2020 IFIP/IEEE 28th International Conference on Very Large Scale Integration (VLSI-SOC 2020)

Salt Lake City, Utah, USA
5 – 7 October 2020

IEEE Catalog Number: CFP20LSI-POD
ISBN: 978-1-7281-5410-7

**Copyright © 2020 by the Institute of Electrical and Electronics Engineers, Inc.
All Rights Reserved**

Copyright and Reprint Permissions: Abstracting is permitted with credit to the source. Libraries are permitted to photocopy beyond the limit of U.S. copyright law for private use of patrons those articles in this volume that carry a code at the bottom of the first page, provided the per-copy fee indicated in the code is paid through Copyright Clearance Center, 222 Rosewood Drive, Danvers, MA 01923.

For other copying, reprint or republication permission, write to IEEE Copyrights Manager, IEEE Service Center, 445 Hoes Lane, Piscataway, NJ 08854. All rights reserved.

*** This is a print representation of what appears in the IEEE Digital Library. Some format issues inherent in the e-media version may also appear in this print version.

IEEE Catalog Number: CFP20LSI-POD
ISBN (Print-On-Demand):   978-1-7281-5410-7
ISBN (Online):   978-1-7281-5409-1
ISSN: 2324-8432

Additional Copies of This Publication Are Available From:

Curran Associates, Inc
57 Morehouse Lane
Red Hook, NY 12571 USA
Phone: (845) 758-0400
Fax: (845) 758-2633
E-mail: curran@proceedings.com
Web: www.proceedings.com

TABLE OF CONTENTS

OPEN-SOURCE EDA: IF WE BUILD IT, WHO WILL COME? 1
    Andrew B. Kahng

SCALABLE OPEN-SOURCE SYSTEM-ON-CHIP DESIGN: (INVITED TALK - EXTENDED
ABSTRACT) 7
    Luca P. Carloni

MULTI-LABEL HD CLASSIFICATION IN 3D FLASH..10
    Justin Morris; Yilun Hao; Saransh Gupta; Ranganathan Ramkumar; Jeffrey Yu; Mohsen Imani; Baris Aksanli;
    Tajana Rosing

EARLY RTL ANALYSIS FOR SCA VULNERABILITY IN FUZZY EXTRACTORS OF
MEMORY-BASED PUF ENABLED DEVICES 16
    Xinhui Lai; Maksim Jenihhin; Georgios Selimis; Sven Goossens; Roel Maes; Kolin Paul

SAT-BASED DATA-FLOW MAPPING ONTO ARRAY PROCESSOR 22
    Yukio Miyasaka; Masahiro Fujita

ABSTRACTPIM: BRIDGING THE GAP BETWEEN PROCESSING-IN-MEMORY
TECHNOLOGY AND INSTRUCTION SET ARCHITECTURE 28
    Adi Eliahu; Rotem Ben-Hur; Ronny Ronen; Shahar Kvatinsky

LAYOUT CONSIDERATIONS OF LOGIC DESIGNS USING AN N-LAYER 3D NANOFABRIC
PROCESS FLOW..34
    Edouard Giacomin; Juergen Boemmels; Julien Ryckaert; Francky Catthoor; Pierre-Emmanuel Gaillardon

SIMULTANEOUS ESTIMATION OF TEMPERATURE AND VOLTAGE FROM DIGITAL
DELAY DIVERSITY 40
    Xiaoyu Lian; Sherief Reda; Jacob K. Rosenstein

EXPLORING THE FPGA IMPLEMENTATIONS OF THE LBLOCK, PICCOLO, TWINE, AND
KLEIN CIPHERS..46
    S. Moraitis; D. Seitanidis; G. Theodoridis; O. Koufopavlou

A 0.8V 875 MS/S 7B LOW-POWER SAR ADC FOR ADC-BASED WIRELINE RECEIVERS IN
22NM FDSOI 52
    David Cordova; Wim Cops; Yann Deval; Francois Rivet; Herve Lapuyade; Nicolas Nodenot; Yohan Piccin

CROSS-LAYER HARDWARE/SOFTWARE ASSESSMENT OF THE OPEN-SOURCE NVDLA
CONFIGURABLE DEEP LEARNING ACCELERATOR 58
    Alessandro Veronesi; Milos Krstic; Davide Bertozzi

X-MAGIC: ENHANCING PIM USING INPUT OVERWRITING CAPABILITIES 64
    Natan Peled; Rotem Ben-Hur; Ronny Ronen; Shahar Kvatinsky

A MINIMALISTIC PERSPECTIVE ON KOBLITZ CURVE SCALAR MULTIPLICATION FOR
FPGA PLATFORMS 70
    Siddhartha Chowdhury; Debapriya Basu Roy; Debdeep Mukhopadhyay

3D LOGIC CELLS DESIGN AND RESULTS BASED ON VERTICAL NWFET TECHNOLOGY
INCLUDING TIED COMPACT MODEL 76
    Chhandak Mukherjee; Marina Deng; François Marc; Cristell Maneux; Arnaud Poittevin; Ian O'Connor; Sébastien
    Le Beux; Cedric Marchand; Abhishek Kumar; Aurélie Lecestre; Guilhem Larrieu

AUTOMATIC TIMING CLOSURE FOR RELATIVE TIMED DESIGNS 82
    Tannu Sharma; Kenneth S. Stevens

MINING HYPERPROPERTIES FROM BEHAVIORAL TRACES 88
    Mayank Rawat; Sujit Kumar Muduli; Pramod Subramanyan

A HYBRID CACHE HW/SW STACK FOR OPTIMIZING NEURAL NETWORK RUNTIME,
POWER AND ENDURANCE 94
    William Andrew Simon; Alexandre Levisse; Marina Zapater; David Atienza

AN ULP SELF-SUPPLIED BRAIN INTERFACE CIRCUIT..100
    Amin Aghighi; Massood Tabib-Azar; Armin Tajalli

ENERGY AND AREA EFFICIENT MIXED-MODE MCMC MIMO DETECTOR 105
    Amin Aghighi; Behrouz Farhang-Boroujeny; Armin Tajalli

MIST: MONITOR GENERATION FROM INFORMAL SPECIFICATIONS FOR FIRMWARE
VERIFICATION 111
    Samuele Germiniani; Moreno Bragaglio; Graziano Pravadelli

BREAKING ACORN AT BITSTREAM LEVEL..117
    Michail Moraitis; Elena Dubrova; Kalle Ngo

BREAKING BARRIERS: MAXIMIZING ARRAY UTILIZATION FOR COMPUTE IN-MEMORY FABRICS 123
Brian Crafton; Samuel Spetalnick; Gauthaman Murali; Tushar Krishna; Sung-Kyu Lim; Arijit Raychowdhury

SANSCRYPT: A SPORADIC-AUTHENTICATION-BASED SEQUENTIAL LOGIC ENCRYPTION SCHEME 129
Yinghua Hu; Kaixin Yang; Shahin Nazarian; Pierluigi Nuzzo

BASIC BLOCK ENCODING BASED RUN-TIME CFI CHECK FOR EMBEDDED SOFTWARE 135
Love Kumar Sah; Srivarsha Polnati; Sheikh Ariful Islam; Srinivas Katkoori

AN OPEN-SOURCE FRAMEWORK FOR AUTONOMOUS SOC DESIGN WITH ANALOG BLOCK GENERATION 141
Tutu Ajayi; Sumanth Kamineni; Yaswanth K. Cherivirala; Morteza Fayazi; Kyumin Kwon; Mehdi Saligane; Shourya Gupta; Chien-Hen Chen; Dennis Sylvester; David Blaauw; Ronald Dreslinski; Benton Calhoun; David D. Wentzloff

ULTRA-COMPACT, SCALABLE, ENERGY-EFFICIENT VO₂ INSULATOR-METAL-TRANSITION OXIDE BASED SPIKING NEURONS FOR LIQUID STATE MACHINES 147
Samiran Ganguly; Nikhil Shukla; Avik W. Ghosh

TESTING THE DIVERGENCE STACK MEMORY ON GPGPUS: A MODULAR IN-FIELD TEST STRATEGY 153
Josie E. Rodriguez Condia; M. Sonza Reorda

A MODEL STUDY OF MULTILEVEL SIGNALING FOR HIGH-SPEED CHIPLET-TO-CHIPLET COMMUNICATION IN 2.5D INTEGRATION 159
Rakshith Saligram; Ankit Kaul; Muhannad S. Bakir; Arijit Raychowdhury

RAT: A LIGHTWEIGHT SYSTEM-LEVEL SOFT ERROR MITIGATION TECHNIQUE 165
Jonas Gava; Ricardo Reis; Luciano Ost

A LOW-POWER 10 TO 15 GB/S COMMON-GATE CTLE BASED ON OPTIMIZED ACTIVE INDUCTORS 171
Amin Aghighi; Armin Tajalli; Mohammad Taherzadeh-Sani

PT CONTROLLED BUCK CONVERTER WITH ADAPTIVE PCCM USING CHARGE MONITORING AND NMOS CURRENT SENSING 176
Yongnan Chen; Yanhan Zeng; Junkai Chen; Hong-Zhou Tan

FAST-TRANSIENT, LIGHT-LOAD EFFICIENT DC-DC CONVERTER USING AN AUXILIARY D-LDO 181
Haochang Zhi; Yanhan Zeng; Wei Zhou; Hong-Zhou Tan

SUBTHRESHOLD-HYBRID SOLUTIONS FOR THERMAL SENSOR AND REFERENCE CIRCUITS IN ADVANCED CMOS 186
Matthias Eberlein; Harald Pretl

TEMPERATURE AND SUPPLY VOLTAGE MONITORING WITH CURRENT-MODE RELAXATION OSCILLATORS 192
Shanshan Dai; Caleb R. Tulloss; Xiaoyu Lian; Kangping Hu; Sherief Reda; Jacob K. Rosenstein

DESIGN, IMPLEMENTATION AND ANALYSIS OF EFFICIENT HARDWARE-BASED SECURITY PRIMITIVES 198
N. Nalla Anandakumar; Somitra Kumar Sanadhya; Mohammad S. Hashmi

MULTIPLE-NOC EXPLORATION AND CUSTOMIZATION FOR ENERGY EFFICIENT TRAFFIC DISTRIBUTION 200
Sonal Yadav; Vijay Laxmi; Manoj Singh Gaur

DESIGN AUTOMATION FOR SIDE CHANNEL RESISTANT LIGHTWEIGHT CRYPTOGRAPHY 202
Rajat Sadhukhan; Debdeep Mukhopadhyay

OPTIMIZATION TOOLS FOR CONVNETS ON THE EDGE 204
Valentino Peluso; Enrico Macii; Andrea Calimera

MEMORY AND ENERGY EFFICIENT METHOD TOWARD SPARSE NEURAL NETWORK USING LFSR INDEXING 206
Foroozan Karimzadeh; Arijit Raychowdhury

ONLINE REWARD-BASED TRAINING OF SPIKING CENTRAL PATTERN GENERATOR FOR HEXAPOD LOCOMOTION 208
Ashwin Sanjay Lele; Yan Fang; Justin Ting; Arijit Raychowdhury

DEVICE MODELING AND CIRCUIT DESIGN FOR SCALABLE BEYOND-CMOS COMPUTING 210
Xuan Hu; Naimul Hassan; Wesley H. Brigner; Maverick Chauwin; Joseph S. Friedman

Author Index

Open-Source EDA:
If We Build It, Who Will Come?

Andrew B. Kahng

Departments of CSE and ECE, UC San Diego
La Jolla, CA 92093-0404 USA
abk@eng.ucsd.edu   https://vlsicad.ucsd.edu/~abk/

Abstract—**The VLSI technology and scaling roadmap has always included Process technology (wrapped as "PDK"), VLSI designs themselves ("System Drivers"), and EDA technology ("Design Technology"). Today, we see an open-source foundry PDK, and we see a vibrant open-source hardware design ecosystem. But what about open-source EDA? The development of open-source EDA technology cannot be separated from the question, "If we build it, who will come?" Today's talk will try to provide some thoughts on this question. What is "it"? Who is "we"? Who is "who"? And in what ways will the "who" come to interact with open-source EDA?**

I. INTRODUCTION

"If you build it, they will come" is a slight variation of the famous line that runs throughout "Field of Dreams" (the film adaptation of W. P. Kinsella's novel, *Shoeless Joe* [16]). In that line, "it" is the field where dreams are realized – and where curing the mistakes of the past brings about a brighter future.

Open-source EDA is truly a field of dreams.

As summarized in [13], a *culture* of open-source EDA brings many clear benefits. It enables the scientific method by bringing transparency and reproducibility to VLSI CAD research. By providing reusable "CAD-IP", it improves research efficiency, thus making the field more attractive. And, when available in an end-to-end flow, it enables research to be evaluated in industry-relevant, flow-scale settings.[1] These are dreams that have been with us for decades, whether as the MARCO GSRC Bookshelf of Fundamental CAD Algorithms in the 1990s [6] [23], or the more recent OpenROAD project [1], [2], [36], [39] in the DARPA IDEA program [26].

Recent years have seen greatly increased complexity of IC design in advanced process technologies. The skyrocketing cost, difficulty and risk of design have put silicon implementation out of the reach of system innovators. This crisis of design and innovation has brought renewed attention to the hardware design process itself, notably since 2017 as one of six main thrusts within the U.S. DARPA Electronics Resurgence Initiative [33].

Recent years have also seen tremendous energy put into reducing barriers – toward a "democratization of hardware

---

[1]Academic end-to-end flows to enable (industry-relevant) research are not necessarily comprised of open-source tools, but are also a well-established goal, e.g., under the leadership of the IEEE CEDA Design Automation Technical Committee (DATC) [7], [8], [11], [12], [27].

design". Today, we see an open-source foundry PDK and design enablement [35] [5], and we see a vibrant open-source hardware ecosystem [32], [34], [45]. But what about open-source EDA? The still-nascent development of open-source EDA technology cannot be separated from the question, "If we build it, who will come?" This paper gives some thoughts on this question, based on the past three years' experience with conception, proposal, and execution of the OpenROAD (Foundations and Realization of Open, Accessible Design) project [29] in the DARPA IDEA program: What is "it"? Who is "we"? Who is "who"? And in what ways will the "who" come to interact with open-source EDA?

II. WHAT IS "IT"

A central goal of the OpenROAD project has always been to deliver "critical mass and critical quality to seed a FOSS EDA ecosystem". This goal has been socialized with a wide range of potential users and stakeholders, at such forums as the DAC-2018 and DAC-2019 "birds of a feather" meetings on Open-Source Academic EDA Software [25], and the 2018 and 2019 workshops on Open-Source EDA Technology (WOSET) [30] – and in numerous public presentations. These discussions, along with the past two years of interactions within the DARPA IDEA program, have made it clear that open-source EDA can mean many things to many people.

Open-source EDA is a moving, many-faceted target.

In July 2018, the community wanted to see a complete RTL-to-GDS flow. In July 2019, with a flow and DRC-clean layout generation in foundry 65nm having been demonstrated, the community wanted to see an integrated tool and tape-out proofs. Today, with an integrated tool and at least one third-party tapeout, community inputs focus on advanced-node foundry support, PPA calibrations, overarching software approach, and a wide range of functionality (DFT, functional simulation, DRC/LVS engine, etc.) that go well beyond the already-ambitious scope of the project.

Even as open-source EDA presents a rapidly moving target, there is also a fundamental tension between OpenROAD's "no human in the loop" goal (no user commands needed, just as a driverless car requires no steering wheel) and many prospective users' typical "here are my 20 favorite Tcl commands that I'd like to see in OpenROAD" request. It has been necessary to continually clarify that while commercial EDA seeks to deliver ultimate quality of results (PPA), OpenROAD seeks to deliver ultimate ease of use. These are two different universes.

978-1-7281-5410-7/20 $31.00 © 2020 IEEE

The ICCAD-2019 invited paper [13] summarized understanding to date of the "table stakes" and "unblocking milestones" needed to attract users to an open-source EDA tool. These included: (1) a unified tool that achieves a full RTL-to-GDS flow; (2) a shared netlist architecture for the tool that enables tight incremental optimization loops; (3) continuous build and integration within a strong software development methodology; and (4) proper open-source licensing to enable unfettered usage across research and commercial settings.

Currently, (1) and (2) are achieved through OpenROAD's integration onto OpenDB [22] [42], a new open-source physical implementation database that holds all essential data for the physical design creation flow (floorplan, global and detailed placement, CTS, and global and detailed routing) as well as timing and power analyses. The underlying data model of OpenDB is similar to that of the LEF/DEF exchange formats, or the well-known OpenAccess database [44]. The open-sourced OpenSTA [43] is intimately connected with OpenDB, such that both timing graph and physical design information are accessible to tools such as a sizing optimizer. The shared netlist data structure enables in-memory communication between tools and the speed improvements that make tight incremental optimization loops feasible.

Since the release of OpenDB, nearly 20 distinct projects in OpenROAD have been integrated into a single binary, referred to as the OpenROAD *top-level app*. Redundancies and inconsistencies such as multiple LEF/DEF readers and writers, as well as file-based or name-based communication between flow steps, have been eradicated. Instead, all projects utilize OpenDB's data structure, via C++ and Tcl APIs. Figure 1 gives a current view of OpenROAD's flow and "v1.0" tool.

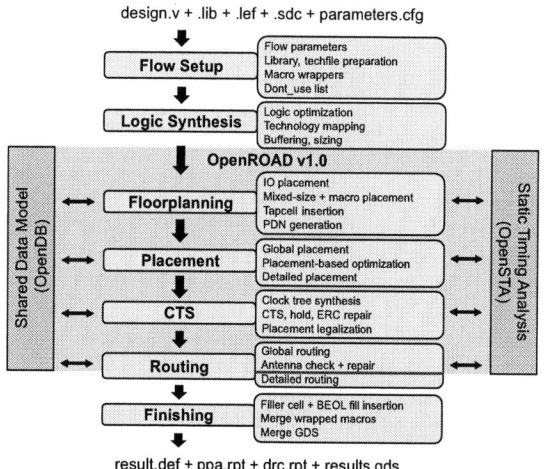

Fig. 1. OpenROAD flow and integrated tool architecture.

III. WHO IS "WE"

OpenROAD was originally proposed to be developed by Ph.D. students and post-docs at four universities. Separately, students and post-docs at a fifth university would serve in an "internal

design advisors" role (see [20]) that was envisioned to span product engineering, expert user testing, and corporate AE-like functions. The clear separation between "internal design advisors" and "tool developers" is built into OpenROAD to avoid improper use of commercial EDA tools. In particular, commercial EDA tools must be used to verify PPA and other calibrations that are required in the project's deliverables; such usage is made by design advisors, not tool developers.

Open-source EDA goes beyond academic research skillsets.

By the project's nine-month mark, it was clear that Open-ROAD needed a dedicated, experienced EDA architect and technical manager from outside the existing team. Voluntary budget reallocations to enable hiring of such a technical project lead were initiated from *within* the OpenROAD team. During the project's second year, these reallocations along with non-DARPA gift funding enabled several industry veterans (Tom Spyrou, James Cherry, Matt Liberty) and additional consulting effort to be recruited into OpenROAD. This has brought much-needed technical leadership and know-how – spanning tool delivery and project management, infrastructure (DB, GUI, build/CI), and key engines (STA, RCX) – on an essentially full-time basis. In the context of this section: *"We" must include professional EDA software developers and architects.*[2]

Additional observations regarding the "We" are as follows.

Strong contributors have software skills *and* the right mindset. Several of OpenROAD's strongest tool developers have been undergraduate and graduate students from outside the U.S. The project has maintained connections with active academic research groups who share the vision of open-source EDA in the RTL-to-GDS space. These groups have been able to identify students who have strong software development skills and who are willing to join the project, particularly when this brings a source of support. Over the past two years, such students have taken on key tool development challenges as well as infrastructure tasks (Jenkins CI [24], measuring and improving test coverage, checking and reconciling Tcl naming [38], etc.); this has led to thesis topics as well as publications along the way. In the forthcoming ICCAD-2020 paper [10], a set of authors from Brazil document their experiences as developers for four separate tools within OpenROAD: global routing, clock tree synthesis, IO placement, and tapcell insertion. In general, the combination of software development skills and willingness to be mentored by industry veterans is very powerful. Physical design understanding is easier to grow than software development maturity.

"We" evolves with "It" (which depends on "Who"). The makeup of an open-source tool development team will depend on feature requirements and roadmap, as well as on the size and sophistication of the user population. At this stage of the OpenROAD project, these "We", "It" and "Who" aspects are

---

[2]Meeting aggressive timelines, and functionality and tapeout-capability requirements, is inconsistent with long learning curves, science experiments, or even publications as an end goal. With recent expansions of OpenROAD's scope for 2020-2022, the addition of EDA industry veterans to the team will almost certainly continue.

978-1-7281-5410-7/20 $31.00 © 2020 IEEE

all very much in flux. This said, it is almost axiomatic that a successful open-source EDA project will see more demands for productization and user support. While success is always welcome, the reality is that it can make project involvement less attractive to academic researchers. This magnifies the need for experienced, full-time architects and developers.

Another reality is that the rigorous software methodology and code organization that improve software quality, maintainability and velocity of development (cf. Google's single-repository approach [19]) can create overheads and barriers – ranging from integration and testing to issues of credit assignment, authorship of publications, etc.[3] An open question is whether "arm's-length development and contribution" might take root in the academic world via such initiatives as the IEEE CEDA DATC Robust Design Flow noted above. For example, when an academic contest is framed in an industry-standard enablement, any winning entry would – if open-sourced by its creators – be available for integration into a more robust, monolithic open-source tool.

"We" must include several types of users. A misconception with open-source EDA is that "users can see the source code, so they can help figure out how to fix bugs or make enhancements." This statement holds for only a very small fraction of OpenROAD's users. In reality, an open-source EDA project such as OpenROAD requires several distinct types of "users" in addition to EDA developers, architects and software engineering infrastructure.

An open-source EDA project requires experienced, *tall-thin tool users* who understand SOC designs as well as their implementation through tapeout. Our "internal design advisors" fit this mold. Such users bring indispensable skills in scripting "make chip" flows, constructing test suites, and bringing up new designs in new enablements. And, experienced EDA tool users have contributed important parts of the OpenROAD flow, notably *pdngen* and the underlying logic of *tapcell*, which are both implemented in OpenDB Tcl.

Importantly, bringup of a new tool *also* requires *application engineers* (AEs), along with expert *power users* who will drive R&D and functional requirements. No successful EDA tool in the history of the field has ever existed without at least a year of intensive "taxicab mode" support delivered by field AEs to key "beta customer" power users. Both power users and AEs find fixes and workarounds and package these up for R&D (developers). Theirs is a much more active, problem-solving mindset compared to mainstream EDA users, who tend to file bugs and wait for fixes.[4] In OpenROAD, an experienced

"power user" and design services consultant was engaged to help accelerate the timeline to a 12nm SOC tapeout milestone. Going forward, it is likely that an additional industry veteran will be recruited to cover the project's current gaps in product engineering and applications engineering.

IV. WHO IS "WHO"

Open-source EDA tools such as Magic, SPICE, MIS/SIS, FastCap and Capo have been in use for decades. Chip tapeouts have been achieved with open-source flows such as qflow [46] and by companies such as efabless [47]. Beyond previous open-source EDA efforts in the RTL-to-GDS space, Open-ROAD has an integrated database and timer, Tcl and python scripting interfaces, GUI, and the "feel" of a commercial back-end EDA tool. OpenROAD *additionally* aims for 24-hour no-human-in-the-loop automation that can directly produce a manufacturable layout database in commercial FinFET technology. But who might come to use OpenROAD in the future? Following are some preliminary thoughts.[5]

Open-source EDA is part of a movement.

The open hardware community will use OpenROAD. A vibrant open-source hardware ecosystem sparked by RISC-V has grown rapidly in recent years [32], [34], [45]. This past June, a SKY130 open-source foundry PDK and design enablement was announced by Google and SkyWater Technology Foundry [35]; the presentation [4] was viewed more than 10,000 times in its first week on YouTube. The OpenLANE flow [15] [21] [31] from efabless.com [47] is built on top of OpenROAD and achieved SKY130 tapeout of the "striVe" SOC in May 2020 (see Figure 2).[6] The coming year is likely to see a number of OpenLANE/OpenROAD SKY130 tapeouts made by researchers, makers and small companies. Broadly, open-source EDA can enable product innovators to perform system ideation and design space exploration in a more friction-free manner, with near-zero overhead.

Fig. 2. Left to right: striVe, strive2 and strive3 SOC designs in SKY130 enablements (source: M. Kassem, efabless.com).

---

[3]This may well lead to more forks and fewer pull requests, i.e., a suboptimal level of "Contribute over Create" (Best Practice #1 in the FOSS102 slides of Ansell [3]). Within OpenROAD, students' natural preferences to maintain independent repositories have slowly faded as the benefits of integration and coding best practices, along with the overheads of keeping functionality in synch between 'integrated' and 'standalone' versions, are comprehended. Particularly during the past half year, the number of independent repositories and the number of submodules in the https://github.com/The-OpenROAD-Project/OpenROAD integrated app have markedly decreased.

[4]As noted in [37], as a non-commercial, academic project, it is not even possible for OpenROAD to receive testcases with bug reports as is the norm for commercial EDA tool suppliers.

[5]It is also important to understand who will *not* use OpenROAD for back-end SP&R implementation. Notably, academic design teams have low-cost licenses to leading-edge commercial tools, so have little reason to explore a raw and limited open-source tool. And, IC product organizations at the bleeding edge will never be able to use an open-source tool: only commercial EDA drives its technology to achieve ultimate quality of (PPA) results in the latest nodes. Recall "two different universes" above.

[6]A family of striVe variants has been designed by efabless.com in SKY130. The figure shows (i) striVe, with PicoRv32 and 1kB of synthesized "Logic RAM", in a high-density library; (ii) striVe2 which swaps in an OpenRAM dual-ported SRAM block; and (iii) striVe3 which uses the OSU scl cell library.

978-1-7281-5410-7/20 $31.00 © 2020 IEEE

Teachers and trainers will use OpenROAD, particularly in contexts where leading-edge back-end implementation tools are unavailable (license limits, tool complexity, etc.). A tool such as OpenROAD is quite simple, yet it gives students insight into back-end database, engine integrations and tight incremental analysis-optimization loops, scripting interfaces, GUI, and other basic aspects of modern P&R tools. For VLSI CAD educators, the transparency of open source enables course assignments that directly delve into tool source code.

Academic EDA researchers will use OpenROAD to improve research efficiency. The reasons for this will only grow stronger over time: bulletproof interfaces, strong test coverage, a user and developer community on GitHub, integration of leading-edge academic methods for easy comparison, etc. Use of OpenROAD as a backplane for academic contests could also increase, for similar reasons.

Mixed-signal SOC designers could use OpenROAD in a "big-A, little-d" context. Digital content on the order of several thousand gates is beyond the reach of manual layout, but a tool such as OpenROAD could suffice to achieve tapeout.

Underserved designers, e.g., in small startups or in government facilities, may be unable to access commercial EDA licenses. Some users may require more complete ownership and control of a transparent chip implementation tool chain, e.g., for reasons of security and trust. Customization opportunities that open source affords may be needed to address rad-hard, 3DIC, trusted IC, or other design applications [14] [48].

V. LOOKING BACK, LOOKING AHEAD

Looking back over the past two years, two high-level takeaways emerge.

First, the passage of time and the accumulation of developed code mean that *some decisions are difficult to revisit*. In other words, "those ships have sailed". OpenROAD's database and timer, Jenkins, coding style, and other project aspects are not likely to change.[7]

Second, OpenROAD's open-source RTL-to-GDS development is challenged by a number of "tensions". (i) Stricter software engineering methodology and tighter integration within a single repository run counter to attracting community participation. (ii) Push-button tapeout capability in advanced-node foundry technology is a requirement that is difficult to "share" with external developers. (iii) Ph.D. research has a mismatch to the demands of EDA tool development and support.[8] More generally, development on an aggressive schedule runs counter to the intermittently available (exams, class projects, winter breaks, summer internships, other research topics, etc.) and

---

[7]Contractual requirements are also constraining. E.g., OpenROAD's requirement to release permissively-licensed open source made the use of OpenAccess [44] impossible. More than a year was spent in pursuit of a database solution; this eventually led to the open-sourcing of Athena Design Systems code and our adoption of OpenDB.

[8]This can be the case even when the Ph.D. student has an EDA R&D career in mind, and despite open-source EDA enabling a new level of EDA and IC design job-readiness. At the same time, all IDEA program performers knew at the outset that "this is not research as usual", and that "the deliverable is working code, not papers".

---

easily-decommitted nature of life in academia. (iv) The IDEA program objective is push-button RTL-to-GDS (a driverless flow that needs no steering wheel), while exploratory use cases and early-adopter users demand much more flexibility and controllability.

These realizations can inform a look ahead, as discussed next.

Looking ahead, two evolutionary directions are of particular interest for open-source EDA.

First, suitable high-value use cases for open-source EDA must be identified. Several are associated with the taxonomy of "Who" above. Beyond these, open-source EDA's low adoption cost and cloud scalability may be well-matched to the long-standing challenge of early design space exploration and pathfinding.

Figure 3 (top) is reproduced from the Design Chapter in the 2009 ITRS Roadmap [41]. The figure's message is that earlier knobs in the flow (system-level, architecture, RTL design) grow relatively more powerful over time. However, while design space exploration should ideally explore more powerful knobs more thoroughly (Ideal "DSE"), attention and iterations still tend to be biased toward the RTL-down flow (Today's "DSE"). This is because optimizing and exploring in early stages has limited value when the back end cannot be predicted accurately enough, and high-level decisions do not correlate to what can actually be closed and signed off. This bespeaks an inability to predict, and an inability to path-find – which are opportunities for leverage of scalable open-source EDA.

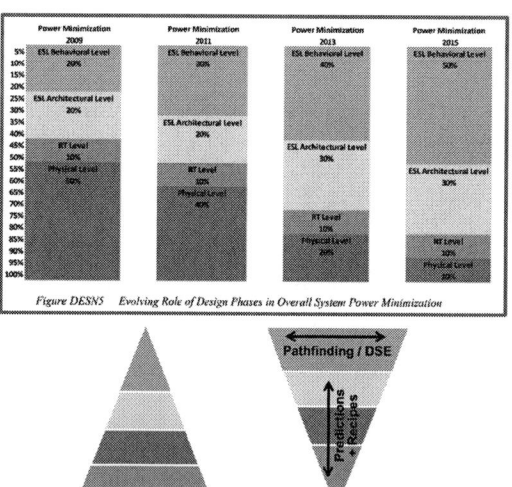

Fig. 3. Top: Growing impact of higher-level design stages on system-level power optimization with advancing technology and system complexity [41]. Bottom: Today's design space exploration (DSE) cannot accurately explore higher-level design stages (left), whereas an ideal DSE would spend effort where it can pay off the most (right).

Second, a sustainable open-source EDA based *ecosystem* must be created that comprehends utility functions and proper incentivizations for *all* stakeholders – spanning IC and system companies, smaller design teams whose needs are ill-served today, open-source hardware and software communities, EDA researchers and professional societies, policy-making bodies and consortia, and commercial EDA vendors. For example, improved separations of "research" (driven by students, professors, enthusiasts and design companies) from "productization and support" (driven by EDA professionals and entrepreneurs who leverage open-source EDA technology to serve a bona fide market) will likely be welcome on multiple fronts.

In his recent 2020 DARPA ERI Summit plenary talk [17], the IDEA program manager, Mr. Serge Leef, posed the question: **Can DARPA *improve access* [to state-of-the-art EDA tools] and *fuel advances* through open-source EDA technologies?** Here, "improve access" was framed as a matter of *economics*, i.e., open-source tools together with cloud deployment can significantly reduce design non-recurring engineering (NRE) costs. And, "fuel advances" was framed as a matter of unleashing hardware *innovation*, i.e., open-source tools can lower barriers and accelerate advances in both hardware design and EDA technology. The overarching challenge is to transition open-source EDA technology from research lab to commercial impact – in a scalable, sustainable way.

Figure 4 presents key elements of three essential pillars that a commercial entity (the Transition "Operator" in the figure) must provide to achieve a scalable, sustainable business based on open-source EDA technology. These pillars – Productization, Support, and Business – exactly correspond to the gaps exposed during OpenROAD's open-source tool development in an academic environment.

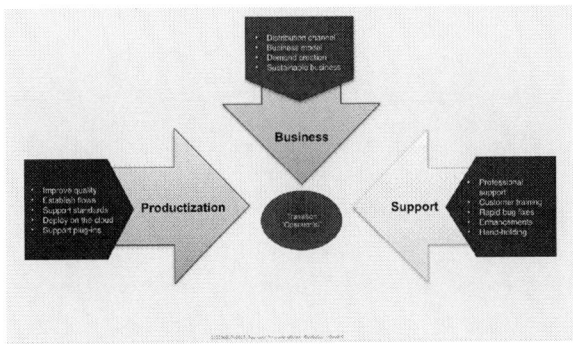

Fig. 4. A potentially scalable and sustainable framework for delivering open-source EDA technology to an IC designer market. (Source: Mr. Serge Leef, DARPA. August 2020.)

Last, if we view open-source EDA as a potential disruptive technology, then the canonical trajectory of adoption is as shown in Figure 5 [9] [28]. A disruptive technology will initially penetrate least-demanding and/or most-underserved market segments. These could correspond to the "Who" in Section IV above.

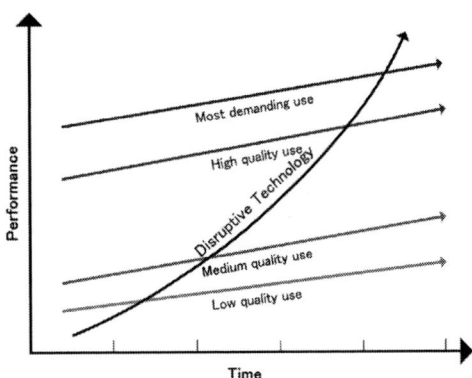

Fig. 5. The trajectory of disruptive innovation.

VI. CONCLUSION

A year ago, [13] considered open-source EDA as a *mirror* that enabled reflections on history, culture, and futures for the research community and the industry. Here, the question of "If we build it, who will come?" also leads to reflections and realizations.

Open-source EDA is truly a field of dreams.

Open-source EDA is a moving, many-faceted target.

Open-source EDA goes beyond academic research skillsets.

Open-source EDA is part of a movement.

The OpenROAD project is still absorbed by many challenges that surround the goal of "critical mass and critical quality to seed a FOSS EDA ecosystem". These challenges include software development infrastructure and robustness; growing available development resources; delivering tool enhancements to satisfy users; adding key missing functionality; improving quality of results; growing a community of users and contributors; and pursuing fundamental EDA research objectives while also helping to accelerate real-world innovation in silicon. As the project continues to execute toward its deliveries, we look forward to the conversations that will illuminate new, sustainable and scalable pathways for the transition of open-source EDA research into real-world impact. And, as developers of open-source EDA, we look forward to continued evolutions: the "It" that OpenROAD project members and many others will build; the community of "We" who contribute; and the "Who" who drive and use open-source EDA to ultimately unleash hardware innovation.

Open-source EDA is a journey.

ACKNOWLEDGMENTS

OpenROAD would not exist without so many contributions from a great team https://theopenroadproject.org/our-team/, as well as the vision and support from the DARPA IDEA program (Serge Leef, Andreas Olofsson and their respective staffs). Sincere thanks are due to everyone who has contributed along this journey so far. Research at UCSD ABKGroup is supported by NSF, DARPA, Qualcomm, Samsung, NXP Semiconductors, Mentor Graphics and the C-DEN center.

REFERENCES

[1] T. Ajayi, D. Blaauw, T.-B. Chan, C.-K. Cheng, V. A. Chhabria, D. K. Choo, M. Coltella, S. Dobre, R. Dreslinski, M. Fogaça, S. Hashemi, A. Hosny, A. B. Kahng, M. Kim, J. Li, Z. Liang, U. Mallappa, P. Penzes, G. Pradipta, S. Reda, A. Rovinski, K. Samadi, S. S. Sapatnekar, L. Saul, C. Sechen, V. Srinivas, W. Swartz, D. Sylvester, D. Urquhart, L. Wang, M. Woo and B. Xu, "OpenROAD: Toward a Self-Driving, Open-Source Digital Layout Implementation Tool Chain", *Proc. GOMACTech*, 2019, pp. 1105-1110.

[2] T. Ajayi, V. A. Chhabria, M. Fogaça, S. Hashemi, A. Hosny, A. B. Kahng, M. Kim, J. Lee, U. Mallappa, M. Neseem, G. Pradipta, S. Reda, M. Saligane, S. S. Sapatnekar, C. Sechen, M. Shalan, W. Swartz, L. Wang, Z. Wang, M. Woo and B. Xu, "Toward an Open-Source Digital Flow: First Learnings from the OpenROAD Project", *Proc. DAC*, 2019, pp. 76:1-76:4.

[3] T. Ansell, "FOSS 101" and "FOSS 102" presentations, July 2019. j.mp/ eri19-foss101 and j.mp/eri19-foss102.

[4] T. Ansell, "Fully open source manufacturable PDK for a 130nm process", *FOSSi Dial-Up presentation*, June 30, 2020. https://youtu.be/EczW2IWdnOM.

[5] T. Ansell and M. Saligane, "The Missing Pieces of Open Design Enablement: A Recent History of Google Efforts", to appear in *Proc. ICCAD*, 2020.

[6] A. E. Caldwell, A. B. Kahng and I. L. Markov, "Toward CAD-IP Reuse: The MARCO GSRC Bookshelf of Fundamental CAD Algorithms", *IEEE Design and Test of Computers*, 2002, pp. 70-79.

[7] J. Chen, I. H.-R. Jiang, J. Jung, A. B. Kahng, V. N. Kravets, Y.-L. Li, S.-T. Lin and M. Woo, "DATC RDF-2019: Towards a Complete Academic Reference Design Flow", *Proc. ICCAD*, 2019, pp. 1-6.

[8] J. Chen, I. H.-R. Jiang, J. Jung, A. B. Kahng, V. N. Kravets, Y.-L. Li, S.-T. Lin and M. Woo, "DATC RDF-2020: Strengthening the Foundation for Academic Research in IC Physical Design", to appear in *Proc. ICCAD*, 2020.

[9] C. Christensen, *The Innovator's Dilemma*, Harvard Business Review Press, 1997.

[10] M. Fogaça, E. Monteiro, M. Danigno, I. Oliveira, P. Butzen and R. Reis, "Contributions to OpenROAD from Abroad: Experiences and Learnings", to appear in *Proc. ICCAD*, 2020.

[11] J. Jung and I. H.-R. Jiang and J. Chen and S.-T. Lin and Y.-L. Li and V. N. Kravets and G.-J. Nam, "DATC RDF: An Academic Flow from Logic Synthesis to Detailed Routing", *Proc. ICCAD*, 2018, pp. 37:1-37:4.

[12] J. Jung, P.-Y. Lee, Y.-S. Wu, N. K. Darav, I. H.-R. Jiang, V. N. Kravets, L. Behjat, Y.-L. Li and G.-J. Nam, "DATC RDF: Robust design flow database", *Proc. ICCAD*, 2017, pp. 872-873.

[13] A. B. Kahng, "Looking Into the Mirror of Open Source: Invited Paper", *Proc. ICCAD*, 2019.

[14] A. B. Kahng and F. Koushanfar, "Evolving EDA Beyond its E-Roots: An Overview", *Proc. ICCAD*, 2015, pp. 247-254.

[15] M. Kassem, T. Edwards and M. Shalan, "Building OpenLane: A 130nm OpenROAD-based Tapeout-Proven Flow". to appear in *Proc. ICCAD*, 2020.

[16] W. P. Kinsella, *Shoeless Joe*, Houghton Mifflin, 1982.

[17] S. Leef, "Open Source Accelerated Chip Design", plenary talk, *DARPA ERI Summit*, August 18, 2020.

[18] G. Moore, *Crossing the Chasm*, Harper Business Essentials, 1991.

[19] R. Potvin and J. Levenberg, "Why Google Stores Billions of Lines of Code in a Single Repository", *Comm. of the ACM*, 2016, pp. 78-87.

[20] A. Rovinski, T. Ajayi, M. Kim, G. Wang and M. Saligane, "Bridging Academic Open-Source EDA to Real-World Usability" *Proc. ICCAD*, 2020.

[21] M. Shalan, "OpenLane, A Digital ASIC Flow for SkyWater 130nm Open PDK", *FOSSi Dial-Up presentation*, July 28, 2020. https://www.youtube.com/watch?v=Vhyv0eq_mLU.

[22] T. Spyrou, "Open-Source EDA Challenges and Architecture", opening talk at DAC2019 Open-Source Academic EDA Software Birds-of-a-Feather meeting.

[23] (MARCO GSRC) VLSI CAD Bookshelf. http://vlsicad.eecs.umich.edu/BK/

[24] Jenkins website, https://jenkins.io/

[25] DAC 2019 Birds-of-a-Feather Meeting: Open-Source Academic EDA Software Continued, https://github.com/The-OpenROAD-Project/Birds-of-a-Feather-Open-Source-Academic-EDA-Software/wiki/DAC-2019-Birds-of-a-Feather:-Open-Source-Academic-EDA-Software

[26] DARPA IDEA, https://www.darpa.mil/program/intelligent-design-of-electronic-assets

[27] DATC, https://ieee-ceda.org/node/2591 and https://github.com/ieee-ceda-datc/RDF2019

[28] Wikipedia, "Disruptive Innovation", https://en.wikipedia.org/wiki/Disruptive_innovation

[29] OpenROAD, https://theopenroadproject.org/

[30] Workshop on Open-Source EDA Technology (WOSET), http://scale.engin.brown.edu/woset

[31] efabless OpenLANE repository, https://github.com/efabless/openlane.

[32] Chips Alliance, https://chipsalliance.org/.

[33] DARPA Electronics Resurgence Initiative, https://www.darpa.mil/work-with-us/electronics-resurgence-initiative

[34] The Free and Open Source Silicon Foundation, https://fossi-foundation.org/.

[35] Google-Skywater, https://github.com/google/skywater-pdk.

[36] OpenROAD: Foundations and Realization of Open and Accessible Design, https://theopenroadproject.org

[37] "OpenROAD Flow Initial Information for Users", November 2018, https://theopenroadproject.org/openroad_event/openroad-flow-initial-information-for-users/

[38] "OpenROAD 'Safe Names' Conventions, v1.0", December 2019, https://theopenroadproject.org/openroad_event/new-openroad-safe-names-conventions-v1-0/

[39] OpenROAD GitHub Repository, https://github.com/The-OpenROAD-Project/OpenROAD.

[40] "International Roadmap for Devices and Systems 2018 Update More Moore", https://irds.ieee.org/images/files/pdf/2018/2018IRDS_MM.pdf

[41] "International Technology Roadmap for Semiconductors 2009 Design", https://www.dropbox.com/sh/ia1jkem3v708hx1/AAB1fo1HrYIKClJNk0dB7YrCa?dl=0&preview=Design.pdf

[42] OpenDB, https://github.com/The-OpenROAD-Project/OpenDB

[43] OpenSTA, https://github.com/The-OpenROAD-Project/OpenSTA

[44] OpenAccess, https://si2.org/openaccess/

[45] OpenHW Group, https://www.openhwgroup.org/

[46] Qflow. https://github.com/RTimothyEdwards/qflow

[47] Efabless.com efabless.com

[48] IEEE CEDA Design Automation Futures Workshop, October 2016. https://ieee-ceda.org/event/design-automation-futures-workshop-2016-dafw

Scalable Open-Source System-on-Chip Design

(Invited Talk - Extended Abstract)

Luca P. Carloni

Department of Computer Science, Columbia University in the City of New York
New York, NY 10027
luca@cs.columbia.edu

Abstract—**The system-on-chip is the dominant architecture in the age of heterogeneous computing, but its energy-efficient performance comes at the cost of higher design complexity. The open-source hardware movement responds to this challenge by promoting design reuse and collaboration. ESP is an open-source research platform for heterogeneous SoC design that combines a scalable tile-based architecture and a flexible system-level design methodology. Conceived as a heterogeneous integration platform, ESP is naturally suited to foster collaborative engineering of SoC designs across the open-source community.**

Index Terms—**system-on-chip (SoC), open-source hardware (OSH), heterogeneous computing, system-level design, RISC-V.**

I. SoC Architectures Are Everywhere

Modern efficient computing is *heterogeneous computing*. The end of Dennard's ideal CMOS scaling [1], the slow-down of Moore's Law [2], [3], and the limits of effective parallelism that can be achieved with homogeneous multi-core processors [4] have pushed designers to achieve performance gains by specializing hardware. *Accelerators*—hardware computing engines that are specialized for a particular application or application domain—provide orders-of-magnitude gains in energy-efficient performance compared to general-purpose processors [5]. To balance specialization and programmability, designers combine accelerators and processors into heterogeneous architectures. Consequently, the *system-on-chip (SoC)* has become the principal computer architecture across the most important classes of computing. The SoC originally emerged in the design of embedded systems, where it continues to be the dominant computing engine for smartphones [6], automotive electronics [7], avionics [8] and the Internet-of-things [9]. Over the last decade, however, the continuous migration of devices from the board to the die [10] and the transfer of critical functionality from software to specialized hardware [4], [11] have made the SoC a popular choice also for personal computers and servers. More recently, as accelerators have gained ground in cloud computing [12], [13], the giants of the information technology industry have started designing their own SoCs [14].

II. SoC Design Is a Challenging Task

Since the quest to maximize energy-efficient performance passes through hardware specialization of computational kernels for critical application workloads, designers combine more and more heterogeneous components in the same SoC.

A state-of-the-art SoC is a highly heterogeneous system that integrates many general-purpose processors, graphics processing units, digital signal processors, and accelerators [15]. Heterogeneity improves energy efficiency and performance, but it increases design complexity. A system consisting in a collection of diverse components is intrinsically more difficult to design, validate, and program than a system made only of homogeneous copies. At design time, the differences among heterogeneous components translate into diminished regularity in chip layout and the need to perform different verification tasks. At runtime, the presence of many heterogeneous components complicates the hardware-software interface and the management of shared resources, such as access to off-chip main memory and use of the tight on-chip power budget. With each SoC generation, the addition of new capabilities is increasingly limited by engineering effort and team sizes [16].

III. Open-Source Hardware to the Rescue

Open-source hardware (OSH) has been proposed as a vehicle to reenergize the innovation in the semiconductor industry in the mold of the proven success of the open-source software ecosystem [17]. The momentum of OSH has been building in recent years [18], [19], thanks particularly to the popularity of the RISC-V project [20]. The number of OSH projects is expected to grow steadily in the upcoming years, fueled by multi-institution organizations [21]–[23], government programs [24], and many diverse contributions from both academia [20], [25], [26] and industry [27], [28]. To date, however, most OSH projects are focused on the development of individual SoC components, such as a processor core or an accelerator. While certainly useful, this leaves open a critical challenge:

How can we realize a complete SoC for a given target application domain by efficiently reusing and combining a variety of independently developed, heterogeneous, OSH components, especially if these components are designed by separate organizations for separate purposes?

While the development of individual OSH components is a necessary precondition, the ultimate goal is the realization of complete SoC designs that leverage these components.

IV. Scaling Up Open-Source Hardware

Achieving this goal requires enabling design reuse and collaboration to directly mitigate the design complexity challenge. A possible path goes through innovations of the SoC architectures as well as the methodologies used to design them. Indeed, architectures and methodologies must be developed together

978-1-7281-5410-7/20 $31.00 © 2020 IEEE

in order to be effective. This approach is captured by the concept of *platform*, which is precisely the combination of an architecture and methodology [29]. Although an architecture limits the space of possible SoC designs, its properties allow for the development of an effective design methodology and supporting CAD tools. In turn, the methodology and tools allow designers to focus on the most important and creative aspects of the design process, while leveraging automation for the repetitive and error-prone tasks.

An SoC architecture enables design reuse when it simplifies the integration of many components that are independently developed. An SoC methodology enables design collaboration when it allows designers to choose the preferred specification languages and design flows for the various components, particularly when these choices provide advantages in the context of important application-specific domains. An effective combination of architecture and methodology is a platform that maximizes the potential of open-source hardware by scaling-up the number of components that can be integrated in an SoC and by enhancing the productivity of the designers who develop and use them.

V. AN OPEN-SOURCE PLATFORM FOR SOC DESIGN

ESP is an open-source research platform for heterogeneous SoC design [30]. The System-Level Design Group at Columbia University has developed ESP by building on the foundations of communication-based system-level design [31] and on years of experience teaching SoC platforms [32]. ESP combines a scalable architecture and a flexible methodology [29]. Just like the architecture simplifies the integration of heterogeneous components developed by different teams, the methodology embraces the use of various design flows for component development [33].

The ESP architecture is structured as a tile grid. The tiles form a distributed system which is inherently scalable, modular and heterogeneous. The main types of tile are three: processor, accelerator and memory. For the processor tile, ESP currently allows a seamless choice between the 32-bit LEON3 SPARC core [34] and the 64-bit ARIANE RISC-V core [35]. An accelerator tile contains one or more loosely-coupled accelerators [36]; these can be accelerators developed with the ESP methodology [37], [38] as well as third-party OSH accelerators like the NVIDIA NVDLA [27]. Processors and accelerators [36] are given the same importance in the SoC. This system-centric view distinguishes ESP from other OSH platforms, most of which take a processor-centric view.

Each tile is encapsulated into a *modular socket* (aka *shell*) that interfaces it to a network-on-chip (NoC), which has a packet-switched 2D-mesh topology with multiple physical planes [39]. Following the principles of the *protocols and shells paradigm* of latency-insensitive design [31], [40], the shell decouples the design of the tile content from the design of the rest of the system, thereby simplifying the integration of an OSH component inside a tile, enabling late-stage optimizations as part of design-space exploration decisions, and promoting its reuse across different SoC designs. Furthermore, the shell

implements a set of *platform services*, which provide pre-validated solutions for common design tasks like accelerator configuration [37], memory management [41], [42], and dynamic voltage-frequency scaling [43].

The ESP methodology guides the choice of the number, mix, and placement of tiles for a target SoC as well as the design of newly-developed components. Third-party IP blocks can be seamlessly integrated [44]. For the development of new components, ESP promotes system-level design [31] and, particularly, the application of high-level synthesis to design-space exploration [45]–[47]. Indeed, the ESP methodology is flexible because it embraces different design flows from specifications written in different languages, including: C with Xilinx Vivado HLS, SystemC with Cadence Stratus HLS, C++/SystemC with Mentor Catapult HLS, as well as SystemVerilog, VHDL, and Chisel. Recently, a flow to design embedded machine learning accelerators with Keras TensorFlow, PyTorch and ONNX through hls4ml [48] became the first example of a domain-specific design flow added to ESP [38].

A graphical user interface allows the selection of the tiles, the number and parallelism of the NoC planes, and the structure of the memory hierarchy, among many other configuration parameters. Once configured, the RTL implementation of the SoC is automatically generated together with all the hardware and software mechanisms for system integration of the chosen processor core. The automatic generation of device drivers from pre-designed templates simplifies the invocation of accelerators from user-level applications running on Linux [41]. The automatic generation of a multi-plane NoC from a parameterized model supports the scaling of the ESP architecture to accommodate multiple cores, many accelerators, and a distributed memory hierarchy [49].

ESP allows SoC architects to rapidly implement FPGA-based prototypes of complex SoCs by combining third-party OSH components that use the AXI protocol (e.g. ARIANE and NVDLA) with newly-designed components. A growing set of tutorials and demos on how to realize these prototypes is available on the ESP website [30]. While the ESP release currently focuses on FPGA-based prototyping, the ESP methodology offers a natural and versatile front end, up to synthesizable RTL, for chip design.

VI. CONCLUSIONS

The concept of platform is the key to handling the complexity of SoC design in the age of heterogeneous computing. By combining a scalable architecture with a flexible methodology, ESP provides the open-source hardware community with a platform for collaborative engineering of SoC designs.

Acknowledgments. This work was sponsored in part by the Army Research Office and was accomplished under Grant Number W911NF-19-1-0476. The views and conclusions contained in this document are those of the authors and should not be interpreted as representing the official policies, either expressed or implied, of the Army Research Office or the U.S. Government. The U.S. Government is authorized to reproduce and distribute reprints for Government purposes notwithstanding any copyright notation herein.

REFERENCES

[1] M. Bohr, "A 30 year retrospective on Dennard's MOSFET scaling paper," *IEEE Solid-State Circuits Society Newsletter*, vol. 12, no. 1, pp. 11–13, Winter 2007.

[2] R. K. Cavin, P. Lugli, and V. V. Zhirnov, "Science and engineering beyond Moore's Law," *Proc. of the IEEE*, vol. 100, pp. 1720–1749, May 2012.

[3] R. Colwell, "End of Moore's law," *IEEE Computer*, vol. 46, no. 12, p. 49, Dec. 2013.

[4] M. Horowitz, "Computing's energy problem (and what we can do about it)," in *ISSCC Digest of Technical Papers*, Feb. 2014, pp. 10–14.

[5] W. J. Dally, Y. Turakhia, and S. Han, "Domain-specific hardware accelerators," *Comm. of the ACM*, vol. 63, no. 7, pp. 48–57, Jun. 2020.

[6] Y. S. Shao, B. Reagen, G. Wei, and D. Brooks, "The Aladdin approach to accelerator design and modeling," *IEEE Micro*, vol. 35, no. 3, pp. 58–70, May-Jun 2015.

[7] G. P. Stein, E. Rushinek, G. Hayun, and A. Shashua, "A Computer Vision System on a Chip: a case study from the automotive domain," in *Conf. on Computer Vision and Pattern Recognition (CVPR'05)*, Sep. 2005, pp. 130–130.

[8] D. Keymeulen, S. Shin, J. Riddley, M. Klimesh, A. Kiely, E. Liggett, P. Sullivan, M. Bernas, H. Ghossemi, G. Flesch, M. Cheng, S. Dolinar, D. Dolman, K. Roth, C. Holyoake, K. Crocker, and A. Smith, "High performance space computing with system-on-chip instrument avionics for space-based next generation imaging spectrometers (NGIS)," in *NASA/ESA Conf. on Adaptive Hardware and Systems*, Aug. 2018, pp. 33–36.

[9] Y. Pu, C. Shi, G. Samson, D. Park, K. Easton, R. Beraha, A. Newham, M. Lin, V. Rangan, K. Chatha, D. Butterfield, and R. Attar, "A 9-mm2 ultra-low-power highly integrated 28-nm CMOS SoC for Internet of Things," *J. of Solid-State Circuits*, vol. 53, no. 3, pp. 936–948, Mar. 2018.

[10] S. Damaraju, V. George, S. Jahagirdar, T. Khondker, R. Milstrey, S. Sarkar, S. Siers, I. Stolero, and A. Subbiah, "A 22nm IA Multi-CPU and GPU System-on-Chip," in *ISSCC Digest of Technical Papers*, Feb. 2012, pp. 56–57.

[11] S. Borkar and A. Chen, "The future of microprocessors," *Communication of the ACM*, vol. 54, pp. 67–77, May 2011.

[12] A. M. Caulfield, E. S. Chung, A. Putnam, H. Angepat, J. Fowers, M. Haselman, S. Heil, M. Humphrey, P. Kaur, J. Kim, D. Lo, T. Massengill, K. Ovtcharov, M. Papamichael, L. Woods, S. Lanka, D. Chiou, and D. Burger, "A cloud-scale acceleration architecture," in *Proc. of the Intl. Symp. on Microarchitecture*, Oct. 2016, pp. 1–13.

[13] N. P. Jouppi, C. Young, N. Patil, and D. Patterson, "A domain-specific architecture for deep neural networks," *Comm. of the ACM*, vol. 61, no. 9, pp. 50–59, Aug. 2018.

[14] E. Jhonsa, "Why tech giants like Amazon are designing their own chips – and who benefits," https://www.thestreet.com/opinion/why-tech-giants-are-designing-their-own-chips-14807638, Dec. 2018.

[15] M. Ditty, A. Karandikar, and D. Reed, "NVIDIA's Xavier SoC," Intl. Symp. on High Performance Chips (HotChips'30), 2018.

[16] B. Khailany, E. Khmer, R. Venkatesan, J. Clemons, J. S. Emer, M. Fojtik, A. Klinefelter, M. Pellauer, N. Pinckney, Y. S. Shao, S. Srinath, C. Torng, S. L. Xi, Y. Zhang, and B. Zimmer, "A modular digital VLSI flow for high-productivity SoC design," in *Proc. of the Design Automation Conf. (DAC)*, Jun. 2018, pp. 72:1–72:6.

[17] G. Gupta, T. Nowatzki, V. Gangadhar, and K. Sankaralingam, "Kick-starting semiconductor innovation with open source hardware," *IEEE Computer*, vol. 50, no. 6, pp. 50–59, Jun. 2017.

[18] B. Bailey, "Open-source hardware momentum builds," https://semiengineering.com/riding-the-risc-v-wave/, Jun. 2020.

[19] The Economist, "The rise of open-source computing," Oct. 2019.

[20] K. Asanovic and D. Patterson, "The case for open instruction sets," *Microprocessor Report*, Aug. 2014.

[21] RISC-V Foundation, https://riscv.org/.

[22] CHIPS Alliance, https://chipsalliance.org/.

[23] OpenHWGroup, https://www.openhwgroup.org/.

[24] S. Moore, "DARPA picks its first set of winners in electronics resurgence initiative," https://spectrum.ieee.org/tech-talk/semiconductors/design/darpa-picks-its-first-set-of-winners-in-electronics-resurgence-initiative, Jul. 2018.

[25] F. Zaruba and L. Benini, "The cost of application-class processing: Energy and performance analysis of a Linux-ready 1.7-GHz 64-Bit RISC-V core in 22-nm FDSOI technology," *IEEE Trans. on Very Large Scale Integration Systems*, vol. 27, no. 11, pp. 2629–2640, Nov. 2019.

[26] J. Balkind, K. Lim, F. Gao, J. Tu, D. Wentzlaff, M. Schaffner, F. Zaruba, and L. Benini, "OpenPiton+Ariane: the first SMP Linux-booting RISC-V system scaling from one to many cores," in *Workshop on Computer Architecture Research with RISC-V (CARRV)*, 2019.

[27] NVIDIA, "NVIDIA Deep Learning Accelerator," www.nvdla.org, 2018.

[28] L. Armasu, "Western Digital bets big on RISC-V with own processor, other innovations," https://www.tomshardware.com/news/western-digital-risc-v-processor-open-source,38200.html, Feb. 2019.

[29] L. P. Carloni, "The case for embedded scalable platforms," in *Proc. of the Design Automation Conf. (DAC)*, Jun. 2016, pp. 17:1–17:6.

[30] Columbia SLD Group, "ESP Release," www.esp.cs.columbia.edu, 2019.

[31] L. P. Carloni, "From latency-insensitive design to communication-based system-level design," *Proc. of the IEEE*, vol. 103, no. 11, pp. 2133–2151, Nov. 2015.

[32] L. P. Carloni, E. G. Cota, G. D. Guglielmo, D. Giri, J. Kwon, P. Mantovani, L. Piccolboni, and M. Petracca, "Teaching heterogeneous computing with system-level design methods," in *Workshop on Computer Architecture Education*, Jun. 2019.

[33] P. Mantovani, D. Giri, G. D. G. L. Piccolboni, J. Zuckerman, E. G. Cota, M. Petracca, C. Pilato, and L. P. Carloni, "Agile SoC development with Open ESP," in *Proc. of the Intl. Conf. on Computer-Aided Design (ICCAD)*, Nov. 2020.

[34] Cobham Gaisler, "Leon3," www.gaisler.com/index.php/products/processors/leon3.

[35] Ariane, www.github.com/pulp-platform/ariane.

[36] E. G. Cota, P. Mantovani, G. Di Guglielmo, and L. P. Carloni, "An analysis of accelerator coupling in heterogeneous architectures," in *Proc. of the Design Automation Conf.*, Jun. 2015, pp. 202:1–202:6.

[37] P. Mantovani, G. D. Guglielmo, and L. P. Carloni, "High-level synthesis of accelerators in embedded scalable platforms," in *Proc. of the Asia and South Pacific Design Automation Conf.*, Jan. 2016, pp. 204–211.

[38] D. Giri, K.-L. Chiu, G. D. Guglielmo, P. Mantovani, and L. P. Carloni, "ESP4ML: platform-based design of systems-on-chip for embedded machine learning," in *Conf. on Design, Automation and Test in Europe*, Mar. 2020, pp. 1049–1054.

[39] Y. Yoon, N. Concer, and L. P. Carloni, "Virtual channels and multiple physical networks: Two alternatives to improve NoC performance," *IEEE Trans. on Computer-Aided Design of Integrated Circuits and Systems*, vol. 32, no. 12, pp. 1906–1919, Dec. 2013.

[40] L. P. Carloni, K. L. McMillan, and A. L. Sangiovanni-Vincentelli, "Theory of latency-insensitive design," *IEEE Trans. on Computer-Aided Design of Integrated Circuits and Systems*, vol. 20, no. 9, pp. 1059–1076, Sep. 2001.

[41] P. Mantovani, E. G. Cota, C. Pilato, G. Di Guglielmo, and L. P. Carloni, "Handling large data sets for high-performance embedded applications in heterogeneous systems-on-chip," in *Intl. Conf. on Compilers, Architecture and Synthesis for Embedded Systems*, Oct. 2016, pp. 1–10.

[42] D. Giri, P. Mantovani, and L. P. Carloni, "Accelerators and coherence: An SoC perspective," *IEEE Micro*, vol. 38, no. 6, pp. 36–45, Nov-Dec 2018.

[43] P. Mantovani, E. G. Cota, K. Tien, C. Pilato, G. Di Guglielmo, K. Shepard, and L. P. Carloni, "An FPGA-based infrastructure for fine-grained DVFS analysis in high-performance embedded systems," in *Proc. of the Design Automation Conf. (DAC)*, Jun. 2016, pp. 157:1–157:6.

[44] D. Giri, K.-L. Chiu, G. Eichler, P. Mantovani, N. Chandramoorth, and L. P. Carloni, "Ariane + NVDLA: seamless third-party IP integration with ESP," in *Workshop on Computer Architecture Research with RISC-V (CARRV)*, May 2020.

[45] H.-Y. Liu, M. Petracca, and L. P. Carloni, "Compositional system-level design exploration with planning of high-level synthesis," in *Conf. on Design, Automation and Test in Europe*, Mar. 2012, pp. 641–646.

[46] C. Pilato, P. Mantovani, G. Di Guglielmo, and L. P. Carloni, "System-level optimization of accelerator local memory for heterogeneous systems-on-chip," *IEEE Trans. on Computer-Aided Design of Integrated Circuits and Systems*, vol. 36, no. 3, pp. 435–448, Mar. 2017.

[47] L. Piccolboni, P. Mantovani, G. D. Guglielmo, and L. P. Carloni, "COS-MOS: Coordination of high-level synthesis and memory optimization for hardware accelerators," *ACM Trans. on Embedded Computing Systems*, vol. 16, no. 5s, pp. 150:1–150:22, Sep. 2017.

[48] hls4ml, https://fastmachinelearning.org/hls4ml.

[49] D. Giri, P. Mantovani, and L. P. Carloni, "NoC-based support of heterogeneous cache-coherence models for accelerators," in *Proc. of the Intl. Symp. on Networks-on-Chip (NOCS)*, Oct. 2018, pp. 1:1–1:8.

978-1-7281-5410-7/20 $31.00 © 2020 IEEE

Multi-label HD Classification in 3D Flash

Justin Morris[*†], Yilun Hao[*], Saransh Gupta[*], Ranganathan Ramkumar[*], Jeffrey Yu[*], Mohsen Imani[‡],
Baris Aksanli[†], and Tajana Rosing[*]

[*]University of California San Diego, La Jolla, CA 92093, USA
[†]San Diego State University, San Diego, CA 92182, USA
[‡]University of California Irvine, Irvine, CA 92697, USA

{justinmorris, yih301, sgupta, rramkuma, jey070}@ucsd.edu, m.imani@uci.edu, baksanli@sdsu.edu, tajana@ucsd.edu

Abstract—Many classification problems in practice map each sample to more than one label - this is known as multi-label classification. In this work, we present Multi-label HD, an in 3D storage multi-label classification system that uses Hyperdimensional Computing (HD). Multi-label HD is the first HD system to support multi-label classification. We propose two different mappings of HD to Multi-label HD. The first, Power Set HD, transforms the multi-label problem into single-label classification by creating a new class for each label combination. The second, Multi-Model HD, creates a binary classification model for each possible label. Our evaluation shows that Multi-Model HD achieves, on average, $47.8\times$ higher energy efficiency and $47.1\times$ faster execution time while achieving 5% higher classification accuracy as state-of-the-art light-weight multi-label classifiers. Power Set HD achieves 13% higher accuracy than Multi-Model HD, but is $2\times$ slower. Our 3D-flash acceleration further improves the energy efficiency of Multi-label HD training by $228\times$ and reduces the latency by $610\times$ vs training on a CPU.

I. INTRODUCTION

The emergence of the Internet of Things (IoT) has created an abundance of small embedded devices [1]. Many of these devices perform classification tasks, such as speech recognition, image classification, etc. More and more devices are now required to perform more complex multi-label classification [2], [3]. However, they have very limited resources such as limited battery lifetime and a small amount of memory, which is not enough to train and run Deep Neural Networks (DNN) [4]. We need a light-weight classification algorithm to perform such tasks on embedded systems.

Brain-inspired Hyperdimensional (HD) computing has been proposed as the alternative light-weight computing method to perform cognitive tasks on devices with limited resources [5]. Inspired by the pattern of neural activity in the brain [6], HD computing maps each data point into high dimension vectors, called *hypervectors* (HVs). HD computing has three main stages, 1) Encoding: mapping data into HVs. 2) Training: combining encoded HVs to create a model representing each class with a HV. 3) Inference: comparing the incoming sample with the trained model to find the most similar class. HD computing shows promising progress for many cognitive tasks such as activity recognition, object recognition, language recognition, and bio-signal classification [7], [8], [9]. However, there has been no work yet on mapping HD computing to multi-label classification tasks.

While HD provides improvements in performance and energy consumption over conventional machine learning algorithms, it still involves fetching each and every data from memory/disk and processing it on CPUs/GPUs. This is exaserbated by the fact that HD expands the dimensionality of the input data into high dimensional space. This massive amount of data needed for HD cannot always fit into the memory. Recent work has introduced computing capabilities to solid-state disks (SSDs) to process data in storage [10], [11], [12], [13], [14]. This not only reduces the computation load from the processing cores but also processes raw data where it is stored. HD computing has compelling properties for efficient hardware acceleration in flash. For instance, HD is highly parallelizable with $D = 10,000$ dimensions where each dimension is independent. Furthermore, HD is comprised of simple operations such as addition, multiplication, and comparisons. With these two properties, HD computing is a prime candidate for acceleration in flash.

In this paper, we design a new Multi-label HD computing in storage system. Our system efficiently accelerates the data-intensive steps of HD, encoding and training, in 3D storage, thus, making it possible to run multi-label classification with HD in the IoT domain. We propose two different mappings of HD to multi-label classification, Power Set HD and Multi-Model HD. Power Set HD, transforms the multi-label problem into classical classification by creating a new class for each label combination. Multi-Model HD that creates a binary classification model for each possible label. Our evaluation shows that Multi-Model HD achieves, on average, $47.8\times$ higher energy efficiency and $47.1\times$ faster execution time while achieving 5% higher classification accuracy as state-of-the-art light-weight multi-label classifiers such as multi-label kNNs. Power Set HD achieves 13% higher accuracy than Multi-Model HD, but is $2\times$ slower. Using our in-3D-flash acceleration, we further improve the energy efficiency of Multi-label HD training by $228\times$ and reduce the latency by $610\times$.

II. RELATED WORK

A. Hyperdimensional Computing

Prior work tried to apply the idea of high-dimensional computing to different classification problems such as language recognition, speech recognition, face detection, EMG gesture detection, human-computer interaction, and sensor fusion prediction [7], [15], [16], [17]. Additionally, work in [9] proposed a new HD encoding based on random indexing for recognizing a text's language by generating and comparing text hypervectors. Work in [18] proposed an encoding method to map and classify biosignal sensory data in high dimensional space. Work in [8] proposed a general encoding module that maps feature vectors into high-dimensional space while keeping most of the original data. There is no work to date that handles multi-label classification in HD.

B. Multi-label Classification

Prior work applied problem transformation methods to transform multi-label classification problems into multiple single-label classification problems [2], [3]. The most widely used transformation method is PT3. PT3 combines each different set of labels into a single label so that the new label set L' is the power set of the old label set L. For a dataset with three binary labels, the new label set would be 000, 001, 010, 011, 100, 101, 110, 111. This causes an exponential increase in the number of labels in the dataset. This transformation method is popular for other light-weight classifiers as their complexities mostly scale with the number of features and not with the number of labels. However, in HD computing, the complexity of inference does scale with the number of labels. Therefore, in this paper we propose a new Multi-Model transformation method that is designed for scalable HD computing.

C. Hardware Acceleration

HD Acceleration on other Platforms: Prior work tried to design different hardware accelerators for HD computing. This includes accelerating HD computing on existing FPGA, ASIC, and processing in-memory platforms [19]. However, these solutions do not scale well with the number of classes and dimensions, primarily due to the data movement issue. Therefore, a new solution is needed that can scale with the dimensionality and number of classes. ISC is a promising acceleration architecture in this aspect. **Computing in 3D Flash:** The current 3D flash-based storage systems suffer from slow flash array read latency and storage to host I/O latency. To alleviate these issues prior work introduced in-storage computing (ISC) architectures [12]. These works exploit the embedded cores present in the SSD controller to implement ISC. Another set of work in [11], [20] used ASIC accelerators in SSDs. The work in [13] proposed a full-stack storage system to reduce the host-side I/O stack latency. While these works propose single-level computing in storage, [14] on the other hand exploited computing at flash die and in top level accelerator to provide multi-layer computing. It also allows for high parallelism in computation. In this work, we adapt the ISC design in [14] to enable multi-label HD in 3D flash.

III. MULTI-LABEL CLASSIFICATION WITH HD

Multi-label classification is the problem of finding a model that maps inputs x to binary vectors y, and each element in

y is a label that is assigned a value either 0 or 1. This is in contrast to single-label classification, where y is a single value, not a vector of labels. Although HD computing performs well for single-label classification tasks, we can't directly apply it to solve multi-label classification problems, as only one label output is chosen. Therefore, we transform the multi-label problem into a single-label problem and then modify the HD computing algorithm to solve the multi-label classification problem. We propose two different mappings of HD to Multi-label HD. The first, Power Set HD, transforms the multi-label problem into single-label classification by creating a new class for each observed label combination. We additionally propose Multi-Model HD that creates a binary classification model for each possible label. By doing this, we can leverage the efficiency of HD computing to complete the multi-label classification task faster and with less energy consumption.

A. Problem Transformation Methods

Power Set: Prior work[2] mapped multi-label classification to single label classification by creating a new label set that was the power set of the multi-labels. For instance, if a multi-label problem had 3 possible labels for every sample, then prior work would transform the 3 multi-labels into 8 single labels. Where each single label represents each possible combination of the 3 individual labels. This exponential increase in the number of labels does not cause challenges for classifiers that do not scale in complexity with the number of labels. However, HD computing complexity does scale with the number of labels. We address this issue with a binary classification transformation for HD computing explained below.

Multi-Model: We propose Multi-Model HD, a method of building a binary classification model for each label as the problem transformation method. Suppose $[l_1...l_h]$ are the labels of the dataset, then after mapping each data point into hypervectors $[v_1...v_n]$, we build h binary classification models, since each label only has a true or false value, i.e., 0 or 1. For example, if a dataset has $h = 3$, we create 3 different HD models, one for each label. Then upon inference, we feed the input data into all 3 of the models, independently checking for the existence of each label. This transformation method is better for HD in multiple ways: 1) HD model size, execution time, and energy scale with the number of classes, so when using Power Set HD, if there is a large number of possible label combinations, Power Set HD will not be as efficient as Multi-Model HD. 2) If a new label is introduced, in Multi-Model HD, we simply need to train a newly added binary classification HD model. However, with Power Set HD, or other models that use the power set transformation method, the entire model needs to be retrained to accommodate the new label combinations. The rest of Section III is mainly focused on Multi-Model HD, while we additionally provide a comparison with Power Set HD in Section V. Now that the problem has been transformed into k binary classification problems, we describe the algorithmic changes to HD computing blow.

978-1-7281-5410-7/20 $31.00 © 2020 IEEE

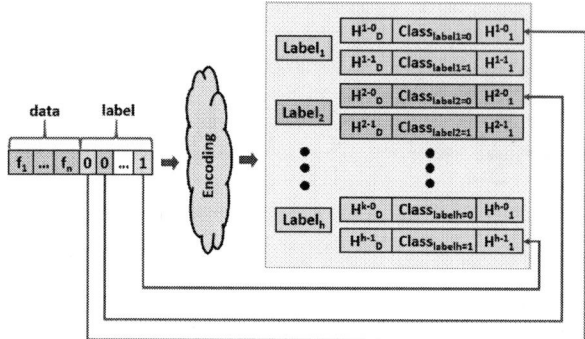

Fig. 1. An example of how the Multi-Model HD model is created

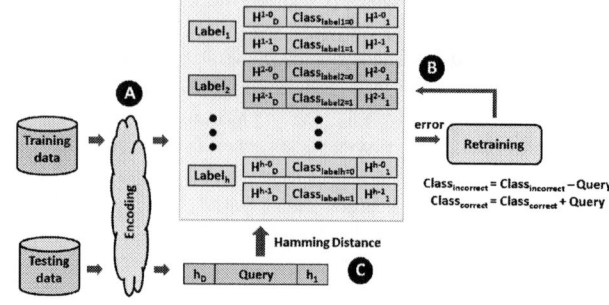

Fig. 2. Overview of how Multi-Model HD is constructed and how Multi-Model HD performs inference.

B. Encoding

HD computing encoding maps each n dimensional feature vector to a D dimensional binary hypervector. We utilize a random projection encoding presented in [21]. Let us assume a feature vector $\mathbf{F} = \{f_1, f_2, \ldots, f_n\}$, with n features ($f_i \in \mathbb{N}$) in original domain. The goal of encoding is to map this feature vector to a D (e.g. $D = 10,000$) dimensional space vector: $\mathbf{H} = \{h_1, h_2, \ldots, h_D\}$. The encoding first generates D dense bipolar vectors with the same dimensionality as original domain, $\mathbf{P} = \{\mathbf{p}_1, \mathbf{p}_2, \ldots, \mathbf{p}_D\}$, where $\mathbf{p}_i \in \{-1, 1\}^n$. Thus, to encode a feature vector into a hypervector, we perform a matrix vector multiplication between the projection matrix and the feature vector using:

$$\mathbf{H} = sign(\mathbf{P}\mathbf{F})$$

Where $sign$ is a sign function which maps the result of the dot product to +1 or -1. In Section IV-A we discuss how we accelerate encoding in flash.

C. Training

In HD computing, the model used in training is initialized through element-wise addition of all encoded hypervectors in each existing class. The result of training are k hypervectors each with D dimensions, where k is the number of classes. For example, the ith class hypervector can be computed as: $\mathbf{C}_i = \sum_{\forall j \in class_i} \mathbf{T_j}$. However, to map this algorithm to Multi-label classification, we need to modify how we create the initial HVs.

As stated in Section III-A since the multi-label classification problem is transformed into multiple binary classification problems with Multi-Model HD, we build two classes for each label (one for value 0 and one for value 1). As shown in Figure 1 after the same encoding process as stated in Section III-B each data point is classified into either $Class_{label_i=0}$ or $Class_{label_i=1}$ for each label i according to the values of its labels $[l_1 \ldots l_h]$. As shown in Figure 2 Ⓐ, for a dataset that has h labels, the binary model of this dataset

would contain $2h$ class HVs in total, one binary classification model for each label where each model contains 2 class HVs.

Unlike in single label classification, in Multi-Model HD, each data point is element wise added to multiple class HVs. For instance, in Figure 1 after the sample is encoded, it is added to the $Class_{label_1=0}$ class HV for the first label, as the first label is 0. It is then additionally added to the $Class_{label_2=0}$ class HV for the second label, as the second label is 0. This is continued for all the labels until it is added to the $Class_{label_h=1}$ class HV for the last label, as the last label is 1. After this procedure is repeated for the entire training set, we are left with k classification models for each label.

This training procedure also results in integer values for the dimensions of the class HVs, requiring the use of a costly cosine similarity during inference to find the best matching class HV to the query HV. We can reduce this computation to a binary operation of Hamming distance by binarizing the model, which is done by changing the class hypervector elements to +1 if they are positive and -1 if they are negative or 0. Hamming distance is desirable because it reduces each multiplication and addition in cosine similarity to a simple bitwise XOR and accumulation, which is significantly more efficient in acceleration circuits. The class with the least mismatching bits to the query is then chosen as the output.

D. Inference

After training, the HD model for single-label classification can now be used for inference. Upon inference, an input data is encoded to a *query* hypervector using the same encoding module used for training. HD Computing then computes the similarity between the *query* hypervector and each class hypervector. It then uses consine similarity to find a class hypervector with the most similarity with the query hypervector.

Multi-Model HD performs inference in a similar way, however, we need to output h labels instead of just 1. Figure 2 Ⓒ shows how inference is performed in Muti-Model HD. Upon inference, an input data is encoded to a *query* hypervector using the same encoding module used for training, just like baseline HD. However, since Multi-Model HD contains h different classification models, the query HV is input into

978-1-7281-5410-7/20 $31.00 © 2020 IEEE

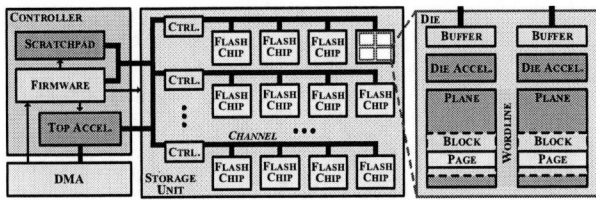

Fig. 3. Overview of Multi-label HD in 3D flash-based storage. ISC enabling components of the design are shown in green.

each classification model independently. For each model, if the query HV is more similar to the 0 label HV, then that label output is chosen as 0, and vice versa if the query is more similar to the 1 label HV. This generates our h different labels for output in a multi-label classification problem. In Multi-label HD, inference is performed on the host CPU.

IV. ACCELERATION WITH 3D NAND FLASH

Here, we present an ISC design that performs Multi-label HD encoding and training completely in 3D flash. Figure 3 shows an overview of the SSD architecture we adopt from THRIFTY [14]. It uses a die-level accelerator (green on the right in Figure 3), in each plane to encode every read page into a hypervector. These hypervectors are then sent to a SSD-level FPGA, which accumulates the hypervectors in the top-level accelerator (green on bottom left in Figure 3) to perform training. The scratchpad (green on top left in Figure 3) in the controller stores the encoding projection matrix, which it receives as an application parameter from the host. The top-level accelerator is an FPGA which uses INSIDER acceleration cluster [13] to implement HDC accumulation and other operations. We utilize THRIFTY's adaptation of INSIDER's software stack to connect our ISC architecture to the rest of the system.

A. Encoding in 3D Flash

As shown in Figure 3, the flash chip may consist of several flash dies which are further divided into flash planes, each plane consisting of a group of blocks, each of which store multiple pages. Each plane has a page buffer to write the data to. Operations in SSD happen in page granularity where the size of the pages usually ranges from 2KB-16KB. Hence, we use accelerators for each flash plane to exploit the flash hierarchy. These accelerators are multiplexed to the page read path.

The die-accelerator in [14] encodes an entire page with raw data to generate a D dimensional hypervector. We assume that the feature vectors are page-aligned, with each page storing one full feature vector. Multi-label HD encoding multiplies an n-size feature vector with a projection matrix containing $D \times n$ 1-bit elements. The accelerator calculates the dot product between the two vectors, one read from the flash array and another being a row-vector of the projection matrix. This involves element-wise multiplication of the two vectors and adding together all the elements in the product. Since the

weights in the projection matrix $\in \{1, -1\}$, we map them to $\{0, 1\}$ respectively. We use 2's complement to break the multiplication into an inversion using XOR gates and then add the total number of inverted inputs to the accumulated sum of XOR outputs. With the assumption that each page consists of a maximum 1K feature elements, the accelerator consists of an array of 32K XOR gates followed by a 1024 input tree adder. It reduces 1024 inputs to 2, which is followed by a carry look ahead addition to get the final dot product. The sign bit (MSB) of the output is the value of one dimension of the encoded hypervector. Complementary to the projection matrix, the output $0 \rightarrow 1$ and $1 \rightarrow (-1)$. The accelerator is iteratively run D times to generate D dimensions. Each encoded hypervector is appended with the corresponding label vector. We write the output of the accelerator to the page buffer of the plane, which serves as the response to the original read request.

B. Training at Top-Level in Storage

The encoded hypervectors from flash chips are input into the top-level accelerator, which is implemented on an FPGA present in the SSD. During training, they are accumulated into the corresponding label hypervectors. At the end of training we obtain two output hypervector for every label ($label_i$), one each for $Class_{label_i=0}$ and $Class_{label_i=1}$.

The design in [14] utilized input queues for each class to increase parallelism between different classes. However in Multi-label HD, each encoded hypervector is added to one of the two classes of each label, i.e. 50% of the classes. Moreover, ideally an encoded hypervector has just one label as '1' while rest are '0's. Hence, all but one classes corresponding to $label_i = 0$ would receive an incoming hypervector. There is negligible parallelism in training between multiple encoded hypervectors. In this case, the input queues of [14] are an overkill. Hence, we remove input queues from the FPGA design of [14]. The label vector of an incoming hypervector is used to input it to the corresponding class ($Class_{label_i=0}$ or $Class_{label_i=1}$) of each label. The inputs to the remaining classes are set to zero. An accumulator is present for each class, which simply needs to read the input and operate on the corresponding data. The accumulators for each class operates in parallel to add an input hypervector to the corresponding class hypervector. While the computation can also be fully parallelized over all dimensions, the large size of hypervectors and the limited read ports of the memory make it impractical. Hence, we utilize the partition-based approach used in [14] to allow partial parallelism. The final class hypervectors are sent to the host.

V. EXPERIMENTAL RESULTS

A. Experiment Setup

We tested Multi-label HD training and inference using an optimized C++ implementation. For comparison, we utilized the open source Mulan multi-label package, which is implemented in Java [22]. We compare Multi-label HD with multi-label versions of k-nearest neighbors (kNN), Sequential

978-1-7281-5410-7/20 $31.00 © 2020 IEEE

13

TABLE I
MULTI-LABEL HD 3D STORAGE PARAMETERS

Capacity	$1TB$	Channels	32
Page Size	$16KB$	Chips/Channel	4
External BW	$3.2GBps$	Planes/Chip	8
BW/Channel	$800MBps$	Blocks/Plane	512
Flash Latency	$53us$	Pages/Block	128
FPGA	$XCKU025$	Scratchpad Size	$4MB$
Avg Power/DA	$8mW$	DA Latency	$1.02ns$

*DA: Die-accelerator

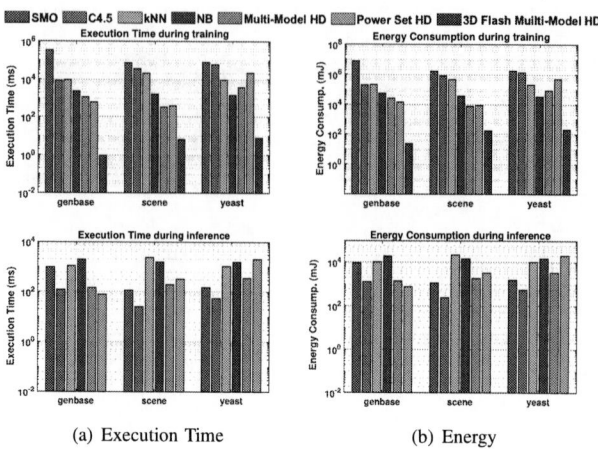

(a) Execution Time   (b) Energy

Fig. 5. Energy consumption and execution time of Multi-label HD during Encoding and Training.

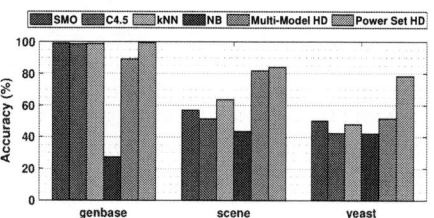

Fig. 4. Classification accuracy of Multilabel HD and other multi-label classification algorithms.

minimal optimization (SMO), C4.5, and Naive Bayes (NB). We also developed a simulator for Multi-label HD in flash which supports parallel read and write accesses to the flash chips. We utilized Verilog and Synopsys *Design Compiler* to implement and synthesize the die-level accelerator at 45nm and scale it down to 22nm. The top-level FPGA accelerator has been synthesized and simulated in Xilinx Vivado. For drive simulation, we assume the characteristics similar to 1TB Intel DC P4500 PCIe-3.1 SSD connected to an Intel(R) Xeon(R) CPU E5-2640 v3 host. The parameters for our 3D flash implementation are shown in Table I. We compare flash implementation with 6th Gen 3.2GHz Sky Lake Intel Core i5-6300HQ CPU with 8GB of RAM and a 256 GB SSD.

We tested our proposed approach on three applications:
Genbase (Genbase) [23]: The protein classes considered are the 27 most important protein families. The training and testing datasets are taken from the Genbase dataset. This dataset consists of 662 samples, each with 1186 attributes.
Scene (Scene) [24]: This dataset contains characteristics about images and their classes. One image can belong to one or more classes. The training and testing datasets are taken from the Scene dataset. This dataset contains 2407 samples, each with 294 attributes.
Yeast (Yeast) [25]: This database contains information about a set of Yeast cells. The task is to determine the localization site of each cell. The training and testing datasets are taken from the Yeast dataset. This dataset consists of 2417 samples, each with 103 attributes.

B. Multi-label HD Comparison with State-of-the-Art

1) Accuracy:

Figure 4 compares the multi-label classification accuracy of current state-of-the-art multi-label classifiers with Multi-label HD. The accuracy for multi-label is calculated by first getting

the accuracy of the model on each label individually. Then to aggregate them, we average each label's accuracy together to get one overall accuracy number for each dataset. As the figure shows, Multi-label HD (Multi-Model HD and Power Set HD) are comparable in accuracy to state-of-the-art multi-label classifiers. In fact, Power Set HD is always better than the state-of-the-art on these three datasets. On the other hand, Multi-model is slightly less accurate than other multi-label classifiers on the genbase dataset by 10%. However, Multi-Model HD is able to achieve higher accuracy on the scene and yeast datasets. This could be attributed to mapping the data into HD space, offering better separability than in the low dimensional data. However, more theoretical analysis on HD Computing is necessary in order to understand why Multi-label HD is more accurate. Overall, on average, Multi-Model HD is 5% more accurate and Power Set HD is 14% more accurate than the highest accuracy state-of-the-art multi-label classifier.

Although Power Set HD achieves higher accuracy than Multi-Model HD, Figure 5 demonstrates that the improvement in accuracy comes at a significant cost in execution time and energy. This is because of the exponential increase in class HVs as discussed in Section III-A. As mentioned before, the exception is the genbase dataset because there is only a small subset of possible combinations that appear in the dataset. On the other hand, when there is a large portion of possible combinations in the dataset, Power Set HD is $3.6\times$ slower than Multi-Model HD. This offers a trade-off between execution time and energy efficiency vs accuracy. If an application requires the highest accuracy, Power Set HD should be used. However, if the key metric is execution time and energy efficiency, for a loss in accuracy compared to Power Set HD, but still comparable with other state-of-the-art multilabel classifiers, Multi-Model HD is the clear choice. If the dataset does not have a diverse combination of labels, such as in genbase, Power Set HD can potentially be more accurate and energy efficient compared to Multi-Model HD.

2) CPU Execution Time and Energy:

Figure 5 compares the execution time and energy consumption of state-of-the-art multi-label classifiers with Multi-label HD on CPU. The data demonstrates that both Multi-Model HD and Power Set HD training are significantly faster than most other multi-label classifiers. On average, Multi-label HD is $60.8\times$ faster and $61.8\times$ more energy efficient than other multi-label classifiers during training. The one exception is Naive Bayes on the yeast dataset, however, although Naive Bayes trains significantly faster than Multi-Model HD on the yeast dataset, Multi-Model HD is $8.6\times$ faster and $8.7\times$ more energy efficient than Naive Bayes during inference. Additionally, Power Set HD is only $3.5\times$ slower than Multi-Model HD on datasets with a large portion of label combinations.

Figure 5 also demonstrates that Multi-Model HD is also significantly faster than kNNs and Naive Bayes multi-label models during inference. Although Multi-Model HD is comparable in execution time and energy efficiency to SMO and C4.5 during inference, Multi-Model HD is $174.4\times(42.8\times)$ faster and $178.1\times(43.1\times)$ more energy efficient than SMO(C4.5) during training. Overall, combining training and one iteration of inference, Multi-Model HD is $47.1\times$ faster and $47.8\times$ more energy efficient than state-of-the-art multi-label classifiers on average, while providing 5% higher classification accuracy. On the other hand, Power Set HD is $24\times$ faster than state-of-the-art multi-label classifiers on average or approximately $2\times$ slower than Multi-Model HD for 13% higher accuracy.

C. Multi-label HD in 3D Flash

Figure 5 also shows the latency and energy consumption of Multi-label HD when accelerated in flash. We implement Multi-label HD encoding and training in flash over the three datasets. We observe that our 3D-flash implementation of Multi-label HD is on average $610\times$ faster and $228\times$ more energy-efficient than CPU. Our evaluations show that the performance and energy consumption of Multi-label HD in 3D-flash increases linearly with an increase in the number of training samples. This happens because more data samples result in more huge hypervectors to generate and process. In conventional systems, this translates to a huge amount of data transfers between the core and memory. In contrast, our 3D-flash implementation generates hypervectors (encoding) while reading data out of the slow flash arrays and processes (training) them on the disk itself, reducing data movement.

VI. CONCLUSION

In this paper, we design the first accelerator for multi-label HD classification in 3D storage. We also propose two different transformation methods to map HD single label classification to multi-label classification: Power Set HD and Multi-Model HD. Overall, combining training and one iteration of inference, Multi-Model HD is $47.1\times$ faster and $47.8\times$ more energy efficient than state-of-the-art multi-label classifiers, while also achieving 5% higher accuracy on average. Power Set HD can achieve 13% higher accuracy than Multi-Model HD, but is $2\times$ slower. We additionally propose in-3D-flash acceleration that

further improves the energy efficiency of Muilti-Model HD training by $228\times$ and speedup by $610\times$.

ACKNOWLEDGEMENTS

This work was supported by NSF grants #1527034, #1730158, #1826967, and #1911095.

REFERENCES

[1] J. Gubbi *et al.*, "Internet of things (iot): A vision, architectural elements, and future directions," *Future generation computer systems*, vol. 29, no. 7, pp. 1645–1660, 2013.

[2] G. Tsoumakas and I. Katakis, "Multi-label classification: An overview," *International Journal of Data Warehousing and Mining (IJDWM)*, vol. 3, no. 3, pp. 1–13, 2007.

[3] T. Durand, N. Mehrasa, and G. Mori, "Learning a deep convnet for multi-label classification with partial labels," in *Proceedings of the IEEE Conference on Computer Vision and Pattern Recognition*, pp. 647–657, 2019.

[4] Y. Sun *et al.*, "Internet of things and big data analytics for smart and connected communities," *IEEE Access*, vol. 4, pp. 766–773, 2016.

[5] M. Imani *et al.*, "Exploring hyperdimensional associative memory," in *HPCA*, pp. 445–456, IEEE, 2017.

[6] P. Kanerva, "Hyperdimensional computing: An introduction to computing in distributed representation with high-dimensional random vectors," *Cognitive Computation*, vol. 1, no. 2, pp. 139–159, 2009.

[7] O. Rasanen and J. Saarinen, "Sequence prediction with sparse distributed hyperdimensional coding applied to the analysis of mobile phone use patterns," *IEEE Transactions on Neural Networks and Learning Systems*, vol. PP, no. 99, pp. 1–12, 2015.

[8] M. Imani *et al.*, "Voicehd: Hyperdimensional computing for efficient speech recognition," in *ICRC*, pp. 1–6, IEEE, 2017.

[9] A. Rahimi *et al.*, "A robust and energy-efficient classifier using brain-inspired hyperdimensional computing," in *ISLPED*, pp. 64–69, ACM, 2016.

[10] I. Jo, D.-H. Bae, A. S. Yoon, J.-U. Kang, S. Cho, D. D. Lee, and J. Jeong, "Yoursql: a high-performance database system leveraging in-storage computing," *Proceedings of the VLDB Endowment*, vol. 9, no. 12, pp. 924–935, 2016.

[11] B. Gu, A. S. Yoon, D.-H. Bae, I. Jo, J. Lee, J. Yoon, J.-U. Kang, M. Kwon, C. Yoon, S. Cho, *et al.*, "Biscuit: A framework for near-data processing of big data workloads," *ACM SIGARCH Computer Architecture News*, vol. 44, no. 3, pp. 153–165, 2016.

[12] G. Koo, K. K. Matam, I. Te, H. K. G. Narra, J. Li, H.-W. Tseng, S. Swanson, and M. Annavaram, "Summarizer: trading communication with computing near storage," in *2017 50th Annual IEEE/ACM International Symposium on Microarchitecture (MICRO)*, pp. 219–231, IEEE, 2017.

[13] Z. Ruan, T. He, and J. Cong, "Insider: designing in-storage computing system for emerging high-performance drive," in *Proceedings of the 2019 USENIX Conference on Usenix Annual Technical Conference*, pp. 379–394, 2019.

[14] S. Gupta, J. Morris, M. Imani, R. Ramkumar, J. Yu, A. Tiwari, B. Aksanli, and T. Rosing, "Thrifty: Training with hyperdimensional computing across flash hierarchy," in *Proceedings of the IEEE/ACM 2020 International Conference on Computer-Aided Design (ICCAD)*, 2020.

[15] Y. Kim *et al.*, "Efficient human activity recognition using hyperdimensional computing," in *IoT*, p. 38, ACM, 2018.

[16] M. Imani *et al.*, "Hdcluster: An accurate clustering using brain-inspired high-dimensional computing," in *DATE*, IEEE/ACM, 2019.

[17] M. Imani *et al.*, "Hdna: Energy-efficient dna sequencing using hyperdimensional computing," in *IEEE BHI*, pp. 271–274, IEEE, 2018.

[18] A. Rahimi *et al.*, "Hyperdimensional biosignal processing: A case study for emg-based hand gesture recognition," in *ICRC*, pp. 1–8, IEEE, 2016.

[19] M. Schmuck, L. Benini, and A. Rahimi, "Hardware optimizations of dense binary hyperdimensional computing: Rematerialization of hypervectors, binarized bundling, and combinational associative memory," *ACM Journal on Emerging Technologies in Computing Systems (JETC)*, vol. 15, no. 4, pp. 1–25, 2019.

[20] V. S. Mailthody, Z. Qureshi, W. Liang, Z. Feng, S. G. De Gonzalo, Y. Li, H. Franke, J. Xiong, J. Huang, and W.-m. Hwu, "Deepstore: In-storage acceleration for intelligent queries," in *Proceedings of the 52nd Annual IEEE/ACM International Symposium on Microarchitecture*, pp. 224–238, 2019.

[21] M. Imani, J. Morris, J. Messerly, H. Shu, Y. Deng, and T. Rosing, "Bric: Locality-based encoding for energy-efficient brain-inspired hyperdimensional computing," in *Proceedings of the 56th Annual Design Automation Conference 2019*, pp. 1–6, 2019.

[22] G. Tsoumakas, E. Spyromitros-Xioufis, J. Vilcek, and I. Vlahavas, "Mulan: A java library for multi-label learning," *Journal of Machine Learning Research*, vol. 12, pp. 2411–2414, 2011.

[23] S. Diplaris, G. Tsoumakas, P. A. Mitkas, and I. Vlahavas, "Protein classification with multiple algorithms," in *Panhellenic Conference on Informatics*, pp. 448–456, Springer, 2005.

[24] M. R. Boutell, J. Luo, X. Shen, and C. M. Brown, "Learning multi-label scene classification," *Pattern recognition*, vol. 37, no. 9, pp. 1757–1771, 2004.

[25] A. Elisseeff and J. Weston, "A kernel method for multi-labelled classification," in *Advances in neural information processing systems*, pp. 681–687, 2002.

Early RTL Analysis for SCA Vulnerability in Fuzzy Extractors of Memory-Based PUF Enabled Devices

Xinhui Lai[1], Maksim Jenihhin[1], Georgios Selimis[2], Sven Goossens[2], Roel Maes[2], Kolin Paul[3]

[1] Department of Computer Systems, Tallinn University of Technology, Estonia
[2] Intrinsic ID, The Netherlands
[3] Department of Computer Science & Engg, Indian Institute of Technology Delhi, India
Email: xinhui.lai@taltech.ee

Abstract—Physical Unclonable Functions (PUFs) are gaining attention in the cryptography community because of the ability to efficiently harness the intrinsic variability in the manufacturing process. However, this means that they are noisy devices and require error correction mechanisms, e.g., by employing Fuzzy Extractors (FEs). Recent works demonstrated that applying FEs for error correction may enable new opportunities to break the PUFs if no countermeasures are taken. In this paper, we address an attack model on FEs hardware implementations and provide a solution for early identification of the timing Side-Channel Attack (SCA) vulnerabilities which can be exploited by physical fault injection. The significance of this work stems from the fact that FEs are an essential building block in the implementations of PUF-enabled devices. The information leaked through the timing side-channel during the error correction process can reveal the FE input data and thereby can endanger revealing secrets. Therefore, it is very important to identify the potential leakages early in the process during RTL design. Experimental results based on RTL analysis of several Bose–Chaudhuri–Hocquenghem (BCH) and Reed-Solomon decoders for PUF-enabled devices with FEs demonstrate the feasibility of the proposed methodology.

Keywords - timing side-channel attack, physical unclonable function, fuzzy extractor, fault-injection attack, error correction code, BCH, Reed-Solomon, RTL analysis.

I. INTRODUCTION

Physical unclonable functions (PUFs) are hardware primitives which derive identifiers and cryptographic keys from the random variations of the silicon manufacturing process. PUFs provide a significantly higher security assurance as keys are volatile and derived only when required. Thus, a PUF can be easily attached or embedded into the cryptographic implementation for authentication and identification [1]. PUF-enabled devices are also an efficient alternative to the expensive conventional measures against the integrated circuit power-off, e.g., by using the Non-Volatile Memory (NVM) for the key storage. The keys generated by PUFs are derived by measurements in the field during the run time and can be saved in a cheaper volatile memory.

PUFs are known to be sensitive to the environmental factors such as the ambient temperature, the supply voltage noise, etc. that may affect the reliability of the response measurement, and ultimately, reduce the reproducibility of the cryptographic key. Along with the external factors, the internal factors of the

PUF's manufacturing technology prevent it from guaranteeing a constant response all the time. This nondeterminism poses issues for applying a PUF as a key generator or identifier [2]. Therefore, for the post-processing, a Fuzzy Extractor (FE) is an essential component to help a PUF generate a reliable key by correcting the errors caused internally or by environmental variations.

Different types of the PUF structure and the environmental conditions imply different requirements for the FE and the corresponding ECC. An example of a silicon PUF is the memory-based PUF, which is widely used in chip-level authentication. FE ECCs such as the Bose–Chaudhuri–Hocquenghem (BCH) [2] or Reed-Solomon [3] are used in memory-based PUF enabled devices.

While FEs with ECCs significantly raise reliability, they can lead to new exploits such as allowing an attacker to extract sensitive information by studying the behavior of ECC. Side-Channel Attacks (SCA) on ECC implementations have attracted particular attention of the research community. In [4], the authors extract the information about the key by non-invasive measurement of electromagnetic radiation together with a differential power analysis of the BCH decoder. In [5], the authors study the simple power analysis of both BCH and Reed-Solomon code and manage to recover the PUF response from the collected power traces. However, there is no research work that refers to attacks that combine timing SCA and fault attacks for FEs, namely targeting to the execution time of the error-correcting code of FE in combination with the insertion of faults to PUF. So in this paper, we address this gap by a study on BCH and Reed-Solomon RTL designs execution time differences as a reaction to intentionally triggered faults inserted to PUF. Specifically, the contributions of the paper include:

- Definition of an attack model based on fault injection and timing analysis of ECC execution that may lead to the secret PUF values extraction.
- An early design stage RTL methodology for verification of an ECC design invulnerability against the proposed attack by employing both structural and simulation-based analysis steps.
- Case studies of Reed-Solomon and BCH based ECC with vulnerabilities identification and exploitation.

978-1-7281-5410-7/20 $31.00 © 2020 IEEE

The rest of the paper is organized as follows. Section II reviews the background of the FE architecture and ECC decoders. The attack model is discussed in Section III. Section IV presents the proposed methodology for verifying invulnerability against the proposed attack. Section V presents a case study for ECC implementations. Section VI concludes the paper.

II. BACKGROUND AND RELATED WORKS

A. Fuzzy Extractor and Secure Sketch

The Fuzzy Extractor [6] is a secure method to generate cryptographic keys from noisy sources. The FE serves as a post-processing unit in memory-based PUF-enabled cryptographic schemes. It is used both in the Generation and Reconstruction Procedures, as illustrated in Fig. 1 and Fig. 2 correspondingly.

In the Generation Procedure case, the fuzzy data from the PUF response W and a random secret S are used to generate the Helper Data by XOR operation on W and $E(S_0)$ which is encoded S_0. The generated helper data is stored in a non-volatile memory. In memory-based PUF-enabled devices, the Generation Procedure happens only once at the first-time power-on of the memory-based PUF.

Generation Procedure

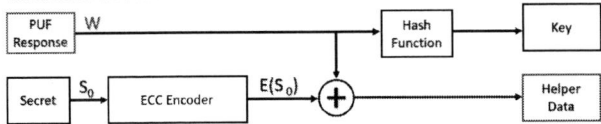

Fig. 1. Generation Procedure in A PUF Fuzzy Extractor

On the contrary to this, the Reconstruction Procedure is executed many times during the product lifetime. Due to the noise and PUF manufacturing randomness, it is difficult to generate the same response consistently. To reproduce the correct cryptographic key, the Helper Data, stored in an NVM, is used in conjunction with the measured PUF response W'. Then with the help of the ECC decoder to detect and correct the divergent bits, the correct W is reproduced. After applying the Hash Function, the expected correct cryptographic key is reconstructed.

Reconstruction Procedure

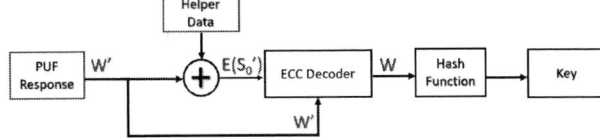

Fig. 2. Reconstruction Procedure in A PUF Fuzzy Extractor

The FE guarantees that the resulting key is consistent while the publicly accessible Helper Data does not leak any information related to the secret of the key. To ensure consistent generation of the correct key, the hamming distance between the measured PUF response W' with the originally measured W in the Generator Procedure should be smaller or equal to the correction capability of the ECC decoder, represented as a constant value t. In this paper, we assume that the measured responses of the memory-based PUF are within this hamming distance constraint.

Recent research works have identified potential attacks on FEs [7]. Most of them target the Reconstruction Procedure. In [8], the authors report on a method to extract the PUF secret by manipulating the Helper Data in the Reconstruction Procedure. In [9], Delvaux et al. provide an in-depth analysis of the Helper Data algorithms, and identify new threats for leaking the Helper Data and the soft-decision coding.

B. ECC decoder

The ECC unit is the main component in a FE. Binary BCH and Reed-Solomon are the two types of ECC that are widely used in PUF-enabled devices. Both codes are cyclic and capable of detecting up to $2t$ and correct up to t errors by adding $2t$ check bits or non-binary values (symbols) to the data. Binary BCH is used for binary error correction, and Reed-Solomon is used for symbol error correction. While both software and hardware implementations of these codes exist, the hardware ones are more adopted. First, this is because the complex algorithms of the decoders require significant computational power along with the real-time constraints. The second difficulty for software implementations is the limited support of the Galois Fields Arithmetic operation in the general-purpose processors [10]. The hardware implementations of binary BCH and Reed-Solomon decoders are discussed in more detail in Section V.

III. ATTACK MODEL

In this paper, we assume an attack combining 1) fault injection to the memory-based PUF with 2) a timing SCA for observing and comparing the different decoding execution times of the ECC unit that is aimed at revealing the correct memory-based PUF data. In case of success, the attack explicitly compromises the core function of the PUF-enabled cryptographic devices, because the attacker can clone the PUF and can steal the secret.

A. Fault Injection Parameters

For the physical fault injection to the memory-based PUF the following fault parameters are assumed.

(a) Granularity: each fault injection results in exactly one fault in one-bit data.
(b) Modification (fault type): after the fault injection, the manipulated data is set to a specified logic value, i.e. either '1' or '0'.
(c) Control: the attacker has a bit-wise precise control of fault injection to the memory-based PUF bits.
(d) Effect of the fault: the injected faults have a transient nature, i.e. the injected values are overwritten by the normal functionality of the device (e.g. the next measurement of the PUF on power-on).

Several studies on laser fault injection [11] have demonstrated similar attack parameters and, therefore, the feasibility of the above assumptions. Technical details of the fault injection attack implementation are out of the scope of this study.

978-1-7281-5410-7/20 $31.00 © 2020 IEEE

B. Attack Assumptions

The following set of assumptions must be satisfied for the success of the attack. The feasibility of the assumptions (iii)-(vi) is supported by several research works in state of the art.

(i) The output of a memory-based PUF measurement in the cryptographic device is processed by a FE with a binary BCH or Reed-Solomon based ECC.

(ii) The ECC implementation leaks exploitable information through the timing-side channel.
Comment: The methodology for identifying the vulnerability enabling this assumption is the core contribution of this paper and presented in Section IV.

(iii) The memory-based PUF is noise-free under stable environmental conditions. The errors in the memory-based PUF are caused by the environment.
Comment: While an ideal noise-free memory-based PUF would not require the FE at all, we assume that the noise is caused by the variations in the external environment while the internal noise is negligible. [12] demonstrated that the external environmental conditions like the ambient temperature, supply voltage, etc. have a significant impact on the error rate of the PUF.

(iv) The generated Helper Data is stored in NVM or the flash memory of the cryptographic devices and remains constant during the Reconstruction Procedure.
Comment: As an added value, this assumption creates an advantage for the proposed attack, compared to alternatives (e.g. [8], [9]), because it does not rely on the attacker being able to modify the Helper Data.

(v) The fault injection parameters (a) to (d) hold (see III.A).
Comment: Several research works proposed bit-wise fault injection in SRAM and other on-chip memories. E.g., in [13], bit-wise faults were successfully injected in a PIC microcontroller through a semi-invasive method and without mechanical damage to the silicon.

(vi) The attacker has a controlled access for measuring the decoding execution time.
Comment: The physical measurement of the ECC decoding execution time can exploit the reflection of timing by the power traces. In [14], the authors analyze use of the AES execution power traces for a SCA. The power traces are represented by changes of power over time, with the timing information embedded. A similar approach is used in [15] for RTL verification of RSA designs against vulnerability to timing SCAs.

C. Attack Procedure

The proposed attack is a combination of fault injection with timing side channel analysis and represented by the following 4 steps. The procedure is illustrated in Fig.3.

1) Power on the device. Measure the initial PUF data. With the above assumptions, this memory value should be error-free, i.e. the same with W generated in the Generation Procedure. Measure and record the reference time T as the number of clock cycles for the execution of the ECC decoding.

2) Inject a fault f at the m_{th} bit of memory-based PUF following the (a) to (d) parameters and generate the new memory data W_{m_f}. W_{m_f} has a one-bit difference value compared to W. E.g, if the f is a set to logic "1" value and $m = 1$ then W and W_{1_f} can be either equal or can be different by exactly one bit at the first position. Then execute the Reconstruction Procedure, measure the decoding execution time $T(m)$.

3) The relation between these two decoding times T and $T(m)$ contains only two possible cases. The PUF's secret single bit m can be revealed by comparing the two decoding times as follows:

- if $T! = T(m)$, then a different value at the m_{th} bit was injected. E.g., for $f = 1$, the original value of the m_{th} bit in memory is '0';
- if $T = T(m)$ then the value at the m_{th} bit was equal to the injected one. E.g., for $f = 1$, the original value of the m_{th} bit in memory is '1';

4) Repeat the steps 1) to 3). of the procedure until the last m_{th} bit of memory-based PUF. The memory-based PUF's secret value is revealed.

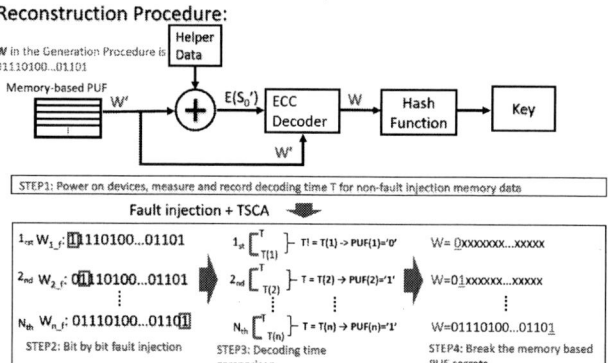

Fig. 3. An illustration of the proposed attack procedure

IV. PROPOSED METHODOLOGY

The precondition for the introduced attack is the non-constant decoding execution time in case of different input data for the ECC unit of the memory-based PUF Fuzzy Extractor. In this section, we propose a methodology to identify this vulnerability in an ECC implementation already at the RTL design phase. The methodology employs both structural and simulation-based analysis for binary BCH and Reed-Solomon algorithms based hardware ECC implementations. In practice, these two algorithms are widely used by the industry in memory-based PUF-enabled devices.

A. Structural Analysis of ECC Decoder

1) Binary BCH Decoder: A general binary BCH decoder hardware implementation has three stages, as shown in Fig.4. The divergent (error) bits are identified by the Syndrome Calculator, Key Equation Solver and the Chien Search. Next, the decoder corrects the error bits by the XOR operation on the stored input with the identified error bits to recover the correct codeword. Let r(x), c(x) and e(x) be the received polynomial, codeword polynomial and error polynomial, i.e. r(x) = c(x) + e(x). Assume the binary BCH decoder can correct t errors. As the structural analysis of the binary BCH, we consider the

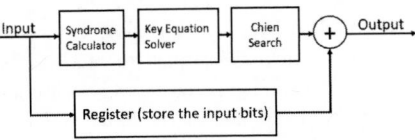

Fig. 4. Binary BCH Decoder Structure

following reasoning.

- **Syndrome Calculator:** It is the first stage in the decoder generates 2t syndromes as defined in (1).

$$S_i = r(x^i) = r_0 + r_1 x^i + r_2 x^{2i} + \ldots + r_{n-1} x^{(n-1)i} \quad (1)$$

where $1 \leq i \leq 2t - 1$. An important feature of the syndromes is that they do not depend on transmitted information but only on error locations. If at position i there is an error then S_i has a non-zero value and it is equal to zero otherwise. For all possible inputs, the decoder always generates $2t$ syndromes. Therefore, the time for the syndrome calculation is constant for the BCH decoder with a fixed error correction capability.

- **Key Equation Solver:** In the second stage, the error location polynomial $\sigma(x)$ is generated. Berlekamp Massey Algorithm (BMA) is one known iterative procedure that determines polynomial equation (2) out of a set of linear equations for the 2t syndromes calculated in the first stage.

$$\sigma(x) = 1 + \sigma_1 x + \sigma_2 x^2 + \ldots \sigma_t x^t \quad (2)$$

BMA can be implemented in parallel or serially. In [16], it is demonstrated that a parallel implementation for a t errors correction BMA needs 2t iterations. A serial implementation implies a significant increase in the number of iterations. According to [17], it needs $2t^2$ iterations. However, for both cases, the total number of iterations is determined only by t, which is the maximum number of errors the decoder can correct.

- **Chien Search:** This stage searches for error locations by checking the roots of $\sigma(x)$. It is a simple trial-and-error procedure. All nonzero elements of the Galois Fields for a binary BCH decoder are generated in sequence and only capture the condition when $\sigma(x_i)$ is equal to zero which the error position. Therefore, in this stage, the total number of nonzero elements depends only on the Galois Field $GF(2^m)$ where $n = 2^m - 1$ and n is the size of codeword.

To conclude, for different binary BCH decoder implementations, the error correction bits and the size of the codeword are the factors which lead to the different decoding execution time. However, for a specific binary BCH decoder, these parameters are fixed at the design phase. Therefore, the structural analysis has not identified timing channels in binary BCH decoder structures.

2) Reed-Solomon Decoder: Reed-Solomon (RS) decoder aimes at non-binary (symbol) error correction. Different from the binary BCH, which needs only to generate error locator polynomial $\sigma(x)$ RS also needs to generate an error value polynomial. Therefore, some RS implementations replace BMA by Euclidean Algorithm (EA) for the Key Equation Solver to calculate the error location polynomial and error value polynomial and add a new component Forney to calculate the error value. The Reed-Solomon decoder structure is illustrated in Fig.5. Here, the differences with the BCH decoder structure are highlighted in red. In the following structural analysis, we focus only on these two different components.

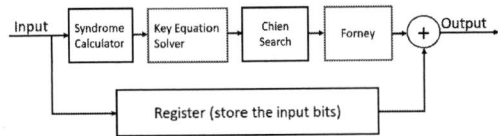

Fig. 5. Reed-Solomon Decoder Structure

- **Euclidean Algorithm (EA):** It is an iterative procedure to generate the *error locator polynomial* and the *error value polynomial* with the 2t syndromes generated by the Syndrome Calculator stage. Particular implementations of EA may prefer a pipelined version with the objective of performance optimization [18]. In EA procedure [18], the error locator polynomial $\sigma(x)$ and the error value polynomial $\omega(x)$ are acquired by solving the equation (3). Equation (3) can be represented in the form of equation (4). The extend Euclidean Algorithm can find a series polynomial by (5). From (4) and (5), $A_i(x) = \sigma(x)$, $R_i(x) = \omega(x)$ and $B_i(x) = -Q(x)$. To solve the Key Equation the EA procedure starts with initiating the values $R_0(x) = x^{2t}$, $Q_0(x) = S(x)$, $L_0(x) = 0$, $U_0(x) = 1$ and then it is followed by interactions of four equations used to calculate $R_i(x)$, $Q_i(x)$, $L_i(x)$ and $U_i(x)$, based on the values from the previous stage, until the degree of $R_i(x)$ gets smaller than the degree of $L_i(x)$ or t. When the iteration is finished, the equation (3) is solved. Because the $R(x)$ starts at the degree 2t, and the iteration can finish at the degree of $R(x)$ equal to t or smaller. Therefore, the EA stage may require a different number of iterations for the different codewords which may introduce different execution times.

$$\omega(x) = S(x)\sigma(x) \mod x^{2t} \quad (3)$$
$$\sigma(x)S(x) = Q(x)x^{2t} + \omega(x) \quad (4)$$
$$A_i(x)S(x) + B_i(x)x^t = R_i(x) \quad (5)$$

- **Forney:** By using the Forney algorithm, the error value e(x) can be acquired by the equation (6).

$$e_j = -\frac{\omega(X_j)}{\sigma'(X_j)} \quad (6)$$

Normally, it is implemented in combinational logic because $\omega(X)$ and $\sigma(x)$ are available. The execution time of this stage is constant.

To conclude, the structural analysis has not identified the

timing channel in the other stages of the Reed-Solomon structure but the second stage. Based on the implementation, the Key Equation Solver stage in the Reed-Solomon based ECC decoder can introduce the vulnerability.

B. Simulation-based analysis of ECC decoder

In an RTL simulation of an ECC decoder implementation, a number of stimuli data parameters may have an impact on the execution time of a decoding iteration. For the proposed simulation-based analysis step, the following parameters are identified:

- $codeword_{value}$: the encoded codeword value
- $error_{value}$: the error value is relevant only for a non-binary (symbol) ECC decoders
- $error_{position}$: the error bit position for a binary ECC decoder or the error symbol position for a non-binary ECC decoder
- $error_{number}$: the number of error bits or symbols for binary or non-binary ECC decoder correspondingly

The structural analysis of binary BCH and RS decoders and the defined attack model allows reducing the search space. Table I presents the relationship of the execution time variation introduced by manipulating a particular decoding parameter and the vulnerability to the proposed attack. The notations C and NC represent constant and non-constant decoding execution time, while V and NV represent vulnerability or invulnerability.

TABLE I
ECC EXECUTION TIME VARIABILITY AND THE SCA VULNERABILITY

ECC Decoding Execution Time/Vulnerability		
Parameters	RS decoder	Binary BCH decoder
$codeword_{value}$	C/NV	C/NV
$error_{value}$	C/V	
$error_{position}$	C/V	C/V
$error_{number}$	NC/V	C/V

In particular, manipulation of the $codeword_{value}$ parameter does not identify the vulnerability of the target decoder. The attacker does not have access to manipulate the predefined correct codeword and can only manipulate the input codeword to cause an error. Based on the structural analysis, it is already known that different codewords do not introduce different decoding time neither in binary BCH nor in RS structures. The $error_{value}$ and $error_{position}$ parameters can be manipulated by the attacker by injecting faults to the input codeword. However, the constant decoding time will not leak information through the timing channel. From Table I, we can conclude that the binary BCH decoder structures are secure with regards to the information leakage through the timing channel. An RS decoder implementation can be vulnerable if the attacker injects a different number of error symbols, i.e. the $error_{number}$. The table guides the designer which simulation campaigns are required to verify a particular implementation against vulnerability to the proposed SCA.

V. CASE STUDY

The feasibility of the proposed methodology was validated by running an exhaustive simulation campaign on 3 case study

ECC designs for memory-based PUF Fuzzy Extractors, i.e. 2 binary BCH and a Reed-Solomon ECC implementations.

A. Binary BCH decoder

The implementation of the binary BCH decoder is an open-source design in RTL Verilog accessible from Github [19]. Its general architecture is illustrated in Fig. 4. The decoder was configured for a 12-bit codeword, 8-bit message and supports two types of BMA, i.e. serial *BMA_serial* and parallel *BMA_parallel* versions. The configuration was set to correct up to two errors, i.e. $t = 2$. Both versions were simulated with an exhaustive set of test vectors to identify the timing information leakage. Only valid values for the 12-bit binary codeword were extracted by running the encoder with all possible inputs. The input for the encoder is 4-bit message and 2-bit error correction capability. Since the number of errors correctable for a given polynomial is sparse, the encoder has the selection algorithm to select suitable polynomial function to meet the provided requirements. Thus the actual message bit might be changed. In our case, the encoder pads 4-bit zeros and makes the input message bit 8-bit. We input all possible 4-bit value into encoder. Then each encoded message value was merged with all possible error combinations considering the injection of 0, 1 or 2 errors at a time, i.e. all combinations of $error_{number}$ and $error_{position}$ were simulated. This means $T_{test\_vectors} = 2^4 * (\binom{12}{0} + \binom{12}{1} + \binom{12}{2}) = 1{,}264$ ECC decoding executions were analyzed for the each design, and the decoding time was measured.

B. Reed-Solomon decoder

The case-study Reed-Solomon decoder implementation is also an open-source design accessible from Github [20] and illustrated in Fig.5. The design was configured for 8-symbol codewords, 4-symbol messages and 8-bit symbols. The error correction capacity was also set to 2 errors, i.e. $t = 2$. By default, the design is pipelined by using registers to extend the execution time for each stage to the worst execution-time case. In practice, for memory-based PUF enabled devices where execution time is a critical factor, a configuration aimed at the decoder speed optimization is often used. This was also applied for the current case study. Different from the binary BCH, the Reed-Solomon decoder uses symbol-based error correction. While the parameter $error_{position}$ represents the position of the error symbol, the $error_{value}$ can take one of the $2^8 = 256$ possible values for an error in each symbol. The number of all combinations for the valid codewords merged with all possible errors for each symbol is $T_{test_{vector}} = \binom{8}{1} * (2^8 - 1) + \binom{8}{2} * (2^8 - 1) + \binom{8}{0} = 1{,}822{,}741$ that represents the number of executions to simulate and analyse per codeword. In the simulation campaign, we limited the analysis to one random valid codeword. Based on the architecture analysis, the other codewords provide the same results.

C. Experiment Results Analysis

Experiment results are shown in Table II. In the list of parameters identified for manipulation by the proposed methodology, the symbols "●" and "-" represent the varied and constant parameters correspondingly. T_d denotes the number of different decoding execution times identified and the corresponding values in clock cycles. For the Binary BCH, the experimental results confirm the conclusions of the structural analysis and do not identify any variations in the execution times. For the Reed-Solomon decoder, the red cells highlight the cases with the varying decoding time. In this experiment, T_d:3 {38, 66, 72} denotes different timing cases in case of the different number of errors to be corrected, i.e. 38, 66 or 72 clock cycles for 0, 1 or 2 errors correspondingly. As shown in the first three rows, different $error_{position}$ and $error_{value}$ can not affect the decoding time, and it remains constant (but can be equal to different values) $T_d : 1$ {38}‖{66}‖{72}.

TABLE II
ECC-BASED FE DECODING TIMING ANALYSIS

Varied Parameters				Decoding time by ECC Implementations (clock cycles)		
$codeword_{value}$	$error_{number}$	$error_{position}$	$error_{value}$	Binary BCH-12-8 BMA_serial	Binary BCH-12-8 BMA_parallel	Reed-Solomon-4-8-8
-	-	-	●			T_d:1 {38}‖{66}‖{72}
-	-	●	-	T_d:1 {28}	T_d:1 {21}	T_d:1 {38}‖{66}‖{72}
-	-	●	●			T_d:1 {38}‖{66}‖{72}
-	●	-	-	T_d:1 {28}	T_d:1 {21}	T_d:3 {38, 66, 72}
-	●	●	-	T_d:1 {28}	T_d:1 {21}	T_d:3 {38, 66, 72}
-	●	-	●	⨉	⨉	T_d:3 {38, 66, 72}
-	●	●	●	⨉	⨉	T_d:3 {38, 66, 72}
●	-	-	-	T_d:1 {28}	T_d:1 {21}	⨉
●	●	-	-	T_d:1 {28}	T_d:1 {21}	⨉
●	●	●	-	T_d:1 {28}	T_d:1 {21}	⨉
●	-	●	-	T_d:1 {28}	T_d:1 {21}	⨉

VI. CONCLUSIONS

Application of Fuzzy Extractors for error correction may enable opportunities to break the secure PUFs if no countermeasures are taken. This paper considers a combined attack model based on fault injection and timing analysis of ECC execution. In the worst case, such an attack may lead to the secret PUF value extraction. An early design stage RTL methodology was developed to verify the ECC design invulnerability against such or a similar SCA.

The methodology involves structural and simulation-based analysis parts. In our study, we targeted at two ECC architectures most widely used in FEs. The structural analysis has not identified vulnerabilities in the considered binary BCH architectures, while the architecture of Reed-Solomon based ECC may be vulnerable in particular implementations. A set of simulation-based experimental results have confirmed the findings and demonstrated the timing information leakage. Under the specified assumptions, the proposed attack procedure is able to exploit this vulnerability and reveal the secret.

The results of the early RTL analysis can guide in the selection of suitable ECC implementation or in the application of design-level countermeasures. To remove the leakage, e.g., a register can be added at the output of the Euclidean Algorithm stage to equalize the timing to the worst-case execution, or

optimizations at the ECC algorithm may be applied. The efficiency of the mitigation solutions can be explored by the proposed methodology at a low cost.

VII. ACKNOWLEDGEMENTS

This research was supported in part by the project H2020 MSCA ITN RESCUE funded from the EU H2020 programme under the MSC grant agreement No.722325 and by European Union through the European Structural, Regional Development and Social Funds.

REFERENCES

[1] R. Maes *et al.*, "Physically unclonable functions: A study on the state of the art and future research directions," in *Towards Hardware-Intrinsic Security*. Springer, 2010, pp. 3–37.

[2] R. Maes *et al.*, "A soft decision helper data algorithm for sram pufs," in *2009 IEEE international symposium on information theory*.

[3] A. R. Korenda *et al.*, "A proof of concept sram-based physically unclonable function (puf) key generation mechanism for iot devices," in *2019 16th Annual IEEE International Conference on Sensing, Communication, and Networking (SECON)*, 2019, pp. 1–8.

[4] L. Tebelmann *et al.*, "Em side-channel analysis of bch-based error correction for puf-based key generation," in *Proceedings of the 2017 Workshop on Attacks and Solutions in Hardware Security*.

[5] D. Karakoyunlu *et al.*, "Differential template attacks on puf enabled cryptographic devices," in *2010 IEEE International Workshop on Information Forensics and Security*. IEEE, 2010, pp. 1–6.

[6] Y. Dodis *et al.*, "Fuzzy extractors: How to generate strong keys from biometrics and other noisy data," in *International conference on the theory and applications of cryptographic techniques*. Springer, 2004.

[7] D. Merli *et al.*, "Side-channel analysis of pufs and fuzzy extractors," in *International Conference on Trust and Trustworthy Computing*. Springer, 2011, pp. 33–47.

[8] G. T. Becker. (2017) Robust fuzzy extractors and helper data manipulation attacks revisited: Theory vs practice.

[9] J. Delvaux *et al.*, "Helper data algorithms for puf-based key generation: Overview and analysis," *IEEE Transactions on Computer-Aided Design of Integrated Circuits and Systems*, vol. 34, no. 6, pp. 889–902, 2014.

[10] M. Riley *et al.*, "An introduction to reed-solomon codes: principles, architecture and implementation," 2003.

[11] C. Roscian *et al.*, "Fault model analysis of laser-induced faults in sram memory cells," in *2013 Workshop on Fault Diagnosis and Tolerance in Cryptography*. IEEE, 2013, pp. 89–98.

[12] Y. Gao *et al.*, "Building secure sram puf key generators on resource constrained devices," in *2019 IEEE International Conference on Pervasive Computing and Communications Workshops (PerCom Workshops)*.

[13] S. P. Skorobogatov *et al.*, "Optical fault induction attacks," in *International workshop on cryptographic hardware and embedded systems*. Springer, 2002, pp. 2–12.

[14] A. Krieg *et al.*, "A side channel attack countermeasure using system-on-chip power profile scrambling," in *2011 IEEE 17th International On-Line Testing Symposium*. IEEE, 2011, pp. 222–227.

[15] X. Lai *et al.*, "Pascal: Timing sca resistant design and verification flow," in *2019 IEEE 25th International Symposium on On-Line Testing and Robust System Design (IOLTS)*. IEEE, 2019, pp. 239–242.

[16] W. Liu *et al.*, "Low-power high-throughput bch error correction vlsi design for multi-level cell nand flash memories," in *2006 IEEE Workshop on Signal Processing Systems Design and Implementation*.

[17] H.-C. Chang *et al.*, "New serial architecture for the berlekamp-massey algorithm," *IEEE transactions on communications*, 1999.

[18] S. Lee *et al.*, "A high-speed pipelined degree-computationless modified euclidean algorithm architecture for reed-solomon decoders," *IEICE Transactions on Fundamentals of Electronics, Communications and Computer Sciences*, vol. 91, no. 3, pp. 830–835, 2008.

[19] "Verilog based bch encoder / decoder," https://github.com/russdill/bch_verilog.

[20] "Freecores reed-solomon codec generator," https://github.com/freecores/reed_solomon_codec_generator.

SAT-based data-flow mapping onto array processor

Yukio Miyasaka
The University of Tokyo
miyasaka@cad.t.u-tokyo.ac.jp

Masahiro Fujita
The University of Tokyo

Abstract—Recently, it has been common to perform parallel processing in machine learning. Reconfigurable array processor is drawing attention in terms of easy custom adjustment and high performance. We propose a method to map a data-flow onto an array processor using a SAT solver. The proposed method is combined with an automatic transformation method, which changes the order of calculations, to generate a more efficient computation scheme. We have solved mapping problems of matrix-vector multiplication. In our experiment, a SAT solver was more scalable than an ILP solver. Our method handled a data-flow of more than a hundred nodes using MAC operation. The automatic transformation under the associative and commutative laws is less scalable but successfully reduced calculation time. We have also mapped sparse matrix multiplication with varying latency and throughput and generated faster schedules utilizing the sparsity.

Index Terms—SAT problem, mapping, data-flow, reconfigurable processor

I. INTRODUCTION

Recently, machine learning with neural networks has played an important roll in image recognition [1] and other applications [2]. However, the calculation of neural network is heavy in that it requires massive numbers of multiplications. There have been many efforts in developing efficient accelerators for neural networks. For example, Google developed TPU [3], which consists of MAC operation units connected in mesh.

On the other hand, fabricating an ASIC for each specific application is not preferable, and reconfigurable devices attract attention these days. FPGA is a well-known reconfigurable device. It can save fabrication time and enables us to modify a design at a fast pace. CGRA is an alternative reconfigurable device, which consists of ALU-like units, whereas FPGA is an array of LUTs. Designed for a domain-specific application, CGRA achieves, with reconfigurability unlike ASIC, higher performance and better energy-efficiency than FPGA [4].

This paper introduces a SAT-based mapping method of a data-flow onto an array processor, which is one of the common architectures for CGRA. Through iterative compilation incrementing the number of cycles from one, it schedules a data-flow in the minimum possible number of cycles. There are several related studies about mapping onto CGRA. A study [5] proposed MRRG, where the computational resources are duplicated by the number of cycles as time-frame expansion, and performed the simulated annealing P&R of a data-flow onto MRRG under the law of casuality. In recent studies, the simulated annealing method was replaced by an ILP problem,

which also proves the mappability in the given number of cycles [6]. Although our SAT problem can be interpreted as an ILP problem without any essential change, SAT solver works faster than ILP solver as shown in the experiment.

We propose a method to transform a data-flow while mapping it. It seems that this is the first trial to automatically transform and optimize a data-flow considering the architecture of array processor. The studies above [5], [6] and others, as far as we know, do not modify the data-flow generated by an architecture-agnostic compiler. A study [7] used an SMT-like approach to rewrite a function during the mapping process, but it is limited to a linear function. Our target is a high-level data-flow consisiting of arithmetic operators as used in [8]. It is also different from the data-flow used in the framework [6], [9], which is at the level of assembly language. We believe that adopting the high-level description makes it easy for users to specify where to parallelize and gives more opportunities for compilers to optimize a data-flow. We have integrated a transformation method under the associative and commutative laws.

In the experiments, we have mapped matrix-vector multiplication and sparse matrix multiplication using the automatic transformation method. The number of cycles has been reduced around 10% by a complicated transformation for matrix-vector multiplication. For sparse matrix multiplication, we used an architecture based on a famous systolic matrix multiplication and successfully saved calculation time according to the sparsity.

This paper is organized as follows. Section 2 explains the basics of the SAT problem. Section 3 defines the mapping problem and shows the simple mapping method. Section 4 proposes the mapping method with the automatic transformation. Section 5 explains the experiments and the results. Section 6 concludes the paper.

II. SAT PROBLEM

A SAT (Satisfiability) problem is a problem to check if a logic formula is always 0 or can be 1. A logic formula is generally expressed as a CNF (Conjunctive normal form). A CNF is AND of clauses, and a clause is OR of literals. A literal is a variable or NOT of a variable. A SAT problem is defined as a problem whether there exists an assignment of the variables such that the logic formula is 1.

In the CNF (1), $(a \vee b)$, $(\overline{a} \vee \overline{c})$, and $(b \vee c)$ are the clauses, a, $\overline{a}$, b, c, and $\overline{c}$ are the literals, and a, b, and c are the variables. This CNF is SAT (can be 1) with the assignment $(a, b, c) =$

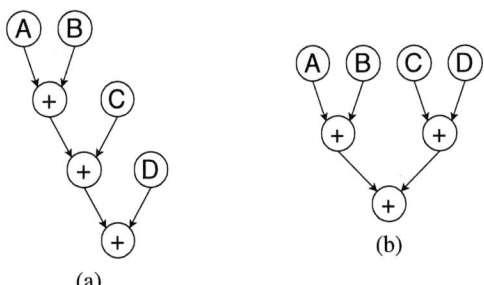

(a)

(b)

Fig. 1. Data-flows summuating 4 variables.

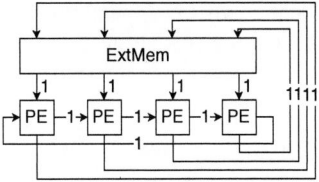

Fig. 2. An array processor consisting of 4 PEs connected in ring.

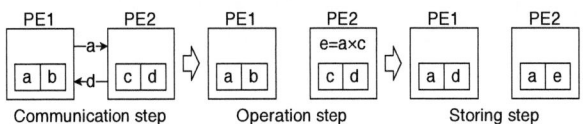

Fig. 3. An example of a cycle.

$(1, 1, 0)$. The CNF (2) is the CNF (1) with a clause $\overline{b}$ added. This CNF is UNSAT (always 0).

It is required in many applications to impose on some boolean variables in a CNF the constraint that at most K variables can be 1 at the same time. (3) is a typical formulation with integer addition (+). This constraint is called Cardinality constraint or At most K constraint. We used the bimander encoding [10] when $K = 1$ and the sequential unary counter encoding [11] otherwise.

$$(a \vee b) \wedge (\overline{a} \vee \overline{c}) \wedge (b \vee c) \qquad (1)$$

$$(a \vee b) \wedge (\overline{a} \vee \overline{c}) \wedge (b \vee c) \wedge \overline{b} \qquad (2)$$

$$a_1 + a_2 + ... + a_N \leq K \qquad (3)$$

III. MAPPING PROBLEM

A. Data-flow

A data-flow is a graph representation of calculation. It consists of input-nodes, operator-nodes, and directed edges connecting these nodes. An input-node corresponds to an input-variable of the calculation. An operator-node expresses an arithmetic operator. It has as many incoming edges as its operands. When the starting point of the incoming edge is an input-node, the value of the corresponding input-variable is used as a operand. When it is an operator-node, the resulting value of the operation is used. If the operator is not commutative, the incoming-edges are weighted to specify the order of the operands.

For example, the calculation summuating A, B, C, and D can be represented as Fig. 1 (a). The order of the operations is specified in a data-flow even if it is arbitrary in the calculation. In this case, Fig. 1 (b) is an alternative data-flow where operands are swapped under the associative and commutative laws.

B. Array processor

We define an array processor as consisting of three kinds of components: PE (Processing Element), Mem (Memory), and ExtMem (External Memory). These components are connected by paths (weighted directed edges) in any topology. We call an edge in an array processor a path to distinguish it from an edge in a data-flow. An example of an array processor is shown at Fig. 2. A PE has operation-units and registers as many as

specified. A Mem is a local memory, which may be shared among PEs. The size of each Mem is unlimited unless given by a user. A Mem can store some specified values in advance of the calculation. If a Mem has no incoming path, it will behave as a ROM. The ExtMem is the external device to supply the values of the input-variables and collect the values of the results. There exists only one ExtMem, while its behavior is the same as a Mem. A user can prohibit the ExtMem from storing values of intermediate variables if necessary.

All components are synchronized. We define a cycle as consisting of three steps: communication step, operation step, and storing step. An example is shown at Fig. 3. The communication step precedes. A path is weighted by the number of the values that the path can communicate in a step. Each PE sends the values in its registers and each Mem sends the stored values to other components through paths simultaneously. Next, in the operation step, each operation-unit perform one operation. The candidates for the operands are the values in the registers in the same PE and the values communicated to the PE. Finally, each register stores a value and each Mem stores values. A register selects the value to store from the same values as the candidates for the operands or the results of the operations in the PE. A Mem selects the values to store from the values in it and the received values.

We assume that a cycle is equal to a clock cycle, where the communication step and the operation step do not take a clock cycle.

C. SAT-based mapping method

We create a SAT problem with the following three kinds of boolean variables. Let i, j, k, and h be integers. Given the number of cycles, N, the range of k is defined as from 0 to $N - 1$, where each cycle is denoted as cycle k in ascending order. The nodes in a data-flow and the components and the paths in an array processor are numbered and referred as node i, component j, and path h respectively. Each range of i, j, and h is restricted to where numbered. In the following, node i also expresses the value of the node (the result of the operation for the operator-node).

978-1-7281-5410-7/20 $31.00 © 2020 IEEE

- $X_{i,j,k}$ ... node i exists in component j at cycle k
- $Y_{i,h,k}$ ... node i is communicated in path h at cycle k
- $Z_{i,j,k}$ ... node i is calculated in component j at cycle k

If component j is a PE, $X_{i,j,k}$ means node i exists in the registers in component j at cycle k.

The CNF of the SAT problem is composed of the following clauses. We denote the set of the incoming paths of component j as H_j, the component at the starting point of path h as s_h, the ExtMem as component e, the set of the nodes which are the starting points of the incoming edges of node i as D_i, and the set of the nodes corresponding to the results as O.

1) $X_{i,j,0}$ for (i,j) where component j is a Mem or ExtMem storing node i, which is an input-node, in advance of the calculation
2) $\neg X_{i,j,0}$ for (i,j) where the condition above is not met
3) $X_{i,e,N-1}$ for $\forall i \in O$
4) $\neg X_{i,j,k} \vee X_{i,j,k-1} \vee \bigvee_{h \in H_j} Y_{i,h,k} \vee Z_{i,j,k}$ for $\forall (i,j,k \neq 0)$
5) $\neg Y_{i,h,k} \vee X_{i,s_h,k-1}$ for $\forall (i,h,k \neq 0)$
6) $\neg Z_{i,j,k} \vee X_{d,j,k-1} \vee \bigvee_{h \in H_j} Y_{d,h,k}$ for $\forall d \in D_i$, for (i,j) where node i is an operator-node and component j is a PE, for $\forall k \neq 0$
7) $\neg Z_{i,j,k}$ for (i,j) where the condition above is not met, for $\forall k \neq 0$
8) At most K constraint on $\{X_{i,j,k}$ for $\forall i\}$ where K is the number of the registers in component j, for j where component j is a PE, for $\forall k$
9) At most K constraint on $\{Y_{i,h,k}$ for $\forall i\}$ where K is the weight of path h, for $\forall (h,k \neq 0)$
10) At most K constraint on $\{Z_{i,j,k}$ for $\forall i\}$ where K is the number of the operation-units in component j, for j where component j is a PE, for $\forall k \neq 0$

The clauses 1 and 2 give the initial condition and the clause 3 gives the condition in the end of the computation. The clause 4 imposes the condition that if a node exists in a component, it must have existed in the component at the previous cycle, been communicated to the component at that cycle, or been calculated (if the node is an operator-node and the component is a PE) in the component at that cycle. The assignment of nodes to the registers in a PE is not performed explicitly, while the number of the nodes in a PE is limited to the number of its registers by the clause 8. The clause 5 imposes the condition that if a node is communicated in a path, the node must have existed in the component at the starting point of the path. The number of the nodes communicated in each path is restricted by the clause 9. The clause 6 and 7 impose the condition that if an operator-node is calculated in a component, the component must be a PE and all of its operands must be available at the PE. The clause 10 restricts the number of the operations performed in each PE by the number of processors. Some other clauses will be added to limit the size of Mems, forbid the ExtMem to have the values of the intermediate variables, and restrict the number of the ports in each of the specified Mems.

TABLE I
THE RESULTS AND RUNTIME TO SOLVE THE PROBLEMS MAPPING
MATRIX-VECTOR MULTIPLICATION

Cycle	Results	Runtime in sec	
		Minisat	CPLEX
9	UNSAT	16.42	1450.99
10	SAT	0.87	1326.04

D. Pipeline

We can incorporate the pipelining constraint with a little modification in the formulation. As proposed in MRRG [5], the same resource must not be shared among the cycles which are congruent modulo T, the number of contexts (the initiation interval), to enable pipelining. Let t be an integer. We modify the clause 8 as follows:

8) At most K constraint on $\{X_{i,j,k}$ for $\forall (i,k)$ **where** $\boldsymbol{k \bmod T = t}\}$ where K is the number of the registers in component j, for j where component j is a PE, for $\forall \boldsymbol{t \in [0, T-1]}$

The changes are written in a bold font. The clause 9 and 10 are modified in the same way

E. Preliminary experiment

We compared a SAT solver (Minisat) and an ILP solver (CPLEX). Each solver worked in a single thread. Regarding the SAT solver, we used the bimander encoding for At most one constraint with two variables in each group. In the ILP problem, Cardinality constraint was directly expressed without encoding, and a parameter "emphasis mip" was set at 1 for the solver to focus on analyzing the feasiblility (satisfiablility).

The experiment was conducted with a data-flow of 4×4 matrix-vector multiplication shown at Fig. 4 and an array processor consisting of 4 PEs connected in ring shown at Fig. 2. The number of the processors and the number of the registers in each PE were 1 and 2 respectively. The number of contexts was not limited where the scheduling was fully sequential without pipelining. The ExtMem was not allowed to store the intermediate variables.

The runtime is shown at Table I. Time out (TO) was set at one day. The mapping problem of size 4 was UNSAT when the number of cycles was 9 and SAT when it was 10. The SAT solver finished 100 times faster than the ILP solver. The runtime of the ILP solver is reasonable because several studies [6], [12], [13] showed that an ILP solver ran out of time (one day) just in mapping a data-flow consisting of dozens of nodes.

IV. TRANSFORMATION

A. Automatic transformation method

The automatic transformation of a data-flow is performed under the associative and commutative laws. We enumerate the order of the operations for each cluster of the operator-nodes which have the same targeted operator. The SAT solver implicitly searches all equivalent data-flows for a data-flow satisfying the mapping requirement, exploring all combinations of the order of the operations in the clusters.

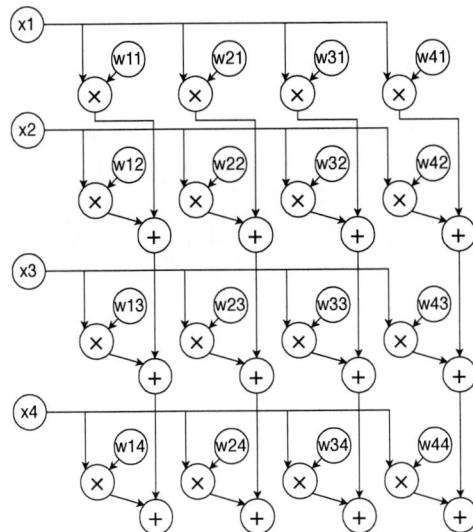

Fig. 4. A data-flow of 4×4 matrix-vector multiplication.

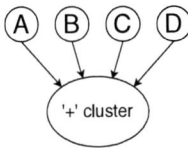

Fig. 5. The data-flow with a cluster-node.

TABLE II
THE LISTS OF THE CANDIDATES FOR THE FINAL OPERATIONS OF THE
CLUSTER-NODE AND THE INTERMEDIATE VALUES

Value	Candidates
A+B+C+D	(A+B+C)+(D), (A+B+D)+(C), (A+C+D)+(B), (B+C+D)+(A), (A+B)+(C+D), (A+C)+(B+D), (A+D)+(B+C)
A+B+C	(A+B)+(C), (A+C)+(B), (B+C)+(A)
A+B+D	(A+B)+(D), (A+D)+(B), (B+D)+(A)
A+C+D	(A+C)+(D), (A+D)+(C), (C+D)+(A)
B+C+D	(B+C)+(D), (B+D)+(C), (C+D)+(B)
A+B	(A)+(B)
A+C	(A)+(C)
A+D	(A)+(D)
B+C	(B)+(C)
B+D	(B)+(D)
C+D	(C)+(D)

following clauses:

6.1) $\neg Z_{i,j,k,l} \vee X_{d,j,k-1} \vee \bigvee_{h \in H_j} Y_{d,h,k}$ for $\forall d \in D_{i,l}$, for $\forall l \in L_i$, for (i,j) where node i is a cluster-node and component j is a PE, for $\forall k \neq 0$

6.2) $\neg Z_{i,j,k} \vee \bigvee_{l \in L_i} Z_{i,j,k,l}$ for (i,j) where node i is a cluster-node and component j is a PE, for $\forall k \neq 0$

The clause 7 is valid for (i,j) where the condition at the clause 6.2 is not met, for $\forall k \neq 0$.

B. Processing multiple operator-nodes

It sometimes reduces the calculation time for a processor to perform an operation with more than three operands. However, such an operation may match not an operator-node but a sequence of operator-nodes in a data-flow. For example, MAC operation will match two contiguous operator-nodes: multiplication and addition. Using the new formulation, we can realize mapping with such operations by modifying the lists of the candidates of the cluster-nodes (an operator-node is treated as a cluster-node here).

We explain the procedure for MAC operation for example. For each node i, if it is a cluster-node of addition, do the following. For each candidate in the candidate-list of node i, let the first operand be node a and the second be node b. If node a is a cluster-node of multiplication, add a new candidate to the candidate-list of node i for each candidate in the candidate-list of node a. The operands of the new candidate are node b and the operands in the candidate in the candidate-list of node a. Do the same for node b. In fact, to avoid interference among new candidates, adding a new candidate is performed simultaneously after all new candidates are enumerated for all nodes and all the operations each of which processes multiple operator-nodes.

We explain the method with the data-flow shown at Fig. 1 (a). First, we create a cluster-node by combining all the contiguous operator-nodes of the same operator, which can be addition, multiplication, or other operators satisfying the associative and commutative laws. The new data-flow is shown at Fig. 5. Next, we enumerate the order of the operations for each cluster-node. In this case, the final operation is addition of two values: the sum of three variables and the other variable, or the sum of two variables and the sum of the other two variables. The list of all such candidates is shown at the first row of Table II. For each intermediate value, we recursively enumerate the candidates for the final operation, which is shown at a subsequent row of the table. The table is shared for all cluster-nodes to avoid duplication of the intermediate values.

Then, we modify the SAT problem. In the following, every operator-node (two-input) is treated as a cluster-node with one candidate. In addition to the nodes in the modified data-flow, each intermediate value in the table is also treated as a cluster-node. We denote the set of the nodes which correspond to the operands in the l-th candidate for the final operation of node i, which is a cluster-node, as $D_{i,l}$ and the set of the possible values of l for such node i as L_i. We add a new kind of variables, $Z_{i,j,k,l}$, meaning all the operands in the l-th candidate of node i, which is a cluster-node, are available at component j at cycle k. The clause 6 is replaced by the

V. EXPERIMENTS

A. Matrix-vector multiplication

We solved the same problem in the preliminary experiment in Section 3 to see the effectiveness of the automatic transformation. We allowed a processor to perform MAC operation, an operation processing multiple operator-nodes. We compared the results with or without the automatic transformation under

978-1-7281-5410-7/20 $31.00 © 2020 IEEE

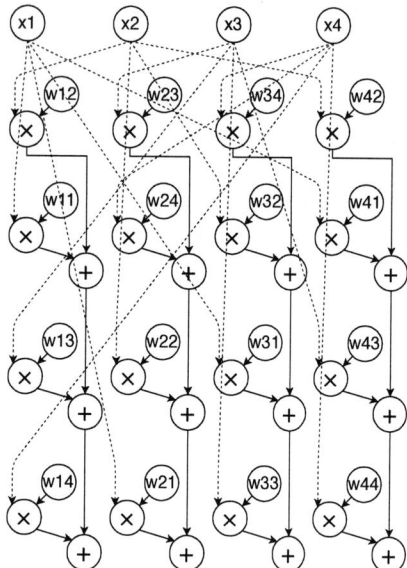

Fig. 6. The transformed data-flow of 4×4 matrix-vector multiplication.

TABLE III
THE RESULTS AND RUNTIME TO SOLVE THE PROBLEMS MAPPING
MATRIX-VECTOR MULTIPLICATION USING MAC OPERATION WITH OR
WITHOUT THE AUTOMATIC TRANSFORMATION

Size	Node		Cycle	Results (Runtime in sec)	
	w/o	w/		w/o	w/
4	48	80	6	UNSAT(0.03)	UNSAT(0.30)
			7	UNSAT(0.20)	SAT(0.65)
			8	SAT(0.08)	SAT(0.24)
5	75	185	7	UNSAT(1.54)	UNSAT(39.14)
			8	UNSAT(7.81)	SAT(793.27)
			9	SAT(1.46)	SAT(1153.01)
6	108	420	8	UNSAT(39.14)	TO
			9	SAT(1411.40)	TO
			10	SAT(1441.67)	TO

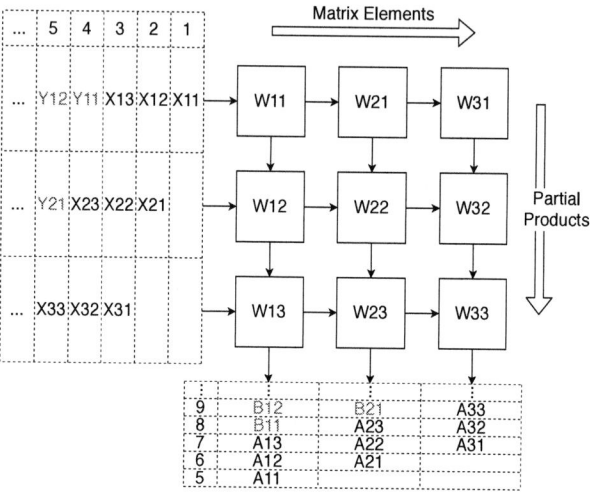

Fig. 7. The algorithm of 3×3 matrix multiplication with PEs in 3×3 mesh.

the associative and commutative laws. We also changed the size of the problem (the size of matrix and the number of PEs) to see the scalability of the method.

The results (SAT, UNSAT, or TO) and runtime are shown at Table III. When the problem size was 4, the minimum possible number of cycles was 8 without the transformation. This number is not more than the number of cycles required to map the data-flow where each set of addition and multiplication is manually converted into a three-input operator-node of MAC operation. On the other hand, with the automatic transformation, the minimum possible number of cycles became 7 through a complex transformation shown at Fig. 6. It spends one cycle for the initial state, one cycle just for reading input-values, four cycles for both reading input-values and performing MAC operation, and one cycle for writing the output-values. The number of cycles was also reduced when the problem size was 5, while we could not get the results with the transformation when the problem size was 6. With the automatic transformation, the number of nodes increases exponentially in proportion to the problem size because there are exponentially more intermediate values.

B. Sparse matrix multiplication

We conducted another experiment for sparse matrix multiplication. We aimed at reducing calculation time of a famous matrix multiplication algorithm, which was used in TPU [3], according to the sparsity of a matrix. The original algorithm is shown at Fig. 7. It calculates $A = W \cdot X$ where W, X, and A are 3×3 matrices. The array processor is composed of 3×3 PEs connected in mesh. The elements of W are distributed among PEs beforehand. The elements at each row of X are supplied one by one while the second row is delayed by one cycle and the third row by two cycles. Each PE multiplies the received element of X and the element of W it has, sends the right-side PE the element of X, and sends the down-side PE the sum of the resulting product and the value which it received from the up-side PE. It takes 9 cycles to calculate one matrix multiplication and works with the initiation interval 3. The figure shows the case that we next calculate $B = W \cdot Y$. The resource used for X will be used again for Y 3 cycles later.

We solved the mapping problems for $W \cdot X$ where some elements in W were fixed at zero and the multiplications using those elements were skipped. The PEs and ExtMem were connected in the same topology with all paths weighted by 1, and for each PE we put a ROM storing one of non-zero elements of W and connected it to the PE. Under this condition, the number of required cycles does not change at least with permutations of rows and columns of W. A permutation of rows is interpreted as swapping output rows and a permutation of columns is interpreted as swapping input rows, where such a swap is arbitrarily done through mapping with the transformation. There are 36 permutationally inequivalent matrices for 3×3 matrices [14]. The number of processors and the number of registers were 1 and 2

978-1-7281-5410-7/20 $31.00 © 2020 IEEE

TABLE IV
THE NUMBER OF MATRICES WHERE THE PROBLEMS FIRST BECAME SAT
WHEN INCREASING THE NUMBER OF CYCLES (THE NUMBER OF CONTEXTS
WAS FIXED AT 3) OR THE NUMBER OF CONTEXTS (THE NUMBER OF
CYCLES WAS FIXED AT 9)

Zeros	Matrices	Cycle						Context		
		4	5	6	7	8	9	1	2	3
0	1	0	0	0	0	0	1	0	0	1
1	1	0	0	0	0	1	-	0	0	1
2	3	0	0	0	0	3	-	0	0	3
3	6	0	0	0	1	5	-	0	0	6
4	7	0	0	0	7	-	-	0	0	7
5	7	0	0	0	7	-	-	0	1	6
6	6	0	0	1	5	-	-	0	1	5
7	3	0	0	3	-	-	-	0	3	-
8	1	0	1	-	-	-	-	1	-	-

respectively. The pipelining was done with the number of contexts 3, where the subsequent calculations like $W \cdot Y$ are considered implicitly as a modulo constraint.

We tried to reduce the number of cycles (latency) and the number of contexts (initiation interval) for each matrix except an empty (all zero) matrix. When we changed the number of cycles, the number of contexts was fixed at 3. On the other hand, when we changed the number of contexts, the number of cycles was fixed at 9.

The results are shown at Table IV. The number of cycles was reduced in proportion to the sparsity, but at least 5 cycles were required. The number of contexts was reduced with more than 4 zeros, but just for one matrix when 5 or 6 zeros.

VI. CONCLUSION

We proposed a mapping method using a SAT solver. We compared a SAT solver with an ILP solver, which was conventionally used, and found the SAT solver works faster than the ILP solver in our formulation. We incorporated an automatic transformation method, which was able to reduce the number of cycles in the mapping for matrix-vector multiplication. In another experiment, a systolic algorithm of matrix multiplication was automatically modified by the mapper, and the sparsity of matrix led the improvement of latency and throughput. Note that ILP solvers can set an objective, minimizing the area usage for example, which SAT solvers cannot deal with.

Our simple mapping method is so scalable that it can be used for a practical example. While this is apart from the main topic (the automatic transformation method) and we do not have enough space to show the details, the problem mapping the AES data-flow consisting of 138 nodes onto a 4×4 mesh array processor was solved to be UNSAT with 51 cycles in 1.32 sec and SAT with 52 cycles in 3.13 sec. It is the theoretical minimum number of cycles because the data-flow has 50 levels and we need 2 cycles for input and output. These are the results without pipelining, and we are now thinking how to pipeline it efficiently.

As future work, formulating a transformation as a swapper (selector) may ease the problem which is now infeasible for large data-flow. It is originally used in [15] for logic factoring

of logic circuits. We are also considering to adopt a rule base transformation, where a rule is a possible transformation defined by a user, to deal with other than the associative and commutative laws. We must take care that the rules should not interfere with each other.

Our method using a SAT solver is not as scalable as the heuristic methods like simulated annealing. We have to consider decomposing a data-flow or imposing some heuristic constraints by generalizing a small mapping result for example. Another direction is to automatically rectify the architecture of array processor. We need a good method to explore the search space: the topology of array processor, the number of processors, the number of registers, and the bandwidth of paths. Such a method is still required just for the problem of adjusting the number of cycles and the number of contexts to maximize resource utilization.

REFERENCES

[1] A. Krizhevsky, I. Sutskever, and G. E. Hinton, "ImageNet Classification with Deep Convolutional Neural Networks," in *Proceedings of International Conference on Neural Information Processing Systems*, 2012, pp. 1097–1105.

[2] D. Silver *et al.*, "Mastering the game of Go without human knowledge," *Nature*, vol. 550, no. 7676, pp. 354–359, 2017.

[3] N. P. Jouppi *et al.*, "In-Datacenter Performance Analysis of a Tensor Processing Unit," *ACM SIGARCH Computer Architecture News*, vol. 45, no. 2, pp. 1–12, 2017.

[4] L. Liu *et al.*, "A Survey of Coarse-Grained Reconfigurable Architecture and Design," *ACM Computing Surveys (CSUR)*, vol. 52, no. 6, pp. 1–39, 2020.

[5] B. Mei *et al.*, "Exploiting loop-level parallelism on coarse-grained reconfigurable architectures using modulo scheduling," *IEE Proceedings - Computers and Digital Techniques*, vol. 150, no. 5, p. 255, 2003.

[6] S. A. Chin and J. H. Anderson, "An architecture-agnostic integer linear programming approach to CGRA mapping," in *Proceedings of Design Automation Conference (DAC)*, 2018, pp. 1–6.

[7] J. W. Greene, "Exact mapping of rewritten linear functions to configurable logic," in *Proceedings of International Workshop on FPGAs for Software Programmers (FSP)*, 2019, pp. 11–18.

[8] M. J. Flynn, O. Pell, and O. Mencer, "Dataflow supercomputing," in *Proceedings of International Conference on Field Programmable Logic and Applications (FPL)*, 2012, pp. 1–3.

[9] S. A. Chin *et al.*, "CGRA-ME: A unified framework for CGRA modelling and exploration," in *Proceedings of International Conference on Application-specific Systems, Architectures and Processors (ASAP)*, 2017, pp. 184–189.

[10] V.-H. Nguyen and S. T. Mai, "A New Method to Encode the At-Most-One Constraint into SAT," in *Proceedings of International Symposium on Information and Communication Technology (SoICT)*, vol. 03-04-Dece, 2015, pp. 1–8.

[11] C. Sinz, "Towards an Optimal CNF Encoding of Boolean Cardinality Constraints," in *Lecture Notes in Computer Science (including subseries Lecture Notes in Artificial Intelligence and Lecture Notes in Bioinformatics)*, 2005, vol. 3709 LNCS, pp. 827–831.

[12] G. Lee, K. Choi, and N. D. Dutt, "Mapping Multi-Domain Applications Onto Coarse-Grained Reconfigurable Architectures," *IEEE Transactions on Computer-Aided Design of Integrated Circuits and Systems*, vol. 30, no. 5, pp. 637–650, may 2011.

[13] J. Yoon *et al.*, "A Graph Drawing Based Spatial Mapping Algorithm for Coarse-Grained Reconfigurable Architectures," *IEEE Transactions on Very Large Scale Integration (VLSI) Systems*, vol. 17, no. 11, pp. 1565–1578, nov 2009.

[14] M. Živković, "Classification of small (0,1) matrices," *Linear Algebra and its Applications*, vol. 414, no. 1, pp. 310–346, apr 2006.

[15] H. Yoshida and M. Fujita, "Exact Minimum Factoring of Incompletely Specified Logic Functions via Quantified Boolean Satisfiability," *IPSJ Transactions on System LSI Design Methodology*, vol. 4, pp. 70–79, 2011.

abstractPIM: Bridging the Gap Between Processing-In-Memory Technology and Instruction Set Architecture

Adi Eliahu, Rotem Ben-Hur, Ronny Ronen, and Shahar Kvatinsky

Andrew and Erna Viterbi Faculty of Electrical Engineering
Technion - Israel Institute of Technology

Haifa, Israel 3200003

{adieliahu, rotembenhur}@campus.technion.ac.il, ronny.ronen@technion.ac.il, shahar@ee.technion.ac.il

Abstract—The von Neumann architecture, in which the memory and the computation units are separated, demands massive data traffic between the memory and the CPU. To reduce data movement, new technologies and computer architectures have been explored. The use of memristors, which are devices with both memory and computation capabilities, has been considered for different processing-in-memory (PIM) solutions, including using memristive stateful logic for a programmable digital PIM system. Nevertheless, all previous work has focused on a specific stateful logic family, and on optimizing the execution for a certain target machine. These solutions require new compiler and compilation when changing the target machine, and provide no backward compatibility with other target machines. In this paper, we present abstractPIM, a new compilation concept and flow which enables executing any function within the memory, using different stateful logic families and different instruction set architectures (ISAs). By separating the code generation into two independent components, intermediate representation of the code using target independent ISA and then microcode generation for a specific target machine, we provide a flexible flow with backward compatibility and lay foundations for a PIM compiler. Using abstractPIM, we explore various logic technologies and ISAs and how they impact each other, and discuss the challenges associated with it, such as the increase in execution time.

Index Terms—Memristor, processing-in-memory, RRAM, stateful logic, ISA

I. INTRODUCTION

In recent years, a plethora of data-intensive applications has been developed. These applications require massive data transfer between the memory and the central processing unit (CPU), and have raised the need for processing-in-memory (PIM) [1, 2]. Various new and emerging memory technologies, *e.g.*, resistive random access memory (RRAM) [3], often referred to as memristors, have been explored lately for the purpose of PIM. By applying voltage across the device, it performs switching between two resistance values, high resistance value (R_{OFF}) and low resistance value (R_{ON}), therefore can function as a binary memory element. In addition to their storage capabilities, memristors can be also used for computation - both application specific and general purpose. Several methods have been proposed to use the memristor as a computation unit for a specific task, *e.g.*, vector-matrix multiplication using analog computation [4]. In this manner, the dual-function memristor can perform efficient computing and reduce data transfer requirements between the CPU and the memory. Numerous accelerators integrating analog memristor-based computations have recently been developed [5].

Together with the approach of using memristors to accelerate application-specific architectures, a different approach, called 'stateful logic', uses memristive memory cells as building blocks to construct logic gates within the memory array. In this paper, we focus on stateful logic rather than analog computation. Stateful logic enables programmable general-purpose architectures since every memristive cell can be used as a storage element, as well as an input, output or a register. Several memristor logic gate families have been designed, including MAGIC [6], IMPLY [7], and resistive majority [8].

Some stateful logic families can be easily integrated within a memristive crossbar array with minor modifications. Designing a functionally complete logic gate set using such a family, *e.g.*, a MAGIC NOR gate, enables in-memory execution of any function. There are many logic gate families which have been explored in the literature, and each of them has different advantages. Previous efforts to execute a function within the memory concentrated on utilizing a specific PIM family and optimizing the latency, area or throughput using this technology, *e.g.*, SAID [9] and SIMPLER [10] for MAGIC technology [6] and K-map based synthesis [11] for IMPLY [7].

While current approaches have substantially improved the latency, area or throughput of a logic function execution, they are strongly dependent on the PIM technique and its basic operations, and therefore are bound to a specific target machine, *i.e.*, the machine on which the logic function is executed. Flexibility in the used PIM technology has many motivations since different logic families have different advantages. For example, MAGIC provides memristive crossbar compatibility and high parallelism, whereas CRS [12] provides flexibility by executing 16 Boolean functions in a single operation.

In this paper, we show a new hierarchical compilation method for PIM which is not restricted to a certain PIM technology by separating the code generation into two components: (1) intermediate code generation using target independent instruction set architecture (ISA), (2) microcode generation for a specific target machine and PIM technology, and executing the code using a third component: (3) runtime execution. In the first component, which is independent of the PIM technology, the compiler generates a compiled program that consists of target independent instructions. In the second

978-1-7281-5410-7/20 $31.00 © 2020 IEEE

component, performed by the PIM technology provider, these instructions are translated into an execution sequence of micro-operations supported by the target machine. In the third component, at runtime, the compiled code instructions are sent from the CPU to the memory controller, which contains the instruction execution sequences from the second component. The controller translates the instructions into micro-operations and sends them to the memory. This third component is similar to an instruction-level opcode being executed using micro-operations in the x86 processors [13].

Figure 1 demonstrates the first and third flow components of a half adder logic for different ISAs and target machines. The first two implementations, shown in Figure 1(a) and 1(b), demonstrate the use of the same target machine while using different ISAs. The code is compiled for a machine that its PIM technology supports only MAGIC NOR logic gates. However, the first example targets a controller which supports only NOR ISA commands, whereas the second example supports all the 2-input and 1-output logic functions as its ISA. In the first component, a netlist and compiled program composed of the ISA commands, dubbed *instructions*, are generated. In Figure 1(a), the netlist is composed of five logic gates that implement the half adder logic, and in Figure 1(b) it is composed of two gates (AND and XOR). The number of gates in the netlist is a representative of both the code size (or number of commands sent from the CPU to the PIM machine), and the control load between the CPU and the memory controller. We will refer to it for the rest of the paper as *code size*. The code size is also a means of estimation of the code abstraction achieved by our flow. In these examples, the code sizes are five and two, respectively. The second component is the microcode generation, where each command is translated to a sequence of MAGIC NOR operations and is embedded in the controller. In the third component, the code is executed. The commands are sent from the CPU to the controller, and then from the controller to the memory; hence, the code size is reduced with minimal changes to the in-memory implementation, namely, adding a few states to the memory controller to support other operations.

Figures 1(b), 1(c) and 1(d) demonstrate the use of the same ISA while using different target machines. These three examples use all 2-input logic functions as their ISA, but first machine uses MAGIC NOR technology, the second uses MAGIC NAND technology and the third uses all MAGIC 2-input logic functions. This example demonstrates the ISA definition flexibility and command hierarchy enabled by our method, and the possible reduction in code size and reduction in the control load between the CPU and the memory controller. It also demonstrates the backward compatibility feature; in Figures 1(c)-(d), machines with technologies which enable lower execution time are used, and yet the generated intermediate code is backward compatible with other PIM technologies. The separation into two independent code generation components also enables the exploration of the impact of the ISA on the used target machine and vice versa.

This paper makes the following contributions:

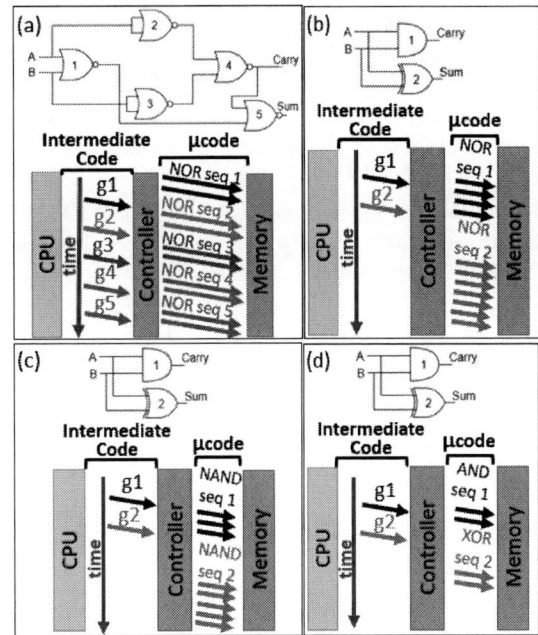

Fig. 1. Compilation example for a half adder using various ISAs and target machines. (a) A NOR ISA and MAGIC NOR target machine. (b) All 2-input and single-output ISA and MAGIC NOR target machine. (c) All 2-input and single-output ISA and MAGIC NAND target machine. (d) All 2-input and single-output ISA and 2-input and single-output MAGIC target machine.

1) Development of technology-independent and ISA-flexible flow for executing any logic function to a memristive crossbar array. Our technique, called abstractPIM, presents a hierarchical view and includes three components. It is a solid foundation for implementation of compilers for general-purpose memristive PIM architectures.
2) Examining the impact of the ISA and the target machine on each other using abstractPIM, in terms of flexibility, performance and code size.
3) A 56% reduction of the control load between the CPU and the memory controller as compared to state-of-the-art solutions [10], demonstrated for different benchmarks.

II. BACKGROUND AND RELATED WORK

A. Stateful Logic

In stateful logic families [14], the logic gate inputs and outputs are represented by memristor resistance. We demonstrate the stateful logic operation using MAGIC [6] gates, which are used as a baseline in this paper. Figure 2(a) depicts a MAGIC NOR logic gate; the gate inputs and output are represented as memristor resistance. The two input memristors are connected to an operating voltage, V_g, and the output memristor is connected to the ground. The output memristor is initialized at R_{ON} and the input memristors are set with the input values. During the execution, the resistance of the output memristor changes according to the ratio between the input values and the initialized value at the output. For example, when one or two inputs of the gate are logical '1', according to the voltage divider rule, the voltage across the output memristor is higher than $\frac{V_g}{2}$. This causes the output memristor to switch from R_{ON}

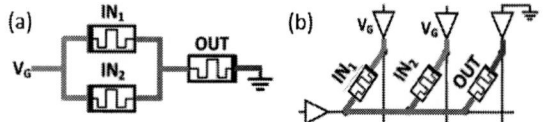

Fig. 2. MAGIC NOR gates. (a) MAGIC NOR gate schematic. (b) MAGIC NOR gate in a crossbar array configuration.

to R_{OFF}, matching the NOR function truth table. The MAGIC NOR gate can be integrated in a memristive crossbar array row, as shown in Figure 2(b). This enables massive parallelism in executing gates in different rows in the same clock cycle.

B. Logic Execution within a Memristive Crossbar Array

Unlike CMOS logic, execution of an arbitrary logic function with stateful logic is performed by a sequence of operations and takes several clock cycles. In each clock cycle, one operation can be performed on a single row, or on multiple rows concurrently. A valid logic execution is defined by mapping of every gate in the desired function to several cells in the crossbar array, and operating it in a specific clock cycle.

Many tools to generate the sequence of operations and map them into the memristive crossbar array cells have been discussed in the literature, *e.g.*, YADAV [15] and SIMPLER [10]. However, the gap between target machine constraints and architectural design choices, *e.g.*, ISA, has never been addressed. Attempts have been made in existing mapping tools to support complex operations in the in-memory execution, *e.g.*, 4-input LUT function [9]. However, their flexibility is limited and they do not completely separate the intermediate code generation and microcode generation, therefore they impose target machine and ISA dependency and do not provide backward compatibility with other target machines.

III. ABSTRACTPIM: THREE-COMPONENT CODE EXECUTION FLOW FOR PIM

The abstractPIM flow includes two code generation components and one execution component. In the first component, *intermediate representation generation*, the program is compiled into a sequence of target independent instructions based on a defined ISA. In the second component, *microcode generation*, each instruction is translated into micro-operations that are supported by the target machine. The translation is performed once per instruction, and is embedded in the controller design. We adopt an existing mapping flow and modify it to support different ISAs and PIM technologies. In the third component, *runtime execution*, the instructions in the compiled code are sent from the CPU to the controller, which translates them into micro-operations and sends them to the memory.

Existing logic execution methods use a set of basic logic operations to implement a logic function. They rely on a memory controller which is configured to perform these operations by applying voltages on the rows and columns of the memory array. In this paper, we assume that the memory controller is configured to perform several logic operations, dubbed *instructions*. Their execution sequence is determined according to a specific target machine and the PIM technology it supports. For example, if the ISA includes an AND instruction and the used technology is MAGIC NOR, 3 computation operations and 1 initialization cycle will be executed one after the other to run the AND instruction, as demonstrated in Figure 1(b), gate 1. An alternative PIM technology that consists of NAND gates will perform the same AND instruction using two NAND computations and one initialization cycle (Figure 1(c)). The instruction execution using different PIM technologies may differ in the execution time and cell usage. Our approach raises the system abstraction level and reduces the flow dependency of the specific PIM technology. It also moves one step closer towards defining a general instruction set to a memristor-based PIM architecture and designing its compiler.

The controller support of complex instructions also reduces the code size and hence the code transfer between the CPU and the memory controller. However, there is a code size and execution time trade-off; the reduction in the code size may cause an execution time penalty. For example, in Figure 1, the first NOR-based implementation takes $5T_{NOR}$ clock cycles to operate, where T_{op} is the number of clock cycles required for execution of an *op* operation. The second implementation, however, takes a total of $T_{AND} + T_{XOR}$ clock cycles. In a machine which supports MAGIC NOR operations, the first implementation takes 10 clock cycles (2 cycles per NOR), and the second implementation takes 11 clock cycles (4 for the AND2 gate and 7 for the XOR2 gate, according to Table I). Some execution cycles are computation cycles and some are initialization cycles, as further elaborated is Section V.

The instruction hierarchy in abstractPIM improves the flexibility of the compilation flow, as demonstrated in Figures 1(b)-(d). This is similar to high-level programming compared to assembly coding, which can improve flexibility at the cost of execution time penalty. While we demonstrate it using MAGIC-based logic families, the flow can be easily used for other target machines and stateful logic families. In our study, we choose different groups of ISAs, and different target machines that support different logic families. We demonstrate how they can be used to execute different benchmarks, and analyze the code size and execution time of the configurations.

IV. CASE STUDY: VECTOR-MATRIX MULTIPLICATION

We showcase our flow with a vector-matrix multiplication (VMM) benchmark (a 5 element vector and a 5×5 matrix with 8-bit elements), which is useful in many applications, *e.g.*, neural networks. The benchmark is tested over a target machine with 1024-sized memristive memory row that supports the MAGIC NOR logic family. The supported set of operations (NOT, NOR2) by the target machine is called *TS0*. Other logic families are discussed in Section VI. We first compile the benchmark for a basic case, where the ISA is also the technology set, *i.e.*, *TS0*. The selection of this ISA enabled a fair comparison between abstractPIM and existing logic execution methods, such as SIMPLER [10], which do not use a two-component code generation process. The used technology sets supported by the target machines we use and their instruction parameters are listed in Table I. Each instruction has three parameters: the number of inputs (*I*), the

978-1-7281-5410-7/20 $31.00 © 2020 IEEE

TABLE I
INSTRUCTION EXECUTION CHARACTERISTICS FOR MAGIC FAMILIES

Instruction	I	O	T_0	T_1	$TS0$	$TS1$	$IS2$	$IS3$
NOT	1	1	1+1	1+1	✓	✓	✓	✓
NOR2	2	1	1+1	2+1	✓	✓	✓	✓
NOR3	3	1	3+1	3+1	-	-	-	✓
NOR4	4	1	5+1	4+1	-	-	-	✓
OR2	2	1	2+1	1+1	-	✓	✓	✓
OR3	3	1	4+1	2+1	-	-	-	✓
OR4	4	1	6+1	3+1	-	-	-	✓
AND2	2	1	3+1	1+1	-	✓	✓	✓
AND3	3	1	6+1	2+1	-	-	-	✓
AND4	4	1	9+1	3+1	-	-	-	✓
NAND2	2	1	4+1	2+1	-	-	✓	✓
NAND3	3	1	7+1	3+1	-	-	-	✓
NAND4	4	1	10+1	4+1	-	-	-	✓
XOR2	2	1	6+1	5+1	-	-	✓	✓
XOR3	3	1	11+1	9+1	-	-	-	✓
XOR4	4	1	16+1	15+1	-	-	-	✓
XNOR2	2	1	5+1	5+1	-	-	✓	✓
XNOR3	3	1	11+1	6+1	-	-	-	✓
XNOR4	4	1	16+1	8+1	-	-	-	✓
IMPLIES	2	1	2+1	2+1	-	-	✓	✓
!IMPLIES	2	1	2+1	2+1	-	-	✓	✓
MUX	3	1	7+1	4+1	-	-	✓	✓
HA	2	2	7+1	6+1	-	-	✓	✓
HS	2	2	6+1	5+1	-	-	✓	✓

The execution time format is $T_c + T_i$, where T_c is the number of computation cycles and T_i is the number of initialization cycles.

number of outputs (O) and the number of execution cycles (T). The first two parameters are technology independent, whereas the last parameter is technology dependent. The parameter corresponding to technology set N is T_N. For example, the OR instruction has two inputs and a single output ($I = 2$, $O = 1$), and requires, when using a target machine that supports $TS0$, three clock cycles for execution ($T_0 = 3$, two computation cycles and one initialization cycle). Using ISA=$TS0$ for the VMM benchmark, there are 25470 execution cycles, out of which, half are initialization cycles and half are computation cycles. Therefore, the code size is 12735 instructions.

In attempt to reduce the code size, we used $IS2$, which contains all the functions with 1 or 2 inputs and 1 output, excluding trivial functions, e.g., constant '0' and identity functions[1]. The set also includes common combinational functions with more than 2 inputs or more than 1 output. Since the number of such functions is large, even for a small number of inputs, we chose three functions which, according to experiments we conducted, were useful in certain benchmarks: half adder [HA], multiplexer [MUX] and half subtractor [HS]. Because of the the circular dependency limitation of our flow, which is further elaborated in Section V, some useful instructions, e.g., 4-bit adder, could not be used. Using $IS2$, code size is reduced by 52%, but execution time is increased by 16%.

To demonstrate the benefit of a larger number of instruction inputs and reduce the execution time, $IS3$ was defined. It contains the $IS2$ instructions, and the 2-input and single output symmetric functions from $IS2$ extended to 3 and 4 inputs. Using $IS3$, lower execution time and code size, as compared to $IS2$, are achieved. The execution time is increased by only 8%, and the code size is reduced by 57%, as compared to $TS0$.

V. ABSTRACTPIM FLOW AND METHODOLOGY

The flow of abstractPIM is composed of three components, as shown in Figure 3. In the first component, the *intermediate*

[1]identity functions, which are in fact copy operations, can be useful in other mapping methods [9, 16], but not in our row-based flow.

representation generation, the input is a Verilog program. The program is synthesized using the Synopsys DC synthesis tool [17], where the synthesis standard cell library includes the ISA in .lib format. The Synopsys DC synthesis tool was chosen since it supports multi-output cell synthesis. Then, a compiled program is generated using a modified and extended version of the SIMPLER mapping tool [10]. This tool builds a directed acyclic graph (DAG). In its original form, every node represents a NOR gate in the netlist, since SIMPLER was designed specifically for the MAGIC NOR family [6]. In the modified mapping tool, each node represents a wider variety of instructions based on the ISA. Using the DAG, the inputs and outputs of the instructions are mapped to row cells in the memristive array, and a compiled program is generated. The I and O parameters are used to build the DAG and are technology-independent. The T parameters (see Table I), which are technology-dependent and determined in the second component, are not used for compilation. Therefore, a complete separation between the code generation components and backward compatibility with other target machines is achieved.

The second component of the abstractPIM flow is *microcode generation*. For each instruction, a microcode is generated by synthesizing the instruction to a micro-operation netlist and then to an execution sequence, which includes mapping to the memristive crossbar array and intermediate computation cell allocation based on specific PIM technology. The second component input is the instruction implemented in Verilog. The instruction is synthesized using the Synopsys DC synthesis tool for a specific PIM technology, described in the synthesis standard cell library. In this paper, we demonstrate the flow with the MAGIC [6] family, and therefore we extended the SIMPLER [10] mapping tool to support different MAGIC operations instead of only MAGIC NOR. The execution times, listed in Table I, were calculated using this flow. The second component of abstractPIM can be replaced by handcrafted execution sequences or other mapping tools, depending on the PIM technology in use, which may produce even faster execution sequences. In the third component, *runtime execution*, the two components outputs are used for full program execution. Instructions are sent from the CPU to the controller, and micro-operations are sent from the controller to the memory.

The SIMPLER mapping tool [10] traces the number of available cells, and when they are all occupied, adds a cycle which initializes several unused cells in parallel. However, not all stateful logic families use initialization, therefore initialization cycles should not be part of the first component of the flow so we remove them. In the second component, since the flow is demonstrated using the MAGIC [6] family, we perform initialization. As opposed to SIMPLER, the second component is not aware of the full program and instruction dependencies, therefore optimized parallel initialization cannot be performed. Instead, output and intermediate computation cell initialization is performed at the first cycle of each instruction execution (if needed, additional initialization cycles can be added to the instruction execution sequence). Overall, the component separation enables flexibility and backward compatibility at

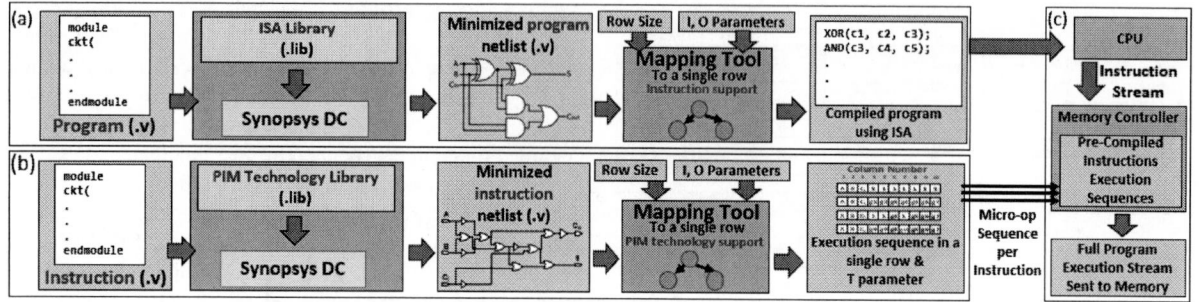

Fig. 3. abstractPIM general flow is composed of three components, two components are for code generation (differences between them are marked with purple.), and the last component is for execution. (a) Intermediate representation generation. (b) Microcode generation. (c) Runtime execution.

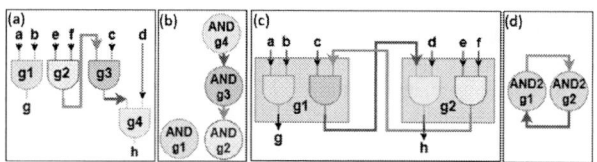

Fig. 4. Compilation with multi-output instructions which creates a circular dependency. (a) Generated netlist using single output gate synthesis. (b) Generated netlist using multi-output gate synthesis. (c) The graph that represents netlist (a), which is a DAG and can be used for the mapping algorithm. (d) The graph that represents netlist (b), which includes a cyclic dependency.

the cost of execution time penalty.

In both code generation components, each standard library cell includes several parameters. Since existing commercial synthesis tools are CMOS-oriented, we set these parameters differently and according to our memristor synthesis flow. Propagation delays, which are relevant for propagating signals in CMOS logic, are irrelevant in the context of memristor logic, where the execution time of each logic operation is a single clock cycle, and are set to 0. The area parameter is set equal for all the library cells, thus the synthesis does not prefer any particular cell, and minimizes the number of cells in the netlist, *i.e.*, minimizes the code size.

AbstractPIM supports multi-output instructions, but not all kinds of multi-output instructions can be used in it, since some may lead to *bogus dependencies* that hinder the execution mapping. Figure 4 shows an example of such bogus dependencies, in which, the input is the function code: $g = ab$, $h = cdef$. In Figure 4(a), the code is compiled using single-output instructions (AND2 instruction), and in Figure 4(b), it is compiled using multi-output instructions (an instruction which computes two AND2 operations, marked in blue). Figures 4(c) and 4(d) show the graphs corresponding to the netlists in Figures 4(a) and 4(b), respectively. While there is no combinational loop in both netlists, a circular dependency was created between the two 2-output AND2 cells. Since abstractPIM relies on the graph acyclic structure, instructions which might cause cyclic dependency cannot be used. A sufficient condition that guarantees no such loops will be created, is to use only cells in which all the outputs depend on all the inputs, *e.g.*, half adder. Future work will ensure support of any multi-output instruction, thus enabling more flexibility in planning the ISA.

After developing abstractPIM and composing the ISAs, the

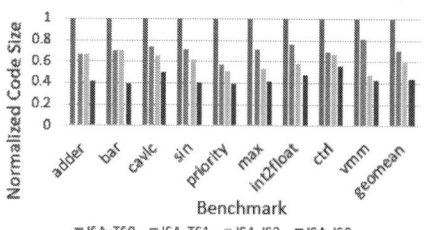

Fig. 5. Normalized code size with respect to TS0 for different ISAs.

code size and execution time were explored. We show the two metrics separately, due to the absence of a natural metric that combines both of them[2]. We used the EPFL benchmark suite [18]. Each benchmark was tested with different technology sets and ISAs, listed in Table I, within a 512-sized row. One benchmark, *max*, could not be mapped to a 512-sized row and was therefore tested with a 1024-sized row.

VI. RESULTS

The abstraction achieved by our flow using different ISAs reduces the code size as compared to an implementation based on a specific PIM technology. Figure 5 shows the code size needed for the execution of each benchmark using different ISAs: *TS0*, *TS1* (used as ISAs and not as technology sets), *IS2* and *IS3*. The code size is determined only by the ISA, and is independent of any target machine. Since the chosen sets are subsets of each other, *i.e.*, $TS0 \subset TS1 \subset IS2 \subset IS3$, then $CS_{TS0} > CS_{TS1} > CS_{IS2} > CS_{IS3}$, where CS_{set} is the code size of *set*. Using *TS1*, *IS2* and *IS3* reduced the code size by 30%, 40% and 56% compared to *TS0*, respectively.

For execution time evaluation, we compiled the benchmarks with the different ISAs and for the different target machines to demonstrate the flexibility and PIM technology independence achieved by our flow. We used two "native" configurations: *TS0/TS0*, *TS1/TS1*, and four "abstract" configurations: *TS0/IS2*, *TS0/IS3*, *TS1/IS2*, and *TS1/IS3*, where the notation is target-machine/ISA. We also compare the results to a single-component target-specific flow, SIMPLER [10].

The results are shown in Figure 6. When comparing *TS0/TS0* with SIMPLER, the execution time is approximately doubled, since in our flow, every NOR or NOT operation takes an additional cycle for initialization. In SIMPLER, which operates at full program context and not at single instruction

---

[2]Weighted product of code-size and execution-time were found misleading.

978-1-7281-5410-7/20 $31.00 © 2020 IEEE

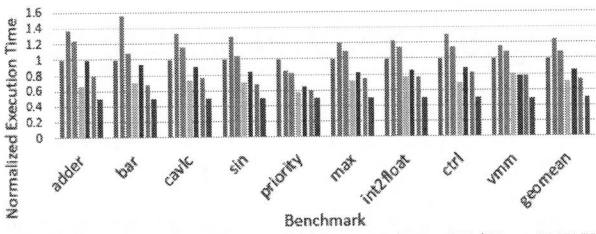

Fig. 6. Normalized execution time with respect to *TS0/TS0* for the different target machines and ISAs.

context, multiple initialization cycles can be combined and therefore the number of initialization cycles is negligible.

When comparing target machines that use native configurations (*TS0/TS0* vs. *TS1/TS1*) we observe that the target machine which is more capable (*TS1*) runs faster (30%). When comparing target machines that use the same abstract configuration (*TS0/IS2* vs. *TS1/IS2* and *TS0/IS3* vs. *TS1/IS3*) we also observe that the target machine which is more capable runs faster (32% and 33%, respectively). When comparing the execution time of a native configuration (*TS0/TS0* and *TS1/TS1*) with that of an abstract configuration using the same target machine, we see that the abstract configuration is slower. *TS0/IS2* and *TS0/IS3* are 24% and 8% slower than *TS0/TS0*, respectively. Comparing the native *TS1/TS1* configuration with the relevant abstract configurations exhibits similar results.

The above observations are quite expected. An important but less obvious benefit of abstractPIM is shown when changing a target machine. For example, when the target machine is upgraded from *TS0* to *TS1*, a program that has been compiled natively (*TS0/TS0*) executes the same number of cycles when running on *TS1* (if $TS0 \subset TS1$, otherwise even slower). However, a program that has been compiled in the first place using *IS3* (*IS2*) runs 27% (16%) faster than on the original machine – no recompilation needed. This is reflected by comparing *TS1/IS3* (*TS1/IS2*) vs. *TS0/TS0*.

Another observation is that among abstract ISAs, higher abstraction usually exhibits better performance, as shown by comparing *TS0/IS3* vs. *TS0/IS2* (13%) and *TS1/IS3* vs. *TS1/IS2* (13%). With higher abstraction it is expected that the execution time will increase compared to native configuration since using basic instructions allows finer granularity. On the other hand, using abstract ISAs reduces the number of instructions, together with the number of initialization cycles. The two opposite trends cause different benchmark behaviors.

The flexibility and code size reduction advantages of abstractPIM come with a cost. The additional execution cycles per benchmark result in proportional additional energy consumption and lower effective lifetime. We believe that higher abstraction is worth the cost of these limitations.

VII. CONCLUSIONS

This paper presents a hierarchical compilation concept and method for logic execution within a memristive crossbar array. The proposed method provides flexibility, portability, abstraction and code size reduction. The abstractPIM flow lays a solid foundation for a compiler for a memristor-based architecture, by enabling automatic mapping and execution of any logic function within the memory, using a defined ISA.

ACKNOWLEDGMENT

This research is supported by the ERC under the European Unions Horizon 2020 Research and Innovation Programme (grant agreement no. 757259).

REFERENCES

[1] S. Hamdioui *et al.*, "Memristor for computing: Myth or reality?," *DATE*, pp. 722–731, Mar. 2017.

[2] D. Ielmini and H.-S. P. Wong, "In-memory computing with resistive switching devices," *Nature Electronics*, vol. 1, pp. 333–343, Jun. 2018.

[3] M. Angel Lastras-Montaño and K.-T. Cheng, "Resistive random-access memory based on ratioed memristors," *Nature Electronics*, vol. 1, pp. 466–472, Aug. 2018.

[4] W. Woods and C. Teuscher, "Approximate vector matrix multiplication implementations for neuromorphic applications using memristive crossbars," *IEEE/ACM NANOARCH*, pp. 103–108, Jul. 2017.

[5] L. Deng *et al.*, "Model compression and hardware acceleration for neural networks: A comprehensive survey," *Proceedings of the IEEE*, pp. 1–48, Mar. 2020.

[6] S. Kvatinsky *et al.*, "MAGIC-memristor-aided logic," *IEEE TCAS II*, vol. 61, pp. 895–899, Nov. 2014.

[7] J. Borghetti *et al.*, "'memristive' switches enable 'stateful' logic operations via material implication," *Nature*, vol. 464, p. 873—876, Apr. 2010.

[8] E. Testa *et al.*, "Inversion optimization in majority-inverter graphs," *NANOARCH*, pp. 15–20, Jul. 2016.

[9] V. Tenace *et al.*, "SAID: A supergate-aided logic synthesis flow for memristive crossbars," *DATE*, pp. 372–377, Mar. 2019.

[10] R. Ben-Hur *et al.*, "SIMPLER MAGIC: Synthesis and mapping of in-memory logic executed in a single row to improve throughput," *IEEE TCAD*, Jul. 2019.

[11] J. Bürger *et al.*, "Digital logic synthesis for memristors," *Reed-Muller*, pp. 31–40, Jan. 2013.

[12] E. Linn *et al.*, "Beyond von neumann - logic operations in passive crossbar arrays alongside memory operations," *Nanotechnology*, vol. 23, p. 305205, Jul. 2012.

[13] "P6 family of processors hardware developer's manual." http://download.intel.com/design/PentiumII/manuals/24400101.pdf.

[14] J. Reuben *et al.*, "Memristive logic: A framework for evaluation and comparison," *PATMOS*, pp. 1–8, Sep. 2017.

[15] D. N. Yadav, P. L. Thangkhiew, and K. Datta, "Look-ahead mapping of boolean functions in memristive crossbar array," *Integration*, vol. 64, pp. 152 – 162, Jan. 2019.

[16] R. Ben Hur *et al.*, "SIMPLE MAGIC: Synthesis and in-memory mapping of logic execution for memristor-aided logic," *IEEE/ACM ICCAD*, pp. 225–232, Nov. 2017.

[17] P. Kurup *et al.*, *Logic Synthesis Using Synopsys*. Springer Publishing Company, Incorporated, 2nd ed., 2011.

[18] L. Amarù, P.-E. Gaillardon, and G. De Micheli, "The EPFL combinational benchmark suite," *IWLS*, 2015.

978-1-7281-5410-7/20 $31.00 © 2020 IEEE

Layout Considerations of Logic Designs Using an N-layer 3D Nanofabric Process Flow

Edouard Giacomin[1], Juergen Boemmels[2], Julien Ryckaert[2], Francky Catthoor[2,3] and Pierre-Emmanuel Gaillardon[1]

[1]University of Utah, Salt Lake City, UT, USA
[2]IMEC, Leuven, Belgium
[3]KU Leuven, Leuven, Belgium
edouard.giacomin@utah.edu

Abstract—In the past few years, novel fabrication schemes such as parallel and monolithic 3D integration have been proposed to keep sustaining the need for more powerful integrated circuits. By stacking several devices, wafers, or dies, the footprint, delay, and power can be decreased when compared to traditional 2D implementations. While parallel 3D does not enable very fine-grained vertical connections, monolithic 3D currently only offers a limited number of transistor tiers due to the high cost of the additional masks and processing steps, limiting the benefits of using the third dimension. In this paper, we introduce an innovative planar circuit netlist and layout approach, which enables a new 3D integration flow called *3D Nanofabric*. The flow, consisting of N identical vertical tiers, is aimed at single instruction multiple data processor *Arithmetic Logic Units* (ALUs). By using a single metal routing layer for each vertical tier, the process flow is significantly simplified since multiple vertical layers can potentially be patterned at once, similar to the 3D NAND flash process. In our study, we thoroughly investigate the layout constraints arising from the Nanofabric flow and the unique metal layer rule and propose several ways to overcome them. We then show that by stacking 32 layers to build a 32-bit ALU, the footprint is reduced by $8.7\times$ when compared to a conventional 7nm FinFET implementation.

I. INTRODUCTION

For many years, the semiconductor industry has continued to scale down the *Metal-Oxide-Semiconductor Field-Effect Transistor* (MOSFET) to increase the number of devices per area unit, thus enhancing the performances of *Integrated Circuits* (ICs). Novel transistor topologies have emerged in the past few years as an alternative to planar transistors, such as FinFETs [1]. They allow better electrostatic control, decreased leakage, and reduced short-channel effects, improving electrical performances. However, FinFETs still suffer from the short-channel effect, as well as other physical limitations, such as quantum effects [2], and can not be scaled indefinitely. Therefore, alternative routes are being investigated to sustain the continuous need for more performant ICs for a given footprint.

In particular, in recent years, three-dimensional integrated circuits (3D ICs) have been proposed [3]–[10], [14]. A 3D IC is an integrated circuit manufactured by stacking silicon wafers, dies, or transistors. They are then interconnected vertically to achieve performance improvements at reduced power, thanks to the shorter interconnects when compared

to conventional 2D approaches. Furthermore, stacked device layers increase the number of transistors per unit footprint without requiring costly feature size reduction. In the past few years, two 3D integration schemes have emerged: parallel 3D [3]–[5] where wafers or dies are stacked and interconnected using *Through Silicon Vias* (TSVs) and bonding techniques, and monolithic 3D [6]–[10], [14], where multiple layers of transistors and/or memory are deposited sequentially on top of one another on the same starting substrate.

While the interconnection density of parallel 3D integration is limited by the large size of the TSVs, monolithic 3D allows a finer integration granularity. However, state-of-the-art monolithic 3D works [6]–[8], [10], [14] are currently constrained by the numbers of active tiers (2-4), limiting the potential offered by 3D integration. In this paper, we introduce a new 3D integration scheme, called *3D Nanofabric*. The *Nanofabric* consists of N identical vertical tiers, each realizing the same logic function. As such, it is aimed at *Single Instruction Multiple Data* (SIMD) processor *Arithmetic Logic Units* (ALUs), where each vertical tier is one ALU bit. We propose here to use a single metal routing layer at each vertical tier, to greatly simplify the process flow, as multiple vertical layers can potentially be patterned at once. While we are aware of the challenges 3D technologies bring, such as thermal aspects including cooling, power distribution, yield, and reliability, those are out of the scope of the paper and are part of ongoing and future work. Instead, the goal of this paper is to first focus on the layout constraints and prove that conventional designs can be integrated into the *3D Nanofabric* flow, given the constraints of a planar graph without crossing wires within a vertical tier. The contributions of this paper are:

- We introduce a novel 3D design style using a very simplified set of masks and describe a possible process flow that could enable a sufficiently high yield across all layers.
- We investigate the physical design constraints arising from the single metal layer rule and propose solutions to planarize standard cells so they can be used in the proposed *Nanofabric*.
- We show that, at the circuit level, by stacking up to 32 layers to build a larger 32-bit ALU, the footprint is reduced by $8.7\times$ when compared to a 2D planar 7nm

978-1-7281-5410-7/20 $31.00 © 2020 IEEE

FinFET implementation.

The rest of this paper is organized as follows: Section II, presents related work. Section III briefly presents the proposed 3D Nanofabric concept and describes a possible technology process flow. Section IV discusses the different physical design constraints and proposes several solutions. Section V shows experimental footprint results. Section VII concludes this paper.

II. RELATED WORK

Several works on parallel 3D integration have been proposed [3]–[5], where devices on separate dies or wafers are fabricated in parallel and followed by a bonding and interconnection step. The stacking can be done with μ-bumps and TSVs [4], which are vertical connections that pass entirely through the wafer or the die. While TSVs allow a fine-grained integration of several dies into a single 3D stack, they also consume a significant area ($\sim \mu m$ pitch), which does not allow them to be used to realize very fine grain interconnects.

Other works focused on monolithic 3D (also called sequential 3D), where multiple transistor tiers and/or memory cells are vertically stacked sequentially on the same starting substrate [6]–[10], [14]. Monolithic 3D opens several opportunities, such as stacking 2 nodes $N-1$ instead of a node N [9], in a Logic-on-Logic or Memory-on-Logic way [6], or more disruptive approaches where emerging technologies can be stacked on top of CMOS [10], [14]. However, only four active tiers have been demonstrated up to this date [14], limiting the benefits of using the third dimension.

On the other hand, 3D NAND flash, consisting of a highly repetitive mask set, has also been introduced [11], [12] for memory applications. Recently, up to 128 vertical layers have been demonstrated for the 3D NAND [12], resulting in a minimal footprint per stored bit.

Our proposed *3D Nanofabric* aims at a similar objective as the 3D NAND, namely, to exploit repetitive vertical layers to decrease the footprint, but is targeted at logic applications. However, this can only be achieved by proposing a circuit netlist topology and layout that relies solely on a single layer where the device channel, poly, and metal wires are all embedded without any other crossing than the gate on top of the device channel. To the best of our knowledge, that is a crucial challenge that has not been enabled by any other proposed netlist approach.

III. PROPOSED *3D Nanofabric* CONCEPT

In this section, we briefly summarize the proposed *3D Nanofabric* concept and then present a possible fabrication flow.

A. General Overview

The proposed *3D Nanofabric* consists of N identical stacked vertical tiers, depicted in Fig. 1 (a). In other words, the *3D Nanofabric* is a 3D ALU where each tier is an ALU bit. Hence, it is aimed at realizing SIMD processor datapaths, where the datapath is composed of an array of 3D ALUs. The way the

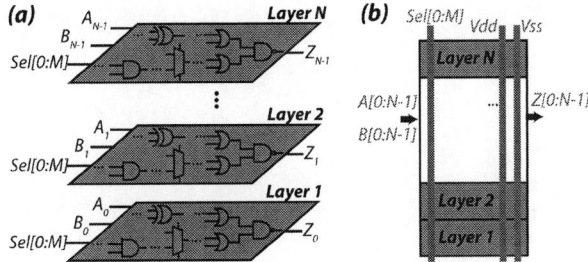

Fig. 1. *3D Nanofabric* concept: (a) Identical transistor tiers; (b) Cross-section general organization.

Nanofabric communicates with the other parts of the processor (control, memory, *etc.*) is out of the scope of this paper and is one of our current studies. To be able to stack many layers, we propose here to use a very restricted set of masks (i.e., only a single metal routing track), which allows multiple layers to be patterned at once during fabrication, as it will be explained in Section III-B. As shown in Fig. 1 (a), the global signals which are shared among all the vertical layers, such as the select signals $sel[0:M]$ (M depending on the number of operations the ALU can realize) or V_{dd} and V_{ss}, are provided through vertical pillars. The other signals (inputs and outputs of each ALU slice) are fed independently to each vertical layer from the side, using staircase-like structures similar to 3D NAND [11] chips.

B. Possible Nanofabric Process Flow

In this Section, we briefly describe a possible technological solution for manufacturing the proposed *3D Nanofabric*. The flow, based on the Coventor® modeling software, has been used to derive the design and layout rules which are presented in this section and which have been employed to obtain the results of Section V. Note that a more complete and thorough process flow study is out of the scope of this paper. While a simple solution would be to create the structure sequentially layer-by-layer, this would not be cost-effective at all as most steps would have to be repeated for each layer. Instead, we propose a solution that only uses a single metal routing layer and patterns multiple vertical layers at once. When patterning multiple layers, special care must be taken about the interaction of the different layers, *e.g.*, we should not destroy any active area by patterning the gate. This means that many of the operations need to happen from the side, as it will be explained later. Furthermore, we need to make sure that the structure always stays mechanically connected to the bulk, and that we never completely undercut a structure.

The processing starts by depositing the layer-stack: for each vertical layer, we deposit an active layer, a sacrificial layer which will become the gate (dummy-Gate), and an interlayer-dielectric. While there are multiple possible options for creating active layers, we propose here to use layer transfer of crystalline silicon, as it is done for *Silicon On Insulator* (SOI) processes. Those SOI-like silicon devices are well understood and have good electrical characteristics. The gate patterning process, which happens from the sides, is

978-1-7281-5410-7/20 $31.00 © 2020 IEEE

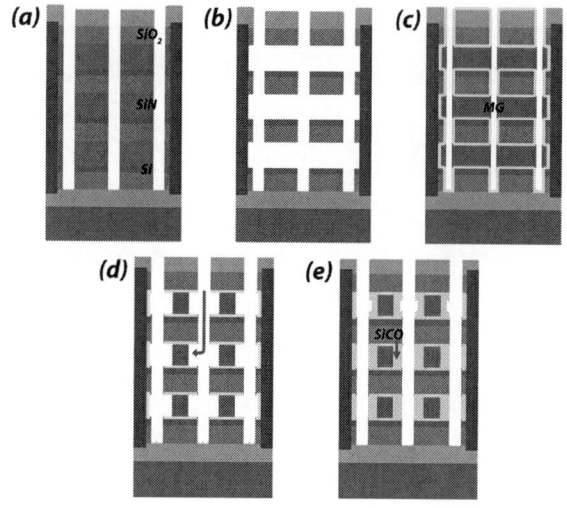

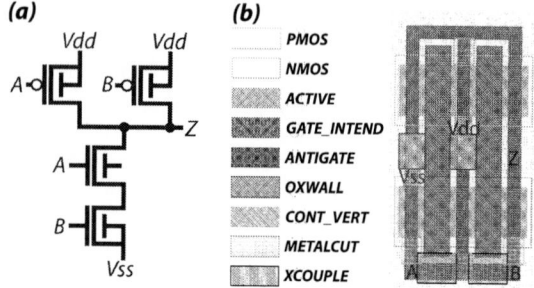

Fig. 3. NAND2: (a) Schematic; (b) Layout with layer legend.

Fig. 2. Cross-section of the gate patterning: a) Gap cut where source and drain will be formed; b) Dummy-gate removal; c) Gate-oxide and metal-gate filling; d) Metal recess; e) Spacer fill and etch-back.

shown in Fig. 2. In fact, there is no gate layer in the design as it is formed in a purely collateral fashion. Instead, the gate is defined by the extended Source-Drain region and is formed indirectly. The layer *ANTIGATE* is surrounding every gate at a fixed distance (in classical terms, the spacer-width). As it is also needed to "repair" the interlayer-dielectric, the *ANTIGATE* layer has a fixed width. As gates will form on both sides of the *ANTIGATE*, a dielectric layer *OXWALL* is used in order to prevent the formation of an unwanted gate but also gives mechanical stability to the structure. First, (Fig. 2 (a)), a gap is cut where the source and drain regions will be formed, by employing a high-aspect-ratio etch. By using a selective isotropic etch process, the dummy-gate material is then removed (Fig. 2 (b)). As we are removing a lot of material between the layers, we need to make sure that every layer is always mechanically supported. This is achieved by the design rule that every gate-island is touching an *OXWALL*. A high-k gate-oxide and then a metal-gate are deposited in the space left by the dummy-Gate (Fig. 2 (c)). To form the spacers, we first recess the metal (Fig. 2 (d)) by an isotropic metal etch and fill the formed cavity with the spacer material. The excess material in the *ANTIGATE*-trenches is removed by an anisotropic high-aspect-ratio etch using the hardmask (Fig. 2 (e)). The active patterning process is not explained here, as it is similar to the gate patterning. The final step is the formation of the metal. For the vertical connection, holes are etched through the layer-stack where *CONT_VERT* layout requests them. Also, for *METALCUT*, holes are formed, which are used as filling ports for the metal lines. The metal lines are filled over the whole length of the line through these filling ports. Therefore, a very conformal deposition is needed to avoid pinch-off. In the last step, the metal is removed from the *METALCUT* plugs and refilled with a dielectric, to cut the metal line at this location.

To illustrate the proposed flow, the layout of a conventional

NAND2 gate is depicted in Fig. 3 (b). For this paper , we consider FDSOI devices using a gate length $L = 24nm$ and a gate pitch $C_{pp} = 48nm$, similar to current 7nm technologies. As discussed, each gate (*GATE_INTEND*) is surrounded by an *ANTIGATE* layer. As such, some metal breakers are required (*METALCUT*) to achieve all the different connections. The *XCOUPLE* layer is used to provide a connection between the gate and the routing layer (*ANTIGATE*). Two *OXWALL* squares can be observed, which are in direct contact with the gates to mechanically support the vertical structure, and also act as metal routing breakersBesideses, the V_{dd} and V_{ss} supply lines are fed through vertical pillars (brown *CONT_VERT* squares) to the logic gate. Note that, as explained earlier, the *GATE_INTEND* layer is not a physical mask as the gates are formed indirectly throughout the flow. This layer is only shown here for layout purposes to ease the design step.

IV. LAYOUT CONSTRAINTS AND SOLUTIONS

In this section, we first describe the different layout constraints arising from the *3D Nanofabric* process flow. We then present solutions to overcome those and produce logic designs using a single metal layer.

A. Layout Constraints

The main layout limitation is that only a single metal routing track can be used within the *Nanofabric*, which considerably restricts the physical design. This means that when designing, no upper metal level layers can be used in case of metal crossing in high congestion areas. While any crossing possibility means that complex gates, such as XOR2 or the FA are challenging to design, some solutions are proposed in the next section. Besides, it is not possible to have the metal routing layer spanning across unrelated gates or active layers, as it is the case in conventional 2D technologies. As explained in Section III-B, the $GATE\_INTEND$ layer is directly derived from the *ANTIGATE* layer, so they are not distinct from a processing point of view. As such, it is strictly impossible to have the *ANTIGATE* layer spanning on the $GATE\_INTEND$ or $ACTIVE$ layers, which adds some restrictions. Finally, the $GATE\_INTEND$ layer has to be surrounded on every four sides by the *ANTIGATE* layer. As a result, for complex cells, some breakers have to be employed

978-1-7281-5410-7/20 $31.00 © 2020 IEEE

to achieve distinct connections on the different source and drain sides.

B. Layout Solutions to Avoid Metal Crossing

Here, we present the algorithm used to overcome the single metal layer and other layout constraints of the *Nanofabric*. We first describe each step with examples and then provide the complete algorithm.

1) Step 1: Resolving Loops at the Cell Level: The first step to resolve metal crossing is to make sure that no metal loop is present within a single logic cell. To do so, several techniques are employed:

Due to the non-conventional way of designing logic cells, there is more freedom to move the transistors vertically and horizontally, instead of having fixed top *p-well* and bottom *n-well* zones as in traditional 2D designs. For complex logic gates like the XOR2, the transistor sharing the same gate signals (mainly A and B) can be stacked on top of each other to relieve congestion within the cell, as shown in Fig. 4 (a). Note that unlike conventional design styles, there is no fixed height for the different logic cells, as complex gates such as the XOR2 will require a larger height due to the transistor stacking. Therefore, more different design styles are possible for a given cell, depending on the desired shape and the internal cell structure. Global signals, including V_{dd}, V_{ss}, or the ALU control signals, which are shared among all the vertical layers to perform the same logic function, are provided to the *Nanofabric* through vertical pillars. In particular, unlike conventional 2D designs, the standard cell power supply grid lines are removed. This relieves metal routability since those signals will not block the metal routing layer. Also, the primary inputs and outputs of the *Nanofabric* are also fed through vertical pillars. However, those are not shared among all the vertical tiers, as each layer requires separate inputs and outputs. As a result, similarly to the 3D NAND process [11], staircase-like structures are employed to convey all the signals to the appropriate tiers independently.

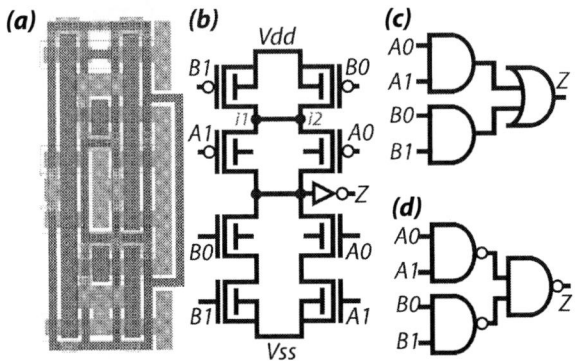

Fig. 4. (a) XOR2 logic gate layout using the proposed *Nanofabric* rules; AO22 gate schematic: (b) Transistor-level based design; (c) Gate-level based design using AND/OR gates; (d) Gate-level based design using NAND gates.

A solution to design complex gates is to use gate-level based designs instead of transistor-level based designs. For the AO22 gate, which transistor level-based design is depicted in Fig. 4 (b), the different connections, notably *i1* and *i2*, make it impossible to be designed using the proposed *Nanofabric* flow. Since each gate has to be surrounded by the metal layer, and there is only a single metal layer, these kinds of connections where 4 transistors share the same drain or source are particularly challenging. However, using the gate-level based design shown in Fig. 4 (c) greatly simplifies the routing and makes it possible, by merely cascading basic gates (NAND2, NOR2, *etc.*). While the gate-level based design uses more transistors (18 instead of 10), it can be rearranged using De Morgan's equation, as shown in Fig. 4 (d), and only uses 2 more transistors than the transistor-level based implementation. Also, due to the boolean commutativity rules, the gate inputs can be re-ordered to facilitate routing.

2) Step 2: Resolving Loops at the Netlist Level: Once all logic gates do not contain any internal metal loop, they are used to build a complete ALU. To resolve any additional metal loop in the netlist when connecting the different gates, duplicated gates can be used. As illustrated in Fig. 5 (a), the input arrangement of the AO22 gate is causing a metal crossing, and there is no way to simply move the gates to overcome this issue. This metal crossing can be resolved by duplicating the OR2 gate (in blue) on the side. As depicted in Fig. 5 (b), its output is now able to be connected to the AO22 gate without being confined, as it was the case before. Note that while it brings an area overhead, duplicating logic gates will always resolve any crossing issue as the gates can be duplicated up to the netlist primary inputs.

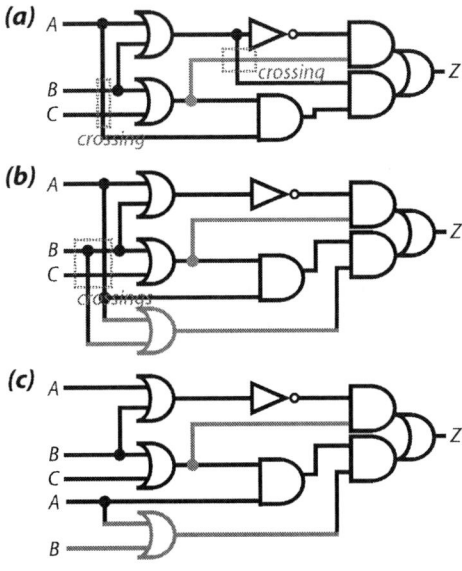

Fig. 5. Logic circuit schematic: (a) Containing 2 metal crossings; (b) Alleviating 1 metal crossing through duplicated inputs from the staircase; (c) Alleviating both metal crossings by using a duplicate gate and duplicated inputs from the staircase.

3) Step 3: Duplicating Signals Through Staircases and Vertical Signals: As explained in Section III-A, each 2D layer

978-1-7281-5410-7/20 $31.00 © 2020 IEEE

will receive its primary inputs from its sides. However, the first logic level of the ALU may require some inputs to be fed to several parallel gates, implying possible metal crossing, as shown in Fig. 5 (b). In this example, input B is driving three parallel gates. However, since there is no way to place them next to each other, the B metal wire has to cross inputs A and C. Since the primary inputs of each 2D layer are provided through a vertical staircase, they can be duplicated to be fed to more gates in the ALU. As depicted in Fig. 5 (c), by duplicating the primary inputs A and B, both metal crossings can be resolved. Besides, as using step 2 might also result in several duplicated primary inputs, the staircase will be able to feed them to the ALU while avoiding metal crossing. As the control signals are provided through vertical pillars, those can also be easily duplicated if they need to control several logic gates.

C. Single Metal Layer Layout Algorithm

In this section, we present the complete algorithm to produce the layout for an ALU netlist while only using a single metal layer. It consists of all the previous layout solutions combined. The algorithm starts from one of the last logic gate (producing an output) and propagates backward through the netlist. For each gate, it first solves the internal gate crossings, before solving the metal loops at the netlist level (between several gates). Once all the gates of a given logic level have been treated, it moves to the previous logic level until it reaches the primary inputs. If necessary, those primary inputs are duplicated through the staircases or the vertical signals. Here, we assume that the netlist does not contain feedback loops. While feedback loops are generally present in sequential circuits, the goal here is to design combinational ALUs for SIMD processors, so it is unlikely to happen. Besides, a proper synthesis of the ALU function would also get rid of the feedback loops within the netlist.

V. EXPERIMENTAL RESULTS

A. Experimental Methodology

For the footprint evaluations, we developed an in-house PDK for the *3D Nanofabric* flow, following the technological assumptions presented in Section III-B. For the 2D baseline, we considered 2 cases: (a) the ASAP 7nm FinFET design kit from ASU [13] and (b) an in-house FinFET IN7 node. For a fair area comparison, transistors are minimum sized in all cases. For all cases, the ALU area values were obtained after synthesis by using the complete available logic libraries. For the 3D case, an extra step is performed to draw the layout by hand following the novel approach described above.

B. Logic Gate Area Comparison

Table I shows the area of a few conventional logic gates, using the proposed *3D Nanofabric* flow when compared to other technologies. As expected, when compared to a highly and aggressively optimized IN7 library, using the *3D Nanofabric* process brings an area overhead (1.8× on average) due to the non-crossing rule, which requires extra transistors or spacing

---

Algorithm 1: *3D Nanofabric* gate placement.

Starts at the output node (last level of logic depth);
$Logic\_level = Get\_Total\_Nb\_Logic\_Levels()$;
while *(Logic_level != 1)* **do**
    $Number\_gates =$
    $Get\_Current\_Logic\_Level\_Nb\_Gates()$;
    while *(Number_gates != 1)* **do**
        if *Current_gate has internal crossings* **then**
            $Duplicate\_Primary\_Input()$;
            $Use\_Gate\_Based\_Logic\_Cell()$;
        else
            $Use\_Transistor\_Based\_Logic\_Cell()$;
        end
        $Number\_gates = Number\_gates - 1$;
    end
    if *Crossing between gates* **then**
        $Use\_Duplicate\_Gate()$;
    end
    $Logic\_level = Logic\_level - 1$;
end
$Duplicate\_Signals()$;

---

for complex gates. In particular, the area overhead is even more important for gate-level based cell such as the AO22 gate due to the additional transistors. Note that the logic gate area is reduced (17% in average) when compared to ASAP7 since the proposed *Nanofabric* allows us to design compact gates, as the *nmos* and *pmos* transistors can be placed closer to each other. In addition, the significant difference between ASAP7 and IN7 is due to the fact that IN7 is equivalent to a commercial foundry 5nm technology node, due to its aggressive dimensions and multiple design boosters enabling a 6-track library, while ASAP7 can only achieve a 7.5 track instance.

TABLE I
LOGIC GATES AREA (IN μm^2) USING ASAP7, IN7 AND THE PROPOSED *3D Nanofabric* PROCESS.

Gate	ASAP7	IN7	*3D Nanofabric*
INVD1	0.044	0.016	0.029
NOR2D1	0.058	0.024	0.041
AO22D1	0.092	0.040	0.127
XOR2D1	0.117	0.072	0.083
NOR3D1	0.073	0.032	0.052
Average	0.077 (-17%)*	0.037 (+1.8×)*	0.066

* *3D Nanofabric* area overhead/reduction, when compared to ASAP7 and IN7 respectively.

C. ALU Footprint Comparison

In this section, we first consider a basic 1-bit ALU aimed at SIMD processor applications, capable of performing the following operations: $A + B + C_{in}$, $A\&B$, $A|B$, $A\char`^B$. Its layout using the proposed *3D Nanofabric* is shown in Fig. 6. Note the presence of several *OXWALL* regions, which fill the extra empty spaces required to route the single metal level layer. Here, there is no need for dummy-poly as in a FinFET technology where the gate is needed to define the Source-Drain. Instead, the empty spaces are filled with the

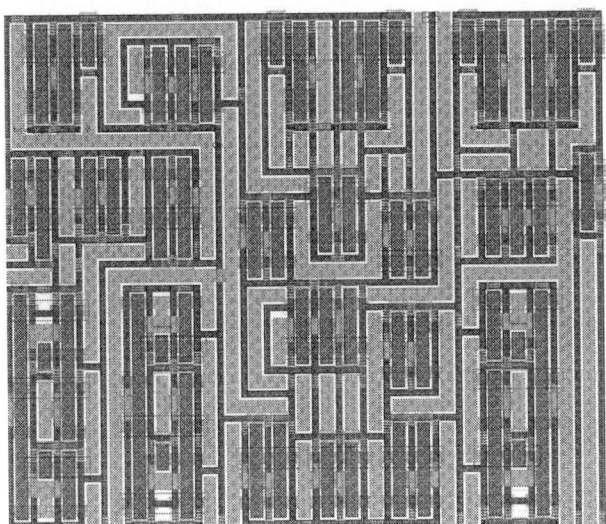

Fig. 6. 1-bit basic ALU layout view using the proposed *3D Nanofabric* rules and process flow.

TABLE II
3D Nanofabric ALU FOOTPRINT COMPARED TO ASAP7 AND IN7 FOR AN
N-BIT ALU.

Number of bits N	Footprint (in μm^2)*		
	ASAP7	IN7	3D
1	0.787 (+1.6×)	0.338 (+3.7×)	1.257
2	1.822 (-1.4×)	0.758 (+1.7×)	1.257
3	2.186 (-1.7×)	1.193 (+1.05×)	1.257
4	2.668 (-2.1×)	1.516 (-1.2×)	1.257
8	4.765 (-3.8×)	2.991 (-2.4×)	1.257
16	11.033 (-8.8×)	5.539 (-4.4×)	1.257
24	16.169 (-12.9×)	8.265 (-6.6×)	1.257
32	21.257 (-16.9×)	10.999 (-8.7×)	1.257

* Also shows the *3D Nanofabric* footprint overhead/reduction, when compared to ASAP7 and IN7 respectively.

are only needed for the primary I/Os of the array, absorbing the area overhead of such structure.

VII. CONCLUSION

In this paper, we introduced a novel 3D design flow called *3D Nanofabric*. The flow consists of several identical stacked logic layers, making it well suited for SIMD processor applications where many basic regular ALUs are repeated. We thoroughly investigated the layout constraints of the *Nanofabric* flow and proposed solutions to overcome them so that basic SIMD ALUs can be designed. We showed that by using 32 vertical layers, the 32-bit ALU footprint is reduced by a factor of 8.7× when compared to a traditional 2D approach using a 7nm FinFET technology. We believe that this novel 3D approach enables cost-effective 3D scaling for SIMD processors, to propose more performant circuits at a smaller footprint.

REFERENCES

[1] S. Natarajan *et al.*, *A 14nm logic technology featuring 2nd-generation FinFET, air-gapped interconnects, self-aligned double patterning and a 0.0588 μm^2 SRAM cell size*, IEDM, 2014.
[2] J.P. Colinge, *FinFET and other multigate transistors*, Springer, 2007.
[3] T. T. Chua *et al.*, *3D interconnection process development and integration with low stress TSV*, ECTC, 2010.
[4] E. Beyne *et al.*, *Through-silicon via and die stacking technologies for microsystems-integration*, IEDM 2008.
[5] W. Ruythooren *et al.*, *Cu-Cu Bonding Alternative to Solder based Micro-Bumping*, EPTC 2007.
[6] P. Batude *et al.*, *Advances, Challenges and Opportunities in 3D CMOS Sequential Integration*, IEDM 2011.
[7] L. Brunet *et al.*, *First demonstration of a CMOS over CMOS 3D VLSI CoolCube™ integration on 300mm wafers*, VLSI 2016.
[8] A. Mallik *et al.*, *The impact of Sequential-3D integration on semiconductor scaling roadmap*, IEDM 2017.
[9] D. Gitlin *et al.*, *Cost model for monolithic 3D integrated circuits*, S3S, 2016.
[10] M. M. Sabry Aly *et al.*, *Energy-Efficient Abundant-Data Computing: The N3XT 1,000x*, in Computer, 48(12): 24-33, 2015.
[11] D. Kang *et al.*, *A 512Gb 3-bit/Cell 3D 6th-Generation V-NAND Flash Memory with 82MB/s Write Throughput and 1.2Gb/s Interface*, ISSCC 2019.
[12] C. Siau *et al.*, *A 512Gb 3-bit/Cell 3D Flash Memory on 128-Wordline-Layer with 132MB/s Write Performance Featuring Circuit-Under-Array Technology*, ISSCC 2019.
[13] L.T. Clark *et al.*, *ASAP7: A 7-nm FinFET Predictive Process Design Kit*, Microelectronics Journal, 53: 105-115, 2016.
[14] M. Shulaker *et al.*, *Three-dimensional integration of nanotechnologies for computing and data storage on a single chip*, Nature 547, 74–78, 2017.

OXWALL dielectric layer. Also, the gate to gate distance is always enforced (36nm) to ensure that all the gate are aligned, so the layout is fully regular. As shown in Table I, ASAP7 and IN7 have a 1.6× and 3.7× smaller area than the proposed *3D Nanofabric*, respectively for the 1-bit ALU as some gates have to be duplicated to avoid crossing. Besides, some extra space is required for routing where 2D processes simply use higher metal layers. However, by going to 3D and stacking several transistor tiers to build larger ALUs, we can observe considerable footprint gains. In particular, when going to 2 and 4 layers, we can already remark some footprint reduction when using the proposed *Nanofabric* flow when compared to ASAP7 (45%) and IN7 (20%), respectively. More importantly, using 32 vertical layers to build a 32-bit ALU reduces the footprint even further by a factor of 16.9× and 8.7× when compared to ASAP7 and IN7, respectively. We believe that stacking 32 vertical layers is a fair assumption, as current 3D NAND processes have demonstrated up to 128 stacked layers [12]. Also, a higher number of vertical layers could be considered once the technology is more mature. Note that while the results presented in this section are for the specific ALU depicted in Fig. 6 (a), similar results are expected when considering different ALU designs.

VI. DISCUSSION AND ONGOING WORK

In ongoing work, we are currently assessing the delay and power benefits of the proposed *3D Nanofabric* netlist and layout approach. Delay and power improvement are expected as each vertical layer is thin (single transistor tier and routing layer). Hence the vertical connections will be short when compared to a 2D implementation. In the context of an adder or multiplier function, the carry propagation path would therefore be shorter. Besides, the 3D staircase area is also being studied. As the idea is to have an array of tiles (each tile being a 3D Nanofabric ALU) with an inter-tile communication, staircases

Simultaneous Estimation of Temperature and Voltage from Digital Delay Diversity

Xiaoyu Lian
School of Enginering
Brown University
Providence, RI, USA
xiaoyu_lian@brown.edu

Sherief Reda
School of Enginering
Brown University
Providence, RI, USA
sherief_reda@brown.edu

Jacob K. Rosenstein
School of Enginering
Brown University
Providence, RI, USA
jacob_rosenstein@brown.edu

Abstract—Semiconductor devices are fundamentally sensitive to process variation, supply voltage, and temperature, which impacts both the design and performance of modern integrated circuits. On-chip sensors are typically designed and calibrated to measure temperature and voltage independently, but designing independent sensors has implicit power and area costs. Here we propose a new simultaneous estimation of temperature and supply voltage using an ensemble of digital logic delays and non-linear regression models. This approach is inherently digital and compatible with modern digital design flows. Using simulations in commercial 12nm FinFET and 65nm bulk CMOS processes, we show the feasibility and efficiency of this approach, and consider its benefits and potential limitations.

Index Terms—temperature, voltage, delay, estimation, regression, FinFET, standard cells

I. INTRODUCTION

Temperature hot spots and localized voltage droops can be difficult to predict at the design phase of modern integrated circuits [1], [2], leading to increased operating margin which carries hidden power and performance costs. Networks of distributed sensors across a high performance chip would provide more precise spatial and temporal information, but increasing sensor density places new demands on the size, power, and design overhead of integrated temperature and voltage monitors.

Historically, many of the highest performance integrated temperature sensors are built around parasitic bipolar junction transistors (BJT) [7]–[9]. Sensors using resistors [4]–[6] have also demonstrated impressive resolution. However, these architectures often rely on analog circuits, such as bandgap references (BGR), which do not necessarily benefit from technology scaling. All-digital temperature and voltage sensors can have advantages in deep sub-micron technologies [1], [3], [6]. Yet existing architectures require multiple constrained supply voltages [1] or non-standard digital logic cells [1], [3], complicating portability and system integration.

In this paper, we present a method for precisely estimating the temperature and supply voltage from the collective response of multiple standard cell logic delays. This architecture offers several important features:

- The approach is fully digital and compatible with commercial standard cells and System-on-Chip (SoC) design

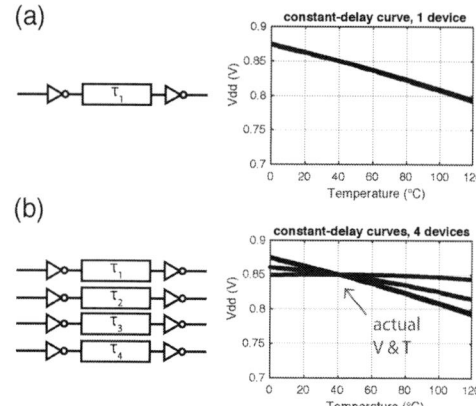

Fig. 1. Transistor logic delays are sensitive to process variation, supply voltage, and temperature (PVT), but one measured delay cannot uniquely report all of these conditions simultaneously. In this paper we explore the idea of using process diversity from several logic delays to perform simultaneous estimation of the temperature and supply voltage. (a) A constant-delay curve for one simulated 12nm FinFET. (b) Constant-delay curves for four different simulated FinFET devices in the same 12nm technology node.

  flows; no custom cells or dedicated power domains are required.
- This approach simultaneously estimates both temperature (T) and power supply voltage (V_{DD}) from the same measurements.
- Low cross-sensitivity: the temperature readout is relatively insensitive to V_{DD}, and vice versa, without using analog compensation schemes or requiring careful device matching at design time.

These methods are quite generalizable across sub-100nm technology nodes, but we present the technique using simplified short-channel device models, along with examples in a commercial 12nm FinFET technology. After introducing the model, we discuss the training requirements, process portability, and simulated performance numbers after implementing this architecture in both 12nm FinFET and 65nm bulk CMOS technology.

II. APPROACH

A. Overview

Digital logic delay is a function of both temperature and V_{DD}, and thus many possible pairwise conditions can produce the same delay (Fig. 1a). In sensor designs based on only one delay line or ring oscillator (RO), measuring the temperature would also require that the voltage is known. However, if we introduce multiple logic paths whose delays have unique temperature and voltage dependence, then the collection of delays can map to exactly one temperature and one V_{DD} (Fig. 1b). Our approach is to produce delay measurements from an ensemble of digital logic paths, built using standard devices with multiple threshold voltages. With some training, we can produce models that estimate the temperature and V_{DD} from the delay measurements.

The advantage of this approach is that it is fully digital, using only standard cells. Its reliance on training rather than analog temperature compensation and device matching schemes make it extremely area-efficient and straightforward to integrate and port between advanced CMOS technology nodes. Possible disadvantages are the computational overhead and training requirements, but these same aspects also offer great flexibility and opportunities for system-level optimization.

B. First-order Device Models of Temperature and Voltage Dependence

A simple model for a digital delay is the charging/discharging of a capacitor [10], which can be expressed as:

$$\tau_{delay} \propto \frac{CV_{DD}}{I_{d,sat}} \tag{1}$$

where V_{DD} is the power supply voltage, C is the load capacitance, and $I_{d,sat}$ is the saturation current of the MOSFET that is charging/discharging the load capacitance.

Accounting for short-channel effects in sub-100nm devices the saturation current can be modeled as:

$$I_{d,sat} = WC_{ox}v_{sat}(V_{gs} - V_{th}(T)) \tag{2}$$

where W is the width of the device, v_{sat} is the saturation velocity, $V_{th}(T)$ is the threshold voltage of the device, and T is temperature. The threshold voltage is also a function of temperature, and $V_{th}(T)$ can be approximated as:

$$V_{th}(T) = V_{th0} - KT - DV_{DD} \tag{3}$$

where V_{th0} is a zero-temperature value of V_{th} and K is the coefficient of temperature. D is a simplified model for Drain Induced Barrier Lowering (DIBL), with $D(DIBL) = \Delta V_{th}/\Delta V_D$. Thus (1) can be expressed as:

$$\tau_{delay} \propto \frac{CV}{WC_{ox}v_{sat}((1+D)V + KT - V_{th0})} \tag{4}$$

C. Discussion of Second-order Effects

Our analysis thus far has been based on first-order short-channel device models. Fortunately, we find that common second-order effects have only small impacts on temperature and V_{DD} estimation. Some of these second-order effects are as follows:

1) Saturation Velocity Temperature Dependence: In Equation (2), the v_{sat} term has a second-order relationship to temperature. The equation is given by:

$$v_{sat} = \frac{v_{sat,300}}{(1 - A_n + A_n \times \frac{T}{300})} \tag{5}$$

where T is temperature, $v_{sat,300}$ is the saturation velocity when $T = 300K$, A_n is a constant that depends on the material. Since A_n is 0.25 for silicon, v_{sat} has a expected variation of $\pm 4\%$ with temperature ranging from $250K$ to $300K$. Therefore, the delay change caused by temperature is still dominated by $V_{th}(T)$ and $v_{sat}(T)$ is a weak function of temperature.

2) Threshold Voltage Roll-off: Threshold Voltage Roll-off is one of the most notable second-order effects brought by short-channel devices. V_{th0} in Equation (3) decreases with the shorter channel length, which can be especially severe for 12nm devices [11]. However, since the roll-off change depends primarily on the channel length, it does not affect either temperature or voltage first-order dependence.

3) Sub-threshold Leakage: Reduced threshold voltages in short-channel devices lead to a smaller I_{ON}/I_{OFF} ratio and increased leakage currents [11]. While leakage current can have significant temperature and supply voltage dependence, I_{ON}/I_{OFF} often remains near 10^3 for short channel devices. Thus leakage is an essential concern for idle logic blocks, but it only has a small contribution to the delays of actively switching logic.

In conclusion, second-order effects have a relatively small impact on the fundamental relationship between τ_{delay} and temperature/voltage. The first-order model is sufficient for the derivations of the estimation model.

D. Mathematical Derivation of the Temperature and V_{DD} Dependence

Using the device models presented in Section II-B, we will next demonstrate that the temperature and V_{DD} can be uniquely represented by measurements of τ_{delay} alone.

By grouping all the constant terms together, (4) can be expressed as:

$$\tau = \frac{AV}{(1+D)V + KT - V_{th0}} \tag{6}$$

$$A \propto \frac{C}{WC_{ox}v_{sat}} \tag{7}$$

For convenience, we can define $f(\tau)$, which has a one-to-one mapping to the delay:

$$f(\tau) = \frac{A}{\tau} - (1+D) \tag{8}$$

Using (6) with (8), we can write expressions for temperature and V_{DD} as:

$$\begin{cases} KT = f(\tau) \times V_{DD} + V_{th0} \\ V_{DD} = f(\tau)^{-1} \times (KT + V_{th0}) \end{cases} \tag{9}$$

If we implement multiple logic paths i and j with different temperature and V_{DD} dependence, then (9) can be expressed as (10) and (11):

$$\begin{cases} KT = f(\tau_i) \times V_{DD} + V_{th0,i} \\ KT = f(\tau_j) \times V_{DD} + V_{th0,j} \end{cases} \quad (10)$$

$$\begin{cases} V_{DD} = f(\tau_i)^{-1} \times (KT - V_{th0,i}) \\ V_{DD} = f(\tau_j)^{-1} \times (KT - V_{th0,j}) \end{cases} \quad (11)$$

After canceling out V_{DD} from (10), and eliminating T from (11), we see that T and V_{DD} can be expressed as a function of logic delays only, as shown in (12) and (13).

$$T \propto \frac{V_{th0,i} \times f(\tau_j) - V_{th0,j} \times f(\tau_i)}{f(\tau_j) - f(\tau_i)} \quad (12)$$

$$V_{DD} \propto \frac{V_{th0,i} - V_{th0,j}}{f(\tau_j) - f(\tau_i)} \quad (13)$$

(12) and (13) also illustrate the nonlinearity between τ_{delay} of logic paths and temperature or V_{DD}.

E. Nonlinear Regression Models

To create nonlinear models which estimate T and V_{DD} from multiple logic path delays, we propose a complete form quadratic model, including cross terms:

$$T = \sum_{i=1}^{N} \alpha_{T,i} \times \tau_i + \sum_{i=1}^{N} \sum_{j=1}^{i} \beta_{T,ij} (\tau_i \times \tau_j) \\ + \sum_{i=1}^{N} \gamma_{T,i} \times \tau_i^2 + C_{T,0} \quad (14)$$

$$V = \sum_{i=1}^{N} \alpha_{V,i} \times \tau_i + \sum_{i=1}^{N} \sum_{j=1}^{i} \beta_{V,ij} (\tau_i \times \tau_j) \\ + \sum_{i=1}^{N} \gamma_{V,i} \times \tau_i^2 + C_{V,0} \quad (15)$$

where N is the number of logic paths measured, α, β, γ are predictor variables for linear, interactive and quadratic terms, and C is the constant term.

However, if we incorporate multiple logic delay paths, the complete quadratic model will produce a large number of variables to train. To reduce training complexity, we implement regularization using lasso regression to try to find a smaller subset of variables that incorporate the strongest effects [13]. In Section III, using 12nm FinFET simulation data, we will use this regularization to reduce the model complexity from 90 terms to 23 terms in (14) and 25 terms in (15).

F. Calibration and Training

After regularization, a regression model is trained using data across multiple temperature and V_{DD} points. During the training process, using multiple linear regression as the solver, predictor variables are fitted by minimizing the loss function, which can be defined as:

$$E(\alpha, \beta, \gamma) = \frac{1}{2n} \sum_{i=1}^{D} (y_i - Y_i)^2 \quad (16)$$

where y_i is the i_{th} recorded value of temperature or V_{DD}, and Y_i is the i_{th} estimated value. After training, the model is tested using a separate subset of the simulated logic delays.

III. RESULTS

A. FinFET Simulations

We tested this approach using SPICE simulations with a commercial 12nm CMOS FinFET design kit. We used only standard cells from a commercial library to ensure that the design is fully synthesizable.

In practice, delays from multiple devices in the same process will have correlated temperature and voltage dependence. To introduce more variation to the training data, we created 12 unique delay lines, with each line containing ten cells of one type of logic gate. Three types of logic gates (NAND4, NOR4, INV) were each implemented with four types of transistors (regular V_{th}, low V_{th}, super-low V_{th}, and thick-oxide extended gate (EG) devices with much higher V_{th}). This set of logic gates aims to represent a range of paths having different delay sensitivity to temperature and voltage.

Simulations were performed across temperature and voltage using a commercial 12nm FinFET design kit at the nominal (TT) process corner. The temperature varies from -20°C to 130°C with a 5°C step size (31 steps in total). The voltage varies from 0.6 V to 0.96 V with a step size of 20 mV (19 steps in total). The simulation results are shown in Fig. 2. Fig. 2(c) presents surface plots of the propagation delay of the regular V_{th}, low V_{th}, super-low V_{th} delay lines. Because of the Inverse Temperature Dependence (ITD) phenomenon in deep sub-micron technology nodes [10], the delay decreases while the temperature increases, as implied in (4). Fig. 2(a) and Fig. 2(b) plot the fractional delay change as the temperature and voltage are varied. It is evident that under these conditions, the delay has a relatively linear temperature dependence, while it varies nonlinearly with supply voltage, as discussed previously.

B. Regularization with 12nm FinFET Simulation Data

The 12 simulated logic paths led to a total of 90 terms in the complete form of the quadratic model for (14) and (15): 12 linear terms, 76 interactive terms and 12 quadratic terms. Including all 90 terms is unnecessary, and it increases training requirements and power overhead. We used lasso regularization to reduce the number of terms required and simplify the function. By sweeping the regularization term, we were able to generate models containing different numbers of terms. During the process, when multiple models share the same number of terms, we chose the one with the least root mean-squared error (RMSE).

Fig. 3 plots the temperature RMSE as a function of the number of terms included. The RMSE improves as the number of terms increases, but there are diminishing returns after it reaches 0.08 °C at 45 terms. Similarly, the supply voltage

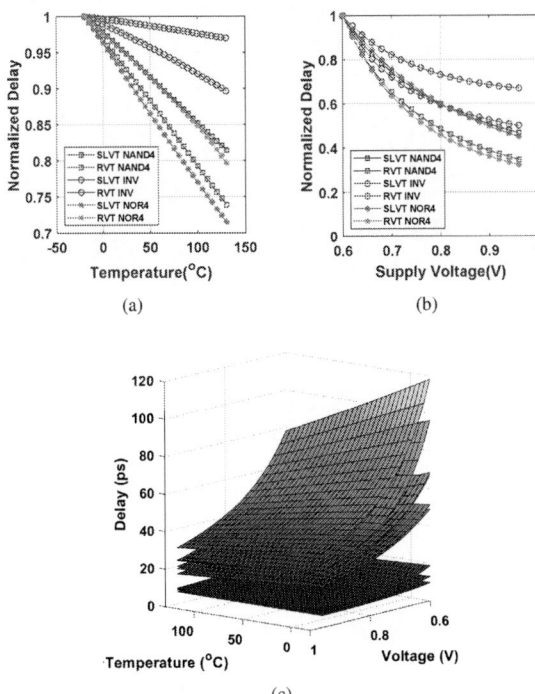

Fig. 2. FinFET simulation results a using commercial 12nm FinFET library; (a) normalized logic delays as a function of temperature at the nominal operating voltage of $0.8V$; (b) delays as a function of supply voltage at room temperature (25°C); (c) surface plots of 9 standard cell delay lines as a function of both temperature and supply voltage.

RMSE stabilizes near $0.12mV$ with 42 model terms. We selected model sizes near the knee of these curves, with 23 terms for the temperature model and 25 terms for the supply voltage model.

As further illustration of the performance of the selected models, Fig. 4 presents residual temperature and V_{DD} errors using a 70%/30% train/test split on 589 simulated data points. In Fig. 4(b) and Fig. 4(d), multiple train/test splits produced similar temperature RMSE of approximately 0.16°C and voltage RMSE of roughly $0.24mV$.

C. Label Noise and Supply Sensitivity

Any real-world implementation will need to tolerate noise and error in its training labels. In this subsection, we introduce noise into the voltage and temperature labels to verify the accuracy and robustness of our model. Fig. 5 and Fig. 6 plot RMS test errors after adding Gaussian noise to voltage or temperature labels separately in the training data. For each noise step, 25 runs were performed, using 412 (70%) data points with added label noise in the training set and 589 (100%) ideal data points in the testing set. In Fig. 5(a) and Fig. 6(b) we see that both voltage and temperature error increase with the corresponding label noise, reaching $10\times$ their baseline errors at roughly 10°C and $25mV$, respectively. We

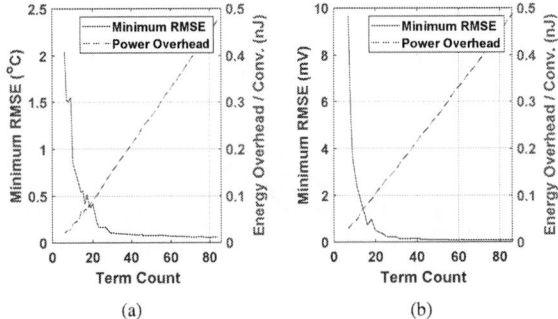

Fig. 3. Lasso regression with reduced number of terms, applied to the 12nm simulated delays; (a) Temperature RMSE and digital computation energy overhead as a function of the number of terms remaining; (b) Voltage RMSE and computation energy overhead per conversion as a function of the number of terms.

note that the resulting model errors are well-behaved and the model output has lower noise than the training labels.

Moreover, Fig. 5b and Fig. 6b reveal that the temperature and supply voltage readouts have very low cross-sensitivity to noise in the other parameter. The estimation models have essentially been trained to perform high-order temperature (or supply voltage) compensation, without requiring analog compensation schemes or device matching at design time.

IV. DISCUSSION

A. Training Requirements

An important consideration is how much production test time would be required to train these models for each chip. Fig. 7 shows the RMS testing error with finite temperature or voltage intervals during training. (Here only one parameter is constrained at a time.) Based on these results, it would be advisable to train the model at $20°C$ temperature intervals and $80mV$ voltage intervals. Several paths may exist to further decrease training requirements, including reducing the supported temperature range or increasing the number of physical delay lines in the sensor.

B. Process Portability

A valuable feature of this system is its excellent process portability and low design effort, since it uses only standard digital logic cells. Supporting circuitry such as communication circuits between the sensor and the processor can similarly be made all-digital. The approach can be synthesized, placed, and routed using standard digital design flows. However, the final accuracy relies on the availability of standard logic cells with multiple V_{th} and/or oxide thickness, which could affect the design cost indirectly. Additionally, the model regularization results and the achievable accuracy may have non-obvious dependence on process parameters.

As a demonstration of its process portability, we implemented the design in 65nm bulk CMOS, resulting in the test accuracy shown in Fig. 8. The temperature RMSE is 0.70°C, and the voltage RMSE is 1.64 mV. Since the relative delay

978-1-7281-5410-7/20 $31.00 © 2020 IEEE

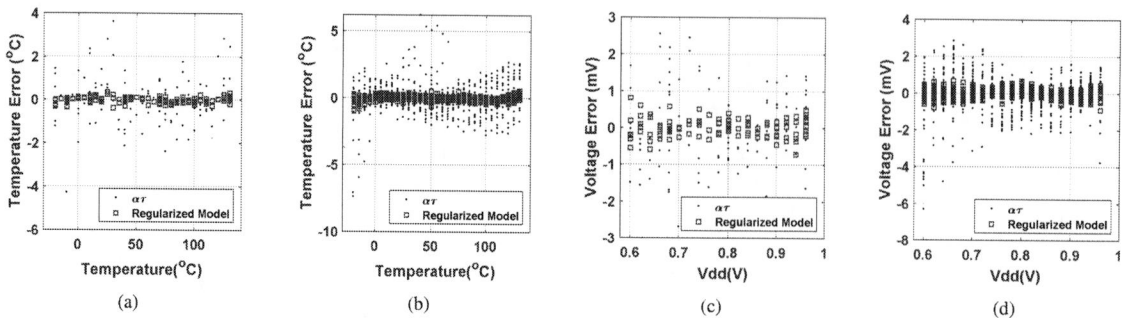

Fig. 4. Residual errors across temperature and voltage for one train/test split of the 12nm simulated data. Also shown for comparison is a linear model (labeled '$\alpha\tau$') which does not include any quadratic terms or interactive terms. (a) temperature training set error; (b) temperature test set error; (c) voltage training set error; (d) voltage test set error.

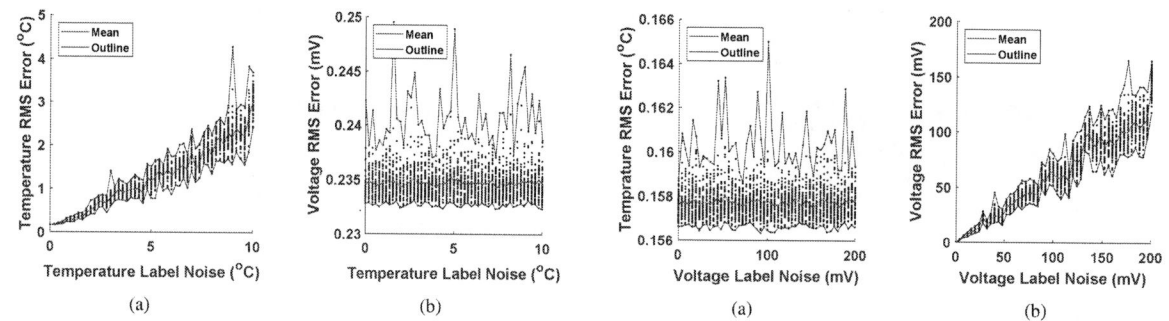

Fig. 5. Testing errors after additional noise was added to the temperature labels. (a) The temperature test error increases with the added noise, and reaches ×10 ideal testing error at approximately $10°C_{RMS}$ label noise. (b) The voltage accuracy does not change after adding temperature label noise.

Fig. 6. Testing errors after additional noise was added to the voltage labels. (a) The temperature test errors does not change after adding voltage label noise. (b) The voltage test error increases with the added noise, and reaches ×10 ideal test error at approximately 10 mV$_{RMS}$ label noise.

TABLE I
TEMPERATURE SENSOR PERFORMANCE OVERVIEW

	This work (simulation only)		Sensors19 [6]	JSSC16 [3]	JSSC10 [2]	CICC15 [1]
Architecture	RO	RO	DLPF	VCO	TDDL	RO
Tech (nm)	12	65	65	65	65	40
Area (um^2)	625†	8,280†	8,400	4,000	600,000	59,000
Supply voltage (V)	$0.6 - 0.96$	$0.8 - 1.2$	$1 - 1.2$	$0.85 - 1.05$	$0.6 - 1$	$0.5 - 1$
Power (μW)	19.6	69.5	35.2	154	0.92	17
Temp. range (°C)	$-20 - 130$	$-20 - 130$	-30 — 100	-30 — 100	0 — 100	-40 — 100
Conv. time (μs)	12‡	11‡	2.5	22	NA	20
Energy/conv. (nJ)	0.3	0.93	0.088	3.4	0.34	0.34

† Post-layout area for sensor logic only, without any computational blocks.
‡ Includes $10\mu s$ (12nm) / $9\mu s$ (65nm) for sensor acquisition and $2\mu s$ computation with a 100 MHz 32-bit RISC-V microcontroller

sensitivity to T and V_{DD} is smaller in 65nm than in 12nm, the RMSE for both temperature and voltage is higher in 65nm bulk CMOS than in the 12nm FinFET process.

C. Simulating Other Performance Metrics

We implemented this approach in both 12nm and 65nm technology, using a ring oscillator (RO) built from each delay type, and a coarse-fine readout consisting of a 16-bit counter plus a 48-step time-to-digital converter (TDC). Table I summarizes the simulated implementation and performance.

1) Conversion Time : The conversion time has two phases, the first phase for digitizing the ring oscillator delays and the

second phase for calculating the output in the microcontroller. We estimated the energy overhead and computation time based on a 100 MHz 32-bit RISC-V microcontroller. The simulated implementation measures the RO one at a time, using the same constant counting window. After regularization, the temperature model requires the measurement of 10 RO. If we allow a counting window of $1\mu s$ and the microcontroller has a clock frequency of 100 MHz, then the conversion time for each temperature readout is $12\mu s$, including $10\mu s$ for digitizing the RO and approximately $2\mu s$ for the calculation. It would be possible to reduce the conversion time by using a faster microcontroller, and it would be possible to acquire

978-1-7281-5410-7/20 $31.00 © 2020 IEEE

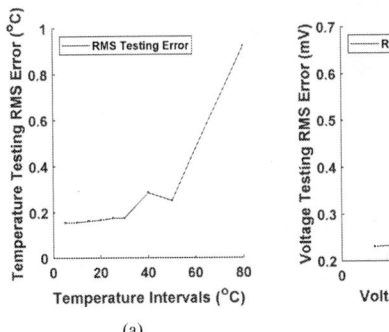

(a)   (b)

Fig. 7. RMS Testing error on different temperature or voltage intervals; (a) temperature RMS error reduces with smaller training interval, being acceptable at 50^oC; (b) voltage RMS error reduces with smaller training interval, being acceptable at $80mV$;

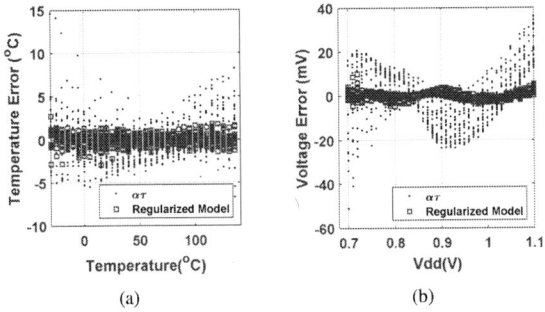

(a)   (b)

Fig. 8. Porting the same model structure and train/test method to a 65nm bulk CMOS process. Included for comparison is a linear model (labeled '$\alpha\tau$') which does not include any quadratic terms or interactive terms. (a) Temperature test set error; (b) voltage test set error

multiple RO in parallel with modified supporting logic, at a cost of more area and higher peak power consumption.

2) Energy per Conversion: The energy consumption can similarly be split into two sections, the energy cost of the sensors and energy cost of the microcontroller computation. The simulated dynamic power for the sensor in 12nm CMOS is $19.6\mu W$, with only one RO running plus the readout circuits. If we once again we allow a $10\mu s$ counting window, then the energy is 0.196 nJ/conv. The computational power cost is estimated in Fig. 3. The simulated energy used by the model calculation is 0.1 nJ for the temperature training model and 0.13 nJ for voltage training model. Hence we come to an estimate of 0.3 nJ per conversion.

3) Area: We have the circuit ready for tapeout both in 12nm and 65nm CMOS. The post-layout area of the sensor only is $625\mu m^2$ in 12nm and $8280\mu m^2$ in 65nm.

V. CONCLUSION

In this paper, we have proposed a novel technique for estimating temperature and voltage based on the digital delay diversity of multiple logic paths, using limited training data. The system is inherently digital and compatible with digital design flows and deep sub-micron technology nodes. A detailed

demonstration using a commercial 12nm FinFET technology has been made to illustrate the practicality of the estimation model. Simulation results show low cross-sensitivity between temperature and V_{DD} readout. Sensors have been built in different technology nodes, occupying $625\mu m^2$ in 12nm and $8280\mu m^2$ in 65nm CMOS. Measured performance numbers will be available after tape-out.

ACKNOWLEDGMENT

The authors thank Sofiane Chetoui for his support with the microcontroller model. This work was supported in part by a grant from the Defense Advanced Research Projects Agency (DARPA), under contract FA8650-18-2-7851.

REFERENCES

[1] M. Saligane, M. Khayatzadeh, Y. Zhang, S. Jeong, D. Blaauw and D. Sylvester, "All-digital SoC thermal sensor using on-chip high order temperature curvature correction," 2015 IEEE Custom Integrated Circuits Conference (CICC), San Jose, CA, 2015.

[2] P. Chen, C. Chen, Y. Peng, K. Wang and Y. Wang, "A Time-Domain SAR Smart Temperature Sensor With Curvature Compensation and a 3σ Inaccuracy of $-0.4^oC \sim +0.6^oC$ Over a 0°C to 90°C Range," in IEEE Journal of Solid-State Circuits, vol. 45, no. 3, pp. 600-609, March 2010.

[3] T. Anand, K. A. A. Makinwa and P. K. Hanumolu, "A VCO Based Highly Digital Temperature Sensor With 0.034 °C/mV Supply Sensitivity," in IEEE Journal of Solid-State Circuits, vol. 51, no. 11, pp. 2651-2663, Nov. 2016.

[4] S. Pan, Y. Luo, S. Heidary Shalmany and K. A. A. Makinwa, "A Resistor-Based Temperature Sensor With a 0.13 pJ · K2 Resolution FoM," in IEEE Journal of Solid-State Circuits, vol. 53, no. 1, pp. 164-173, Jan. 2018.

[5] W. Choi, Y. Lee, S. Kim, S. Lee, J. Jang, J. Chun, K. A. A. Makinwa, and Y. Chae. "A Compact Resistor-Based CMOS Temperature Sensor With an Inaccuracy of 0.12 °C (3 σ) and a Resolution FoM of 0.43 pJ · K ^{2} in 65nm CMOS," in IEEE Journal of Solid-State Circuits, vol. 53, no. 12, pp. 3356-3367, Dec. 2018.

[6] A. Wang, C. Chen, C. Liu and C. -J R. Shi, "A 9-Bit Resistor-Based Highly Digital Temperature Sensor With a SAR-Quantization Embedded Differential Low-Pass Filter in 65nm CMOS With a 2.5- μ s Conversion Time" in IEEE Sensors Journal, vol. 19, no. 17, pp. 7215-7225, 1 Sept.1, 2019.

[7] J. S. Shor and K. Luria, "Miniaturized BJT-Based Thermal Sensor for Microprocessors in 32- and 22-nm Technologies," in IEEE Journal of Solid-State Circuits, vol. 48, no. 11, pp. 2860-2867, Nov. 2013.

[8] B. Yousefzadeh and K. A. A. Makinwa. "A BJT-based temperature sensor with a packaging-robust inaccuracy of $\pm0.3^oC(3\sigma)$ from -55^oC to $+125^oC$ after heater-assisted voltage calibration." In 2017 IEEE International Solid-State Circuits Conference (ISSCC), pp. 162-163. IEEE, 2017.

[9] A. Heidary, G. Wang, K. Makinwa and G. Meijer. "12.8 A BJT-based CMOS temperature sensor with a 3.6pJ·K2-resolution FoM." In 2014 IEEE international solid-state circuits conference digest of technical papers (ISSCC), pp. 224-225. IEEE, 2014.

[10] M. Cho, M. Khellah, K. Chae, K. Ahmed, J. Tschanz and S. Mukhopadhyay. "Characterization of inverse temperature dependence in logic circuits." Proceedings of the IEEE 2012 Custom Integrated Circuits Conference. IEEE, 2012.

[11] K. Roy, S. Mukhopadhyay and H. Mahmoodi-Meimand. "Leakage current mechanisms and leakage reduction techniques in deep-submicrometer CMOS circuits." Proceedings of the IEEE 91.2, 2003.

[12] A. A. Abidi. "Phase noise and jitter in CMOS ring oscillators." IEEE journal of solid-state circuits 41.8, 2006.

[13] R. Tibshirani. "Regression shrinkage and selection via the lasso." Journal of the Royal Statistical Society: Series B (Methodological) 58.1, 1996.

978-1-7281-5410-7/20 $31.00 © 2020 IEEE

Exploring the FPGA Implementations of the LBlock, Piccolo, Twine, and Klein Ciphers

S. Moraitis, D. Seitanidis, G. Theodoridis*, O. Koufopavlou
Electrical Computers Engineering Department*)
University of Patras
Patras, Greece
*theodor@ece.upatras.g

Abstract—**In this work, the implementations of the LBlock, Piccolo, Twine, and Klein lightweight ciphers in FPGA technology are studied in terms of area, frequency, throughput, and throughput/area. To accomplish this, loop unrolling and pipelining were employed in two phases. In the first phase, different loop unrolling factors were used to implement the round function of each cipher, while in the second phase, 2-stage pipelining with loop unrolling per stage was applied. The produced designs were implemented in Xilinx (Kintex-7) FPGA technology. Based on the implementation results, a detailed study on the above-mentioned design metrics was performed and important outcomes were derived.**

Keywords— *Lightweight ciphers, FPGA, loop unrolling, pipeline, area, frequency, throughput*

I. INTRODUCTION

Lightweight ciphers are developed for a large number of emerging and rapidly growing application fields such as wireless body area network (WBAN), wireless sensor network (WSN), internet of things (IoT), radio-frequency identification (RFID) and smartcards [1]-[4]. The main feature of these applications is that they are implemented on computing platforms that usually employ a limited number of resources. So, the implementations of these ciphers must be characterized by low area cost. On the other hand, these applications demand high processing rate.

To improve the throughput of the implementations of ciphers in hardware, loop unrolling and pipelining are among the most widely applied techniques. In general, as loop unrolling and pipeline stages increase, both throughput and area increase but not necessarily equally. So, there is an exploration space that must be investigated to find a good balance between these metrics. Also, many surveys have been published in which the performance of the implementations of lightweight encryption algorithms in software and hardware platforms was studied in terms of several metrics such as throughput, area, execution time and program memory, while different design techniques were applied [2]-[7]. In all these studies, when hardware implementations are mentioned, they concern ASIC technology. However, in ASIC technology, the area overhead can be estimated with acceptable accuracy since it is mainly dominated by the increase of the datapath's area due to the applied unrolling. Similar estimations can be also made for frequency.

On the other hand, when FPGA technology is employed, the situation becomes more complex. Both area cost and clock period depend strongly on the type, number, and organization of the resources of the FPGA device, as well as the way the synthesis tool exploits them to produce the final design. For instance, when the unrolling factor increases from 2 to 4, the datapath's area is not necessarily doubled. This may happen because the synthesis tool may be able to use some available resources (e.g. LUTs) of the slices which were used when unrolling factor was two. Moreover, the situation is further complicated since these algorithms are characterized by low complexity, making difficult to estimate easily and with adequate accuracy the throughput and area values. Hence, when FPGA technology is used for the implementation of lightweight ciphers, it is not so easy to estimate the impact of unrolling and pipeline in terms of area and throughput and a detailed study is demanded.

In this paper, four lightweight encryption algorithms specifically the LBlock, Piccolo, Twine and Klein are implemented in FPGA technology (Xilinx, Kintex-7) and they are studied in terms of area, frequency, throughput, and throughput/area metrics. For the algorithms' implementation, loop unrolling and pipelining were applied in two phases. Firstly, the encryption round of each algorithm was unrolled by 1, 2, 4 and 8 and the corresponding implementations were studied in terms of the metrics mentioned above. Then, a 2-stage pipeline architecture was developed, where in each pipeline stage the above unrolling factors were also applied.

Based on the experimental results, several important outcomes are derived. Firstly, as unrolling increases, both area and clock period increase, but at smaller rates than those expected due to the applied unrolling. Secondly, when the unrolling factor is kept low, the critical path lies in the control circuit. However, as it increases, the critical path is moved to the combinational logic that implements the rounds. Also, both throughput and throughput/area increase as the applied unrolling increases, but not the same for each algorithm. Concerning area, LBlock is the best, while LBlock and Twine are the best in terms of frequency. On the other hand, Klein is the best in terms of throughput, while Klein and LBlock are the best in terms of throughput/area metrics.

The paper is organized as follows. In Section II the target algorithms are described, while in Section III the adopted architectures are presented. The experimental results and their discussion are provided in Section IV, while Section V concludes the paper.

II. DESCRIPTION OF THE ALGORITHMS

In this section the studied algorithms are presented. Due to space limits, their description is not detailed. For more details, the reader may read the corresponding reference of each algorithm.

A. LBlock

LBlock receives as inputs a 64-bit plaintext, P, and a 80-bit master key, K, and produces a 64-bit ciphertext, C, after 32 rounds [8]. Its encryption procedure employs a variant of the Feistel structure and is described by the following procedure

For $i = 0$ to 31 **do**
$$X_{i+2} = F(X_{i+1}, RK_{i+1}) \oplus (X_i <\!\!<\!\!< 8)$$
end for
$C = X_{32} \| X_{33}$

978-1-7281-5410-7/20 $31.00 © 2020 IEEE

where RK_i is the 32-bit round key of the i-th round, $\oplus$ is the bitwise XOR operation, $<<<8$ is the 8-bit left cyclic shift operation, and $\|$ is the concatenation of two or more vectors.

The round function, F, implements the expression $F=P(S(X \oplus RK_i))$, where S is a confusion function consisting of eight 4-bit S-boxes in parallel, and P is a diffusion function, which corresponds to a permutation of eight 4-bit words.

Initially, the plaintext is divided into two 32-bit vectors, X_0 and X_1. Then, in each round, the vector X_i is being left cyclic shifted by 8 bits, while the vector X_{i+1} along with the round key, RK_{i+1}, become inputs to the round function, F. Next, the outputs of the above two operations are driven to a 32-bit XOR to form the output X_{i+2} of the current round. Finally, the results of the last two rounds are concatenated to produce the 64-bit ciphertext, C. Concerning the 32-bit round keys RK_i, they are produced by the key scheduling procedure, which uses the 80-bit input master key K and produces the round keys as described in [8].

B. Piccolo

Piccolo gets as input a 64-bit plaintext, P, and produces a 64-bit ciphertext, C. The master key K, can be 80 or 128 bits, while the executed rounds, N_R, are 25 when K is 80 bits and 31 when K is 128 bits [9]. In this work we study the version in which K is 80 bits. The algorithm's encryption procedure is the following

$X_0 \| X_1 \| X_2 \| X_3 = P$
$X_0 = X_0 \oplus wk_0, \quad X_2 = X_2 \oplus wk_1$
For i = 0 to N_R - 2 **do**
$\quad X_1 = X_1 \oplus F(X_0) \oplus RK_{2i}$
$\quad X_3 = X_3 \oplus F(X_2) \oplus RK_{2i+1}$
$\quad X_0 \| X_1 \| X_2 \| X_3 \leftarrow RP(X_0 \| X_1 \| X_2 \| X_3)$
end for
$X_1 = X_1 \oplus F(X_0) \oplus RK_{2r-2}, \quad X_3 = X_3 \oplus F(X_2) \oplus RK_{2r-1}$
$X_0 = X_0 \oplus wk_2, \quad X_2 = X_2 \oplus wk_3$
$C = X_0 \| X_1 \| X_2 \| X_3$

where wk_i are the whitening keys produced by the master key. Also, F is the round function, which is a substitution network consisting of two S-box layers separated by a multiplication process of the first layer with a diffusion matrix, M, over $GF(2^4)$ field defined by the x^4+x+1 polynomial [9].

Initially, the plaintext is divided into four 16-bit vectors X_0, X_1, X_2 and X_3. Then, the XOR operations $X_0 = X_0 \oplus wk_0$ and $X_2 = X_2 \oplus wk_1$ take place. Next, the X vectors are configured in the way shown above, using two 16-bit round keys in each one of the 24 rounds and four whitening keys, two at the beginning and two at the end of the process. All 50 round keys, RK_i, are produced after 25 rounds by the key scheduling procedure [9].

C. Twine

Twine gets as input a 64-bit plaintext, P, and produces a 64-bit ciphertext, C. Also, the master key K, can be 80 or 128 bits, while the executed rounds, N_R, are 35 for both key versions [10]. In this work, we study the version in which K is 80 bits. The encryption process is a generalized Feistel structure, which is described in the following

$X^1 = P$
$RK^0 \| \dots \| RK^{35} = K$
For i = 0 to 34 **do**

$X_0^i \| \dots \| X_{15}^i = X^i$
$RK_0^i \| RK_1^i \| \dots \| RK_6^i \| RK_7^i = RK^i$
For j = 0 to 7 **do**
$\quad X_{2j+1}^i = S(X_{2j}^i \oplus RK_j^i) \oplus X_{2j+1}^i$
end for
For h = 0 to 15 **do**
$\quad X_{\pi[h]}^{i+1} = X_h^i$
end for
$X^{i+1} = X_0^{i+1} \| X_1^{i+1} \| \dots \| X_{14}^{i+1} \| X_{15}^{i+1}$
end for
For j = 0 to 7 **do**
$\quad X_{2j+1}^{35} = S(X_{2j}^{35} \oplus RK_j^{35}) \oplus X_{2j+1}^{35}$
end for
$C = X^{35}$

In each round, the plaintext, P, is divided into sixteen 4-bit vectors, X_0^i - X_{15}^i, while the 32-bit round key is also divided into eight 4-bit sub-keys, RK^i. The odd vectors of each round are produced from a XOR operation of themselves with the result of the S function, which substitutes the XOR process of the even vectors and the sub-keys RK^j. Then, a diffusion process permutes these sixteen vectors according to the index produced from the function π, whose outputs are extracted from a look-up table. After the completion of the last round, the procedure is performed once again without the diffusion process and next, the concatenation of the 16 vectors forms the 64-bit ciphertext C.

The key scheduling of this version divides the 80-bit master key K into twenty 4-bit vectors. In each one of the 35 rounds these vectors form the round keys, through a process using constant values, S-boxes and XOR gates [10].

D. Klein

Klein takes as input a 64-bit plaintext, P, and produces a 64-bit ciphertext, C. Its master key, K, can be 64, 80 or 96 bits, while the executed rounds, N_R, may be 12, 16, or 20, respectively [11]. In this work, we use the version of the algorithm in which the master key is 80 bits, while the rounds are 16. The structure of the encryption round is a typical Substitution-Permutation Network (SPN), which is described in the following

$RK_0 = K$
$STATE = P$
For i = 1 to N_R **do**
$\quad AddRoundKey(STATE, RK_i)$
$\quad SubNibbles(STATE)$
$\quad RotateNibbles(STATE)$
$\quad MixNibbles(STATE)$
$\quad RK_{i+1} = KeySchedule(RK_i, i)$
end for
$C = AddRoundKey(STATE, RK_r)$

Initially, the master key, K, is assigned to the round key, RK_0, while the plaintext is assigned to the vector $STATE$. The round function, F, consists of four steps, which are described in the following. In the $AddRoundKey$ step, the XOR operation is applied in vector $STATE$ and the RK_i. Then, in the $SubNibbles$, the result of the previous step is divided into 16 nibbles that are substituted according to a Substitution Network. Next, in the $RotateNibbles$, the results of the previous step are permuted to each other. Afterwards, the $MixNibbles$ is applied, which is a modular multiplication between the results of the previous step with a fixed polynomial. At the end of the round, the round key, RK_{i+1}, of the next round is produced using the *KeySchedule* function. Finally, after the completion of the rounds, the *AddRoundKey* is applied to produce the ciphertext, C. More details of these steps are provided in [11].

III. STUDIED ARCHITECTURES

To study the performance of the above algorithms in terms of important design metrics (e.g. area, frequency, throughput), two basic architectures are considered. These are the Unrolled architecture and the 2-stage Pipelined one with unrolling per pipeline stage. The description of them follows.

A. Unrolled Architecture

As it is depicted in Fig. 1, the Unrolled architecture consists of two main modules, which are the Round Module and Keys Module. It also includes a Control Module that is omitted for clarity reasons.

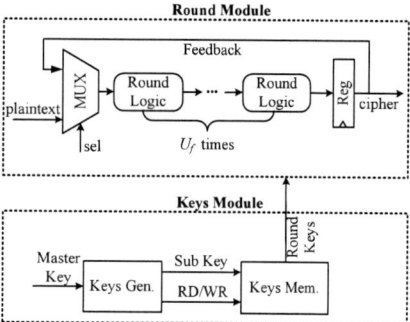

Fig. 1.   Unrolled Architecture

The Round Module is responsible for the execution of the algorithm's encryption rounds. It consists of a two-input multiplexer (MUX), an output register (Reg) and Uf serially connected copies of the Round Logic, where Uf is the applied unrolling factor, which depicts the number of Round Logic blocks between the multiplexer and the register that are being unrolled. Initially, the plaintext is selected by the MUX and it is driven to the first Round Logic component, while for the remaining execution cycles, the Feedback signal is selected by the MUX. After executing a specific number of iterations, which depends on the applied unrolling, the encryption process completes and the ciphertext is produced. This architecture is suitable for algorithms with identical or almost identical encryption rounds, as it happens in our case.

Regarding the Key Module, it receives the Master Key, spends some cycles to produce the required Round Keys using the Keys Generation component and stores them in the Keys Memory. Afterwards, the encryption process starts. This approach is widely used to alleviate the circuit complexity required to generate on-the-fly the Round Keys, for the cases where multiple keys are required per execution cycle. Also, in real-life applications, when a connection has been established between the transmitter and the receiver, the Master Key remains the same for a long time so, there is not a strong need to generate on-the-fly the Round Keys.

The main benefit of the unrolled architecture is that it allows to study the trade-offs between area and throughput. When Uf is kept low, the number of the included Round Logic components is small, leading in high hardware reuse and consequently in low area cost. For instance, when $Uf = 1$, only one Round Logic component is included, which results in the lowest area cost.

Throughput is computed by

$$Throughput_{unroll} = \frac{bits \times F_{Uf}}{exec.cycles} = \frac{bits \times F_{Uf}}{N_R/Uf} \qquad (1)$$

where *bits* is the algorithm's block size, Uf is the unrolling factor, F_{Uf} is the frequency for the given Uf value, *exec. cycles* are the cycles that are spent to complete the encryption process, and N_R is the algorithm's number of rounds.

By increasing Uf, the execution cycles are reduced. However, this may result in a frequency reduction, since Uf Round Logic modules are connected serially, which may increase the critical path. Hence, if the frequency reduction ratio is lower than the increase of Uf then, throughput is improved.

However, the studied algorithms are characterized by low computational complexity. So, it is not certain that the increase of the critical path will be equal to the increase of Uf. This may happen in the cases where the value of Uf is small. In these cases, the critical path may be included in the control circuit and not in the serially connected Round Logic blocks. On the other hand, for large unrolling values, the critical path may be moved to the Round Module. Moreover, this issue becomes more complex in FPGA technology where the performance of the implementation strongly depends on the type, amount and organization of the resources of the target FPGA device, as well as the way the synthesis tool exploits them to improve frequency.

In addition, when unrolling is applied, area increases because Uf Round Logic components are included in the design. As in the case of frequency, when the algorithms of our study are implemented in FPGA technology, it is not certain that area will increase equally to the unrolling factor.

To study the above topics, the target encryption algorithms implemented with $Uf = 1, 2, 4, 8$ and the corresponding designs are denoted as U1, U2, U4, and U8, respectively.

B. Pipelined Architecture

To study the impact of pipeline, a second architecture was studied in which the Round Module of the Unrolled architecture was replaced by a new one, where a 2-stage pipeline was applied. Also, in each pipeline stage, unrolling was applied. The new Round Module is depicted in Fig. 2.

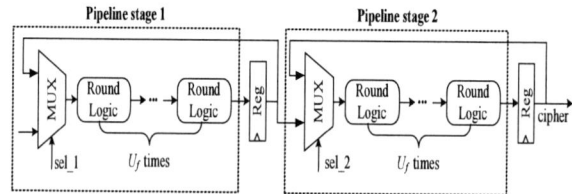

Fig. 2.   Pipelined Round Module

Thus, the above Round Module is a 2-stage pipeline design and the throughput is computed by

$$Throughput_{pipe\&unroll} = \frac{bits \times F_{Uf}}{exec.cycles} = 2 \times \frac{bits \times F_{Uf}}{N_R/Uf} \qquad (2)$$

Concerning the variations of area and frequency with respect to the applied unrolling, the same discussion as in the case of the unrolled architecture holds. Also, for the case of the pipeline designs, the same unrolling was applied as in the case of the unrolled architecture. That is, $Uf = 1, 2, 4, 8$ and the corresponding designs are denoted as PU1, PU2, PU4, and PU8, respectively.

C. Particularities in Piccolo and Twine Implementations

To perform a fair comparison, we had to implement the algorithms by applying the same unrolling values. However, Piccolo executes 24 rounds and it cannot be implemented in a PU8 architecture. This happens because a 2-stage pipeline is applied, and each stage would require executing 12 rounds. Also, the rounds of Twine are 35, so it cannot be divided evenly with the applied unrolling values. To overcome this, the following modifications were applied. The first 32 rounds were implemented using the above architectures and unrolling values, while the last 3 rounds were implemented by an extra Round Logic unit. For the cases of U1, U2, PU1 and PU2 designs, in order to not affect the critical path, the applied unrolling factor of the extra units was set to one ($Uf = 1$). Also, in the cases of U4, U8, PU4 and PU8 designs, the unrolling factor of the extra unit was equal to three ($Uf = 3$), so that their execution lasts for one clock cycle.

IV. EXPERIMENTAL RESULTS AND DISCUSSION

The above designs described in VHDL while for their simulation, synthesis and implementation, the Xilinx Vivado framework was used. The correct functionality of the implementations was verified by performing extensive functional and post-implementation simulations, using as references the test vectors provided in the algorithms' description documents [8]-[11]. Moreover, reference software was developed for each algorithm to produce additional input and output test vectors, which are used to further verify the correct functionality of the designs. Also, all the designs were implemented in a Kintex (xc7k70tfbg484) device using the default options of Vivado. So, all the presented results have been produced after completing the placement and routing steps.

The metrics that were studied are the area, frequency, throughput, and throughput/area. The values of each metric for each design are shown in TABLE I. and the discussion on each of them is provided in the following.

A. Area

Concerning the unrolled U1, U2, U4, and U8 designs, it is concluded that for all cases, the designs of the LBlock and Klein are the best and worst in terms of area, respectively.

Comparing Piccolo and Twine, it is also observed that Piccolo is the best in U1 and U2 designs while Twine is the best in U4 and U8 designs. A second important remark is that for all cases, the area does not increase equally with respect to the increase of the unrolling factor. This was expected because these algorithms are characterized by low complexity. Studying in detail the area increases of the U2, U4 and U8 designs over the U1 ones, we observe the following. For the case of LBlock, the values of the area ratios $U2_A/U1_A$, $U4_A/U1_A$, $U8_A/U1_A$ are 1.1, 1.3 and 1.8, respectively. Also, the corresponding area increases are 1.4x, 2x and 3x for Piccolo, 1.4x, 1.7x and 2.3x for Twine and 1.3x, 1.9x and 2.6x for Klein. Hence, LBlock exhibits the smallest area increase even when the unrolling factor equals to 8.

Regarding the pipelined PU1, PU2, PU4, and PU8 designs, it is concluded that LBlock is again the best in terms of area, but it is not clear which algorithm is the worst. Also, as in the case of the unrolled designs, the area does not increase equally with respect to the unrolling factor. Particularly, the area ratio values $PU2_A/PU1_A$, $PU4_A/PU1_A$ and $PU8_A/PU1_A$ are 1.2, 1.6 and 2.4, respectively for LBlock, 1.5 and 1.7 for Piccolo, 1.1, 1.2 and 1.8 for Twine and 1.3, 2.0 and 2.8 for Klein. Thus, among all algorithms, Twine exhibits the lowest area increase with respect to the applied unrolling.

To study the overhead (extra registers and slightly more complex controller) due to pipeline, we compare the $PU1_A/U2_A$, $PU2_A/U4_A$ and $PU4_A/U8_A$ area ratios. The corresponding values are 1.12, 1.06 and 1.07 for LBlock, 1.18, 1.20 and 0.91 for Piccolo, 1.37, 1.23 and 0.99 for Twine and 0.93, 0.85 and 0.91 for Klein. Thus, it is concluded that area overhead due to the pipeline is also low in all cases.

Summarizing, for all algorithms, both in the unrolled and pipelined designs, the area ratio increase is kept much lower than the increase of the unrolling factor. Also, LBlock is the best in terms of area in both architectures. Finally, the area overhead due to pipeline is small in all cases.

B. Frequency

For the unrolled U1, U2, U4, and U8 designs, the outcome is that LBlock and Twine achieve similar frequency values and they are the best in all cases, except for U1 designs in which Klein is the best. Moreover, the frequency reduction is lower than the one that was expected due to the applied unrolling. This was also expected due to the low complexity of the algorithms.

TABLE I.   IMPLEMENTATION RESULTS: AREA, FREQUENCY, THROUGHPUT, THROUGHPUT/AREA

Alg.	Area (Slices)								Freq (MHz)							
	$U1_A$	$U2_A$	$U4_A$	$U8_A$	$PU1_A$	$PU2_A$	$PU4_A$	$PU8_A$	$U1_F$	$U2_F$	$U4_F$	$U8_F$	$PU1_F$	$PU2_F$	$PU4_F$	$PU8_F$
LBlock	579	612	770	1030	684	816	1098	1666	323	343	301	190	342	318	291	187
Piccolo	630	892	1291	1902	1049	1548	1739		332	224	132	67	341	220	125	
Twine	699	945	1169	1627	1290	1443	1607	2310	322	320	307	189	332	317	295	183
Klein	732	945	1368	1930	881	1156	1764	2432	352	275	160	89	326	276	165	92

Alg.	Throughput (Mbps)								Throughput/Area (Mbps/Slice)							
	$U1_T$	$U2_T$	$U4_T$	$U8_T$	$PU1_T$	$PU2_T$	$PU4_T$	$PU8_T$	$U1_{T/A}$	$U2_{T/A}$	$U4_{T/A}$	$U8_{T/A}$	$PU1_{T/A}$	$PU2_{T/A}$	$PU4_{T/A}$	$PU8_{T/A}$
LBlock	647	1372	2405	3034	1369	2543	4656	5990	1.12	2.24	3.12	2.95	2.00	3.12	4.24	3.60
Piccolo	849	1102	1207	1066	1678	2013	2006		1.35	1.24	0.94	0.56	1.60	1.30	1.15	
Twine	588	1078	2186	2417	1118	1844	3778	3895	0.84	1.14	1.87	1.49	0.87	1.28	2.35	1.69
Klein	1408	2202	2568	2862	2612	4411	5291	5889	1.92	2.33	1.88	1.48	2.96	3.82	3.00	2.42

Calculating the frequency ratios $U2_F/U1_F$, $U4_F/U1_F$ and $U8_F/U1_F$, we observe the following. The values of these ratios are 1.06, 0.93 and 0.59, respectively for LBlock, 0.67, 0.40 and 0.20 for Piccolo, 0.99, 0.95, and 0.59 for Twine and 0.78, 0.45 and 0.25 for Klein. Hence, the first outcome is that for LBlock and Twine, concerning the designs up to U4, the frequency variation is not large. Also, for the case of U8 designs, frequency is reduced by 40% on them; however, this is much better than the 87% that was expected due to the unrolling by 8. On the other hand, in Piccolo and Klein, frequency starts reducing from U2 designs, reaching an 80% reduction for Piccolo in U8 design. Exploring the reports provided by Vivado, we found the following. For all algorithms, the critical path lies in the controller for the case of the U1 designs, while it lies in the Round Unit for the cases of the U4 and U8 designs. Also, for the case of U2 designs, the critical path lies in the controller for LBlock and Twine, while for Piccolo and Klein, it is included in the Round Unit. Considering the above, the frequency behaviour for each case can be easily understood.

Regarding the pipelined PU1, PU2, PU4, and PU8 designs, the conclusions are the same as in the case of the unrolled ones. Again, LBlock and Twine are the best in most cases. Also, the frequency reduces as the unrolling factor increases but the reduction is smaller than the expected one due to the applied unrolling. Specifically, compared to PU1 designs, the frequency of PU2, PU4 and PU8 designs is reduced by 7%, 15% and 45%, respectively in LBlock, 35% and 63% in Piccolo, 4.5%, 11% and 45% in Twine and 15%, 49% and 72% in Klein

Summarizing, LBlock and Twine achieve similar frequency values and they are the best both in the unrolled and pipelined designs. Also, in both architectures, the frequency reduction ratio is lower than the increase of the unrolling factor. In addition, frequency does not vary when pipeline is applied. Finally, for LBlock and Twine, both in the unrolled and pipelined designs, frequency is kept almost intact except for the cases where the unrolling factor equals to 8.

C. Throughput

For the unrolled designs, it is concluded that Klein achieves the highest throughput, except for the case of U8 in which LBlock is slightly better. Concerning the remaining algorithms, their throughput values vary per case making difficult to determine a ranking of them for all cases. The second topic is to study the increase of throughput with respect to the applied unrolling. Thus, for LBlock, comparing the U2, U4, and U8 designs against the U1 one, it is derived that throughput improves by 2.12x, 3.72x and 4.69x, respectively. Also, the corresponding throughput improvements are 1.3x, 1.42x and 1.26x for Piccolo, 1.83x, 3.72x and 4.11x for Twine and 1.56x, 1.82x and 2.03x for Klein. Therefore, it is concluded that for LBlock and Twine and up to the case of U4 designs, throughput improves almost equally with respect to the applied unrolling. However, for the case of U8 designs, throughput improves about 4x. On the other hand, for both Piccolo and Klein, throughput improves with the increase of the unrolling factor. However, they do not exhibit a significant improvement of throughput with respect to the applied unrolling

Regarding the pipelined designs, the outcomes are similar to the ones of the unrolled designs. Again, Klein is the best except for the PU8 designs where LBlock is slightly better. Also, for LBlock and Twine and up to the case of PU4 designs,

the throughput improvement follows the increase of the unrolling factor at a very good level. However, in the case of PU8 designs, compared to the PU1 ones, throughput is improved by 4.38x for LBlock and 3.48x for Twine.

Also, to compare the unrolled and pipelined designs in terms of throughput, we calculate the corresponding throughput ratios namely $PU1_T/U1_T$, $PU2_T/U2_T$, $PU4_T/U4_T$ and $PU8_T/U8_T$. Thus, the corresponding values of these ratios are 2.12, 1.85, 1.94 and 1.97 for LBlock, 1.98, 1.83 and 1.66 for Piccolo, 1.90, 1.71, 1.73 and 1.61 for Twine and 1.86, 2.00, 2.06 and 2.06 for Klein, respectively. Hence, the outcome is that it is worthy to apply pipeline when high throughput is required. Specifically, due to the applied two-stage pipeline, the throughput is almost doubled for LBlock and Klein. However, this is not happening for Piccolo and Twine.

Summarizing, both in the unrolled and pipelined designs, Klein achieves the highest throughput in all cases, except for the PU8 designs, in which LBlock is slightly better. Also, for LBlock and Twine, in both architectures, the throughput improvement follows the increase of the unrolling factor. Finally, the applied two-stage pipeline improves the throughput, as it was expected and especially for LBlock and Klein, it is doubled over the unrolled designs.

D. Throughput/Area

For the case of the unrolled designs, it is derived that Klein is the best in terms of throughput/area for the U1 and U2 designs, while LBlock is the best for the U4 and U8 ones. Also, the values of the $U2_{T/A}/U1_{T/A}$, $U4_{T/A}/U1_{T/A}$ and $U8_{T/A}/U1_{T/A}$ ratios are 2.00, 2.79 and 2.63 for LBlock, 0.92, 0.70 and 0.41 for Piccolo, 1.36, 2.23 and 1.77 for Twine and 1.21, 0.98 and 0.77 for Klein. It is clear, that LBlock and Twine benefit from the increase of the unrolling factor. On the other hand, this does not happen for Piccolo and Klein, in which the increase of the unrolling factor is not beneficial. The reason for the above is mainly based on the achieved throughput values per case and the variation of throughput with respect to the unrolling factor, as well as the area study, which has been discussed in the previous sub-sections.

Concerning the pipelined designs, it is concluded that, onace more, Klein is the best in terms of throughput/area for the U1 and U2 designs, while LBlock is the best for the U4 and U8 designs. Also, compared to the U1 designs, the corresponding improvements of throughput/area metric of the U2, U4 and U8 designs are 1.56x, 2.12x and 1.80x for LBlock, 0.81x and 0.72x for Piccolo, 1.47x, 2.70x and 1.94x for Twine and 1.29x, 1.01x and 0.82x for Klein. Thus, as in the case of the unrolled designs, LBlock and Twine benefit from the increase of the unrolling factor, which does not happen for Piccolo and Klein.

Summarizing, for the unrolled and pipelined designs, Klein is the best in terms of throughput/area for the U1 and U2 designs, while LBlock is the best for the U4 and U8 ones. Also, LBlock and Twine benefit from the increase of the unrolling factor. On the other hand, this does not happen for Piccolo and Klein, where the increase of the unrolling factor is not beneficial.

V. CONCLUSIONS

The implementation of encryption mode of the LBlock, Piccolo, Twine and Klein lightweight ciphers in FPGA technology was studied in terms of area, frequency, throughput, and throughput/area factors. To achieve this, the techniques of loop unrolling and pipelining were used for the implementation of the encryption rounds of these ciphers. As a future work, we plan to perform a similar study concerning the decryption and the combined encryption-decryption modes of these ciphers, as well as to study additional metrics such as power and energy consumption.

REFERENCES

[1] S. Singh, P.K. Sharma, S.Y Moon, et al., "Advanced lightweight encryption algorithms for IoT devices: survey, challenges and solutions," Journal of Ambient Intelligence and Humanized Computing, SpringerLink, 2017.

[2] J. H. Kong, Li-Minn Ang, K. P. Seng, "A comprehensive survey of modern symmetric cryptographic solutions for resource constrained environments," Journal of Network and Computer Applications, Elsevier, Vol. 49, pp. 15-50, March 2015.

[3] B. J.Mohd, T. Hayajneh and A.V.Vasilakos, "A survey on lightweight block ciphers for low-resource devices: Comparative study and open issues," Journal of Network and Computer Applications, Elsevier, Vol. 58, pp. 73-93, December 2015.

[4] G. Hatzivasilis, K. Fysarakis, I Papaefstathiou, C. Manifavas, "A review of lightweight block ciphers," Journal of Cryptographic Engineering, SpringerLink, vol. 8, pp. 141–184, 2018.

[5] J. Hosseinzadeh, M. Hosseinzadeh, "A comprehensive survey on evaluation of lightweight symmetric ciphers: hardware and software implementation," Journal on Advances in Computer Science, vol. 5, No. 4, pp/31-41, 2016.

[6] S. Sallam; B. and D. Behesht, "A Survey on Lightweight Cryptographic Algorithms," Proceedings of TENCON, 1748-1789, 2018.

[7] J. Mohda,et. al, "Hardware design and modeling of lightweight block ciphers for secure communications," Future Generation Computer Systems, Elsevier, vol. 83, pp. 510-521, 2018.

[8] W. Wu and L. Zhang, "LBlock: A Lightweight Block Cipher," Int. Conf. on Applied Cryptography and Network Security (ACNS), Lecture Notes in Computer Science, Springer, vol. 6715, pp. 327-344, 2011.

[9] K. Shibutani, T. Isobe, H. Hiwatari, A. Mitsuda, T. Akishita and T. Shirai, "Piccolo: An Ultra-Lightweight Blockcipher,"Int. Workshop on Cryptographic Hardware and Embedded Systems (CHES), Lecture Notes in Computer Science, Springer, vol. 6917, pp. 342-357, 2011.

[10] T. Suzaki, K. Minematsu S. Morioka and E. Kobayashi., "TWINE: A Lightweight, Versatile Block Cipher," ECRYPT Workshop on Lightweight Cryptography, 2011.

[11] Z. Gon., S. Nikova and Y.W. Law, "KLEIN: A New Family of Lightweight Block Ciphers," Int. Workshop on Radio Frequency Identification: Security and Privacy Issues (RFIDSec), Lecture Notes in Computer Science, Springer, vol. 7055, pp. 1-18, 2012.

A 0.8V 875 MS/s 7b low-power SAR ADC for ADC-Based Wireline Receivers in 22nm FDSOI

David Cordova[1,2], Wim Cops[2], Yann Deval[1], Francois Rivet[1], Herve Lapuyade[1],
Nicolas Nodenot[2], and Yohan Piccin[2]

[1]IMS Lab., Univ. Bordeaux, Bordeaux INP, CNRS UMR 5218, Talence, France
[2]MACOM Technologies Solutions, Sophia Antipolis, France

Abstract—This paper presents a very low-power 875 MS/s 7b single-channel high-speed successive approximation register (SAR) analog-to-digital converter (ADC) that achieves a SNDR/SFDR at Nyquist rate of 41.46/55.01 dB. The use of an integer-based split CDAC combined with an improvement for the LSB capacitor allows a substantial improvement in the SNDR. A simple and accurate calibration procedure for the ADC is presented thanks to body biasing. The ADC is designed in 22 nm FDSOI while consuming 1.65 mW from a 0.8 V supply with a core chip area of 0.00074 mm^2. The Walden figure-of-merit of 19.5 fJ/conversion-step at Nyquist rate making it one of the lowest among recently published medium resolution SAR ADCs.

Keywords—Analog-to-digital conversion, analog-to-digital converter (ADC), low power, single channel, successive approximation register (SAR).

I. INTRODUCTION

EMERGING standards (200 - 400G Ethernet and others) in wireline communications will demand higher data rates. A shift towards more sophisticated encoding schemes that require less bandwidth will be favored. For example, PAM4 encoding, using 4 levels, could reduce the bandwidth by a factor of 2. These encoding schemes will be harder to support by purely analog solutions, so a trend towards mixed-signal architectures is underway.

An ADC-based solution seems to be a robust implementation over channels with high losses (> 20dB), because it could take advantage of technology scaling and most of the equalization can be done in the digital domain [1], [2].

Such ADCs are implemented using time-interleaving: identical sub-ADCs multiplexed in time, operating in parallel to achieve a higher sampling rate, [3]. A suitable sub-ADC should not magnify the complexity and the achievable speed and resolution versus power performance should be optimized in the interleaved system.

A very low-power 7b sub-ADC that is suitable for integration into such a time-interleaved (TI) receiver is presented in this paper. The potential integration of this ADC is illustrated in Fig 1. The paper is organized as follows: Section II presents the system level design of the sub-ADC. In Section III, the circuit description is presented. Simulation results are shown in Section IV and Section V draws the main conclusions from this work.

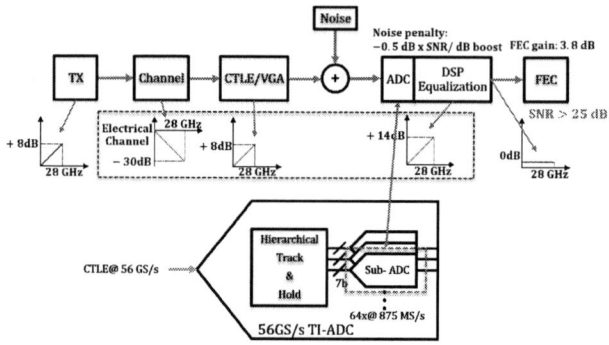

Fig. 1. sub-ADC in a Time Interleaved System

II. SYSTEM LEVEL DESIGN: RESOLUTION OF SUB-ADC

The proposed circuit is designed to meet the SNR requirements for 56 GBaud PAM4 signaling. Considering the same amplitude swing for NRZ and PAM4 encodings, the signal-to-noise ratio (SNR) in PAM4 is lower than NRZ. Forward error correction (FEC) is used to improve link integrity and counteract physical layer level errors introduced by reduced SNR in PAM4 signals [4].

At 56 GBaud, the pre-FEC raw BER of $< 10^{-4}$ is a typical target. In our design the receiver is designed to achieve BER$= 10^{-6}$ over a 30dB loss channel at 28GHz. Fig. 1 shows the link budget, we start from the output SNR (post-FEC BER$= 10^{-15}$) and work backwards from there.

$$\text{BER}_{\text{post,FEC}} = \left(\frac{M-1}{2M}\right) \times \text{erfc}\left(\sqrt{\frac{3 \cdot \text{SNR}}{2(M^2-1)}}\right) \quad (1)$$

where is M=4, for PAM4 modulation and *erfc* is the complementary error function, which yields a SNR$\approx$25dB.

For a FEC gain of 3.8dB, the pre-FEC BER is

$$\underbrace{25\text{dB}}_{\text{BER}_{\text{post,FEC}}} - \underbrace{3.8\text{dB}}_{\text{Gain}_{\text{FEC}}} = \underbrace{21.2\text{dB}}_{\text{BER}_{\text{pre,FEC}}<10^{-6}} \quad (2)$$

The DSP equalization boost is calculated as

$$\underbrace{30\text{dB}}_{\text{channel loss}} - \underbrace{8\text{dB}}_{\text{Tx}_{\text{gain}}} - \underbrace{8\text{dB}}_{\text{Rx}_{\text{gain}}} = 14\text{dB} \quad (3)$$

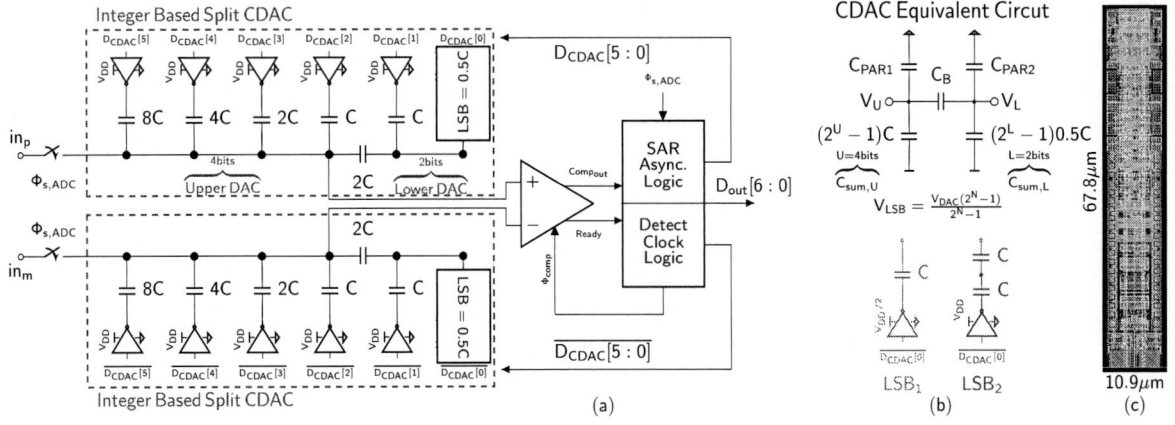

Fig. 2. (a) Architecture of the proposed single-channel 7b SAR ADC; (b) LSB Capacitor Variation: LSB_1 (red) and LSB_2 (blue) ; (c) Layout

Considering a SNR noise penalty of -0.5dB per dB of DSP boost and a margin of 6dB, the SNR requirement at the input of the ADC yields:

$$SNR_{ADC} = \underbrace{21.2dB}_{BER_{pre,FEC}<10^{-6}} + \underbrace{14 \times 0.5dB}_{N_{o,DSP}} + \underbrace{6dB}_{Margin} = 34.2dB \tag{4}$$

which roughly corresponds to a resolution of ≈ 5.4b. Our design targets a resolution of 7b as a trade-off of speed and power for a single-channel ADC.

III. CIRCUIT DESCRIPTION

The proposed SAR ADC uses asynchronous design to improve the speed conversion, an integer-based split capacitive DAC (CDAC) and the monotonic switching principle.

Fig. 2a illustrates the top-level ADC architecture. The front-end T&H circuit consists of a sampling bootstrapped switch with feed-through and charge-injection compensation. The sampling clock ($\Phi_{s,ADC}$), with a 25% duty cycle, drives the T&H and triggers the asynchronous internal conversion logic, which is responsible for generating the bit-cycle phases, controlling the comparator, storing its decisions and switching the CDAC.

The ADC comprises two identical 6b CDACs to accommodate differential operation. Since the MSB decision is the sign bit, it can be decided without changing the state of the CDAC. Thus, 6b CDAC instead of 7b CDAC is sufficient for this 7b SAR ADC design [5].

A. Capacitive DAC

The 6b CDAC, depicted in Fig. 2a, is based on an integer-based split capacitive DAC. The CDAC is divided in two sides: the lower side and the higher side of 2b and 4b, respectively. The split capacitor with a value 2C (C=unit capacitor) is placed just after the smallest unit capacitor of the 4b array, thus avoiding the non-linearity issues found in the common split capacitor array [6]. In the lower side two variations for the LSB capacitor were evaluated, Fig. 2b. Instead of switching a fraction of the unit capacitor (0.5C) during the

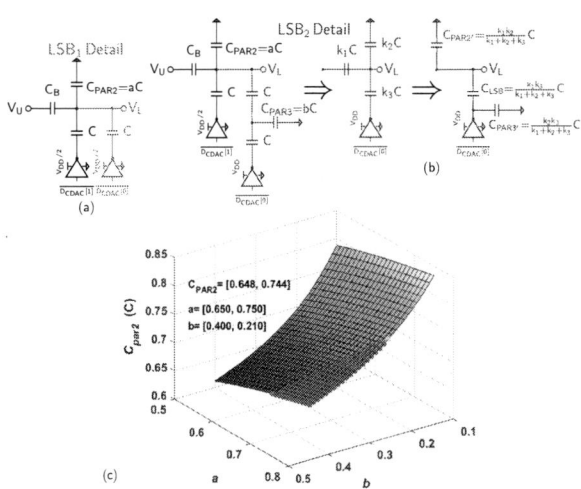

Fig. 3. LSB equivalent circuit with parasitics: (a) LSB_1, (b) LSB_2. (c) C_{PAR2} grid for a,b values

LSB conversion by the full difference of the reference voltage, $V_{REF}=V_{DD}$, a C unit capacitor is switched by a fraction of the reference voltage, $V_{REF}=V_{DD}/2$, which will be referred as LSB_1,[7]. And the other LSB_2 is implemented switching two unit capacitors (1C) in series by $V_{REF}=V_{DD}$.

The unit capacitors of the CDAC are designed to achieve best area efficiency using alternate-polarity metal-finger capacitors (APMOM) which offer high density, good matching characteristics and low parasitics.

For this 7b SAR ADC, the total capacitance of this capacitor array is $(2^{7-3}+3)$C=19, [6]. A 70% area reduction compared to a conventional CDAC with monotonic switching is obtained, thus reducing settling time between conversion cycles and minimizing area and power consumption. The designed one-side total capacitance is 45.6fF, which produces an equivalent kT/C thermal noise of 301.4μV. The least significant bit (LSB) value for a 7b ADC at a reference voltage of 0.8V is 6.25mV so the thermal noise introduced by this capacitor array is not a limiting factor.

978-1-7281-5410-7/20 $31.00 © 2020 IEEE

Following the equivalent circuit shown in Fig. 2b, the CDAC voltage (V_{DAC}) and LSB voltage (V_{LSB}) are

$$V_{DAC} = \frac{C_B(C_{VREF}^U + C_{VREF}^L) + C_{VREF}^U C_{sum,L}}{den} \cdot V_{REF}$$
$$+ \frac{C_{VREF}^U C_{PAR2}}{den} \cdot V_{REF} \quad (5)$$

$$V_{LSB} = \frac{V_{DAC}(2^N - 1)}{2^N - 1} =$$
$$\frac{C_B(C_{sum,U} + C_{sum,L}) + C_{sum,U}C_{sum,L} + C_{sum,U}C_{PAR2}}{(2^N - 1)den} \cdot V_{REF}$$
$$(6)$$

where N=U+L, den=$(C_{sum,U} + C_{PAR1})(C_{sum,L} + C_{PAR2}) + C_B(C_{sum,U} + C_{sum,L} + C_{PAR1} + C_{PAR2})$, C_B=2C. $C_{sum,U-L}$ and C_{VREF}^{U-L} denotes the total capacitance and total capacitors connected to V_{REF} in the Upper and lower DACs, respectively.

The parasitic capacitance C_{PAR2} in the numerator of Eq. 6 contributes to a code dependent error, degrading the DAC linearity [8]. C_{PAR2} is defined by the total parasitic capacitance of the top plates in the Lower CDAC and the metal interconnection. Fig. 3a shows the Lower DAC section for the LSB_1 capacitor. C_{PAR2} is denoted as a fraction of the unit capacitor (aC), for this condition the V_{LSB} error can be reduced by minimizing C_{PAR2} contribution.

In Fig. 3b it is shown a variation of LSB capacitor (LSB_2) to reduce the V_{LSB} error. C_{PAR2} is the same as in LSB_1, but with the addition of C_{PAR3} as the parasitic capacitance of the bottom plates of the two unit capacitors in series. C_{PAR3} is represented as a fraction of of the unit capacitor (bC). Through a delta-star transformation LSB_2 is shaped with the same arrangement as for LSB_1. Fig. 3c shows the new values of C_{PAR2} and C_{LSB}. A range for C_{PAR3} between [0.21 0.4]C yields the reduction of C_{LSB} and C_{PAR2} simultaneously. In 22nm FDSOI, the APMOM capacitors present a parasitic $\approx$0.13C per capacitor yielding a value of b around 0.26.

B. Comparator

Since the comparator determines the accuracy and speed of the ADC, special care has to be taken for its design [9]. A strong-ARM comparator is chosen for its superior decision speed enabled by the single-stage design. It is optimized for low noise and low power.

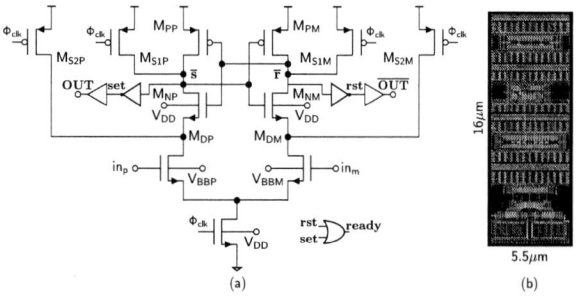

Fig. 4. Strong ARM Comparator: (a) Schematic; (b) Layout

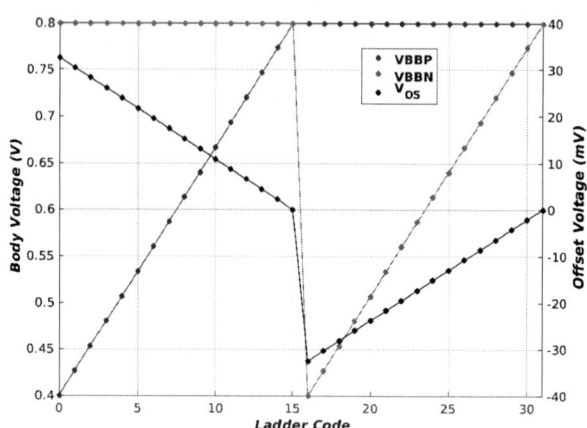

Fig. 5. Body biasing DAC: body bias and offset voltage

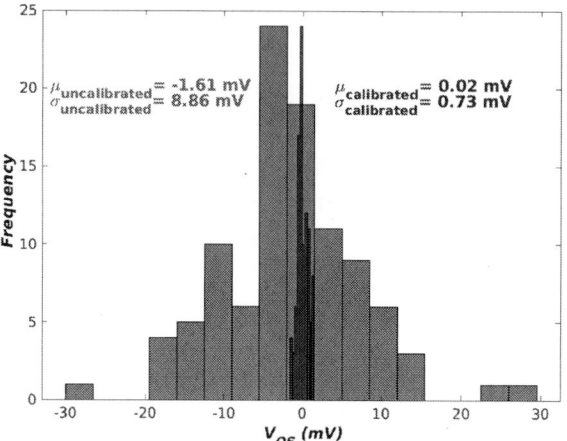

Fig. 6. Histogram Strong ARM V_{OS}: Uncalibrated; Calibrated for 100 MC runs

Fig. 4a shows the schematic, the latch regeneration forces one of the signals, *rst* and *set*, to high and the other to low, depending on the comparison result. As a result, the output of the OR gate (*ready*) is pulled high to enable the asynchronous control clock and to facilitate the progression to the next step in the SAR conversion.

The offset voltage V_{OS} introduced by the input differential pair is calibrated using a simple and accurate procedure. The calibration operates in two phases: V_{OS} extraction and V_{OS} calibration. The extraction is a fast loop operating at the comparator's clock frequency Φ_{CLK} and follows the smart resettable SAR (SR-SAR) technique presented in [10]. Here, the voltage offset is calculated. The second phase is a slow loop (below 1 MHz), due to the low frequency nature of the transistor's body. After extracting the V_{OS}, the control block begins to adjust the threshold voltages (V_{TH}) of the differential pair through a body biasing DAC, which consists of a resistor ladder controlled by a 5 bit digital word.

The nominal voltage supply for the comparator is 0.8V, the same for the body bias at nominal operation. In Fig. 5 is

978-1-7281-5410-7/20 $31.00 © 2020 IEEE

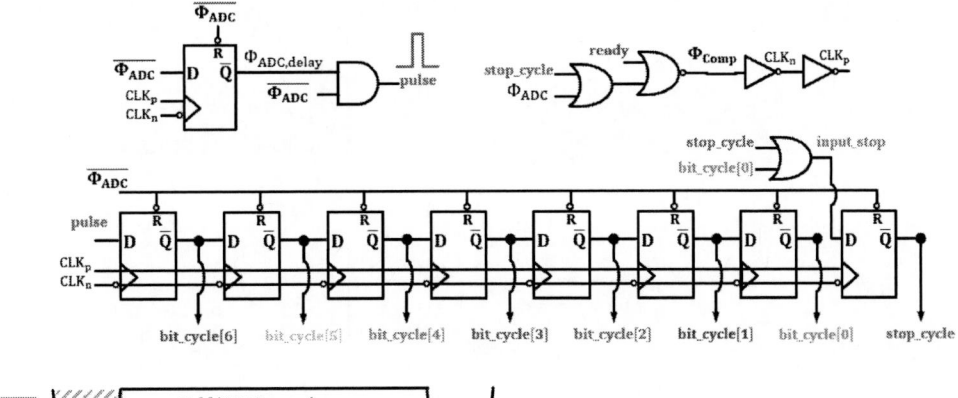

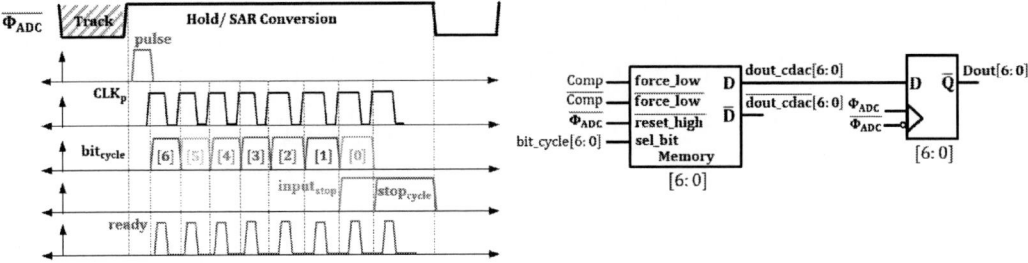

Fig. 7. SAR Logic and timing

shown the body bias control. The ladder was designed using a resolution of 5bits, with the body voltage varying between 0.4 to 0.8V. An artificial offset voltage range of +/- 32mV ($\approx$ 2.1mV steps) was created. For example if a negative V_{OS} is extracted, the calibration controller will only choose from the positive range to counterbalanced the negative V_{OS} generated by the differential pair.

Post-layout Monte Carlo (MC) simulations were performed to calculated the statistical distribution of V_{OS} and the histograms were plotted in Fig. 6. The V_{OS} mean was almost zeroed and the standard deviation was reduced by a factor of 12.

C. SAR Logic

Besides the advantage of faster bit-cycle conversion, an additional benefit of asynchronous SAR logic is that it does not require an external high-frequency clock, and thus saves the power needed to generate and distribute it. This is extremely important, since this ADC will be part of a time-interleaved system.

The logic of the SAR ADC can be divided into two parts: 1) the clock generation, which provides the clock for the comparator and the bit-cycle phases and 2) the state memory, in charge of controlling the CDAC based on the comparator decision in each of the bit-cycle phases, Fig 7. The Track duration is set at 25% of the 875MHz clock period.

The clock generation combines the clock Φ_{ADC}, ready and stop$_{cycle}$ signals to generate the comparator clock (Φ_{Comp}) and bit-cycle clocks (CLK$_{p,n}$) with simple combinational logic. At the same time, a sampling *pulse* is generated from $\overline{\Phi_{ADC}}$ and its delay. This *pulse* propagates sequentially, as controlled by

the CLK$_{p,n}$ signals, generating each of the bit-cycle phases, Fig. 7. The Track duration is set at 25% of the 875 MHz clock period. Additionally, the stopcycle signal is used to indicate the end of the conversion cycle.

The state memory part connects directly to the differential output of the comparator. A dynamic register is used as memory to optimize loop delay and enable fast settling upon comparator decision. One cell is activated during every comparison by its corresponding bit$_{cycle}$[i] and provides Dout$_{CDAC}$ and $\overline{Dout_{CDAC}}$ as control signals for the CDAC. Finally, Dout$_{CDAC}$ is retimed by Φ_{ADC} to create the output data.

IV. SIMULATION RESULTS

This circuit has been designed using the 22FDX^{TM} platform in 22nm FDSOI CMOS of GLOBALFOUNDRIES, [15]. One of the most differentiated features of the 22FDX platform is the capability of effective body biasing. Body biasing applies a positive or a negative voltage to the back gate of the transistor, which allows the transistor V_{TH} to be tuned, and can be done statistically or dynamically.

The nominal full-scale ADC input is 800mV$_{pp,diff}$ with a common mode of 400mV. The ADC operates from a core 0.8 V supply and the simulated, under typical conditions, power consumption is 1.65mW based on post-layout simulation results at 875MS/s. This overal power consists of 0.08mW for the bootstrapped input switch, 0.04mW for the CDAC, 0.68 mW for the comparator, and 0.85mW for the phase and SAR logic.

Although, the calibration for the comparator was implemented to extract the V_{OS}, it can also be used to calibrate the ADC. Fig. 9 shows the MC simulation results, the ADC

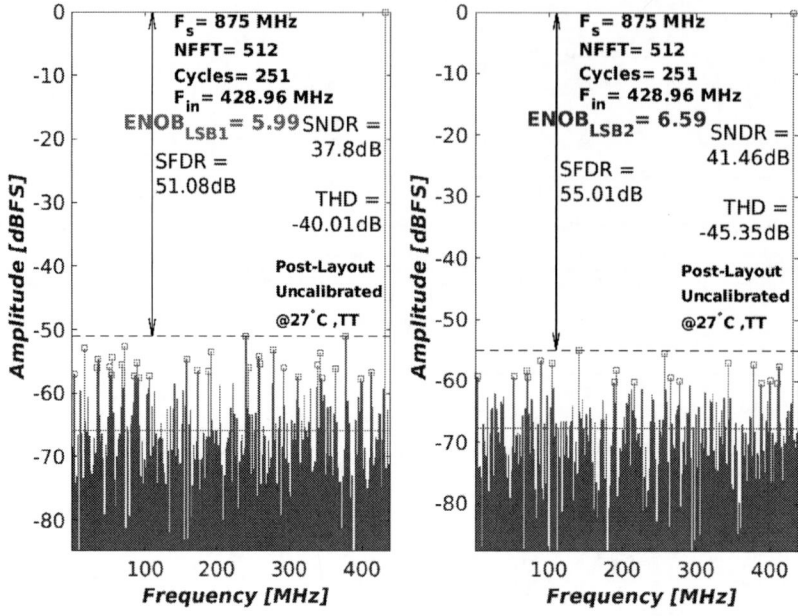

Fig. 8. Output Spectrum with 0 dBFS signal applied at 428.96 MHz, sampling frequency is 875 MHz. LSB$_1$ (red) and LSB$_2$ (blue)

TABLE I
PERFOMANCE SUMMARY AND COMPARISON WITH SINGLE-CHANNEL STATE-OF-THE-ART SAR ADCs

	This Work*	[12]+	[13]+	[7]+	[14]+
Technology [nm]	22nm FDSOI	28nm CMOS	65nm CMOS	32nm SOI	40nm CMOS
Architecture	SAR	SAR	2b/c SAR	2 comp. SAR	ci-SAR
Calibration	YES	NO	YES	YES	YES
Resolution [bits]	7	7	8	8	6
Supply [Volts]	0.8	1.0	1.2	1.0	1.0
Samplig Rate [GS/s]	0.875	1.25	0.4	1.2	1
Power Consumption [mW]	1.65	3.56	4.0	3.1	1.26
Active Area [mm^2]	0.00074	0.0071	0.024	0.0031	0.00058
SFDR@ Nyq. [dB]	55.01	52	53	49.8	49.7
SNDR@ Nyq. [dB]	41.46	40.1	40.4	39.3	34.6
FoM@Nyq. [fJ/conv-step]	19.5	34.4	116.9	34	28.7

+ Measured; *Post Layout

was calibrated at the middle of the range (64LSB) and a 4x reduction of the output code σ was achieved. It is worth to mention that the limiting factor for the ADC calibration is the resolution, LSB = 6.25mV (7 bits).

It is worth to mention, that the calibration procedure for the comparator and ADC was performed at low frequency, due to the time step of the transient simulation is coarse enough to allow a much faster convergence solution. For that reason, the results for static and dynamic characteristics shown are just for nominal conditions without calibration.

The simulated DNL and INL results at 875 MS/s for a ramp up input for the two LSB capacitor versions are depicted in Fig. 10. It can be seen that the second version (LSB$_2$) is more robust to implement the LSB capacitor of the lower side of the differential DAC. LSB$_2$ presents a DNL within 1.08/-0.51LSB, and the INL within -0.78/+0.6LSB.

The output spectrum at 428.96MHz input frequency is shown in Fig. 8. The ADC achieves a SNDR/SFDR of 41.46/55.01dB at Nyquist frequency for the LSB$_2$ version, presenting an overall improvement of 0.6ENOB with respect to the LSB$_1$ version, showing the advantage of the proposed modification for the LSB capacitor.

Table I shows a performance summary and comparison with recent state-of-the-art SAR ADCs of similar performance, Fig. 11. The presented ADC achieves a comparable SNDR and lower FoM for a similar sampling rate (between 400MS/s ⇔ 1.25GS/s) and resolution (6-8b) while having a lower complexity and a smaller area.

V. CONCLUSION

A 7b 875MS/s single-channel SAR ADC has been presented. The integer-based split CDAC combined with an improvement for the LSB capacitor allows a substantial improvement in the SNDR. A simple and accurate calibration

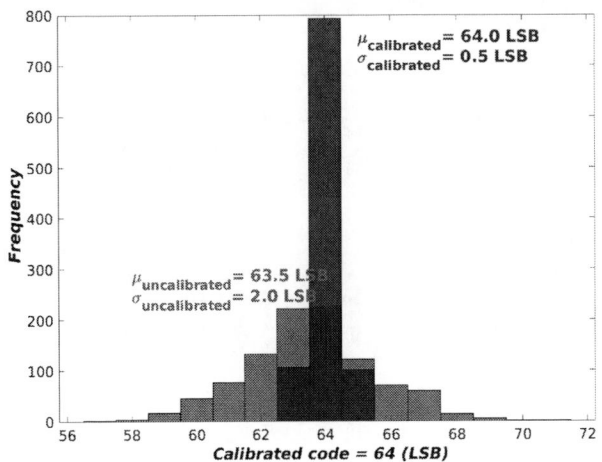

Fig. 9. Histogram of ADC Output Code: Uncalibrated, Calibrated for 1000 MC runs

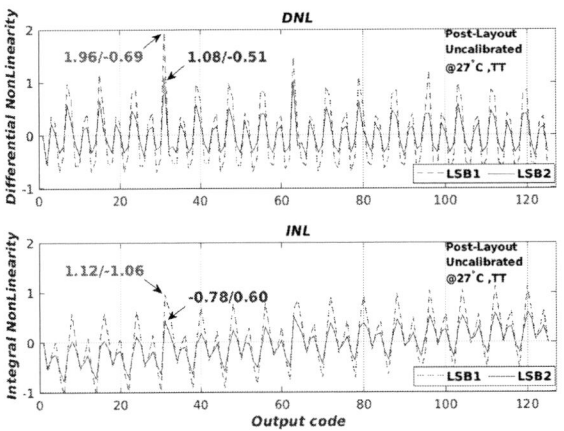

Fig. 10. DNL and INL. LSB$_1$ (red) and LSB$_2$ (blue)

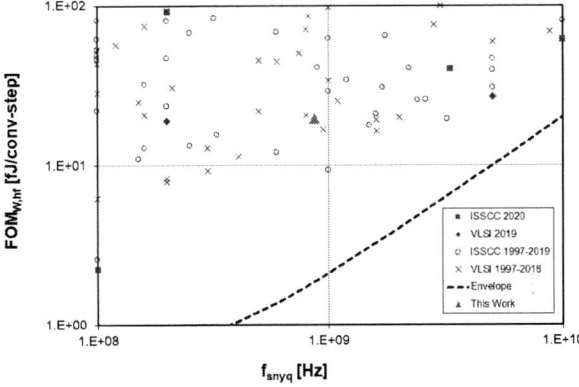

Fig. 11. Comparison with state-of-the-art single-channel SAR based on [11]

procedure for the ADC is presented thanks to body biasing. The use of a dynamic register in the SAR logic yields a shorter and more uniform settling time per cycle and therefore a faster ADC. The circuit in 22nm FDSOI achieves a Nyquist

Walden FoM of 19.5fJ/conversion-step, which is the lowest FoM among previously medium resolution (5.5 ⇔7.5ENOB) SAR ADCs with sampling rates greater than 0.8GS/s/channel.

ACKNOWLEDGMENT

The authors thank the MACOM High-Performance-Analog design team for contributing to the circuit design and GLOB-ALFOUNDRIES for technology access.

REFERENCES

[1] J. Hudner, D. Carey, R. Casey *et al.*, "A 112GB/S PAM4 Wireline Receiver Using a 64-Way Time-Interleaved SAR ADC in 16NM FinFET," in *2018 IEEE Symposium on VLSI Circuits*, June 2018, pp. 47–48.

[2] K. Sun, G. Wang, Q. Zhang, S. Elahmadi, and P. Gui, "A 56-GS/s 8-bit Time-Interleaved ADC With ENOB and BW enhancement Techniques in 28-nm CMOS," *IEEE Journal of Solid-State Circuits*, vol. 54, no. 3, pp. 821–833, March 2019.

[3] L. Kull, J. Pliva, T. Toifl, M. Schmatz, P. A. Francese, C. Menolfi, M. Brndli, M. Kossel, T. Morf, T. M. Andersen, and Y. Leblebici, "Implementation of Low-Power 6-8 b 30-90 GS/s Time-Interleaved ADCs With Optimized Input Bandwidth in 32 nm CMOS," *IEEE Journal of Solid-State Circuits*, vol. 51, no. 3, pp. 636–648, March 2016.

[4] D. Cordova, W. Cops, Y. Deval, F. Rivet, H. Lapuyade, N. Nodenot, and Y. Piccin, "A Hierarchical Track and Hold Circuit for High Speed ADC-Based Receivers in 22nm FDSOI," in *2019 26th IEEE International Conference on Electronics, Circuits and Systems (ICECS)*, 2019, pp. 358–361.

[5] C. Liu, S. Chang, G. Huang, and Y. Lin, "A 10-bit 50-MS/s SAR ADC With a Monotonic Capacitor Switching Procedure," *IEEE Journal of Solid-State Circuits*, vol. 45, no. 4, pp. 731–740, April 2010.

[6] L. Deng, C. Yang, M. Zhao, Y. Liu, and X. Wu, "A 12-bit 200KS/s SAR ADC with a Mixed Switching Scheme and Integer-Based Split Capacitor Array," in *2013 IEEE 11th International New Circuits and Systems Conference (NEWCAS)*, June 2013, pp. 1–4.

[7] L. Kull, T. Toifl, M. Schmatz, P. A. Francese, C. Menolfi, M. Brndli, M. Kossel, T. Morf, T. M. Andersen, and Y. Leblebici, "A 3.1 mW 8b 1.2 GS/s Single-Channel Asynchronous SAR ADC With Alternate Comparators for Enhanced Speed in 32 nm Digital SOI CMOS," *IEEE Journal of Solid-State Circuits*, vol. 48, no. 12, pp. 3049–3058, Dec 2013.

[8] Yan Zhu, U-Fat Chio, He-Gong Wei, Sai-Weng Sin, Seng-Pan U, and R. P. Martins, "A power-efficient capacitor structure for high-speed charge recycling SAR ADCs," in *2008 15th IEEE International Conference on Electronics, Circuits and Systems*, 2008, pp. 642–645.

[9] H. Xu and A. A. Abidi, "Analysis and Design of Regenerative Comparators for Low Offset and Noise," *IEEE Transactions on Circuits and Systems I: Regular Papers*, vol. 66, no. 8, pp. 2817–2830, Aug 2019.

[10] H. Omran, "Fast and accurate technique for comparator offset voltage simulation," *Microelectronics Journal*, vol. 89, pp. 91 – 97, 2019.

[11] B. Murmann. ADC Performance Survey 1997-2020. [Online]. Available: http://web.stanford.edu/ murmann/adcsurvey.html

[12] A. T. Ramkaj, M. Strackx, M. S. J. Steyaert, and F. Tavernier, "A 1.25-GS/s 7-b SAR ADC With 36.4-dB SNDR at 5 GHz Using Switch-Bootstrapping, USPC DAC and Triple-Tail Comparator in 28-nm CMOS," *IEEE Journal of Solid-State Circuits*, vol. 53, no. 7, pp. 1889–1901, July 2018.

[13] H. Wei, C. Chan, U. Chio, S. Sin, U. Seng-Pan, R. Martins, and F. Maloberti, "A 0.024mm2 8b 400MS/s SAR ADC with 2b/cycle and resistive DAC in 65nm CMOS," in *2011 IEEE International Solid-State Circuits Conference*, Feb 2011, pp. 188–190.

[14] K. D. Choo, J. Bell, and M. P. Flynn, "27.3 Area-efficient 1GS/s 6b SAR ADC with charge-injection-cell-based DAC," in *2016 IEEE International Solid-State Circuits Conference (ISSCC)*, Jan 2016, pp. 460–461.

[15] R. Carter *et al.*, "22nm FDSOI technology for emerging mobile, Internet-of-Things, and RF applications," in *2016 IEEE International Electron Devices Meeting (IEDM)*, Dec 2016, pp. 2.2.1–2.2.4.

Cross-Layer Hardware/Software Assessment of the Open-Source NVDLA Configurable Deep Learning Accelerator

Alessandro Veronesi
IHP - Leibniz Institut für Innovative Mikroelektronik
veronesi@ihp-microelectronics.com

Milos Krstic
IHP - Leibniz Institut für Innovative Mikroelektronik, also with University of Potsdam
krstic@ihp-microelectronics.com

Davide Bertozzi
Engineering Department University of Ferrara
brtdvd@unife.it

Abstract—The Nvidia Deep Learning Accelerator (NVDLA) is a free and open architecture that aims at promoting a standard way of designing deep neural network (DNN) inference engines. The analogy between open-source software and hardware points to FPGAs as ideal implementation platforms for open hardware accelerators. However, the instantiation flexibility enabled by reconfigurable logic should be correlated to the capacity of cost-effective devices. This paper explores the resource utilization-performance trade-offs spanned by the main precompiled NVDLA accelerator configurations on top of the mainstream Zynq UltraScale+ MPSoC. For the sake of comprehensive end-to-end performance characterization, the inference rate of the software stack is matched to that of the accelerator hardware, thus identifying current bottlenecks and promising optimization directions.

Index Terms—Deep-Learning, Reconfigurable Logic, Bare-Metal Software, Open Hardware, Configurable Accelerator

I. INTRODUCTION

Designing efficient hardware systems to support deep learning is an important step towards enabling its wide deployment, particularly for embedded applications such as mobile, Internet of Things (IoT), and drones [1], [2]. Field-programmable gate arrays (FPGAs) are gaining momentum in assisting CPUs for deep neural network (DNN) acceleration [3]. In fact, they have higher energy efficiencies than graphics processing units (GPUs) and shorter development time, lower cost as well as higher flexibility than their ASIC counterparts, while still achieving moderate performance figures.

The rise of FPGAs not only for prototyping but also as mainstream computing acceleration platforms has fundamental synergies with another emerging trend: open hardware. The latter follows the successful trajectory of open-source software and typically refers to the public availability of RTL models of IP components that allow any potential user to run them through a physical implementation flow.

The relevance of open hardware initiatives becomes higher when they receive active industrial support [4]. Industry itself may lead open hardware initiatives as a way of speeding up the adoption of disruptive technologies, thus triggering a virtuous cycle that is likely to end up in wider market opportunities. One recent and relevant example is represented by the Nvidia Deep Learning Accelerator (NVDLA) [5], an open-source industry-grade inference engine that has also been integrated into the Jetson Xavier SoC platform [6]. NVDLA is an end-to-end software/hardware stack from the high-level deep learning framework to the actual hardware implementation through the runtime environment. The accelerator is highly configurable

so to adapt to a wide range of Edge computing applications with different resource budgets.

The hardware-software open source analogy applies to different aspects of the development process and points to reconfigurable logic as the ideal implementation platform for open-source DNN accelerators. First, the released hardware models typically undergo an intensive customization effort by adopters, which may not be effectively performed through simulation for complex architectures. Second, the business model of open hardware cores may envision FPGA prototyping boards that let the users download them, just as computer users would download open-source software and run it on general purpose processors. Third, reconfigurable fabrics may turn out to be already a good compromise between power, performance and development cost as an implementation platform for open hardware.

Unfortunately, building a high-performance FPGA-based DNN accelerator while fitting the resource availability of the reconfigurable platform at hand is a challenging task, and typically implies to trade architectural complexity and performance for lower resource utilization. This trade-off is exacerbated by the recent improvements of DNN algorithms (e.g., Winograd and FFT convolutions, weight compression) [7]–[9], which enable the achievement of higher performance than conventional designs at the cost of more complex design flows and expensive hardware extensions.

This paper moves from the observation that the flexible NVDLA accelerator has mainly undergone virtual or physical prototyping so far, while validating only single design points. Therefore, its wide configuration range has never been correlated to the capacity of mainstream reconfigurable fabrics, and to achievable performance.

At the same time, like many other open hardware cores, NVDLA comes with testbenches for direct hardware performance evaluation. However, they may turn out to be misleading since they ignore the overhead of the software stack, including user-mode and kernel-mode drivers, the runtime environment and portability layers for compatibility across compute platforms.

This paper reports on the mapping of the main pre-compiled configurations of the NVDLA accelerator on the mainstream Zynq UltraScale+ MPSoC, with a threefold objective:

- performing a relative comparison of the performance-resource utilization trade-offs spanned by such configurations;

978-1-7281-5410-7/20 $31.00 © 2020 IEEE

- achieving a stripped-down bare-metal porting of the software stack onto the board host processor, resulting in a bare sequence of lightweight function calls;
- assessing the software stack and correlating its sustainable performance to that of the accelerator hardware to derive end-to-end performance figures.

The paper is structured as follows. After presenting related work in Section II, the NVDLA hardware/software stack is presented in Section III, while our approach to prototyping is discussed in Section IV. Hardware and software characterization results are reported in Section V and VI, respectively, while aggregate performance figures are given in Section VII. Finally, conclusions are drawn in Section VIII.

II. RELATED WORK

There is a surge of interest in FPGAs as implementation platforms for DNN accelerators [10]–[12]. The availability of high-level synthesis (HLS) tools from FPGA vendors lowers the programming hurdle and allows sophisticated end-to-end hardware-software co-design flows [13], [14]. Flexibility of reconfigurable logic is typically exploited to directly map hardwired DNN models for better performance [11], [12], [15].

A more general approach consists of supporting the mathematical operations at the core of deep learning inference in the accelerator hardware. This approach is better suited for virtualized environments and frequent model updates. Along this direction, the NVIDIA deep learning accelerator project promotes interoperability with the majority of modern deep learning networks [5]. NVDLA has undergone an intensive virtual prototyping effort [16], while further research focuses on integrating it with different MCU architectures [17].

Several porting frameworks of the accelerator to FPGA have been documented [18], but most of them only report resource utilization without in-depth analysis. Only the work in [19] provides performance evaluation. Unfortunately, a single design point is investigated (e.g., there is no discussion on the feasibility of fitting larger NVDLA configurations into the target board), and no insight is provided on the overhead of the runtime. Overall, previous prototyping frameworks mainly aim at functional validation.

To date, the most informative parametric study is still the one reported on the NVDLA website [5], which is however referred to ASIC technology, is far from exhaustive, and ignores the overhead of the software stack.

This paper aims to be a milestone in correlating the configuration space of the NVDLA accelerator to the performance-resource utilization trade-offs spanned on a commercial FPGA platform. Breadth and depth of analysis are pursued by combining the hardware and the software execution times to quantify achievable end-to-end performance.

III. THE NVDLA ARCHITECTURE

The NVDLA project is composed of three main GitHub repositories, providing the source code of:

- The Virtual Prototyping platform;

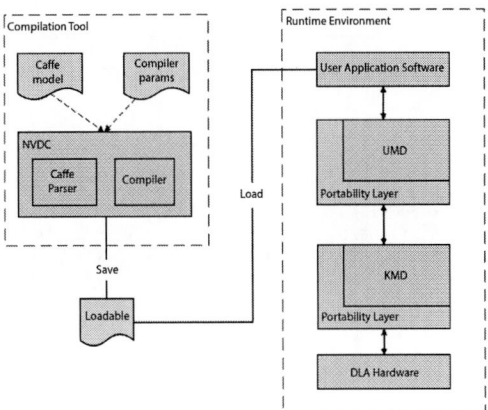

Fig. 1. NVDLA software stack.

- The Software Stack, composed of a Runtime Driver and a Neural Network Compiler;
- The Hardware RTL and SystemC models.

For our purpose, we focus on the software stack and the Verilog RTL model.

A. Software Environment

The software environment is composed by a Network Compiler and by a Runtime Environment. The Compiler acquires a pre-trained Neural Network written in *Caffe* as well as an NVDLA configuration and produces a *".loadable"* file, containing the list of hardware layers to be executed (see Fig. 1). A hardware layer is a set of operations to be scheduled on each functional unit of the NVDLA architecture. The loadable file is given to the Runtime Environment, which is composed by a *User Mode Driver (UMD)* and a *Kernel Mode Driver (KMD)*. The former loads the network into the main memory (which has to be shared between the two drivers and the hardware), reads and unpackages the image file and executes data preparation. Next, the UMD sends an NVDLA Task to the KMD, which contains the functional unit scheduler and the hardware abstraction layer. The inference is performed in a single NVDLA Task, through a layer-by-layer execution. Both the UMD and the KMD are originally written for Linux. The OS-driver interactions are wrapped into a Portability Layer structure and hidden to the drivers' main routines. Currently, no bare metal implementation exists for the NVDLA drivers, a gap that this paper aims to bridge not to include the OS overhead into the NVDLA performance evaluation.

B. Hardware Architecture

The NVDLA architecture is composed of several functional units. As shown in Fig. 2, NVDLA revolves around a sophisticated Convolution Pipeline, which is augmented by an activation engine and a pooling engine. There are also more specialized hardware units to extend the range of compatible deep learning applications. They include a Cross-Channel Data Processor (CDP) for local response normalization and a Reshape Engine (RUBIK) for simple image manipulation.

978-1-7281-5410-7/20 $31.00 © 2020 IEEE

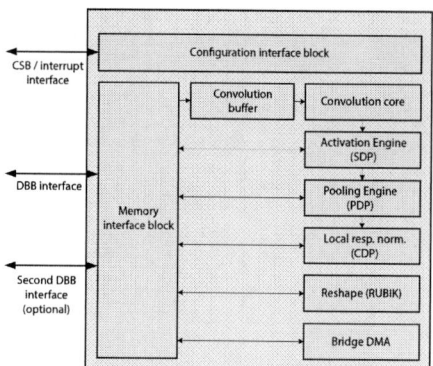

Fig. 2. NVDLA hardware block diagram.

The accelerator has two interfaces to the outside world. On the one hand, the Control Space Bus (CSB) is a slave interface connected to the host processor, and is used for accelerator configuration together with an interrupt signal. On the other hand, the data backbone (DBB) is a link with the main system memory, which is shared with the host processor. It is possible to instantiate an optional DBB interface, which typically consists of a high-speed dedicated SRAM, onto which a Bridge DMA moves data from the main memory in a software-controlled way.

Beyond supporting the processing of a wide range of DNNs, NVDLA targets instantiation flexibility: in fact, most of the above hardware units are optional and highly configurable (e.g., number of MACs, convolution buffer size, batch size, number of operations per convolution round, activation function, etc.). Even when all functional units are supported, there are still significant degrees of freedom as to the deep learning processing features, such as the convolution algorithm and the weight compression option. The list of configurations addressed in this work is reported in Section IV-B.

The most documented operating mode of the released NVDLA HDL model is named *independent mode*. Each functional block is independently configured and performs memory-to-memory operations. This causes a per-block round-trip through main memory instead of having blocks communicate with each other directly through small FIFOs (*fused mode*). This paper will adopt the independent operating mode, since fused mode is not an available option in the accelerator configuration file yet.

IV. METHODOLOGY

Our prototyping platform is the industry-standard Zynq UltraScale+ MPSoC (ZCU102) from Xilinx. A bare-metal version of the NVDLA firmware has been derived and ported to one core of the 64-bit ARM Cortex A53 host processor on this platform, running at 1.2 GHz. The DRAM memory subsystem revolves around a 4 GB Micron MTA4ATF51264HZ device and has been configured to reach a peak bandwidth of 21.3 GB/s. This is enough for the requirements of the DNNs that will be tested later [5]. Without lack of generality, we use

the board SD Card as the repository for the input loadable file and for the images to be processed.

A. Bare-Metal Runtime Implementation

The portability layers of the whole NVDLA runtime environment have been largely re-written in order to come up with a low-overhead bare-metal realization. We relied on the Xilinx "standalone" libraries for file system support (to read images and loadable file from the I/O unit), interrupt control (to interact with NVDLA hardware) and dynamic memory management (extensively used by the UMD).

The original Linux driver strongly relies on file descriptors to manage the communication between the UMD and the KMD, using virtualized memory regions for data exchange. As a workaround, we have instantiated specific data structures as an exchange area between UMD and KMD. The actual exchange takes place via *memcpy* operations, while the KMD functions are invoked by the UMD via a set of KMD APIs based on the original *ioctl()* procedures.

After the bare-metal porting of the runtime, the latter consists of a unique execution flow where we can focus on key runtime operations rather than on the driver entities that perform them. Therefore, key runtime operations include:

- **Network loading**. It transfers a pre-compiled loadable file from an I/O device (in our case, the SD Card) to main memory.
- **Test operation**. It performs image reading, data preparation and fills up a data structure with task scheduling information.
- **Submit operation**. It schedules operations to the accelerator functional units, with which interaction takes place through the hardware abstraction layer.

In order to assess the software stack of the NVDLA framework in isolation, the hardware execution has been at first assumed to occur in zero time. This was achieved by disabling write/read operations to/from the accelerator registers in the KMD, and by virtually driving interrupt responses of the accelerator to the hardware abstraction layer, under the assumption of instantaneous execution of scheduled commands.

Regarding the tested DNNs, we compiled their Caffe models with the baseline NVDLA Compiler provided by NVIDIA. In order to characterize the software stack, all the DNNs under test have been compiled for the NVDLA "Large" configuration of the accelerator hardware.

B. Tested NVDLA Hardware Configurations

NVDLA currently comes with several pre-compiled hardware configurations. We tested all the stable configurations at the time of the work reported in Tab. I. They encompass "Small", "Medium" and "Large" or even "Full" instances. All of them share a common baseline feature: they instantiate all the functional units of the accelerator, except for RUBIK. Moreover, the usage of the SRAM inteface is configuration-specific.

Instead, they mainly differ by the target data type, number of instantiated MACs and the buffer size in the convolutional unit,

978-1-7281-5410-7/20 $31.00 © 2020 IEEE

TABLE I
TESTED HARDWARE CONFIGURATIONS

	Data Type	MACs	CBUF (KB)	SDP	PDP	CDP	Winograd Support	Weight Compression	Batch Size	SDP LUT Support	SDP Throughput (op/cycle)	PDP Throughput (op/cycle)	CDP Throughput (op/cycle)	RUBIK	SRAM Mem If
nv_small	Int8	64	128	Yes	Yes	Yes	No	No	1	No	1	1	1	No	No
nv_small_256	Int8	256	128	Yes	Yes	Yes	No	No	1	No	1	1	1	No	No
nv_medium_512	Int8	512	512	Yes	Yes	Yes	No	No	1	No	4	2	2	No	Yes
nv_large	Int8	2048	512	Yes	Yes	Yes	Yes	Yes	32	Yes	16	8	8	Yes	Yes
nv_full	Int8/Int16/Fp16	2048	512	Yes	Yes	Yes	Yes	Yes	32	Yes	16	8	8	Yes	Yes

TABLE II
SYNTHESIZED HARDWARE RESOURCE UTILIZATION

	LUTs	LUTRAMs	FFs	BRAMs	DSP slices	CLB occupation
nv_small	68927	196	69347	100	41	43%
nv_small_256	90871	196	92830	165	41	57%
nv_medium_512	130532	200	116364	187	75	84%

batch size, and by DNN algorithmic features such as weight compression or Winograd convolution (which correspond to matching hardware units).

The "Full" version includes, among the other things, Winograd convolution, weight compression, a batch size of 32, 2048 MACs, INT8/INT16/FP16 data path and a dedicated SRAM interface. In this configuration, internal optimizations are enabled in the SDP, PDP and CDP in order to maximize throughput.

The "Small" configuration exhibits a baseline convolutional pipeline without the optional memory interface, and a tighter resource budget for reduced area and power requirements, including 64 MACs, INT8 data path, and a batch size of 1 (see Tab. I for all details).

In all configurations, the accelerator is inferred as a single clock domain, and no specific mapping optimization to heterogeneous FPGA resources has been applied. Thus, absolute performance measurements reported in this paper have to be considered as pessimistic.

V. HARDWARE CHARACTERIZATION RESULTS

A. FPGA Resource Utilization

In Tab. II, FPGA resource utilization is reported for the configurations under test. Interestingly, even the "Small" accelerator instance takes as much as 43% of the available CLBs, an utilization that grows to 84% for the "Medium" configuration. From our tests, the "Large" and "Full" architectures are so overprovisioned that they do not fit the Programming Logic of the Zynq UltraScale+ platform, which questions their applicability to cost-effective applications. More in detail, the second column of Tab. II shows an extensive logic LUT utilization, which grows from 25% in the "Small" configuration to 47% in the "Medium" one.

MAC units occupy only 10% of the LUTs in the "Small" baseline configuration, and roughly add from 60 to 90 additional LUTs for each increment in the number of MACs, which depends on the specific configurations. At the same time, the number of instantiated DSP slices is quite limited: from 1.63% in "Small" to 2.98% in "Medium". This points to an inherent limitation of the released source code of the accelerator: it natively targets ASIC implementation and is not optimized to make an extensive use of DSP slices of the target FPGA. We identified the reason in the logic-level

specification of multipliers in the Convolutional unit, which biases the synthesis tool toward a LUT-based implementation.

Allocated DSP slices are for the Convolution DMA, which uses them for address manipulation, and especially for the CDP. As a result, when target CDP performance is boosted in the transition from "Small 256" to "Medium" configurations, hardware redundancy is used to meet the requirement, which results in a doubled allocation of DSP slices. Source code optimizations for better use of DSP slices in the Convolutional pipeline is left for future work.

For each hardware configuration under test, Fig. 3 illustrates breakdown of LUT occupation into the individual functional units. Only interfaces to external busses are omitted, since they are constant throughout the configurations and lightweight with respect to reported results (e.g., 300 LUTs for the CSB interface).

We can first notice that the LUT occupation of the MAC array significantly changes across configurations. In the "Small" one, the MAC array area is irrelevant compared to contributions of other functional units, while in the "Medium" configuration it becomes as large as 30% of the total LUT count.

Besides the MAC array, the CDP makes the most extensive usage of LUTs, followed by the SDP engine. In particular, the number of LUTs in both CDP, PDP and SDP increases from "Small 256" to "Medium" due to the boost in the throughput requirements of these functional units.

Observing the Convolutional Buffer (CBUF), we can notice a counterintuitive trend in LUT (Fig. 3) and BRAM allocation (Tab. II). Starting from the "Small" configuration, the CBUF LUT utilization decreases with the growth of the CBUF size (from 128kB to 512kB), while the BRAM utilization has a significant increase in the transition from "Small" to "Small 256" architectures (i.e., from 64 to 128 BRAMs, which determines a similar trend also for the total number of BRAMs allocated by the design as a whole). However, the allocated BRAMs remains stable in the transition from "Small 256" to "Medium" configurations.

The explanation can be found in the way the CBUF is managed. In all the tested configurations, NVDLA comes with 32 SRAM banks, but in the "Small" one they have a width of 8-Byte, which grows to 32-Byte data in the "Small 256" and "Medium" configurations.

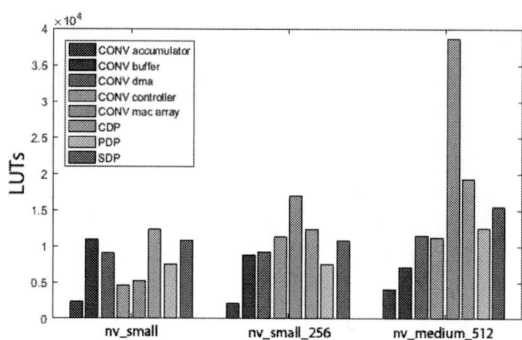

Fig. 3. LUT occupation for each NVDLA functional unit.

TABLE III
SYNTHESIZED HARDWARE TIMING

	Clock Speed	ResNet-50 (frame/sec)
nv_small	130 MHz	0.95
nv_small_256	130 MHz	5.98
nv_medium_512	80 MHz	7.44

This has implications over the CBUF organization. The latter consists of a set of SRAM macros with a bitwidth of 64 in the two "Small" architectures. Therefore, when the MAC array needs to be fed with a 64-bit input ("Small" case), one macro would be enough. However, in order to increase the buffer size, the "Small" configuration combines two macros into a unique aggregate macro, while leaving the output bitwidth unaffected.

When the bitwidth increases to 256 bits ("Small 256" and "Medium" cases), the CBUF aggregates 4 64-bit macros to provide it in "Small 256", and 2 128-bit macros in "Medium".

The mapping of these convolution buffer configurations to FPGA turns out to be sub-optimal. In fact, both the 16kb SRAM macros in the "Small" configuration and the 8kb macros of the "Small 256" configuration are exclusively mapped to 36kb BRAMs of the FPGA. The result is twofold. On the one hand, this explains the increase in allocated BRAMs in Tab. II when moving from "Small" to "Small 256" despite the convolutional buffer size stays the same. On the other hand, this mapping results in a total of 2304Kb occupied memory for the "Small" and 4608Kb for the "Small 256" instances versus a request of only 1024Kb in both cases.

When moving to the "Medium" configuration, the SRAM macro size grows to 64Kb, which exceeds the BRAM size, hence justifying Vivado's choice to map each macro to two different 36Kb BRAMs. Thus, the number of occupied BRAMs is the same as in "Small 256" (see Tab. II). However, memory utilization efficiency is better: 4608Kb of memory is occupied versus the requested 4096Kb.

Last but not least, the different convolution buffer organization leads to different complexity of the multiplexing and control logic. From Fig. 3, the "Small" configuration takes more LUTs in its CBUF due to an additional multiplexing layer to aggregate SRAM macros. Moreover, "Medium" reduces the used LUTs in CBUF with respect to "Small 256" because only two SRAM macros have to be combined to provide the target bitwidth instead of 4.

B. Inference Performance

Clock speed and inference performance results are reported in Tab. III. We observe a 38% reduction of the clock speed only for the "Medium" configuration.

The frame rate has been computed for the ResNet-50 DNN benchmark. From the "Small" to the "Small 256" configuration, an allocation of 4x the number of MACs results in an inference rate improvement by 6x.

When we move from "Small 256" to "Medium", the benefits of doubling the number of MACs, improving the convolutional buffer size and speeding up the functional unit throughput are partly offset by the more complex hardware and the lower operating speed. Therefore, the inference rate speedup is only 1.2x (instead of the theoretical 2x), achieved with a number of CLBs which is 1.4x, thus giving rise to an unbalanced cost-benefit trade-off for "Medium".

VI. SOFTWARE RESULTS

As anticipated in Section IV-A, the key runtime operations include compiled Network Loading, Test operation and task Submit. Next, we assess performance of these operations.

As shown in Fig. 4, the network Loading time dominates over that of other operations, due to reading of the loadable file from SD Card. On the target board, the loading time turns out to be slightly better than a similar operation performed from Flash memory, given the bandwidths of the devices under test: 100 MByte/sec for the SD card vs. a peak bandwidth of up to 90 MByte/sec for the Flash memory. It is worth recalling that network loading takes place only occasionally for DNN model update, while Test operation and Submit are performed at each inference. Therefore, from now on, we will focus only on Test and Submit.

In contrast, the scheduler and the hardware abstraction layers (see "Submit" column) run much faster, mainly because the DNN interpretation has already been realized during the compilation phase, thus runtime scheduling restricts only to task-to-functional unit assignment over time. As a result, the "Submit" time grows with the number of schedulable "hardware layers" the DNN is broken into by the compiler, and that the KMD scheduler will process.

While the "Submit" time scales with the computational cost of the DNN under test, the Test operation time shows a tighter correlation with the format of the input dataset (MNIST for LeNet-5 vs. ImageNet for the remaining DNNs). This results from the breakdown in Fig. 5, which shows the contribution of the image loading time from SD card (in yellow). Clearly, the Test execution time is dominated by data preparation on the ARM and DRAM subsystems, and not by I/O unit.

A. Software throughput

Next, we characterize software sustainable throughput, under the assumption of instantaneous hardware execution. In order to get realistic performance estimates, we also assume that the image to be processed is already in DRAM. Under

978-1-7281-5410-7/20 $31.00 © 2020 IEEE

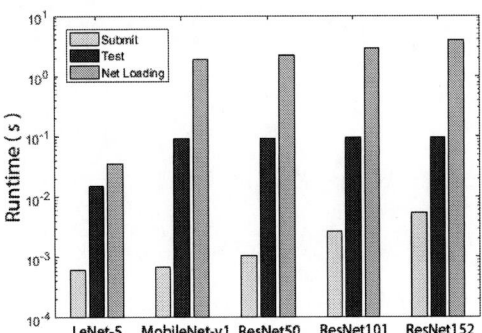

Fig. 4. Execution time of the key operations of the NVDLA runtime for different DNNs under test. Image format: 28 × 28 pixels for LeNet-5, 224 × 224 pixels for the others.

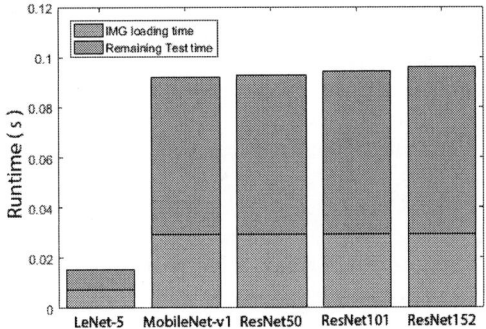

Fig. 5. Breakdown of the Test execution time.

these assumptions, the Test routine remains the throughput bottleneck, and does not enable a frame rate higher than roughly 20 frames-per-second for ImageNet-processing DNN models, and 120 fps for LeNet-5. The Submit operation alone would enable up to 1600 fps for LeNet-5 and slightly more than 200 fps for the most complex Resnet152.

VII. END-TO-END HW/SW PERFORMANCE

By combining hardware and software execution times for a single inference of the ResNet-50 DNN, we need to budget 63ms for the software stack (the Test operation takes two orders of magnitude longer than Submit) and 134ms for hardware inference in the "Medium" configuration. Overall, the resulting frame rate amounts to 5.13 fps[1]. For the "Small" configuration, the frame rate can be as small as 1 fps.

As a future optimization, we could think of parallelizing data preparation with hardware operation. Since the total frame rate is currently hardware-dominated, the expected upper bound consists of the 7.44 fps derived in Tab. III for "Medium", while no noticeable improvement is expected for "Small". Future performance optimizations may come from a better exploitation of FPGA's DSP slices in the Convolutional pipeline, and from enabling the fused operating mode.

---

[1]The software stack has been characterized for the "Large" configuration, but it is matched to the performance of a "Medium" accelerator instance since the only difference, i.e., the Submit time, would be negligible.

VIII. CONCLUSIONS

In this paper, we have mapped the industrial-strength NVDLA DNN accelerator onto the commercial Xilinx UltraScale+ MPSoC, with integrated reconfigurable logic. Our analysis reveals that: (i) the pre-compiled "Large" and "Full" configurations are overprovisioned for cost-effective Edge computing platforms; (ii) the "Medium" configuration makes an effective use of FPGA memory resources and provides the best performance, but exhibits an unfavourable cost-benefit trade-off with respect to the "Small 256" configuration; (iii) the frame rate on the target platform is currently hardware-dominated, and achieves roughly 5 fps for ResNet-50, which can be potentially extended up to 8 fps; (iv) future optimizations of hardware performance will bring a software-limited throughput of 20 fps to the forefront.

REFERENCES

[1] V. Sze, Y. H. Chen, T. J. Yang, and J. S. Emer, "Efficient processing of deep neural networks: A tutorial and survey," *Proceedings of the IEEE*, vol. 105, no. 12, 2017.

[2] J. Chen and X. Ran, "Deep learning with edge computing: A review," *Proceedings of the IEEE*, vol. 107, no. 8, pp. 1655–1674, 2019.

[3] A. Shawahna, S. M. Sait, and A. El-Maleh, "FPGA-based accelerators of deep learning networks for learning and classification: A review," *IEEE Access*, vol. 7, pp. 7823–7859, 2019.

[4] [Internet], "RISC-V Foundation, available at: riscv.org."

[5] [Internet], "NVDLA open source project, available at: nvdla.org."

[6] [Internet], "NVIDIA Jetson modules, available at: nvidia.com/autonomous-machines/embedded-systems."

[7] L. Lu, Y. Liang, Q. Xiao, and S. Yan, "Evaluating fast algorithms for convolutional neural networks on FPGAs," in *Proceedings of the 25th IEEE International Symposium on FCCM*, pp. 101–108, 2017.

[8] R. DiCecco, G. Lacey, J. Vasiljevic, P. Chow, G. Taylor, and S. Areibi, "Caffeinated FPGAs: FPGA framework for convolutional neural networks," in *Proceedings of the 2016 International Conference on FPT*, pp. 265–268, 2016.

[9] C. Zhuge, X. Liu, X. Zhang, S. Gummadi, J. Xiong, and D. Chen, "Face recognition with hybrid efficient convolution algorithms on FPGAs," in *Proceedings of the 2018 GLSVLSI Great Lakes Symposium on VLSI*, pp. 123–128, 2018.

[10] J. Qiu, J. Wang, S. Yao, K. Guo, B. Li, E. Zhou, J. Yu, T. Tang, N. Xu, S. Song, Y. Wang, and H. Yang, "Going deeper with embedded FPGA platform for convolutional neural network," in *Proceedings of the 2016 ACM/SIGDA International Symposium on Field-Programmable Gate Arrays*, pp. 26–35, 2016.

[11] J. Zhang and J. Li, "Improving the performance of OpenCL-based FPGA accelerator for convolutional neural network," in *Proceedings of the 2017 ACM/SIGDA International Symposium on Field-Programmable Gate Arrays*, p. 25–34, 2017.

[12] X. Zhang, J. Wang, C. Zhu, Y. Lin, J. Xiong, W. M. Hwu, and D. Chen, "DNNBuilder: An automated tool for building high-performance DNN hardware accelerators for FPGAs," in *Proceedings of the 2018 International Conference on Computer-Aided Design*, pp. 1–8, 2018.

[13] T. Moreau, T. Chen, Z. Jiang, L. Ceze, C. Guestrin, and A. Krishnamurthy, "VTA: An open hardware-software stack for deep learning," 2018.

[14] D. Wang, K. Xu, and D. Jiang, "PipeCNN: An OpenCL-based open-source FPGA accelerator for convolution neural networks," in *Proceedings of the 2017 International Conference on FPT*, pp. 279–282, 2017.

[15] N. Suda, V. Chandra, G. Dasika, A. Mohanty, Y. Ma, S. Vrudhula, J. S. Seo, and Y. Cao, "Throughput-optimized OpenCL-based FPGA accelerator for large-scale convolutional neural networks," in *Proceedings of the 2016 ACM/SIGDA International Symposium on Field-Programmable Gate Arrays*, p. 16–25, 2016.

[16] G. Zhou, J. Zhou, and H. Lin, "Research on NVIDIA deep learning accelerator," in *2018 12th IEEE International Conference on Anti-counterfeiting, Security, and Identification*, pp. 192–195, 2018.

[17] F. Farshchi, Q. Huang, and H. Yun, "Integrating NVIDIA deep learning accelerator (NVDLA) with RISC-V SoC on FireSim," 2019.

[18] [Internet], "GitHub issue 110: NVDLA running on a FPGA platform, available at: github.com/nvdla/hw/issues."

[19] S. Luo, "Customization of a deep learning accelerator," in *Proceedings of the 2019 International Symposium on VLSI-DAT*, pp. 1–2, 2019.

X-MAGIC: Enhancing PIM using Input Overwriting Capabilities

Natan Peled, Rotem Ben-Hur, Ronny Ronen, and Shahar Kvatinsky

Andrew and Erna Viterbi Faculty of Electrical Engineering
Technion - Israel Institute of Technology
Haifa, Israel 3200003
{natanpeled, rotembenhur}@campus.technion.ac.il, ronny.ronen@ef.technion.ac.il, shahar@ee.technion.ac.il

Abstract—Processing-in-memory (PIM) using memristive technologies is an attractive solution for the memory wall problem. PIM can improve the performance and energy efficiency of computing systems by reducing the data transfer between the memory and the processor. Memristor Aided loGIC (MAGIC) is a popular memristive PIM technique that can perform any combinational logic as a sequence of atomic NOR/NOT operations. These NOR/NOT operations rely on initializing their output cell prior to computation. In this paper, we explore input overwriting: the use of the MAGIC gate output cell as an additional input without initializing it. We extend MAGIC and introduce X-MAGIC (eXtended MAGIC) which uses input overwriting, and demonstrate it by two gates, $A \cdot (\overline{B+C})$ and $A \cdot \overline{B}$, where A is an overwritten input. We show that input overwriting improves functionality, performance, and effective lifetime of the system.

Due to algorithmic difficulties, available PIM synthesis tools do not support input overwriting. We address these difficulties by modifying an existing synthesis tool for MAGIC (SIMPLER), and presenting several general principles and methods for supporting input overwriting. We examine two configurations of the modified synthesis tool using X-MAGIC gates, differing in their performance/area trade-off. Both configurations achieve a geomean improvement of over 16.5% in performance, and over 20% in effective lifetime compared to standard MAGIC.

Index Terms—memristor, logic synthesis, PIM, memory

I. INTRODUCTION

In recent years, data transfer between the computation unit and the memory has become the main bottleneck in computer architecture [1]. One possible approach to alleviate this bottleneck, so to improve performance and reduce energy consumption, is the decades-old concept of processing-in-memory (PIM). Recent advances in memory technologies, and specifically, the emergence of new electronic devices, have brought PIM to the center-stage of contemporary computer systems research.

One example of such an emerging electronic device is the memristor [2]. In memristors, the resistance can be varied by applying current or voltage across the device. Hence, memristors can modulate data into resistance. Memristors can be used to build standard memories [3], and can even implement logic gates [4]–[6]. Furthermore, memristors can combine both capabilities [7], [8], *i.e.*, they can integrate data storage and computation to perform PIM.

Memristor aided logic (MAGIC) [4] is a technique which uses memristors to perform logic, and is compatible with

memristive crossbar memory arrays. The inputs and outputs of MAGIC gates are stored within the memory array, in the same cells that are used for computation; that is, the entire computation is performed within the memory cells. MAGIC gates can form a complete logic structure using NOR/NOT operations, which means that any function can be broken down into a sequence of MAGIC operations.

MAGIC gates rely on initialization of their output cells to a specific logical state prior to the computation. Then, based on the input values, the output cell either switches or retains its value. In fact, MAGIC gates can skip the initialization step and use their output cells as an additional input variable. FELIX [9] presented a two-cycle MAGIC-based XOR gate. Both cycles use the same cell as an output, without a cell initialization between the cycles. By avoiding this initialization, the output value generated in the first cycle serves as an input for the second cycle, and the result of the second cycle overwrites the value stored in the output cell in the first cycle. By referring to the second cycle operation as a new MAGIC gate, *i.e.*, a MAGIC NOR gate without initialization, the idea of input overwriting with MAGIC can be leveraged.

Input overwriting, as referred in this paper, is the ability of a logic operation to replace one of its inputs with the operation result. We show that input overwriting provides additional functionality that improves the performance and extends the effective lifetime of the system.

The integration of input overwriting into the execution sequence of a full and complex combinational logic raises several challenges and difficulties. Existing PIM synthesis tools do not support the integration of input overwriting and therefore face these challenges. In this paper, we propose several principles and methods to address these challenges for different tools and algorithms. To demonstrate the impact of input overwriting in PIM, we introduce two eXtended MAGIC (X-MAGIC) gates with input overwriting capability, and modify an existing MAGIC synthesis tool by adding input overwriting support to it. We test two synthesis configurations, differing in their performance/area trade-off, using the modified synthesis tool and compare the results to a state-of-the-art synthesis process that does not support the integration of input overwriting.

This paper makes the following contributions:

- Identification of principles and development of methods to support input overwriting in PIM.

978-1-7281-5410-7/20 $31.00 © 2020 IEEE

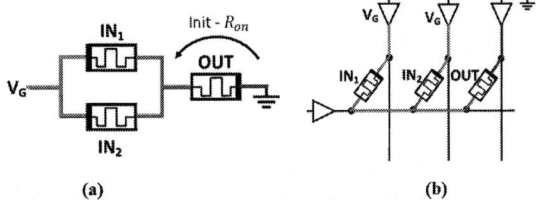

Fig. 1. (a) Schematic of a MAGIC NOR gate and (b) its mapping into a memory crossbar array. To perform a logical NOR operation using the gate, the output cell is initialized to R_{on}. Then, a voltage is applied at the negative terminals of input memristors and the negative terminal of the output memristor is grounded. The operation relies on a voltage divider between the input and output memristors.

- Proposing two new MAGIC gates that overwrite one of their inputs and store their outputs instead of it.
- Demonstration of the performance and effective lifetime benefits of the proposed gates.

The rest of the paper is organized as follows. Section II provides the background about MAGIC and the relevant synthesis tool. Section III introduces the concept of input overwriting, describes its general principles, and shows how to apply it to extend MAGIC. The methodology and evaluation are presented in Section IV, and Section V concludes the paper.

II. BACKGROUND

The concept of input overwriting may be useful for a diversity of technologies, algorithms, and tools. In this work, we focus on the MAGIC technique and the synthesis tool called SIMPLER.

A. MAGIC

Memristor aided logic (MAGIC) [4] is a technique which uses memristors to perform logic operations, including logical NOR and NOT. Since NOR constitutes a functionally complete set, MAGIC gates can implement any combinational logic sequence. MAGIC gates work with the same data representation of the stored data in memristive memories, where logical '1' is represented as a low resistive state (R_{on}) and logical '0' is represented as a high resistive state (R_{off}). Fig. 1 shows the schematic of a single MAGIC NOR gate and its mapping into a memory crossbar. The operation of this gate (as well as the NOT gate) consists of two steps. At the first step, the output memristor is initialized to R_{on}. At the second step, a voltage is applied to the negative terminals of the input memristors, while the negative terminal of the output memristor is grounded. By applying the voltage, the resistance of the output memristor is changed depending on the ratio between the resistances of the input and the output memristors.

Operation of MAGIC gates is compatible with memristive memory crossbars, allowing for performance of single instruction multiple data (SIMD) operations. By selecting the same columns as inputs and output in multiple rows, all the selected rows behave as the same logic gate, but with different input values. Furthermore, the initialization stage can be performed simultaneously on multiple columns within all selected rows.

B. SIMPLER

To perform different logical functions with MAGIC, the MAGIC gates must be sequenced and then mapped into specific locations within the memory. SIMPLER [10] is a state-of-the-art synthesis and mapping tool for the conversion of combinational netlist files into PIM execution sequences, where each sequence can be executed within a single row in the crossbar array. The execution sequence created by SIMPLER is composed of MAGIC instructions, where each instruction includes the operation type (NOR or NOT), the location of its inputs, and the location where its result should be placed within the memory row. The tool also minimizes the number of memory cells participating in the computation.

III. INPUT OVERWRITING

As explained above, input overwriting is the ability of an operation to replace one of its inputs with its result, *i.e.*, the ability to overwrite an input with the operation output. Input overwriting introduces new functionality based on the same operation, which may improve the performance, area, and effective lifetime of the system. However, using input overwriting raises several limitations and implications. In this section, we discuss some of the challenges caused by input overwriting, and describe principles and methods to handle them. Then, we describe how to use input overwriting with MAGIC and how to integrate it in logic synthesis.

For our discussion, we describe combinational logic as a dependency Directed Acyclic Graph (DAG) [10]. The combinational logic is synthesized using an existing CMOS synthesis tool that generates a list of logic gates and their connections. For a given netlist, each logic gate is represented as a node (where the node is associated with the gate's logic operation), and a wire between logic gates is represented as an edge. The inputs of each gate are referred as its child nodes, while the gates that are connected to its output are referred to as its parent nodes. The number of parents a node has is termed as the *Fan Out* (FO) of the node. To perform the logic operation associated with a node, all the operations associated with its child nodes must be performed prior to the node operation, *i.e.*, the node depends on its children in term of execution order. Wires (*i.e.*, edges) describe true dependencies between the nodes they connect. The roots of each DAG are the outputs of the netlist, and the leafs of the DAG are the inputs of the netlist. Besides a logic operation, each node is associated with the operation value (its result). In this paper, an overwriting relation (dependency) between two nodes means that the result of the parent node operation **replaces** the original value that is associated with the child node.

To distinguish between overwriting dependencies and non-overwriting dependencies, we use two types of edges in the dependency DAG. The first edge type, named *regular edge*, represents a non-overwriting dependency. The second edge type, named *overwriting edge*, represents an overwriting relation between a child node and a parent node, namely, the value stored in the child node is overwritten by the parent operation. Fig. 2(a) shows the two different types of edges.

978-1-7281-5410-7/20 $31.00 © 2020 IEEE

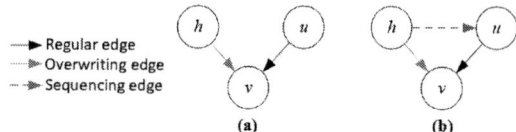

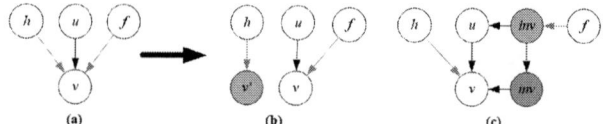

Fig. 2. DAG of three logic gates, demonstrating the different types of edges. Nodes u and h consume the output of node v. Node u is connected to v via a regular edge (black), and node h is connected to v via an overwriting edge (green). (a) Execution order impact. If h is executed before u, the original value of v is overwritten; hence it is invalid as an input of node u. (b) Use of a sequencing edge. To ensure u is executed before h, a sequencing edge (red) is added.

Note that in this work, a parent node can have, at most, one child connected to it via an overwriting edge.

A. Input Overwriting Principles

In this section, we describe several principles that an input overwriting DAG should follow, explain why these principles are needed, and describe methods to ensure that a given DAG is following these principles.

The first principle for working with input overwriting is to check that **each node has no more than one parent connected to it via an overwriting edge**. Otherwise, each one of its overwriting parents will try to overwrite the value stored in the node. Let h, u, f, and v be nodes in the DAG, as shown in Fig. 3(a). If h is executed before f, the value that is stored in v will be overwritten, hence one of inputs that belongs to f will contain an incorrect value. The same phenomenon will occur if the execution order of h and f is reversed. This case can be further generalized by having more than two parents connected to v via overwriting edges. Our method for overcoming this challenge is based on the relation between v and its children. If v is not connected to any of its own children via an overwriting edge, we duplicate v into a new node, v', where v and v' have the same children. Then, we choose one of v's overwriting parents and move the connecting overwriting edge to connect between v' and the parent instead, as shown in Fig 3(b). We repeat this process until v has only a single parent connected via an overwriting edge. If v is connected to one of its own children via an overwriting edge, the duplication of v will also require duplication of the overwritten child of v. That is, at least two nodes will be duplicated, and even more than two if v's child is also connected to one of v's grandchildren via an overwriting edge. Hence, instead of duplication, a buffer is inserted between v to all but one of its overwriting parents. Fig. 3(c) demonstrates the insertion of the buffer nodes.

Another principle to follow is that **the overwriting parent should be the last parent that is executed**. Assume there are two parent nodes with a shared child node, where the first parent node is connected to the child node via an overwriting edge, and the second node is connected via a regular edge, as shown in Fig. 2(a). If the overwriting parent node is executed before the other parent node, the value of the child will be changed, and hence, it will hold an incorrect value as an input for the second parent node. This ordering problem may become even worse if there are several parents that are connected to the same child via regular edges. The method

Fig. 3. Example of several overwriting parents. The nodes in the DAG are v, h, u and f. Node u is a parent of v connected via a regular edge (black). Nodes h and f are parents of v connected via overwriting edges (green). (a) Several overwriting parents. Without intervention, f and h will overwrite the value stored in v, where the first node that is executed will make the value invalid for the other. (b) The solution if v is not connected to any of its children via an overwriting edge. v is duplicated into v' (blue node), and the edge from h to v is replaced by an edge from h to v'. (c) The solution if v is connected to one of its own children via an overwriting edge. Two inverter nodes in a row (a buffer) are inserted between v and f ($f \rightarrow$ blue node $\rightarrow$ red node $\rightarrow v$). If u is associated with an inverter, u operates as the first buffer's inverter, and only the second inverter is inserted ($f \rightarrow$ blue node $\rightarrow u \rightarrow v$).

to solve this issue consists of defining a third type of edge - a *sequencing edge*. The purpose of this edge is to ensure that the parent that is connected to the mutual child via an overwriting edge, is the last one to be executed. Sequencing edges create artificial dependencies among all the parents that are connected to the child via regular edges and the parent node that is connected to the child via an overwriting edge, as illustrated in Fig. 2(b). Sequencing edges are added only after it was ensured that each node has no more than one parent connected to it via an overwriting edge (as described in the first principle).

The third principle is that **sequencing edges must not create cycles in the DAG**. Let v, h, and u be nodes in the DAG, as shown in Fig. 4(a). The addition of the sequencing edge from h to u creates a cycle, since a path between u and h exists. The existence of this cycle contradicts the fact that the dependency graph is a DAG and creates a cyclic dependency. We call this problem the *cyclic dependency problem*. To avoid the *cyclic dependency problem*, we check each node for a connection to one of its parents via an overwriting edge. If an overwriting edge exists, we search for a path from any of the other parents (u) to the one which overwrites (h) the mutual child. If a path is found, we distinguish between the same cases as we did in the single overwriting parent check, as shown in Fig. 4(b) and Fig. 4(c).

The last principle is that **overwriting the inputs of the original netlist is prohibited**. If one of the input nodes of the netlist is connected to its parent via an overwriting edge, the execution of this parent will change the input value, and it will not be available to other netlists which use this input or as general data. Hence, it is necessary to protect the value of this input. To support this, we use an approach similar to those shown in Fig. 3 and Fig. 4.

We combine all the above methods into a single process, and apply it to the initial DAG. We name the entire process the *Pre-Processing Stage*. To analyze the time complexity of the *Pre-Processing Stage*, assume the number of nodes in the initial DAG is $|V|$, and the number of edges is $|E|$. The time complexity of assuring that each node has at most one parent connected to it via an overwriting edge, adding

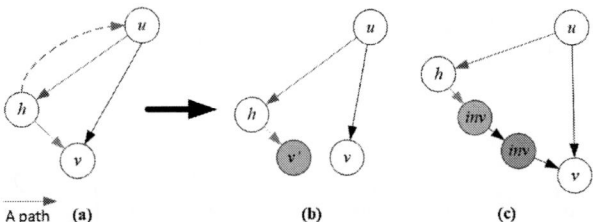

TABLE I
X-MAGIC NOR TRUTH TABLE

A	B	C	Output (A^*)
$0\ (R_{off})$	ϕ		$0\ (R_{off})$
$1\ (R_{on})$	$0\ (R_{off})$	$0\ (R_{off})$	$1\ (R_{on})$
$1\ (R_{on})$	$0\ (R_{off})$	$1\ (R_{on})$	$0\ (R_{off})$
$1\ (R_{on})$	$1\ (R_{on})$	$0\ (R_{off})$	$0\ (R_{off})$
$1\ (R_{on})$	$1\ (R_{on})$	$1\ (R_{on})$	$0\ (R_{off})$

Fig. 4. The cyclic dependency problem and the proposed solution. The nodes in the DAG are v, h, and u. Node u is a parent of v and is connected to v via a regular edge (black), h is a parent of v and is connected to v via an overwriting edge (green). Let p be a path from u to h, represented by the blue edge. (a) A cyclic dependency. Adding a sequencing edge (red) from h to u creates a dependency cycle, composed of p and the edge. (b) The solution if v is not connected to any of its children via an overwriting edge. v is duplicated into v' (blue node), and the overwriting edge from h to v is replaced by an edge from h to v'. (c) The solution if v is connected to one of its own children via an overwriting edge. Two inverters (as a buffer) are inserted between v and h. If v has one inverter as a parent (red node), only one inverter is inserted (blue node).

the sequencing edges, and protecting the netlist's inputs, is $O(|V| + |E|)$. To solve the *cyclic dependency problem*, we use the *DFS* algorithm for checking the existence of paths to each overwriting node, which makes the time complexity of this method to be $O(|V| \cdot (|V| + |E|)) = O(|V|^2 + |V||E|)$. Hence, the time complexity of the entire *Pre-Processing Stage* is $O(|V|^2 + |V||E|)$.

B. Input Overwriting using MAGIC

As described in Section II-A, to perform a NOR/NOT operation using MAGIC, the output cell must be initialized to R_{on} prior to computation. FELIX [9] introduces a two-cycle MAGIC XOR gate, which uses the same output cell for both cycles without a re-initialization between the cycles. By avoiding this initialization, FELIX makes use of the input overwriting concept, but only as an intermediate computation step. FELIX does not show the potential of MAGIC-based input overwriting beyond small and handcrafted logic circuits, nor does it address the challenges introduced by input overwriting. FELIX does not discuss how to integrate input overwriting in a full synthesis flow. In FELIX, the intermediate value is unavailable as an input for other MAGIC gates, which may result in some re-computations. By defining new MAGIC-based gates which use the input overwriting capacity, applying the *Pre-Processing Stage* to any given combinational netlist that uses the new gates, and adding the relevant support to SIMPLER, we present a full synthesis flow for any function, that supports both the new and the old MAGIC gates.

Assume a MAGIC NOR operation is performed on inputs B and C, and the operation result is stored in an uninitialized cell, A. Table I lists the possible results for different input combinations. If the value of A before the computation is logical $'0'$ (R_{off}), the output value, *i.e.*, the value stored in A after the computation, marked as A^*, is logical $'0'$ (R_{off}) regardless of the values of B and C. The value that is stored in A does not change owing to the output memristor polarity and under the voltage constraints for proper operation [4]. Conversely, if the value of A prior the computation is logical

$'1'$ (R_{on}), the gate operates as a standard MAGIC NOR gate. Overall, A can be viewed as an input of a gate that implements the $A \cdot (\overline{B+C})$ logic function, where the output value (A^*) and input A must share the same memory cell. In other words, the result value overwrites the value stored in A. Similarly, performing a MAGIC NOT operation with an uninitialized output cell implements the $A \cdot \overline{B}$ logic function. We named these overwriting MAGIC gates as X-MAGIC (eXtended MAGIC) gates.

X-MAGIC gates provide additional capable logic on top of the standard MAGIC NOR/NOT gates; the enhanced functionality of the new gates may reduce the number of total gates needed to implement a given combinational function. To perform the same logic operation of X-MAGIC gates with MAGIC NOR/NOT, four clock cycles are required instead of the single clock cycle required in X-MAGIC. Hence, using X-MAGIC may lower the number of gates and reduce execution time. Furthermore, since X-MAGIC gates avoid the initialization of the output cell, and since fewer gates are used to execute a function, the number of write operations to the memory cells is also lowered, lengthening the effective lifetime of the system due to endurance limitations. The effective increase in lifetime is proportional to the relative reduction in the number of writes.

IV. EVALUATION AND RESULTS

To evaluate the impact of X-MAGIC gates, we developed X-SIMPLER, a modified version of the Python-based implementation of SIMPLER to support the new gates. We also developed and integrated the *Pre-Processing Stage* as the first stage of X-SIMPLER flow, that is, before the synthesis and mapping stages. We examined the outputs produced by X-SIMPLER by comparing the performance (execution latency), area (required memory row size), and effective lifetime (correlates with the number of write operations per cell) of each tested benchmark to those of the of SIMPLER (using only NOR/NOT gates). In this paper, we evaluated the impact of the $A \cdot (\overline{B+C})$ gate only, since its functionality includes the functionality of the second X-MAGIC gate. This section describes the evaluation methodology and results.

A. Environment and Methodology

Synthesis constraints and optimization. The first stage of the X-SIMPLER flow is synthesis based on a standard synthesis flow (typically used for CMOS logic). To generate a netlist that contains X-MAGIC gates, the X-MAGIC gate was modeled as a three-input gate. A specific input port was defined as the overwriting port (corresponds to *A*). Since the number of gates directly determines the latency, we attempted to reduce the number of nodes added during the *Pre-Processing Stage* by using the optimization capabilities of the synthesis tool. The synthesis was optimized for area, where all gates were assigned equal area. To ensure that each gate drives, at most, one overwriting input port, we use Fan-Out (FO) constraints during the synthesis process through the gates definition file. The X-MAGIC overwriting input port was assigned a high FO load value, while all the other input ports (whether they belong to an X-MAGIC or NOR/NOT), were assigned a low FO load value. The output ports of all gates were assigned a FO value that ensured that no more than a single overwriting input port will be connected to each gate output.

CMOS synthesis tool. SIMPLER used the ABC synthesis tool (ABC) [11]. Since *ABC* cannot apply any FO constraints, this work uses the *Synopsis Design Compiler* (DC) [12] as the synthesis tool. The gates definition file used by DC is written in *Liberty* format. To make a fair comparison, all workloads [13] were re-synthesized using DC. For a given set of design constraints, DC makes the "best-effort" for applying them. The constraints defined in the *Liberty* file may therefore be violated in the netlist created by the tool. The *Pre-Processing stage* was built in a way that checks if violations exist, and solves these violations during its different sub-stages.

Synthesis configurations. We examined two synthesis configurations for the integration of the X-MAGIC gates (each configuration is defined by a different *Liberty* file). The first configuration allows each gate to drive, at most, a single X-MAGIC node via an overwriting port (edge) and to drive several additional gates via regular ports (including X-MAGIC gates, to their regular ports), as shown in Fig. 5(a). In the second configuration, each node can only drive either a single overwriting port or several regular ports, as shown in Fig. 5(b). The configurations are named *Different-Edge Types* (DET) and *Same-Edge Types* (SET), respectively.

Benchmarks, baseline, and comparison. To evaluate the impact of the X-MAGIC gates in both configurations, we used the EPFL benchmark suit [13]. We applied SIMPLER on the re-synthesized files (without X-MAGIC), where the results serve as the baseline for the comparison between the two X-MAGIC synthesis configurations. To evaluate the potential effective lifetime improvement, we assumed the operations are uniformly distributed among all cells within each memory row (as proposed in, *e.g.*, *RRAM Endurance Resiliency* [14]). Under this assumption, we measured only the reduction in the number of write operations for each tested benchmark. To assess the impact of X-MAGIC on area, we measured, for each configuration, the minimum row size needed for the computation of the different benchmarks. To make a fair latency comparison between SIMPLER and X-SIMPLER (for both SET and DET), the memory row size each tool uses should be the same. For the comparison between each X-MAGIC configuration (SET **or** DET) and SIMPLER, we

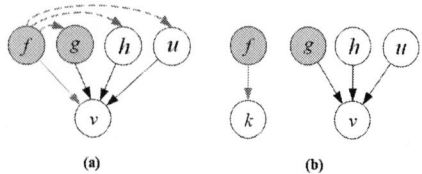

Fig. 5. The two tested synthesis configurations. f, g, h, u, k, and v are nodes in the DAGs. Nodes h, u, k, and v are regular MAGIC nodes (*i.e.*, NOT or NOR), and f and g are X-MAGIC nodes (marked in orange). (a) Different-Edge Types (DET). Each child can be connected to both regular and overwriting edges. (b) Same-Edge Types (SET). Each child can be connected to only a single edge type. After the *Pre-Processing Stage*, the child node can be connected to several regular edges, or to a single overwriting edge.

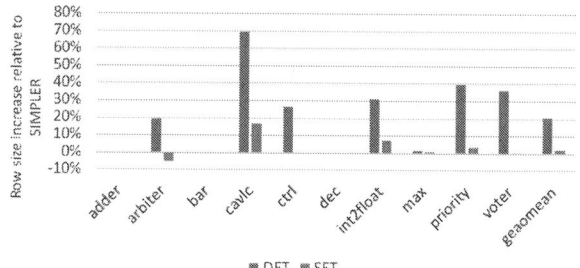

Fig. 6. Row size increase relative to SIMPLER (lower is better).

set the size to the highest of the two minimum row sizes SIMPLER and X-SIMPLER require, where SIMPLER serves as the baseline. To make a fair latency comparison between SET and DET, we set the row size to the highest of the three minimum sizes SIMPLER and X-SIMPLER (SET and DET) require, where SIMPLER serves as the baseline.

B. Results

Area. SIMPLER uses a heuristic technique to reuse cells and reduce the row size. This heuristic is not fully compatible with the case where the execution order influences the correctness of the logic. Hence, X-MAGIC increases the row size. Fig. 6 shows the increase in the minimum required row size for each configuration compared to SIMPLER minimum row size. The geomeans of DET and SET configurations, relative to SIMPLER, are 20.5% and 2.1%, respectively. The increase in the row size for SET is relatively low, and is even lower than the baseline in one case (arbiter). Conversely, the DET configuration incurs a non-negligible area increase. The zero row-size increase in adder and dec stems from the fact that these two benchmarks do not use X-MAGIC gates at all.

Latency. A netlist consisting of both X-MAGIC gates and standard MAGIC gates has fewer gates than a netlist consisting of MAGIC gates only. Since the latency is proportional to the number of gates, the latency is lower when X-MAGIC gates are used. Fig. 7 shows the latency relative to the baseline. Note that for benchmarks which do not use X-MAGIC gates (adder and dec), there is no improvement. The comparisons between each configuration and its SIMPLER baseline are shown in blue for DET and in orange for SET. The geomean latency improvements are 18% for DET and 16.2% for SET. We select the row size to be the maximum between the sizes

978-1-7281-5410-7/20 $31.00 © 2020 IEEE

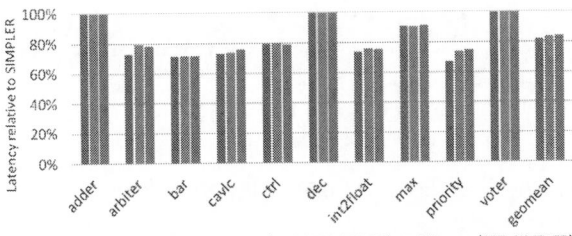

Fig. 7. Latency relative to SIMPLER (lower is better).

■ DET - max{DET,SIMPLER} ■ SET - max{DET,SET,SIMPLER} ■ SET - max{SET, SIMPLER}

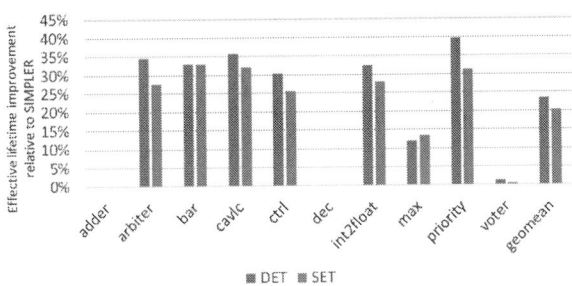

■ DET ■ SET

Fig. 8. Effective lifetime improvement relative to SIMPLER, measured as reduction in the number of write operations (higher is better).

of SIMPLER and the compared SET or DET configuration. To compare between DET and SET, we set the same row size for both configurations. Since the highest size is always the size in DET, only SET is re-executed using the row size of DET. The results for SET with this row size are shown in gray, where the geomean latency improvement is 16.5%.

Lifetime. We approximate the lifetime improvement by counting the number of write operations associated with each benchmark. We count each occurrence of a standard MAGIC gate (NOR/NOT) twice, once for the output initialization and once for the gate operation. Each X-MAGIC gate is counted once, as they do not require initialization. Fig. 8 shows the effective lifetime improvement (measured as reduction in the number of write operations) relative to SIMPLER. The geomean improvement for DET is 23.4% and for SET is 20.2%. This shows that both configurations fulfill our expectations for a nice improvement in the effective lifetime of the system.

The differences between the DET and SET configurations relative to SIMPLER are clear. Both configurations significantly improve latency and reduce number of write operations, but need a larger row. Among the DET and SET configurations, when both are allocated the same row size, DET is slightly better than SET in both latency and number of writes, but DET needs a larger minimal row size to operate. Based on these findings, it is reasonable to conclude that DET is likely a better choice in general, unless area is so scarce that saving several cells makes a difference.

Finally, the computational complexity of the *Pre-Processing stage* is $O(|V|^2 + |V||E|)$, higher than SIMPLER computational complexity of $O(|V| + |E|)$. This increase is less significant in practice since (1) the modified tool is still quite fast, taking less than one minute to process the largest benchmark (which contains over $14K$ gates), (2) the Pre-

Processing stage is only a small part of the entire synthesis flow, hence, its longer runtime does not significantly affect the entire synthesis runtime.

V. CONCLUSION

This paper presents several principles and methods for working with input overwriting in PIM. It defines X-MAGIC, two MAGIC-based logic gates which overwrite one of their inputs. X-MAGIC gates implement the $A \cdot (\overline{B + C})$ and $A \cdot \overline{B}$ logic functions. In both functions, the operation result overwrites the same memory cell used to store the input value A. To demonstrate the impact of input overwriting on PIM, we added support for X-MAGIC gates in the synthesis of combinational logic for PIM. The results show a decent improvement in the latency and the effective lifetime, while the minimal area is slightly higher in one of the configurations. We examined two X-MAGIC synthesis configurations, and our experimental results show a geomean latency improvement of over 16.5% and a geomean improvement of over 20% in the effective lifetime for the same row size as the baseline.

ACKNOWLEDGMENT

This research is supported by the ERC under the European Unions Horizon 2020 Research and Innovation Programme (grant agreement no. 757259).

REFERENCES

[1] Pedram *et al.*, "Dark Memory and Accelerator-Rich System Optimization in the Dark Silicon Era," *IEEE Design Test*, vol. 34, pp. 39–50, April 2017.

[2] L. Chua, "Memristor-The Missing Circuit Element," *IEEE Transactions on Circuit Theory*, vol. 18, pp. 507–519, September 1971.

[3] C. Xu *et al.*, "Overcoming the Challenges of Crossbar Resistive Memory Architectures," *2015 IEEE 21st International Symposium on High Performance Computer Architecture (HPCA)*, pp. 476–488, February 2015.

[4] S. Kvatinsky *et al.*, "MAGIC-Memristor-Aided Logic," *IEEE Transactions on Circuits and Systems II: Express Briefs*, vol. 61, pp. 895–899, November 2014.

[5] Borghetti *et al.*, "'Memristive' Switches Enable 'Stateful' Logic Operations via Material Implication," *Nature*, vol. 464, pp. 873–876, April 2010.

[6] S. Kvatinsky *et al.*, "Memristor-Based Material Implication (IMPLY) Logic: Design Principles and Methodologies," *IEEE Transactions on Very Large Scale Integration (VLSI) Systems*, vol. 22, pp. 2054–2066, October 2014.

[7] Linn *et al.*, "Beyond von Neumann—Logic Operations in Passive Crossbar Arrays Alongside Memory Operations," *Nanotechnology*, vol. 23, July 2012.

[8] Y. Levy *et al.*, "Logic operations in memory using a memristive Akers array," *Microelectronics Journal*, vol. 45, no. 11, pp. 1429 – 1437, 2014.

[9] S. Gupta *et al.*, "FELIX: Fast and Energy-Efficient Logic in Memory," in *2018 IEEE/ACM International Conference on Computer-Aided Design (ICCAD)*, pp. 1–7, 2018.

[10] R. Ben-Hur *et al.*, "SIMPLER MAGIC: Synthesis and Mapping of In-Memory Logic Executed in a Single Row to Improve Throughput," *IEEE TCAD*, July 2019.

[11] A. Mishchenko, "ABC: A System for Sequential Synthesis and Verification," *Berkeley Logic Synthesis and Verification Group*, 2012.

[12] H. Bhatnagar, *Advanced ASIC Chip Synthesis: Using Synopsys' Design Compiler and PrimeTime*. USA: Kluwer Academic Publishers, 1999.

[13] L. Amarù *et al.*, "The EPFL Combinational Benchmark Suite," in *Proceedings of the 24th International Workshop on Logic Synthesis (IWLS)*, 2015.

[14] M. M. Sabry Aly *et al.*, "The N3XT Approach to Energy-Efficient Abundant-Data Computing," *Proceedings of the IEEE*, vol. 107, no. 1, pp. 19–48, 2019.

A Minimalistic Perspective on Koblitz Curve Scalar Multiplication for FPGA Platforms

Siddhartha Chowdhury[1], Debapriya Basu Roy[1,2], Debdeep Mukhopadhyay[1]

[1]*Department of Computer Science and Engineering*
Indian Institute of Technology Kharagpur, West Bengal, India
[2]*Technical Univeristy of Munich, Department of Electrical and Computer Engineering*
Chair for Security in Information Technology
siddhartha.chowdhury92@gmail.com[1], debapriya.basu-roy@tum.de[1,2], dmcseiitkgp@gmail.com[1]

Abstract—Koblitz Curves offer excellent optimization opportunities for characteristic-2 Elliptic Curve Cryptosystems (ECC). However, porting such choices onto a lightweight and cost-effective FPGA platform is a major challenge. The underlying characteristic-2 algebra is not aligned with the on-chip components like DSP multipliers, which if not utilized leads to large LUT counts of the designs. In this work, we develop several techniques to propose a minimal instruction set, centered around ADDN (Add and Branch if less than zero) based OISC (One-Instruction-Set-Computing) coupled with a lightweight comba characteristic-2 finite field multiplier to ensure the aggressive utilization of the FPGA resources. The paper develops an ADDN based scalar multiplier for Koblitz curves which has a minimal footprint on the LUTs, leveraging methods for utilizing the underlying DSP blocks in FPGAs for performing $\mathbb{GF}(2)$ operations, in conjunction with proposed reduced variants of the ADDN instruction to perform the computations. The architecture uses FPGA resources like BRAMs, DSPs effectively to have a minimal requirement for FPGA LUTs, thus leaving room for other peripheral designs to be hosted in a single FPGA. The same design can be configured to realize scalar multiplications by two popular strategies, namely due to Solinas and Montgomery, to demonstrate the modular nature of the design. The minimalistic design style leads to a resource-constrained architecture performing scalar multiplication in less than 1000 slices, needing less than 0.5 ms on a cost-effective Artix-7 FPGA. The design has been compared with reported literature to highlight the area efficiency of the design, however ensuring a competitive AT (area-time) product.

I. INTRODUCTION

In recent years due to the ubiquitous presence of embedded devices in our everyday life, with devices ranging from smart cards, mobile phones to modern cars, particular attention is imperative for security features of these devices. To fulfill the demands of these resource constrained applications, continuous efforts are needed to improve the design and implementation of the hardware root of trust which comprises complex cryptographic primitives, however, ensuring that the area and energy are minimally consumed. More specifically, efforts have been made to develop compact implementations for public-key cryptographic algorithms which have been a principal ingredient in assuring integrity and non-repudiability in security subsystems. Strategies for porting the complex arithmetic operations of the state-of-the-art public-key algorithms like ECC onto a lightweight hardware platform to accommodate other peripheral components on a single chip has always been a challenge. Field Programmable Gate Array

(FPGA) is well suited to this purpose due to attributes like in-house reconfigurability, shorter design cycles, etc. Additionally, modern FPGAs are coupled with multiple high-performance modules like digital signal processing (DSP) blocks and block RAMs (BRAMs). Developing a lightweight implementation of ECC with a balanced consumption of these different modules along with standard FPGA resources (slices and LUTs) is a challenging design problem. A complete SOC (system-on-chip) design will contain multiple hardware modules supporting different operations along with the ECC accelerator. Therefore, to develop single chip FPGA SOCs housing the ECC accelerator, a judicious design choice is needed which utilizes uniformly the various FPGA primitives, like LUTs, BRAMs, DSPs, etc. This becomes even more imposing when the developed lightweight architecture needs to exhibit a competitive area-time product value, which is a standard performance metric used for evaluation of any design. Our design strategy involves a minimalistic single instruction computing framework to design an ECC processor on FPGA with the objective of minimalistic consumption of different FPGA primitives (DSPs, BRAMs, LUTs, registers, slices), however with competitive area-time performance.

In our current work, we focus on the implementation of a characteristic-2 Koblitz NIST curve. We develop a programmable architecture to perform scalar multiplication using two different scalar multiplication algorithms. Based on τNAF proposed by Solinas in [1] to reduce the number of non-zero terms in the scalar, we have explored the scalar multiplication formula proposed in [1]. On the other hand, while using the Montgomery ladder technique proposed in [18], owing to x-coordinate only operations, we can compute using less number of field multiplications. In this paper, we target to explore these trade-offs using a Turing complete `URISC` instruction, ADDN (Add the operands and branch if the answer is less than zero) to design a lightweight ECC scalar multiplier with few necessary modifications. The idea of designing an ECC processor using a single instruction-based approach was first proposed in [5], where the authors had used SBN (Subtract the operands and branch if negative) instruction to design an ECC scalar multiplier for NIST prime field curve. But such prime field arithmetic can be mapped to on-chip DSPs of FPGAs due to their conventional arithmetic and thus can drastically reduce LUT consumptions. However, characteristic-2 algebra on

978-1-7281-5410-7/20 $31.00 © 2020 IEEE

which Koblitz curve is defined cannot be directly implemented on DSPs, and thus requires dedicated techniques as proposed in this work. We have also implemented Itoh–Tsujii [15] for a fast inversion operation without any additional hardware. The experimental results show the proposed scalar multiplier exhibits a competitive area-time metric and consumes least LUT resources among reported literature making the design ideal for even using cost-effective Artix-7 FPGA platforms for developing crypto-SOCs. It may be emphasized that on such an FPGA 96.79% LUTs, 96.3% BRAMs, 98.75% DSPs are free for realizing other peripheral circuits required in a SOC, thus showing the minimalistic perspective of the design approach.

The remaining paper is arranged as follows: A brief background on the literature is presented in Section II. Section III demonstrates the design challenges and components. Section IV presents the complete working of the proposed architecture. Section V presents the results and comparison of the proposed design followed by a conclusion in Section VI.

II. BACKGROUND

In this section, we will give a brief overview of elliptic curve algorithms used by the processor.

A. Koblitz Curves

Koblitz curves [11], also known as anomalous binary curves, are elliptic curves defined over $\mathbb{F}_2$ by

$$E_a : y^2 + xy = x^3 + ax^2 + 1 \tag{1}$$

where $a \in (0, 1)$. Due to the definition of the curve over $\mathbb{F}_2$, it encompasses a special property of Frobenius endomorphism. According to Frobenius map, since Koblitz curves are defined over $\mathbb{F}_{2^m}$: if $P = (x, y)$ is a point on E_a, then so is the point (x^2, y^2). The mapping can be represented as $\tau(x, y) := (x^2, y^2)$. With this form of endomorphism, Solinas proposed τ-adic non-adjacent form known as τNAF [1] in order to reduce the number of non zero terms in the scalar. But the length of τNAF is almost twice the length of binary expansion of the integer scalar which negates the advantage offered by Frobenius endomorphism. In order to overcome the drawback, Meier and Staffelbach proposed a solution in [2]. They established the fact that $kP = \gamma P$, where k and $\gamma \in \mathbb{Z}[\tau]$, such that $\gamma \equiv k \mod (\tau^m - 1)$):

$$\gamma P = kP + \lambda(\tau^m - 1)P = kP + \lambda\mathcal{O} = kP \tag{2}$$

,where λ is some integer. Brumley and Järvinen in [3] based on this reduction mechanism devised an algorithm commonly known as lazy reduction which was the first hardware friendly scheme proposed for the reduction of τNAF. In the case of Koblitz curves, division by τ only involves shifts and additions. The approach involves repeated division by τ for m times. The length of the τNAF may be at most $m + 4$ as stated in [3]. Lazy reduction was further improved by double lazy reduction [12] scheme. In double lazy reduction scheme, division by τ^2 is performed $(m - 1)/2$ times followed by a division by τ in the final step. But due to the complexity in the algorithm we have observed that latency for performing

double lazy reduction is more than lazy reduction using our single instruction architecture framework.

B. Scalar multiplication

In the proposed design, we have explored two widespread scalar multiplication techniques used for Kobliz Curve. Solinas in [1] proposed a scalar multiplication algorithm that replaces point doubling with Frobenius map operation which can be performed with few shifts and additions in binary field arithmetic. In our proposed scheme we have coupled the scalar multiplication methodology in [1] with lazy reduction algorithm, to obtain a reduction in scalar at the same time.

A different approach to compute kP was introduced by Montgomery in [17]. The same was optimized and proposed in [18] along with the usage of projective coordinates for point addition and doubling. The approach uses a x-coordinate only formulation which reduces the total number of field multiplication thus achieving a reduction in overall latency of the design. We have a programmable framework to implement both methodologies using the same datapath with two different instruction sets.

C. The Coordinate system and field inversion

In the proposed design while implementing the algorithm proposed by Solinas in [1] we have used the mixed affine addition formula where one point is taken as affine coordinates and one in LD projective coordinates. The formula for $(X_3, Y_3, Z_3) = (X_1, Y_1, Z_1) + (x_2, y_2)$ are as follows:

$$A = Y_1 + y_2 Z_1^2 \qquad\qquad B = X_1 + x_2 Z_1$$
$$C = BZ_1 \qquad\qquad\qquad Z_3 = C^2$$
$$D = x_2 Z_3$$
$$X_3 = A^2 + CA + B^2(C + aZ_1^2) \quad Y_3 = (D + X_3)(AC + Z_3)$$
$$+ (y_2 + z_2)Z_3^2$$

, where a is a curve constant.

But while computing the point multiplication proposed in [18], we have used standard projective coordinates and used the point addition and doubling formula as given in [18]:

$$x(2P_i) = X_i{}^4 + Z_i{}^4 \qquad Z_3 = (X_1 Z_2 + X_2 Z_1)^2$$
$$z(2P_i) = X_i{}^4 Z_i{}^4 \qquad X_3 = xZ_3 + (X_1 Z_2)(X_2 Z_1)$$

, where $i \in \{1, 2\}$. These require only 5 field multiplications and 5 squarings in each step compared to 9 field multiplications in mixed affine addition formula.

For field inversion we have implemented Itoh–Tsujii inversion [15] instead of Fermat's Little theorem. Using the addition chain for $\mathbb{GF}(2^{163})$ field we could avoid 153 field multiplications which helped us reduce a significant amount of clock cycles. While implementing the method proposed by Solinas in [1], two finite field inversions are required. On the contrary, a single finite field inversion is required when the algorithm proposed in [18] is implemented owing to the x-coordinate only formulation.

D. Single Instruction Processor Architecture

The first single instruction processor coined as URISC (Ultimate Reduced Instruction Set Computer) was proposed in [4].

978-1-7281-5410-7/20 $31.00 © 2020 IEEE

Applying the idea of URISC, an elliptic curve scalar multiplier was proposed in [5]. In that paper, modified SBN (Subtract the operands and branch if the answer is less than zero) instruction had been used to implement scalar multiplication for NIST P-256 curve using less than 100 slices on FPGA. But as scalar multiplication was implemented for prime curve in [5], DSP blocks could directly be used for arithmetic operations which is a challenge in case of characteristic 2 curve.

In our proposed design we have chosen ADDN (Add the operands and branch if the answer is less than zero) for implementing ECC scalar multiplication on K-163 curve. The operation of ADDN instruction is shown in Algo. 1. In the

---

Algorithm 1: ADDN Operation

Data: A,B,C
Result: ADDN A,B,C
$D.Mem[A] \leftarrow D.Mem[A] + D.Mem[B]$;
if *D.Mem[A]<0* **then**
 | jump to C;
else
 | jump to next instruction;

---

next section, we will present the detailed idea of implementing the point multiplication using our framework. The operation of ADDN instruction is shown in Algo. 1.

III. ADDN INSTRUCTION AND ELLIPTIC CURVE SCALAR MULTIPLICATION

In this section, we will have a detailed discussion regarding the challenges a designer may face along with the components required while implementing the proposed single instruction architecture.

A. Single Instruction Processor Architecture

Fig. 1 depicts the general architecture of the proposed single instruction processor.

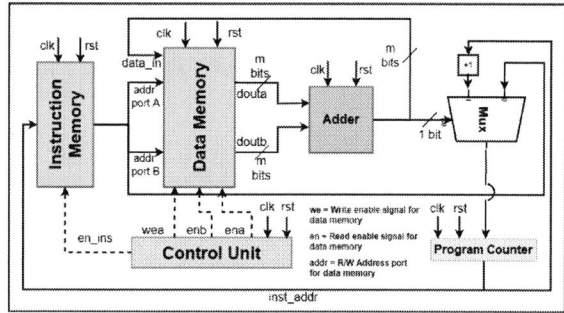

Fig. 1: Architecture of ADDN processor .

The three main components of the generic processor as shown in Fig. 1 are Instruction Memory, Data Memory, and Arithmetic and Logic Unit (ALU). The above-mentioned architecture has been modified based on our requirements and implemented on Artix-7 FPGAs.

B. Challenges for Finite Field Hardware Design

While inspecting the hardware design strategy for a finite field architecture we can visualize certain challenges:

· **Finite Field Addition/Subtraction**: Characteristic 2 field addition/subtraction is equivalent to bitwise XOR. Hard-IPs like DSP cannot perform XOR or AND as they are meant to operate on integer values. Thus LUT overhead will increase if we want to perform the same operation using FPGA logic units.
· **Right Shift Operation**: Right shift operation using ADDN instruction is impractical to realize as even by repeated subtraction, in worst case scenario we would require exponential time (2^{163} operations) for $\mathbb{GF}(2^{163})$.
· **Field Multiplication and Squaring**: Field multiplication and squaring are similarly impractical to realize using ADDN instruction. In the worst-case scenario, we may need to run a loop of ADDN instruction for 2^{163} times.

C. Optimization of ADDN Instruction

The traditional ADDN instruction takes in 3 parameters as described in Algo.1. But our modified ADDN instruction takes in 4 parameters as described in Algo. 2. This modification

---

Algorithm 2: Addition using Modified ADDN

Data: A,B,C,D
Result: D.Mem[C]=D.Mem[A]+D.Mem[B]
$D.Mem[C] \leftarrow 0$;
ADDN C,A,B,2; // 2→ Next instruction

---

in the ADDN instruction allows us to write the result of the addition in a separate memory address directly. As mentioned in [5] we have also used two variants of the modified ADDN instruction: $ADDN_{nw}$ and $ADDN_w$. $ADDN_{nw}$ does not allow the data to be written back into the data memory while $ADDN_w$ switches on the data write-back mode of the data memory. In the proposed architecture we have used 2 data memories configured as true dual port block RAM. Along with these modifications, we have connected the Adder module with the second output port of the first data memory and first output port of the second data memory as in Fig. 2. The output of

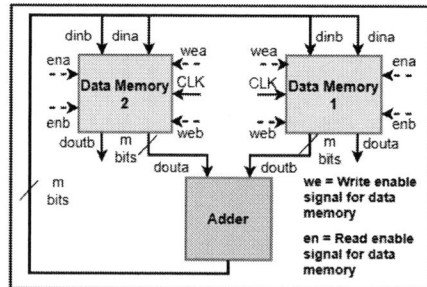

Fig. 2: Modified structure of ADDN processor

the Adder module travels to all the data input ports of the data memories. This arrangement allows the concurrent reading and writing of data from the same data memory, thus allowing the pipe-lining of several instructions that could not have been possible using a single data memory. Fig. 3 demonstrates the pipe-lining as below:

1) **Clock Cycle 1:** Data 1 and Data 2 read from Data Memory 1 (doutb port) and Data Memory 2 (douta port) and gives input to adder

978-1-7281-5410-7/20 $31.00 © 2020 IEEE

2) **Clock Cycle 2:** Data 3 and Data 4 read from Data Memory 1 (doutb port) and Data Memory 2 (douta port) and gives input to adder. The adder performs addition operation on Data 1 and Data 2

3) **Clock Cycle 3:** Data 5 and Data 6 read from Data Memory 1 (doutb port) and Data Memory 2 (douta port) and gives input to adder. The output of the adder is written to dina port of Data Memory 1.

If we note the operations in clock cycle 3, the data is read and written into Data Memory 1 via different ports. This contribution has significantly helped in the reduction of latency of the architecture. This concurrent operation will not be possible with a single data memory as in traditional ADDN architecture.

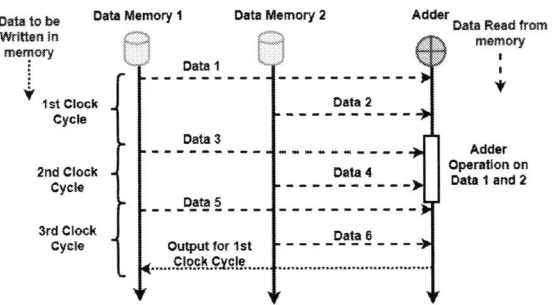

Fig. 3: Modified ADDN processor Operation

D. Finite Field Addition and AND operation

In the proposed design, we are mostly confined to characteristic 2 arithmetic. In characteristic 2 field, addition is identical to XOR operation. AND operation is required while implementing lazy reduction [3]. In contrast to integer addition and subtraction, which can be carried out using DSP cores in a FPGA, the same cannot be done in case of characteristic 2 field. To realize XOR and AND operation using DSP adder, we need to perform the following as depicted in Fig. 4. The rewiring operation in Fig. 4 expands the integer by inserting zeros in alternate bit positions. The sum part of the addition results in XOR of the two numbers and the carry part results in AND of the two numbers. Using a flag (**XORCon**) we can retrieve the desired AND/XOR output into the data memory. Hence, performing these operations using DSP cores in a FPGA helps in the reduction of LUTs in the architecture.

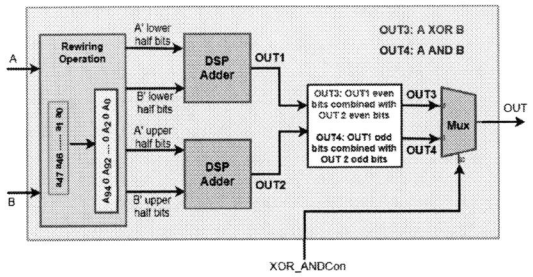

Fig. 4: XOR and AND operation using ADDN

E. Finite Field Multiplication

Finite field multiplication using only ADDN instruction is impractical for a real-life scenario as, we may need to run a loop for 2^{163} times for elements $\in \mathbb{GF}(2^{163})$.

To solve this problem, we have designed a lightweight field multiplier with **rstMul** flag to reset the multiplier core. We have used the Comba multiplication algorithm as mentioned in [8] with word size 48 bits by pipe-lining certain operations. Fig. 5 shows the architecture of the lightweight field multiplier used in the proposed design.

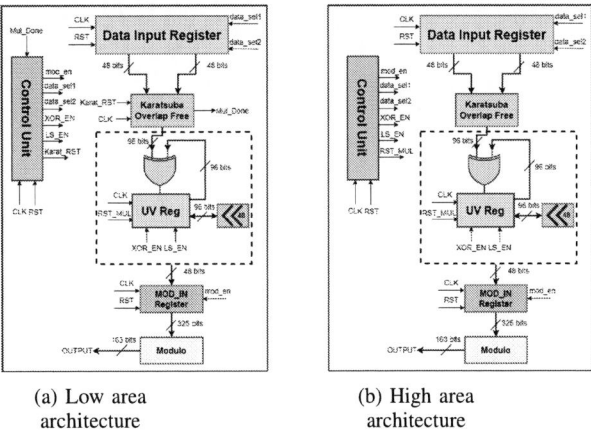

(a) Low area architecture   (b) High area architecture

Fig. 5: Lightweight Field Multiplier

The core 48 bit field multiplier can be implemented in two different ways as depicted in Fig. 5a and 5b. The 48 bit overlap-free combinational Karatsuba multiplier [14] in Fig. 5b produces output in single clock cycle while the sequential one in Fig. 5a requires 5 clock cycles to produce an output. A similar hardware implementation of overlap-free Karatsuba was done in [16]. Thus overall for a 163 bit field multiplication, the multiplier in Fig. 5a requires 112 clock cycles while Fig. 5b requires 32 clock cycles. Both the multipliers are followed by a similar combinatorial modulo operation block. The one in Fig. 5a consumes much lesser area compared to Fig. 5b owing to the difference in the architecture of the Karatsuba multiplier block. The XOR operation is performed using a combinatorial XOR block which can be implemented in a similar way as mention in Sec. III-D but considering the latency of the field multiplier we have opted for the combinatorial XOR block. The operations are pipelined using the control path designed within the field multiplier. For field squaring we have implemented a separate combinatorial squarer block which requires total 9 clock cycles including data loading for a single squaring. Owing to the total number of squaring operations required due to Frobenius map, point addition and doubling and Itoh-Tsuji inversion, a separate squarer block reduces the execution time of the complete scalar multiplication operation.

F. Right Shifter on ADDN Processor

Right shift operation is required several times while implementing scalar multiplication operation on the Koblitz curve.

Shifting key bits or division by 2, both can be performed using right shift operation. Executing these operations by repeated subtraction using ADDN is not feasible. So with a zero LUT overhead, we can implement a right shifter module in the design enabled by a flag bit. The flag-bit **rsCon** is used to write the right-shifted output into the data memory.

G. Instruction Memory and Control Path

The instruction memory is configured as single port ROM can hold up to 2^{10} instructions. The instructions are 60 bits wide. A single instruction consists of the 22 flag bits, four 7 bits operand address corresponding to the four ports of the data memory, and a 10-bit jump address. The flag bits consists of:

- · Read-Write Operations: It controls the read/write operations for all the input ports of the two data memories
- · ALU Operation: It controls which of the ALU block provide output to the data memory

The instruction format of the modified design is as shown in Fig. 6. Instead of having a hardwired control path we have micro-coded the control signals within the processor instructions which helps us in saving few resources.

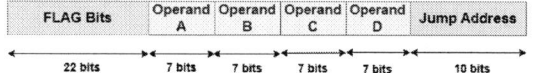

Fig. 6: Instruction structure of the ADDN processor

IV. ECC CO-PROCESSOR USING ADDN

In this section, we will present a detailed description of our proposed ADDN ECC Processor. The DSPs in the architecture are configured for 48 bits operands. The instruction format of the whole design is as shown in Fig. 6. The flag bits in Fig. 6 controls the operation of the dedicated ALU blocks and the read/write operations of the data memories. The multiplier core is initiated by **rstMul** flag bit. The key is stored in the data memory in an inverted manner. The right-shifter helps in traversing through the key which is controlled via **rsCon** flag bit. The MUX in the program counter sections serves the purpose of selecting the next line address or the jump address based on the MSB of the output ALU blocks. The output of the ALU blocks are written into data memory by selecting the correct output via the select signal of the DIN Select module and the write control bits of the data memory. The complete ADDN processor is shown in Fig. 7. The output port of the DIN Select module is connected to all the data input ports of the two data memories used. This allows the parallelization of data read and write cycle. The instruction memory is put to a halt while field multiplication is performed. This helps in avoiding no operation instructions thus saving instruction memory. We are able to implement two different methodologies of scalar multiplication using architecture specified in Fig. 7 with only two different instruction sets.

V. RESULT AND COMPARISON

In this section, we will analyze the performance of the proposed ADDN processor for the ECC Scalar Multiplication.

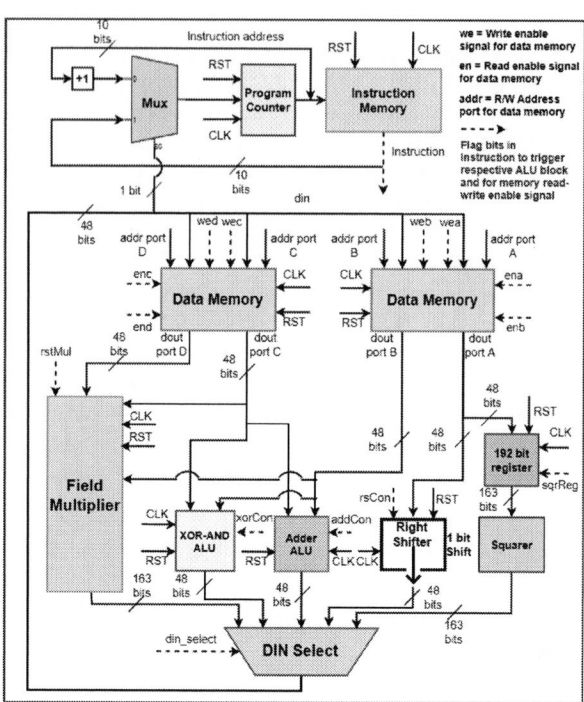

Fig. 7: Architecture of ADDN processor

The design goal is to develop a design with minimal footprint on the slices and properly utilizing the BRAMs and DSPs judiciously to ensure that the FPGA can also house other circuits needed in an application. Table I shows the statistics of resource utilization of the proposed design with respect to the available resources in Artix-7 (XC7A100TCS) FPGA board. Table II reflects 4 proposed designs where the data paths of all the designs are the same except for 2 different multiplier cores; one combinational and another sequential showing two design data-points each for Solinas and Montgomery techniques. Combining the results of Tables I and II we can see that, the most competitive result in terms of Area-Time metric (**0.269**) is obtained for the design with the multiplier core depicted in Fig. 5b and the scalar multiplication implemented using Montgomery ladder. Thus the $x-$coordinate only formulation with no additional operations to reduce the scalar has proved to be more advantageous with respect to the method in [1]. The design consumes only 657 slices, leaving 96.79% LUTs, 96.3% BRAMs, 98.75% DSPs free, while requiring 0.410 ms to execute the complete scalar multiplication.

Platform	Slice	LUT	Register	DSP	BRAM
Artix-7	4.15% (657)[1]	3.21% (2033)	0.98% (1237)	1.25% (3)	3.7% (5)
(XC7A100TCS)	2.64% (419)[2]	1.92% (1220)	1.15% (1458)		

TABLE I: Resource utilization details of the ECC processor

[1]48 bit combinational Karatsuba Multiplier which produces output in 1 clock cycle
[2]48 bit sequential Karatsuba Multiplier which produces output in 5 clock cycles

Most of the existing designs implement an ECC scalar multiplication using a high-performance architecture where the latency of the design is crucial. The designs in Table II are mostly implemented in Virtex-4 or older Xilinx platforms and none of them utilizes DSPs for finite field arithmetic. It would be unfair to directly compare the statistics of our design to these as they are on different families. However to demonstrate the minimalistic perspective of the design we choose a modern low-end Artix-7 board. A high-performance ECC scalar multiplier on K-163 with a digit-level finite field multiplier over Gaussian normal basis is presented in [6]. The target of the design is to achieve low latency at the cost of area wherein it employs four parallel field multipliers for speed-up. The designs in [7] and [13] supports all 5 Koblitz curves recommended by NIST. The design of [7] had tried to reduce latency by the use of the Karatsuba-Ofman field multiplier. However, the slice consumption of the architecture is 10 times more than even our high area architecture. In [13] several parallelization techniques have been applied in the field multiplier section. Additionally, an improved τNAF point multiplication algorithm is proposed to improve the efficiency of elliptic curve point multiplication. Despite having an impressive area-time metric, the architecture consumes significant resources compared to ours. A scalable architecture with improved point addition algorithm and use of comba multiplication algorithm is described in [8]. They have also targeted a low latency, high-performance design with high LUT and BRAM consumptions. An architecture for $GF(2^{163})$ is shown in [9], where a digit-serial multiplier with improved latency is used. The design has a fairly high resource consumption which is 10 times more than ours. A low area overhead scalable design is shown in [10] using a micro coding technique. The control path of the design is coded within the instruction memory. This way they have achieved a fairly low area consumption of the overall architecture. But the latency of the design is 10 times more than the present proposed work.

VI. CONCLUSION

In this work, we develop an ADDN-instruction based ECC processor for Koblitz curves. Several design choices for judicious usage of FPGA hard IPs have been proposed for reducing the area overhead while ensuring a competitive area-time product. The design is reconfigurable to support two different methodologies for scalar multiplication. We furnish experimental results to show that the design consumes minimal resource on cost-effective FPGAs, leaving ample room for hosting other important SOC circuitry.

VII. ACKNOWLEDGEMENT

Authors would like to thank the Defence Research & Development Organisation, India for partially funding the work through the project entitled, "Secure Resource-constrained Communication Framework for Tactical Networks using Physically Unclonable Functions".

REFERENCES

[1] Solinas, J.A, "Efficient Arithmetic on Koblitz Curves," Designs, Codes and Cryptography, vol. 19, pp. 195—249, 2000.
[2] Meier W., Staffelbach O., "Efficient Multiplication on Certain Nonsupersingular Elliptic Curves," CRYPTO, pp. 161–177, 1992.
[3] B. B. Brumley and K. U. Jarvinen, "Conversion Algorithms and Implementations for Koblitz Curve Cryptography," IEEE Transactions on Computers, vol. 59, pp. 81–92, 2010.
[4] Mavaddat F., Parhamt B. "URISC: the ultimate reduced instruction set computer," Int. J. Electr. Eng. Educ, pp. 327–334, 1988.
[5] D.B. Roy, Poulami Das, D. Mukhopadhyay "ECC on Your Fingertips: A Single Instruction Approach for Lightweight ECC Design in GF(p)," SAC , pp. 161–177, 2015.
[6] R. Azarderakhsh and A. Reyhani-Masoleh, "High-Performance Implementation of Point Multiplication on Koblitz Curves," IEEE Transactions on Circuits and Systems II: Express Briefs, vol. 60, no. 1, pp. 41–45, Jan. 2013
[7] K. C. C. Loi, S. An and S. Ko, "FPGA implementation of low latency scalable Elliptic Curve Cryptosystem processor in GF(2m)," 2014 IEEE International Symposium on Circuits and Systems, 2014, pp. 822–825
[8] K. C. C. Loi and Seok-Bum Ko "High performance scalable elliptic curve cryptosystem processor for Koblitz curves," Microprocessors and Microsystems, 2013, pp. 394–406
[9] Lutz J., Anwarul Hasan M., "High Performance Elliptic Curve Cryptographic Co-processor," International Conference on Information Technology: Coding and Computing , pp. 486–492, 2004
[10] Mohamed N. Hassan and Mohammed Benaissa , "Efficient Time-Area Scalable ECC Processor Using μ-Coding Technique," ITCC, pp 250–268, 2010
[11] Koblitz N., "CM-curves with good cryptographic properties," CRYPTO, pp. 279-287, 1991.
[12] J. Adikari, V. S. Dimitrov and K. U. Jarvinen, "A Fast Hardware Architecture for Integer to τNAF Conversion for Koblitz Curves," IEEE Transactions on Computers, vol. 61, no. 5, pp. 732-737, 2012.
[13] K. C. C. Loi , Seok-Bum Ko, "Parallelization of scalable elliptic curve cryptosystem processors in $GF(2^m)$," Microprocessors and Microsystems, 2016.
[14] H. Fan , J. Sun , M. Gu , K.-Y. Lam , "Overlap-free Karatsuba-Ofman polynomial multiplication algorithms," IET Information Security, 2010.
[15] T. Itoh, S. Tsujii, "A fast algorithm for computing multiplicative inverses in $GF(2^m)$ using normal bases," Information and Computation, pp. 171-177, 1988.
[16] D.B. Roy, D. Mukhopadhyay, "An Efficient High Speed Implementation of Flexible Characteristic-2 Multipliers on FPGAs," Progress in VLSI Design and Test, pp. 99-110, 2012.
[17] P. L. Montgomery, "Speeding the Pollard and Elliptic Curve Methods of Factorization," Mathematics of Computation, Jan 1, 1987.
[18] López J., Dahab R., "Fast Multiplication on Elliptic Curves Over GF(2^m) without precomputation," CHES, pp 316-327, 1999.

Works	Platform	LUTs	Area (Slices)	Max Freq (MHz)	Latency (μ-sec)	$\frac{AT}{10^6}$
[6]	ALTERA STRATIX II	18964 ALM	-	-	9.84	-
	ALTERA STRATIX II	23084 ALM	-	-	9.15	-
[7]	Virtex-4	-	7427	-	29.28	0.217
[8]	Spartan-3 XC3S400	3850	1232 (8 BRAMs)	93.084	456	0.561
	Virtex-4 XC4VFX12	3815	1219 (8 BRAMs)	155.376	273	0.332
	Virtex-E XCV2000E	4320	1641 (18 BRAMs)	43.983	964	1.58
[9]	Xilinx XCV2000E	7362 (g = 14)	-	-	135	0.993
	Xilinx XCV2000E	10017 (g = 41)	-	-	75	0.751
[10]	Spartan 3 XC3S200	2220	1180 (4 BRAMs)	70.5	2700	3.18
	Spartan 3 XC3S50	847	417 (4 BRAMs)	75	17517	7.304
	Spartan 3 XC3S50	1056	543 (4 BRAMs)	76	5360	2.91
[13]	Virtex-5 XC5LX110T	7073	2199 (5 BRAMs)	223.46	68	0.149
		8609	2708 (5 BRAMs)	222.67	55	0.148
Proposed[1] (Montgomery)	Artix-7 (XC7A100TCS)	2033	657 (5 BRAMs, 3 DSPs)	109.5	410	0.269
Proposed[2] (Montgomery)	Artix-7 (XC7A100TCS)	1220	419 (5 BRAMs, 3 DSPs)	120.25	919	0.385
Proposed[1] (Solinas)	Artix-7 (XC7A100TCS)	2033	657 (5 BRAMs, 3 DSPs)	109.5	663	0.435
Proposed[2] (Solinas)	Artix-7 (XC7A100TCS)	1220	419 (5 BRAMs, 3 DSPs)	120.25	1049	0.439

TABLE II: Comparison of ECC ADDN Processor with Existing Designs

[1] 48 bit combinational Karatsuba Multiplier which produces output in 1 clock cycle
[2] 48 bit sequential Karatsuba Multiplier which produces output in 5 clock cycles

3D logic cells design and results based on Vertical NWFET technology including tied compact model

Chhandak Mukherjee, Marina Deng,
François Marc, Cristell Maneux
IMS Laboratory
University of Bordeaux, CNRS UMR
5218, Bordeaux INP
Talence, France
firstname.name@u-bordeaux.fr

Arnaud Poittevin, Ian O'Connor,
Sébastien Le Beux, Cedric Marchand
Lyon Institute of Nanotechnology
University of Lyon, CNRS UMR 5270,
Ecole Centrale de Lyon
Ecully, France
firstname.name@ec-lyon.fr

Abhishek Kumar, Aurélie Lecestre,
Guilhem Larrieu
LAAS-CNRS
Université de Toulouse, CNRS, INP
Toulouse, France
firstname.name@laas.fr

Abstract— Gate-all-around Vertical Nanowire Field Effect Transistors (VNWFET) are emerging devices, which are well suited to pursue scaling beyond lateral scaling limitations around 7nm. This work explores the relative merits and drawbacks of the technology in the context of logic cell design. We describe a junctionless nanowire technology and associated compact model, which accurately describes fabricated device behavior in all regions of operations for transistors based on between 16 and 625 parallel nanowires of diameters between 22 and 50nm. We used this model to simulate the projected performance of inverter logic gates based on passive load, active load and complementary topologies and carry out an performance exploration for the number of nanowires in transistors. In terms of compactness, through a dedicated full 3D layout design, we also demonstrate a 48% reduction in lateral dimensions for the complementary structure with respect to 7nm FinFET-based inverters.

Keywords—Vertical NWFET technology, compact model, VNWFET DC measurements, 3D logic circuit cell, circuit simulation results.

I. INTRODUCTION

Data size and functionality requirements for computing are increasing, according to the expectation that hardware performance will continue to improve, irrespective of the actual implementation. This is particularly true for emerging computing paradigms such as Edge Computing which is placing extraordinarily stringent constraints on computing hardware performances. However, the end of the roadmapped technological scaling is anticipated in a few technology nodes, mainly for cost reasons down to the 7nm FinFET gate length node. In this context, vertical integration is an attractive approach to fully take advantage of 3D integration and scale pitch between contacts. Huge gains in silicon area are expected through the combination of extremely small elementary device footprint and minimal device usage with MIG and PTL design styles, for instance. This paper is the first attempt to quantify the gains in terms of compactness and energy efficiency of 3D logic blocks based on actual fabricated p-type VNWFET devices.

The paper is organized as follows: section II recalls the VNWFET technology main features while detailing its associated scalable compact model. In particular, the unified charge-based control model has been self-consistently modified to take into account depletion and accumulation regimes, electrostatic control, short-channel effects (SCE), drain-induced barrier lowering (DIBL) and band-to-band tunneling (BTBT) contributions through gate-induced drain

leakage (GIBL). Simulated results are compared to measurements to illustrate the p-type VNWFET model versatility in terms of dimensions. An n-type VNWFET model has also been delivered using the electron mobility value from the literature. These scalable compact models have been implemented in Verilog-A, and subsequently implemented in a dedicated circuit design workspace. In section III, we demonstrate the efficiency of this design workspace to simulate and quantify the 3D logic blocks. 3D layouts implement inverter functions with various topologies: (i) passive load, (ii) active load and (iii) complementary. Their static and dynamic energy consumption and delays are given. In section IV, we propose a layout footprint comparison between the 7nm FinFET and the VNWFET through conventional λ-rules. Going beyond this approach, section V deals with large-scale integration considerations suitable for a fully 3D logic block architecture.

II. VNWFET DEVICES

A. Technology description

The VNWFET technology has a junction-less architecture composed of a homogenous highly doped nanowire channel, patterned into boron doped (2×10^{19}cm^{-3}) Si substrate. The current flows between silicided source/drain contacts and is controlled by a gate-all-around structure with a physical channel length of 14nm (Fig. 1). More details on the fabrication steps can be found in [2].

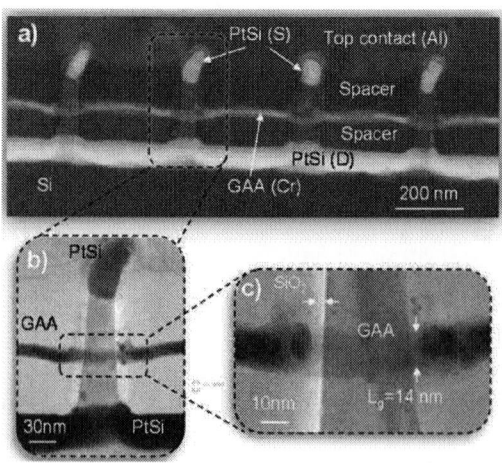

Fig. 1: VNWFET device [1]: (a) STEM image in cross section of the vertical transistor implemented in nanowire arrays, (b) single nanowire showing its (c) gate formation.

B. Compact Model

The model formulation is based on the unified charge-based control model (UCCM) elaborated in [3] for long-channel devices, which furthers the physical basis of the junctionless nanowire transistor (JLNT) model presented in [4] and adapted in [5] for the JLNT technology under test [2]. The limitations of the model in [4] is mainly the piece-wise continuous drain-current model which requires additional smoothing functions and fitting parameters to smooth the transition between depletion and accumulation modes of operation. In order to overcome this, the explicit and non-piece-wise solution in [3] treats the mobile charge (Q_m) to be decoupled between the depletion (Q_{DP}) and complementary (Q_C) components. In the depletion mode the UCCM expression has been formulated as [3]:

$$Q_{DP} = Q_{eff} L W \left\{ \frac{Q_{sc}}{Q_{eff}} \exp\left(\frac{V_g - V_{th} - \eta V}{\eta \phi_T} + \frac{Q_{dep}}{Q_{sc}} \right) \right\} \tag{1}$$

with the depletion charge, $Q_{dep}=qN_DR/2$, the effective charge during depletion, $Q_{eff}=Q_{sc}\eta C_{ox}\varphi_T/(Q_{sc}+\eta C_{ox}\varphi_T)$, $Q_{sc}=2\varepsilon_{Si}\varphi_T/R$, R being the nanowire diameter, η an interface trap parameter, φ_T the thermal voltage and V the potential along the channel. A Lambert W function has been used in both [3] and [4] to develop the solution of total mobile charge in the JLNT. While the expression for Q_{DP} predicts the depletion contribution correctly (for $V_g<V_{th}$), it underestimates the value of the drain current above the flat-band condition. So in accumulation mode, especially in high accumulation with $Q_C \geq Q_{dep}$, the charge Q_C has been derived to act complementary to Q_{DP}, considering that the threshold voltage is pinned at V_{FB} in the accumulation region, in order to avoid using additional smoothing functions and improve simulation time. Under high accumulation $Q_C \geq Q_{dep}$ and Q_C is simplified using another Lambert function as follows [3]:

$$Q_C = \eta C_c \phi_T L W \left\{ \frac{Q_{sc}}{\eta C_c \phi_T} \exp\left(\frac{V_g - V_{FB} - \eta V}{\eta \phi_T} \right) \right\} \tag{2}$$

with corrected electrostatic control through $C_c=C_{ox}-C_{eff}$, $C_{eff}=1/C_{ox}+R/2\varepsilon_{Si}$. Having evaluated both the depletion and complementary parts of the mobile charge, one can formulate the non-piece-wise continuous model of the total drain current in terms of Q_{DP} and Q_{DC} at the source and the drain end, Q_{DP0}, Q_{C0} and Q_{DPL}, Q_{CL}, respectively:

$$I_{DS,0} = \mu_{eff} \frac{2\pi R}{L_{eff}} \phi_T \left[\frac{Q_{DP}^2}{2\eta C_{ox}\phi_T} + Q_{DP} + \frac{Q_C^2}{2\eta C_c \phi_T} + 2Q_C \right]_{Q_{DPL}, Q_{CL}}^{Q_{DP0}, Q_{C0}} \tag{3}$$

The drain current expression is free of any fitting parameters and can be evaluated based on the physical device parameters such as that of geometry and doping. Additionally, short channel effects were taken into account considering velocity saturation, an effective mobility, μ_{eff}, and incorporating an effective gate length, $L_{eff}=L-\Delta L$, where L is the physical device gate length and ΔL is calculated following the expression in [6]. Considering that the source and drain access region resistances degrade the drain current above threshold, the final expression of the drain current can be written as a function of the long channel current ($I_{DS,0}$), using (3), taking into account the corrections due to short-channel effects [5], as follows [6]:

$$I_{DS} = \frac{I_{DS,0}NF}{1 + 2\pi \frac{R}{L_{eff}} NF \mu_{eff}(R_s + R_D)\left[(Q_{DP0}+Q_{C0}) - \eta_1\left(Q_{DP0}+Q_{C0} - (Q_{DP,Vdeff}+Q_{C,Vdeff})\right)\right]} \tag{4}$$

Here, R_S and R_D are the source and drain series access resistances, respectively; NF is the number of nanowires in parallel, η_1 is a fine tuning parameter to take into account the drain-voltage dependence of the series access resistances and $Q_{DP,Vdeff}+Q_{C,Vdeff}$ is the total mobile charge at the drain end (pinch-off) of the channel.

Additionally, considering formation of Schottky contacts at the source and drain access regions, the subthreshold leakage currents are also taken into account. Consequently, thermionic (I_{th}), tunneling (I_{tun}) and band-to-band tunneling (BTBT) contributions through gate-induced drain leakage (GIDL) are added as separate branch currents [7] to the total drain current, in order to model the subthreshold behavior of the drain current. The expression used in the compact model for the BTBT current at the drain end reads [7]:

$$I_{GIDL} = 2\pi R L_{Access} NF \cdot A_{GIDL} V_{DS} E_{segd}^2 \exp\left(-\frac{B_{GIDL}}{E_{segd}} \right) \tag{5}$$

wher L_{Access} represents the lengths of the source and drain access regions outside the channel, B_{GIDL} is a physics-based parameter with a theoretical value of 21.3MV/cm [7] and E_{segd} is the electric field in the drain overlap region, given as

$$E_{segd} = \frac{C_{ox}\sqrt{V_{segd}^2 + (C_{GIDL}V_{DS})^2}}{\varepsilon_0 \varepsilon_{Si}} \tag{6}$$

Here, V_{segd} is the gate-drain voltage across the oxide and A_{GIDL}, C_{GIDL} are two GIDL fitting parameters. Lastly, additional model improvement has been achieved compared to the model reported in [5], in the subthreshold regime. In order to improve model accuracy, the accurate extraction of the parameter η is ensured in order to correctly adjust the subthreshold slope. Moreover, the effect of drain-induced barrier lowering (DIBL) is also taken into account in the compact model by a modification of the threshold voltage through the following equation,

$$V_{th} = V_{FB} - \frac{Q_{dep}}{C_{ox}} - DIBL(V_{DSmax} - V_{DSmin}) \tag{7}$$

with $DIBL$ being the drain-induced barrier lowering in mV/V.

C. Measured and simulated results

For the validation of the compact model against measurement results, we chose a wide range of geometries where test structures had diameters (D) ranging between 22-50nm with 16-625 nanowires in parallel (NF). Figs. 2(a) and (b) show the transfer characteristics, I_D-V_{GS}, of the JLNTs with D=22nm/NF=16 and D=50nm/NF=36, respectively. The model simulation results show very good agreement with the measurements over the entire bias range, indicating accuracy of individual modules of the compact model. Particularly, the improvement of the model accuracy in the subthreshold region is observable compared to the results reported in [5], leveraging eqs. (5) and (7) as well as the parameter η. The improvement in drive current and subthreshold leakage with a higher number of nanowires in parallel is obvious from Fig. 2(b), which however suffers from a more pronounced DIBL induced V_T-shift. This is most likely due to quantum confinement effects in smaller nanowire diameters [8].

978-1-7281-5410-7/20 $31.00 © 2020 IEEE

Nevertheless, the compact model captures these effects with sufficient accuracy. A second order validation is performed in Figs. 3 (a) and (b) depicting the output characteristics, I_D-V_{DS}, of the JLNTs, further affirming model accuracy.

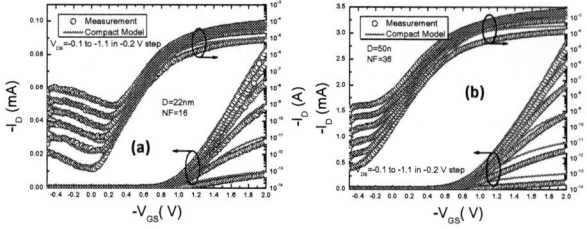

Fig. 2: I_D-V_{GS} of JLNTs with (a) 22nm diameter and 16 nanowires in parallel and (b) 50nm diameter and 36 nanowires in parallel.

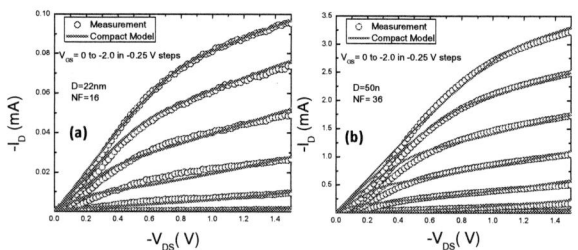

Fig. 3: I_D-V_{DS} of JLNTs with (a) 22nm diameter and 16 nanowires in parallel and (b) 50nm diameter and 36 nanowires in parallel.

III. LOGIC PERFORMANCE ASSESSMENT

In this section, we leverage the developed compact model to assess the performance metrics of various topologies of an elementary logic gate in the VNWFET technology. While it is possible to simulate logic gates implementing multiple Boolean operations, we focus in this paper on the comparison between several topologies implementing a single inverter operation. This is partly due to the lack of experimental devices and consequently measurements with which the compact model parameters can be defined; but it also targets a full understanding of the relative merits and drawbacks of the device itself, minimizing design-specific issues. In a first exploration we assess the simulated performance of p-type only inverters, while in a second exploration, using a literature survey, we extrapolate the model to n-type VNWFETs in order to explore a simple complementary inverter structure. Finally, we establish a comparison with the 7nm FinFET technology node using typical values, in preparation for further analysis in section IV.

A. Simulated structures

This work focuses on inverter structures implemented with passive- and active-load topologies, as well as with complementary topologies.

1) P-type only inverters

P-type only structures use a p-type device as conventional pull-up, and implement the pull-down branch as a resistance, with either a passive (Fig. 4(a)) or an active (Fig. 4(b)) load. For the latter, we use a p-type device configured as current source. In both cases, the pull-down load is responsible for the low logic state output. This type of structure is known to be less efficient than their complementary counterparts, but firstly enables validation of the use of experimental data for designing logic, and secondly gives first insights into the use of such devices.

2) Complementary inverter

For the complementary circuit, based on [9][10] and shown in Fig. 4(c), we conjecture a value for the carrier mobility in the n-type VNWFET channel and consequently its drive current. This value is 3x that of the p-type VNWFET. Hence to balance the circuit for a switching input voltage value halfway between the supply rails and for roughly equivalent noise margins, we target identical currents in both devices. To achieve this, we set the NF (number of nanowires) per device in the P-type equal to 3x that of the n-type.

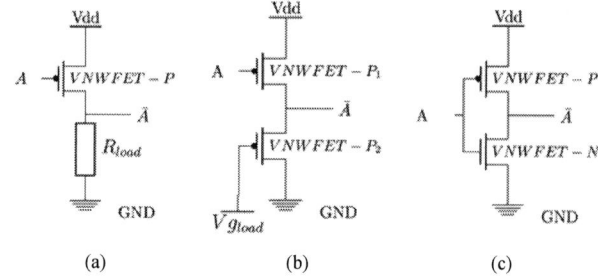

Fig.4: Schematics of the studied inverter structures, (a) passive load, (b) active load and (c) complementary

B. Results

The goal of the following simulation-based exploration is to study the impact of using a large range of nanowires per transistor on typical static and dynamic logic performance metrics. In the simulation protocol, we assume that the gate capacitance behaves in the same way for both p- and n-type VNWFETs, and that the capacitive load on the output of each structure is equivalent to its own theoretical input capacitance (i.e. fanout of 1). Since the p-type VNWFET gate capacitance with NF=16 is experimentally determined to be 50aF, and assuming that its evolution with NF is linear, we deduce a capacitance contribution per nanowire of 3.25aF. Measurements were performed using the model described in section II as implemented in Verilog-A and simulated using the Spectre™ commercial simulator.

1) Static performance

DC simulation points enable the extraction of typical static characteristics of the p-type VNWFET transistor.

a) I_{on}/I_{off} ratio

In this analysis, we characterize the p-type VNWFET characteristics in terms of I_{on}/I_{off} ratio for values of NF ranging from 3-300. To measure I_{on} (resp. I_{off}), we set input A=0 (resp. A=1) such that the pull-up device is on (resp. off) in all structures.

We observe a linear increase in the leakage current with NF at a rate of 61pA per nanowire. Given the 16nm nanowire diameter, this translates to 0.3μA/μm^2 leakage current density in the p-type VNWFET. However, device I_{on} does not increase linearly with NF – in fact, the rate of increase slows down when using large values of NF. As a result, device I_{on}/I_{off} ratio *decreases* with increasing NF, from $15 \times 10^3 |_{NF=10}$ to $6.5 \times 10^3 |_{NF=300}$.

b) Logic level degradation

The load in the pull-down branch of both passive- and active-load inverters is a major factor both for logic '1' level degradation and for high-low propagation delay. Its value is a tradeoff: increasing the load decreases logic level degradation but increases propagation delay. For the studied structures the best compromise, as shown in Fig. 5, gives a 15% logic '1' level degradation and a 15% logic '0' level overshoot during high-low transitions at the output for a 1GHz input signal.

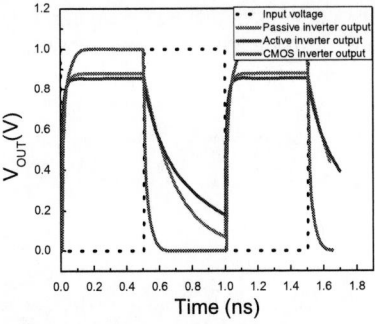

Fig. 5: Voltage across the load capacitor for the studied structures

2) Dynamic performance

As shown in Fig. 5, we carried out transient simulations to extract relevant dynamic performance metrics, using a 1GHz data input with rise and fall times equal to 10ps. As previously indicated, each inverter shown in Fig. 4 was simulated with a fan-out of 1.

a) Propagation delay

For small values of NF, we observe a delay (measured as $t|_{Vout=50\%Vdd} - t|_{Vin=50\%Vdd}$) ranging from 5-10ps according to the type of inverter (the lowest delay is achieved by the passive load inverter). When increasing NF, the gate delay increases. This result can be linked principally to the sublinear increase in I_{on} with NF, and the linear increase in gate capacitance with NF.

b) Dynamic energy consumption

We also measure the energy required to transit through the transistor channel when changing state. We calculate the amount of charge for a low-high transition at the output for the self-loaded complementary inverter. This value varies linearly with NF and works out to 11aC per nanowire. With a 1V supply voltage this gives us an energy consumption of 11aJ per nanowire for a low-high transition at the output.

3) Fanout analysis

Due to the sublinear variation of I_{on} with NF, self-loaded logic cells with high values of NF cannot charge completely in the available time (Fig. 6). At 1GHz and for NF>300, an inverter cannot cascade with an identical logic cell. Similarly, when increasing the fan-out (number of cells controlled by the inverter), this boundary reduces until fan-out = 5, where a single nanowire transistor cannot drive 5 identical cells simultaneously. This information is crucial regarding power and delay management when designing larger cells.

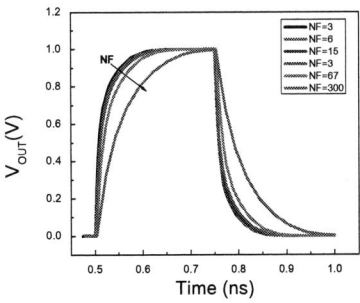

Fig. 6: Voltage across the load capacitance of the self-loaded complementary inverter according to NF (p-type VNWFET)

C. Comparison with FinFET and conclusion

Based on the previous results, there is a clear advantage for using logic cells with low NF, both for power consumption concerns and for fan-out. The values obtained are compared in

Table I to FinFET values from the literature, both for static values [11] and for propagation delay [12]. Note that while being an academic non-optimized technology, the VNWFET shows similar order of magnitude compared with the industrial mature process 7nm FinFET.

TABLE I. VNWFET / FINFET COMPARISON

Metric	VNWFET	7nm FinFET
Static leakage current density ($\mu A/\mu m^2$)	0.3	1
I_{on}/I_{off} ratio ($*10^4$)	[1.5 – 0.65]	~8
Propagation delay (ps)	[5 - 10]	2.2

IV. 3D LOGIC CELLS

A. Going vertical: Implications on physical circuit design

1) Paradigm change

Vertical transistor channels lead to a paradigm change in the design of logic cells. Source and drain contacts, separated by the vertical channel, can occupy the same lateral space. Stacked series transistors further improve the gain in circuit density. Further, the additional dimension enables numerous spatial configurations for the same logic functionality [13]. However, careful evaluation of gate contacts and routing is necessary to ensure the best tradeoff between density and performance. In this section, we identify critical dimensional constraints, formulate λ-rules for the VNWFET technology and leverage them to compare footprint to lateral FinFET technology.

Although this article does not aim to explore complex logic structures using this technology, it lays the foundations for carrying out a complete and exhaustive study with this objective. For this reason, in order to deal with the significant differences between planar FinFET technology and vertical nanowire technology, initial designs must share as much common ground as possible with a tried-and-tested yet cutting-edge technology. Based on the comparison results we extrapolate the comparison metrics to projected figures considering using the potential of the VNWFET technology to its full extent.

2) Comparison basis

A planar FinFET channel is composed of a number of fins, according to the desired transistor characteristics (Fig. 7). Similarly, a VNWFET channel is composed of several vertical nanowires. In this work, we aim to compare the footprint of VNWFET-based logic cells with respect to FinFET-based logic cells. We take as baseline reference λ-rules for elementary standard cells based on the 7nm FinFET technology [14] established in the context of exhaustive layout and performance benchmarking. Lambda-based rules (λ-rules) constitute a simple tool that allows first order scaling by linearizing the resolution of the complete wafer implementation. While modern processes rarely shrink uniformly, λ-rules remain useful to make first-order cross-technology spatial comparisons.

The principle of λ-rules is to decorrelate characteristic sizes from absolute dimensions by expressing them as a function of some reference length unit (λ). The λ value used for the FinFET represents twice the fin thickness (T_{si} –shown Fig. 7), which represents the smallest mask dimension (oxide thickness, established through epitaxial growth, is not correlated to lithography or mask limitations). Correspondingly, the smallest dimension in the VNWFET transistor is the nanowire diameter D (Fig. 8) and is accordingly used to define λ for the VNWFET technology. An

978-1-7281-5410-7/20 $31.00 © 2020 IEEE

important observation in both technologies is that dimensions in the transistor zone are comparable to λ, while dimensions comparable to 3λ are used in the routing and contacting of the transistor. In the baseline reference, FinFET planar transistors are at the 7nm node, such that λ=3.5nm, while the current state of VNWFET technology allows a minimal nanowire diameter such that λ=16nm. It should be stressed that this is representative of an emerging research technology under development rather than an inherent limitation to the technology.

Table II shows the λ-rules as established in [14] for 7nm FinFETs as well as those chosen in this paper for VNWFETs.

TABLE II. LAMBDA-RULE COMPARISON BETWEEN FinFET AND VNWFET

Parameter	Value in 7nm FinFET (nm)	Value in projected VNWFET (nm)	Comment
T_{fin} / D	$3.5 = \lambda_{fin}$	$11 = \lambda_{NW}$	Fin thickness / nanowire diameter
T_{si}	$2*\lambda_{fin}$		Fin length
H_{fin} / H_{NW}	$4*\lambda_{fin}$	30	Height
T_{ox}	1.55	5	Oxide thickness
P_{fin} / P_{NW}	$2*\lambda_{fin} + T_{fin}$	$2*\lambda_{NW} + D$	Pitch
W_C	$3*\lambda_{fin}$	$3*\lambda_{NW}$	Contact size
W_{M2M}	$2*\lambda_{fin}$	$2*\lambda_{NW}$	Gate to contact space

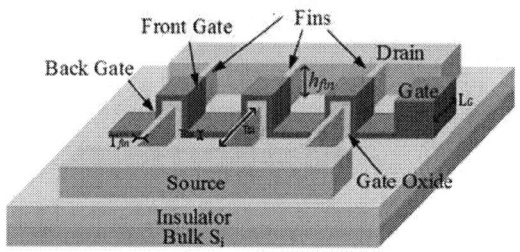

Fig. 7: Perspective view of a 7nm node FinFET transistor [14]

B. VNWFET technology and device structure

There are 3 metallic contacts along the transistor's vertical channel. This structure is shown in Fig. 8 [2]:

- The bottom PtSi contact surrounds the bottom of the nanowire and establishes a first access to the transistor channel. This contact is ultimately used as drain or source.
- A top Al contact covers the top of the nanowire and establishes a second access to the transistor channel. This contact is similarly used as a drain or source.
- A Cr layer in the middle surrounds the center of the nanowires and is separated from the silicon by a gate oxide. This metal contact acts as the gate.

It is worth noticing that the gate structure surrounds the channel, thus categorizing this type of transistor as a Gate-all-around (GAA) FET. Moreover, as compared to FinFET technologies, the silicon in the region of the drain and source is doped in the same way as for the channel zone. Specifically, the nanowire to which the drain, gate and source are attached is etched in a uniformly doped silicon bulk [1]. This transistor is thus also a junction-less transistor.

The perspective view of the nanowire transistor provides insights into the tridimensional structure of the device. Since we focus on the lateral footprint and for the sake of improved visibility, vertical dimensions are not to scale in this view. A single transistor may comprise multiple (NF) nanowires in its channel and each nanowire is surrounded by gate oxide before

any contacts other than the bottom contact is deposited. Spacers made of oxide are represented between top contact and gate and between gate and bottom contact to isolate those metals.

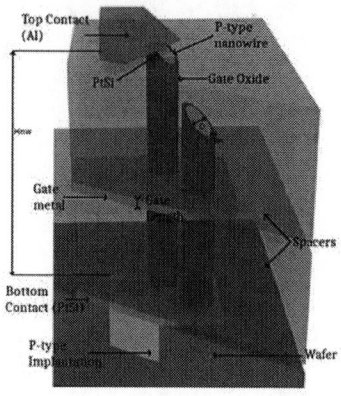

Fig. 8: Perspective and cut view of a projected VNWFET transistor

In order to facilitate the differentiation between p- and n-type, and for the sake of clarity, the figures representing VNWFET structures in the remainder of this paper will show neither the gate oxide surrounding the silicon nanowire nor the insulating spacers for each metallic layer.

C. Footprint estimation

As indicated previously, we focus on the footprint (lateral area) in order to keep common grounds with the FinFET technology. The vertical height of the logic cell is considered unimportant in this comparison and unrelated to any FinFET dimension. The same is true for gate length and contact thickness.

In the standard cell approach, the lateral "height" of the cell (i.e. the distance between supply voltage and ground) is constant for all standard cells in a technology. Data inputs and outputs are typically located in the middle of the cell. Their position is not constrained and their access is not taken into account in the cell design. This type of layout allows the designer to assemble each logic gate on the same level with only inputs and outputs to route properly, usually through a dedicated routing channel.

Several important points mentioned in section III are taken into account in order to implement logic cells in the context of standard cell design. The current technology is used for characterization and trials on vertical nanowires. It is thus unable to sustain the requirements of the λ-rule constraints we introduced in Table II. Indeed, the 16nm diameter value is not the main concern, since the dimensions of the metallic layers are much larger for electrical characterization purposes and manufacturing process limitations. The lithography equipment used in the process is not intended for this scale of precision.

These manufacturing changes, as compared to [1], remain credible in a foreseeable future. In a nutshell, the main assumption is that contact and gate dimensions are in the same range as the nanowire dimensions.

D. Layout footprint comparison example

The comparison method explained above is applied to a CMOS inverter structure with balanced switching corresponding to nanowire mobility [9].

The difference between the mobility of both transistor channels suggests that the n-type and p-type transistors respectively possess one fin or one nanowire and one fin or

978-1-7281-5410-7/20 $31.00 © 2020 IEEE   80

three nanowires. In Fig. 9, the layout is composed of large voltage supply extensions for both V_{DD} and GND and an active zone where the transistors are connected together. The layout footprint is the product of the width and length given in λ values.

In both Figs. 9(a) and (b), we notice a similar inverter structure, where the position of the supply voltages and input/output are indicated.

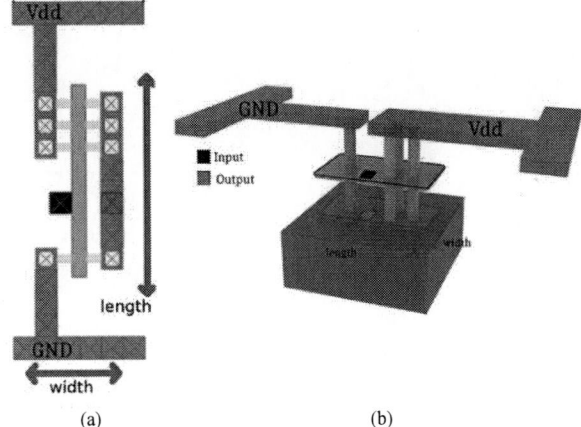

Fig. 9: (a) Layout (top view) of a 7nm FinFET inverter gate [Cui14], (b) Perspective view of a VNWFET inverter gate according the projected technology

The FinFET inverter footprint is 48λ long and 18λ wide while the VNWFET inverter footprint is 31λ long and 15λ wide. This represents a 48% footprint area reduction. If we choose a less restrictive comparison criterion and consider the active part alone (removing the 12λ supply contacts for both layouts), we observe an 84% footprint area reduction.

V. INSIGHTS FOR LARGE SCALE INTEGRATION

While devices based on vertical nanowires have been compared to lateral GAA devices in the past [15], this work used apparent mobility differences between both device channels to justify the difference in gate lengths. In order to achieve similar drive strength for both devices, the vertical device gate length is around 2× that of the lateral device. Such electrical considerations help to set the vertical dimensions for the VNWFET. Gate length values in the referenced article are in the 10-20nm range. This value fits our designs without any impact on the device footprint area and its impact on overall performance will be studied in the future. The fact that VNWFET dimensions such as gate, spacer and channel lengths are decorrelated from the lateral footprint allows electrical parameters to be tuned without touching the cell design. This favors the standardization of simple cells, as well as the achievement of complex logic design and scalable electronics. In this work, the comparison method separated electrical behavior concerns from device layout to establish a workflow and to enable the future consideration of stacked-gate vertical devices [16]. Stacking gates requires device I_{on} to be high enough to drive the whole common channel. We demonstrate in section III that with current technology, the fan-out limit is 4 for a very low number of transistors. Thus, we can expect at least a functional 4-stacked gates transistor, which already enables significant opportunity for disruptive logic designs.

VI. CONCLUSION

This work considers the use of VNWFETs as a means to implement 3D logic blocks. We have built a technology

scalable physics-based compact model and implemented it in Verilog-A as incorporated in a dedicated circuit design workspace. This environment has been used to simulate innovative 3D layouts of inverter cells. The layout of the complementary inverter has been compared with projected 7nm FinFET technology through the use of λ-rules. We showed that the VNWFET-based approach achieves 48% footprint reduction and can reach 84% if only the active part is considered. Beyond λ-rule comparisons, we presented another physical layout implementation that leverages the unique features of VNWFETs, where dimensions such as gate, spacer and channel lengths are de-correlated from the transistor footprint. This important property allows electrical parameters to be tuned without any impact on cell design. The standardization of such simple logic cells will pave the way for more complex VNWFET logic cell designs.

ACKNOWLEDGMENTS

This work was supported by the French RENATECH network (French national nanofabrication platform) and by the LEGO project through ANR funding (Grant ANR-18-CE24-0005-01).

REFERENCES

[1] G. Larrieu and X.-L. Han, "Vertical nanowire array-based field effect transistors for ultimate scaling", *Nanoscale*, vol. 5, pp. 2437-2441, 2013

[2] Y. Guerfi and G. Larrieu "Vertical Silicon Nanowire Field Effect Transistors with Nanoscale Gate-All-Around", *Nanoscale Research Letters*, vol. 11, pp. 210, 2016

[3] A. Hamzah, R. Ismail, N E. Alias, M. L. Peng Tan and A. Poorasl, "Explicit continuous models of drain current, terminal charges and intrinsic capacitance for a long-channel junctionless nanowire transistor", *Phys. Scr.* vol. 94, pp. 105813, 2019

[4] F. Lime, O. Moldovan and B. Iñiguez, "A Compact Explicit Model for Long-Channel Gate-All-Around Junctionless MOSFETs. Part I: DC Characteristics", *IEEE Trans. Electr. Dev.*, vol. 61, no. 9, pp. 3036-3041, Sept. 2014

[5] C. Mukherjee, G. Larrieu and C. Maneux, "Compact Modeling of 3D Vertical Junctionless Gate-all-around Silicon Nanowire Transistors", *EUROSOI ULIS*, 2020 (accepted)

[6] F. Lim, F. Ávila-Herrera, A. Cerdeira and B. Iñiguez, "A compact explicit DC model for short channel Gate-All-Around junctionless MOSFETs", *Solid State Electron*, vol. 131, pp. 24-29, 2017

[7] G. Zhu *et al.*, "Subcircuit Compact Model for Dopant-Segregated Schottky Gate-All-Around Si-Nanowire MOSFETs", *IEEE Trans. Electr. Dev.*, vol. 57, no. 4, pp. 772-781, April 2010

[8] S. Sahay, M. J. Kumar, "Junctionless Field-Effect Transistors: Design, Modeling, and Simulation". *IEEE Press Series on Microelectronic Systems*, John Wiley & Sons, 2019

[9] O. Gunawan *et al.* "Measurement of Carrier Mobility in Silicon Nanowires", *Nano Letters*, vol. 8, pp. 1566-1571, 2008

[10] J-P. Colinge *et al.*, "Nanowire transistors without junctions", *Nature Nanotechnology*, vol. 5, pp. 225-229, 2010

[11] L.T. Clark, V. Vashishtha, L. Shifren, A. Gujja, S. Sinha, B. Cline, "ASAP7: A 7-nm FinFET predictive process design kit", *Microelectronics Journal*, vol. 53, pp. 105-115, 2016

[12] P. Raghavan *et al.*, "Holistic device exploration for 7nm node", *IEEE Custom Integrated Circuits Conference (CICC)*, San Jose (CA), USA, 28-30 Sept. 2015

[13] V. Moroz *et al.*, "Power-performance-area engineering of 5nm nanowire library cells", *Int. Conf. Simulation of Semiconductor Processes and Devices (SISPAD)*, Washington (DC), USA, Sep. 9-11 2015

[14] T. Cui, Q. Xie, Y. Wang, S. Nazarian and M. Pedram, "7nm FinFET Standard Cell Layout Characterization and Power Density Prediction in Near- and Super-Threshold Voltage Regimes", *Int. Green Computing Conf. (IGCC)*, Dallas (TX), USA, Nov. 3-5 2014

[15] D. Yakimets, G. Eneman, P. Schuddinck, T.H. Bao and Al. "Vertical GAAFETs for the Ultimate CMOS Scaling", *IEEE Trans. Electr. Dev.*, vol. 62, pp. 1433-1439, 2015

[16] J. Shi, M. Li, M. Rahman, S. Khasanvis, C. A. Moritz, "NP-Dynamic Skybridge: A Fine-Grained 3D IC Technology with NP-Dynamic Logic", *IEEE Trans. Emerg. Topics Comput.*, vol. 5, pp. 286-299, 2017

Automatic Timing Closure
for Relative Timed Designs

Tannu Sharma and Kenneth S. Stevens
Email: kstevens@ece.utah.edu
Electrical and Computer Engineering, University of Utah

Abstract—**Relative timing is a universal representation of the sequencing property of time. Relative timing can be applied to specify the correctness and performance properties of digital circuits in the form of a set of timing constraints. These constraints are passed to commercial synthesis, place and route, and timing verification tools to optimize the robustness, power, and performance of digital integrated circuits. This paper presents an algorithm for the automatic convergence of a set of relative timing constraints in the synthesis and place and route flows to create a timing closed design. The algorithm can perform timing closure for a design starting with zero maximum delay targets. This tool produces improved results compared to a start point with delay values close to final closed values.**

I. INTRODUCTION

Time delays are manifested in the components and wires of an integrated circuit (IC). These delays are evaluated based on a timing path between two points in a circuit, which consists of a sequence of components a signal must pass through. They dictate the robustness, performance, and power of a system. Static timing analysis is employed to evaluate and optimize delays and to close timing during synthesis and layout [1].

The traditional techniques employed by commercial EDA tools are insufficient to close timing on a relative timed (RT) design. RT timing paths can be cyclical, and may be controlled by state bits in sequential logic implemented as combinational gates with feedback. To improve quality of relative timed designs, an engine that understands relative timing is required to obtain delay target value and sign-off timing in the current commercial framework.

A. Relative Timing

Relative timing (RT) is a universal representation of the sequencing property of time [2]. Sequencing in the time domain is a common correctness requirement used in integrated circuits. An example of a critical well known correctness requirement in the time domain is the storing of data in a flip-flop. The data must arrive at a flop earlier than the clock.

The simplified universal specification of a relative timing constraint is shown in Eqn. 1 [2]. There must be a common timing start point called a point of divergence (pod) that has causal paths to two timing endpoints called points of convergence (poc_0, poc_1). To ensure the delay on the early path is always less than the delay of the late path requires taking the maximum delay from pod to poc_0 (plus margin m) and

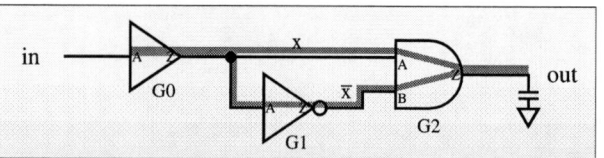

Fig. 1: Pulse generator circuit. The maximum delay path from the rising edge of in (pod) to the rising edge of out (poc_0) of Eqn. 2 is highlighted in red, and the minimum delay path from the rising edge of in (pod) to the falling edge of out (poc_1) is highlighted in blue. The capacitor models wire and gate capacitance of a latch array.

the minimum delay from pod to poc_1. RT expressions employ unbounded delays, and thus are a property of the circuit structure. Therefore, RT constraints are agnostic to specific implementation details of a design that affect circuit delays such as technology node, device sizes, or standard cell layout. Specific path delays are not known until design instantiation and timing closure.

$$\text{pod} \mapsto \text{poc}_0 + m \prec \text{poc}_1 \qquad (1)$$

Assume one needs to synthesize and verify a circuit that generates a pulse. Such a circuit could be used to pulse clock a latch bank. The circuit in Fig. 1 generates a pulse on the out net upon a rising edge of the in net when proper circuit delays are employed. The delay path through the inverter can be designed to generate the required minimum pulse width. The relative timing constraint (RTC) to correctly realize the pulse generator of Fig. 1 is shown in Eqn. 2. This produces a pulse on net out with a minimum width m. The '+' or '-' appended to a net name indicates a rising or falling transition respectively on the net. The causal path through the circuit to create a rising edge on out is through the A pin of gate G2; the causal path for the falling edge is through pin B.

$$\text{in+} \mapsto \text{out+} + m \prec \text{out-} \qquad (2)$$

A relative timed design can contain as many as several million RT constraints. Many of these paths conflict as maximum and minimum delay path segments can partially or completely overlap [3]. Manually converging timing by modifying the timing constraints of individual paths is not a feasible option for such designs, and it may not be possible to resolve the large number of violating paths through mere post-layout ECO [4]. As such, an automated aid to produce timing closed designs is required.

978-1-7281-5410-7/20 $31.00 © 2020 IEEE

The timing on a relative timed design is complex, where the optimization of one path may affect timing of other paths (associated or non-associated). The entire problem is an interaction of non-convex optimization algorithms across often competing timing path constraints. To model the delay target value for each path, the algorithms consider device sizes, drive strength, transition capacitance, fanout, derating factors, EDA uncertainty along with the affects of other paths on a path.

The key contribution of this paper describes a heuristic based timing closure (HBTC) CAD tool that automatically generates functional delay targets for a complete RT constraint set. The algorithm updates relative timing targets based on the results of synthesis or place and route EDA tools to produce a design with high quality of results in terms of circuit power, performance, and area (PPA). The goal of the HBTC algorithm is to produce high quality PPA in as few a number of EDA tool iterations as possible because EDA tools such as Design Compiler (DC) and Integrated Circuit Compiler (ICC) are in the inner loop. The iteration terminates when there is no negative slack on any relative timing constraint.

Iterations start with a set of RTCs where delay targets have been assigned to timing constraints (such as y in Eqn. 3). The HBTC engine presented here is capable of generating a timing closed design with arbitrary initial delay targets, including initial maximum delay targets as zero (0 ps) (e.g. $y = 0$ is the starting condition). The tool is usually able to produce a completely closed set of constraints with no negative slack violations.

The heuristic based timing closure tool presented here is evaluated on a few designs for run time, power, performance, and design robustness at different technology nodes. Various types of RT constraints are employed in these designs. Pipelined designs employ the bundle data design style with handshake controllers [5]. In general fully automated timing closure often shows better results than designs which have manually been brought to near convergence.

II. BACKGROUND AND RELATED WORK

A. Specifying RT Constraints

Relative timing constraints can be mapped to Synopsys design constraints (sdc) to drive timing verification and timing driven synthesis and place and route [6], [7]. An example of a relative timing constraint expressed in sdc format that specifies timing of the pulse circuit of Fig. 1 with the RTC of Eqn. 2 is shown in Eqn. 3 and 4. The early arriving path – with a maximum constrained delay of y – is highlighted in red in Fig. 1. The late arriving path – with a minimum delay target of z – is highlighted in blue. RT constraints assume causal paths, but static timing only employs net connectivity. Therefore, when mapping the early and late RTC paths to sdc format the causal paths through this circuit must explicitly specified with the -through commands.

$$\begin{aligned} &\texttt{set\_max\_delay } y \texttt{ -rise\_from in} \\ &\qquad \texttt{-rise\_through G2/A -rise\_to out} \end{aligned} \quad (3)$$

$$\begin{aligned} &\texttt{set\_min\_delay } z \texttt{ -rise\_from in} \\ &\qquad \texttt{-fall\_through G2/B -fall\_to out} \end{aligned} \quad (4)$$

Specific delay values must be assigned to RTC paths for synthesis and evaluation with tools such as DC, ICC, and PrimeTime. These are the y and z values in Eqn. 3 and 4. Given a specific timing delay target for y and z, the synthesis and place and route (PnR) tools will optimize the design, attempting to create path delays that meet the path constraints.

Assume $y = 0.100$ ns and $z = 0.200$ ns in Eqn. 3 and 4. If the synthesis and PnR tools are able to create a design without negative slack for these sdc constraints, then the margin m, and thus the pulse width, will be at least 100 ps.

We assume that an sdc representation has been created for the design for which timing closure is to be applied. The relative timed design flow consists of an independent but related flow to characterize a set of sequential or timed design macros and also to implement these macros in a system design. The characterized macros contain sdc representations of relative timing constraints to direct the timing driven optimization of commercial EDA tools. These constraints are used by HBTC that is integrated with the commercial EDA tools to verify and/or update the target values.

B. Relative Timed Circuits

Relative timed circuits are an interesting class of circuits because of their ability to leverage the advantages of both clocked and asynchronous design styles. They can also provide a significant improvement over power, performance, and area than a clocked design.

An implementation of a C-Element with combinational feedback is shown in Fig. 2. The output c asserts high when both inputs are high, and goes low when both inputs are low [5]. The [c (ac)'] cycle remembers that a has risen, and the [c (bc)'] cycle remembers that b has risen. The set of RT constraints that make the design conform to its specification is given in Eqn. 5–8. Eqn. 5 and 7 ensure that the circuit remembers that a+ occurred by asserting the [c+ (ac)'-] cycle before the environment observes c+ and lowers a- (path [c+ a-]), and before b- unasserts cycle [c+ (bc)'-] through timing path [c+ b- (bc)'+]. The second symmetrical pair of RTCs enforces the same constraints on cycle [c (bc)'].

$$c+ \mapsto (ac)\text{'-} + m \prec a\text{-} \quad (5) \qquad c+ \mapsto (ac)\text{'-} + m \prec (bc)\text{'+} \quad (7)$$

$$c+ \mapsto (bc)\text{'-} + m \prec b\text{-} \quad (6) \qquad c+ \mapsto (bc)\text{'-} + m \prec (ac)\text{'+} \quad (8)$$

In order to employ commercial EDA tools, combinational loops in the C-element circuit are cut using sdc commands

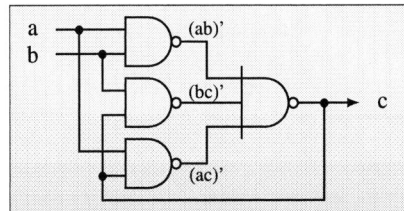

Fig. 2: C-Element NAND Implementation.

978-1-7281-5410-7/20 $31.00 © 2020 IEEE

`set_disable_timing` to create a DAG. This DAG must preserve RT paths specified in the `sdc` constraint file.

C. Related Work

Much work has been invested to develop CAD to support asynchronous design as a viable design choice [8], [9], [10], [11]. Also, there has been work to advance timing flows and apply clocked CAD to timed asynchronous methodologies [12], [13], [14], [15], [16]. The absolute timing analysis tool for asynchronous designs like ATACS does well on small designs. Its extension was limited for complex designs. The ACDC synthesis flow for RT designs works with commercial EDA framework, however, the flow is also suitable for small subsystems [14]. This flow doesn't ensure convergence or optimization for better PPA. To overcome the gaps of conventional modeling, the timing loops are systematically removed without disabling timing arcs. The flow works well for WCHB QDI asynchronous circuits within traditional EDA flow [16]. An iterative approach to converge timing on a RT design by gradually increasing delay on failing paths has proven effective to sign-off timing on complex designs [15]. However, it also suggests incremental ECO runs to overcome non-convergence on designs with large number of failing RT constraints.

The iterative approach presented in this paper is applicable for both small and large designs with millions of RT constraints. It employs cycle cutting to create a DAG to perform timing analysis using conventional EDA. To converge timing on violating paths, the delay targets are increased. Min-delay path targets in an RTC are also increased when necessary to ensure that they remain greater than the max-delay path plus the assigned margin when the associated max-path delay has been increased. In addition, optimizations exist to reduce power consumption by relaxing timing to reduce the drive strength on some paths.

III. LANGUAGE

This work leverages commercial tools and languages whenever available. However there are some properties of a RT constraint which are not supported by modern circuit design specification languages such as `sdc`. Thus we have created a simple language to represent relative timing constraint relationships that can be embedded into an `sdc` constraint set.

While we can specify and constraint independent paths in a circuit as done in Eqn. 3 and 4, there are no semantics in the `sdc` specification language to express relationships between constraints. Thus we have added a `#margin` pragma to the language. This identifies the two timing paths in an RTC, as well as their margin of separation. The semantics are:

$$\texttt{\#margin} \;\; \langle\langle \texttt{margin} \rangle\rangle \;\; \langle\langle \texttt{maxpath} \rangle\rangle, \; \langle\langle \texttt{minpath} \rangle\rangle; \quad (9)$$

An example of a complete RTC constraint for the pulse circuit of Fig. 1 expressed in our extended `sdc` notation is shown in Eqn. 10.

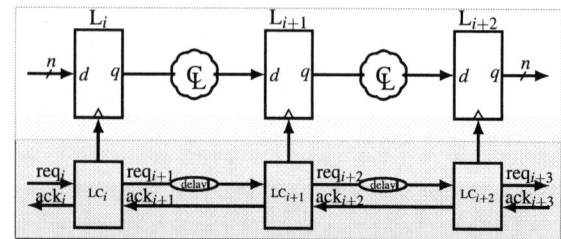

Fig. 3: Timed bundled data handshake design.

```
set_max_delay y -from in -through NAND/A -to out
set_min_delay z -from in -through NAND/B -to out
#margin m -from in -through NAND/A -to out , \
        -from in -through NAND/B -to out ;
```
$\qquad(10)$

Note that EDA tools optimize against specific delay values, rather than relative values. When mapping to a specific technology node or design implementation, the path delays y and z in Eqn. 10 must be replaced with specific values. For the circuit to be properly specified, $y + m < z$. Timing closure consists of adjusting the values of y and z such that each path has no negative slack, and $y + m < z$. In the current implementation of HBTC, the margin m is specified by the designer and remains constant.

When constraints such as those expressed in Eqn. 10 are passed to a commercial tool, a specific delay value is required. Specific timing values are therefore provided for each `sdc` constraint for timing closure. HBTC will iterate across the design until the maximum delay plus a margin in an RTC constraint is less than the minimum delay. This is achieved by modifying the delay variables passed to each tool until the synthesis or place and route tool are able to create a design that meets all of the margins. Because of this, nearly any initial maximum delay value is okay, including zero delay targets.

Many designs have the best PPA quality of results when multiple delay targets are equivalent or optimized in concert with each other. For example, a linear pipeline may be best optimized when each pipeline stage has the same cycle time targets. Identical delay targets can be achieved by using a tcl variable to identify a delay or wildcards to specify timing path end points. For example, delay z in Eqn. 10 can be expressed as a tcl variable, such as `$pd0`, rather than a specific scalar value. All RT constraints using a similar delay target are updated concurrently. The constraint in Eqn. 11 is an example that employs both a tcl variable and wild cards to create uniform delays between pipeline controllers.

```
set pd0 0.200
set_min_delay $pd0 -from pipe/*/rr -to pipe/*/lr
```
$\qquad(11)$

RT constraints are associated to each module master in an object oriented fashion. For example, each linear controller (LC) in the design of Fig. 3 will contain a set of RT constraints that are used to direct EDA tools in performing timing driven synthesis and place and route. However, some timing endpoints may lie outside of the LC controller itself – such as as the `d` and `clk` ports of a flip-flop. We have adopted a language extension to express endpoints outside of the module that contains the relative timed point of divergence (pod) [17]. Endpoints outside of the current module are referenced relative to adjacent pipeline stages upstream or downstream from the standard direction of the data flow.

IV. ALGORITHM

The EDA timing synthesis and PnR optimization algorithms do not result in a convex search space when provided a set of relative timing constraints. A change in any RT constraint in the system can trigger a change in the delay of associated or non-associated RT constraint path, be the min or max delays.

A number of different algorithmic approaches have been tried to converge timing closure on a set of RT constraints. Gradient descent approaches were explored during the implementation stage. The Newton Raphson method and Brent algorithm were implemented to calculate delay target values for each path. The cost function evaluates delay target value for each path while optimizing the performance of the design. Newton Raphson results are too dependent on the initial value selected, whereas with Brent algorithm the range has to be known. Defining the values between zero and a large number wasn't sufficient to obtain a value for which that design would achieve timing convergence.

We were not able to converge timing on a relative timed circuits using these convex optimization techniques. We collaborated with mathematicians to find an algorithm to model this problem. However, none was obtained before we expanded to implement the greedy heuristic algorithm described here.

A. Heuristic Based Timing Closure

A greedy search algorithm, presented as Algorithm 1, has been employed to quickly converge timing on a design with good PPA. EDA tools must support the synthesis and optimization of minimum delay constraints because the delay of the min-delay path must be larger than the delay of the max-delay path in an RT constraint.

The heuristic algorithm is based on two assumptions. (a) The delay target of a maximum delay path can be increased until the tools can implement a design where the path meets timing. (b) We can maintain fidelity of a relative timing constraint by adding delay to the late path (which is the minimum delay path) in order to ensure that the margin of the relative timing constraint continues to hold.

The heuristic based search engine modifies a subset of the design's timing constraints by monotonically increasing max and min delay constraints such that the new constraint set has a higher probability of reaching a timing converged design. This iteration continues until the circuit meets timing or no progress is made. It is possible to create designs which have no consistent solution, such as designs where a min-delay path with a large delay is entirely covered by a max-delay path with a small delay [3]. In such circumstances, there is no solution.

Based on this approach, one could very quickly create a timing closed design by directly applying path delay results of synthesis or place and route to each delay target. However, the PPA of the design will be poor. Instead, the heuristic algorithm

slowly "cools" the design until no outliers remain. HBTC is also capable of performing a second iteration of the algorithm that evaluates and relaxes timing on paths with large device sizes as a power saving optimization.

The commercial EDA tools, in general, do not directly support minimum delay constraints as a part of their central timing driven optimizations; the tools generally provide "min-delay fixing". Path delays are padded to fix min delay violations after the design has been optimized for maximum delays. Likewise, this HBTC tool does not repair minimum delay violations until after the maximum delay violations have all been met. At this point both minimum and maximum timing violations will be concurrently repaired. However, when a max-delay target is increased the associated min-delay path in the RTC will be updated to ensure that the RTC margin holds.

This algorithmic approach provides RT designs that better meets performance targets given the current EDA optimization algorithms. However, designer intervention may still be required to fully close a design as timing and routeability are significantly influenced by area and design density, which is not adjusted as part of this algorithm.

Algorithm 1 Heuristics Based Algorithm for each path.

1: $checkMin \leftarrow false, failures = \{0,0,0\}, W_p \leftarrow 0.8, D_m \leftarrow 100ps$
2: **while** 1 **do**
3:    $failures[2] \leftarrow failures[1], failures[1] \leftarrow failures[0], failures[0] \leftarrow 0$
4:    Run Synthesis or PnR
5:    **begin forall** $n \in$ max $\cup$ min timing constraints   ▷ Check all paths
6:       $s_n \leftarrow$ slack(n)
7:       $t_n \leftarrow$ getTargetDelay(n) ▷ Get path slack
8:       **if** (isMaxPath$(n) \lor checkMin) \land s_n < 0$ **then**
9:          ▷ Update target delay on max paths or all paths with negative slack
10:          $t_n \leftarrow t_n +$ min$((0 - s_n) \times W_p, D_m)$ ▷ Increase path target delay
11:          setDelayTarget(n, t_n)
12:          $failures[0] \leftarrow failures[0] + 1$ ▷ Increment failed path count
13:          $t_m \leftarrow$ getTargetDelay(minPath(n))
14:          **if** (isMaxPath$(n) \land t_m - t_n <$ margin(n)) **then**
15:             ▷ RTC margin doesn't hold, fix to retain RTC margin
16:             $t_m \leftarrow t_n +$ margin(n)
17:             setDelayTarget(minPath$(n), t_m$)   ▷ increase min path target
18:          **end if**
19:       **end if**
20:    **end forall**
21:    **if** $failures[0] = 0 \land checkMin = false$ **then**
22:          ▷ No max path violations, now check min paths too
23:       $checkMin \leftarrow true$
24:       goto 5
25:    **end if**
26:    **if** $failures[0] = 0 \lor (failures[0] = failures[1] = failures[2])$ **then**
27:       Break   ▷ Exit on no negative slack or non-convergence
28:    **end if**
29: **end while**

B. Implementation

Algorithm 1 summarizes HBTC where n is the index number of a timing path, $failures$ is an integer array identifying the number of paths with negative slack in the last three runs, $checkMin$ controls whether minimum delay paths are evaluated, and W_p and D_m are user defined values used to control convergence time and quality of results of the algorithm. Their values vary based on design type and technology node. Function isMaxPath(n) returns $true$ if path number n is a maximum delay path, delay(n) returns the maximum delay for early RTC path n and minimum delay if n is a late path, minPath(n) returns the path index number of the minimum delay path associated with maximum delay path n in a relative timing constraint, and margin(n) returns the margin between the two paths of a RTC for which n is one of the two paths. Slack(n) returns the slack of path n, and the functions getTargetDelay(n) and setDelayTarget(n, t) retrieves the delay target for the RTC associated with delay path n, or sets it to value t.

Each path n has delay target t_n which is initially assigned by the user (e.g. y in Eqn. 3). In the case of negative slack, delays are updated and the margin between the minimum and maximum delay paths of an RTC are maintained. To fix a path with negative slack, the target delay is updated based on the size of the negative slack and the convergence values W_p and D_m. The delays for a single path between synthesis runs can vary greatly, particularly when some paths are poorly placed. Converging too quickly allows the algorithm to settle to a poor local minimum. By slowly converging to a closed solution, poor placement is significantly reduced and PPA improves. In Algorithm 1 the target delay will be updated by either 80% of the negative slack, or 100ps, whichever is smaller. The cost is increased iterations and run time.

The results of the last three iterations are compared to detect non-convergence (normally related to the min delay constraints). If they are equal, no progress has been made toward convergence and the program terminates with paths that have negative slack.

Several extensions to the algorithm have been implemented. To bypass non-convergence, margin relaxation is performed, where margin for critical paths is raised to accommodate variations and timing. Timing relaxation can be performed to improve power based on cell size on critical paths. Timing targets for multiple paths can be shared with a single target variable. This can be used to mitigate cycle time variation between pipeline stages and to reduce tool run times.

A design can have thousands of relative timing constraints. Thus, employing an SQL database to search for and update RT constraints and delay targets substantially improves the run time performance of HBTC. The constraint data for each design is managed under delay and margin tables, with separate tables for both synthesis and physical design runs. The delay table contains rows for both minimum and maximum delay constraints. HBTC verifies constraint data for missing maximum or minimum delay paths entries and duplicate constraint entries.

C. Timing Violations After Timing Closure

HBTC assumes that synthesis and PnR is performed by the commercial EDA tools in a single pass. This requires representing the circuit as a directed acyclic graph (DAG) [7]. Many relative timed designs will have sequential circuits built from combinational gates similar to the circuit in Fig. 2. In such designs, full timing paths of an RTC will often contain cycles. The process of creating a DAG representation of cyclic timing paths will result in an incomplete timing

path specification for these RTCs. Therefore, when the HBTC tool closes timing, it is doing so on a segment of the full RTC timing path. In such cases a closed design, with no negative slack on the path segments, can still fail full path timing because the full cyclic path cannot be provided to the commercial EDA tools. Also in these cases, simulation through ModelSim may identify the violation, as reflected by sim-errors column in Table I. The means of formally identifying full-path cyclical timing verification are outside the scope of this work, but will be included in future implementations.

V. EVALUATION

This paper reports on applying the HBTC tool to converge timing on designs with varying complexity at the 40 nm cmos technology node. These include a watchdog timer, wakeup timer, timer, FFT_64, and peripheral system design which contains the timers, a network-on-chip, and scratchpad register file. The three timers are all "clocked" designs that have been optimized using relative timed techniques to significantly reduce power. The FFT-64 and network-on-chip used in the peripherals design are bundled data asynchronous handshake designs. The small designs contain approximately 100 RT constraints, with the large designs containing approximately 50,000 RT constraints and over 30,000 pipeline latches.

Synopsys Design Compiler is used for synthesis, ModelSim for simulation, and PrimeTime for path delays and power analysis. The results are compared for power, performance, run time, and simulation errors that occur after the timing closure run. Two starting points are employed, one with zero maximum delay targets and a second set using user-defined delay targets that are close to a timing closed solution.

VI. RESULTS

Table I shows results obtained in two scenarios. The first set is where a designer specified critical path delays and manually drove the timing targets to get close to a closed solution. The second set allows HBTC to fully close the design by assigning zero delay targets to all max delay paths. These two approaches are compared with manual design effort used as the baseline. This shows that HBTC tool can produce faster and lower power designs when no targets are provided. However, this comes at a considerable run time cost for larger designs. CPU run time for the HBTC algorithm is less than 2 seconds for all examples.

The number of iterations is critical for run time, because synthesis or place and route is in the inner loop. When starting with timing constraints dictating desired performance, the small watchdog timer converged in two cycles. The more complicated peripheral and FFT-64 designs required 8 or 9 cycles to converge. Many more iterations are required when HBTC closes a design when no timing targets are provided. The timer designs took between 10 and 25 iterations to converge. The iterations for the peripherals more than tripled. The FFT-64 design employed tcl variables for the zero delay start point to help identify frequency domains, so that design converged in 18 iterations. Most of the designs simulated

TABLE I: HBTC results for two configurations and their comparison with zero-max delay target design as baseline.

Designs	Close to Convergence Target						Zero Max Delay Target						Compare – Zero / Close			
	child process run time (s)	No. of iterations	Power (uW)	Energy (pJ)	sim. errors	$e\tau^2$	child process run time (s)	No. of iterations	Power (uW)	Energy (pJ)	sim. errors	$e\tau^2$	Run time (x)	Power	Energy	$e\tau^2$
Watchdog	1,646.48	2	56.88	71.19	1	11.15	1,735.86	20	52.37	64.87	0	9.95	1.05	0.92	0.91	0.89
Wakeup	634.84	7	54.56	30.58	0	0.96	1,972.70	25	52.53	28.80	0	0.87	3.11	0.96	0.94	0.91
Timer	842.65	3	41.38	87.52	2	40.64	2,073.17	10	39.70	85.55	2	38.27	2.46	0.96	0.98	0.94
Peripherals	2,811.87	9	135.80	147.90	1	175.40	25,230.97	33	132.30	143.10	3	167.41	8.97	0.97	0.97	0.95
FFT-64	1,972.70	8	12,100	20.43	0	58.21	32,758.92	18	12,400	20.93	0	59.64	16.61	1.02	1.02	1.02

without timing violations. To completely close a design, a full path timing validation must also be applied which supports cyclic timing paths.

Most of the designs demonstrated improved energy efficiency and performance when the HBTC tool was allowed to perform timing closure from zero delay targets. In general, the smaller designs provided greater improvements than the larger designs. The FFT-64 design is a multirate design, and used a different design target approach as the zero max delay target design employed tcl variables for common frequency targets which reduced the iterations. The user specified frequencies helped the design PPA. The number of iterations required to timing close a system design like peripherals is reduced by employing HBTC on individual hierarchical blocks.

The HBTC tool provided a rapid approach to converging timing for designs which otherwise could only be closed manually at great time and effort. It is well integrated in commercial EDA flow and scalable to different process technologies that could be represented in relative timed methodology.

VII. CONCLUSION

Relative timing is a method for specifying and optimizing digital integrated circuits. A heuristic algorithm was developed to automate timing closure of a complex set relative timing constraints. The HBTC tool provides a fully automated approach to converge timing for designs which otherwise could only be closed by hand.

The heuristic based timing closure tool is implemented and applied to designs reported in this paper. The tool provides reasonable run times with a synthesis tool, such as Design Compiler, in its inner loop. The tool can begin the closure process agnostic of technology node delay information by using zero delay starting point targets. Starting from this point, this tool produced results for several designs showing 5% power reduction and about 8% better performance in our design suite compared to designs that started with a near-converged values provided by designers. This relative timing closure tool is designed to work well with the commercial EDA tools and is well integrated with the design flow.

VIII. ACKNOWLEDGMENTS

This material is based upon work supported by the National Science Foundation under Grant No. 1111533 and a grant from Granite Mountain Technologies, in which Kenneth S. Stevens declares financial interest.

REFERENCES

[1] R. Nair, C. L. Berman, P. S. Hauge, and E. J. Yoffa, "Generation of Performance Constraints for Layout," *IEEE Transactions on Computer-Aided Design*, vol. 8, no. 8, pp. 860–874, Aug 1989.

[2] K. S. Stevens, R. Ginosar, and S. Rotem, "Relative Timing," *IEEE Transactions on Very Large Scale Integration (VLSI) Systems*, vol. 1, no. 11, pp. 129–140, Feb. 2003.

[3] J. V. Manoranjan and K. S. Stevens, "Qualifying Relative Timing Constraints for Asynchronous Circuits," in *International Symposium on Asynchronous Circuits and Systems*, May 2016, pp. 91–98.

[4] T. Sharma and K. S. Stevens, "Physical Design Variation in Relative Timed Asynchronous Circuits," in *IEEE Computer Society Annual Symposium on VLSI (ISVLSI)*, July 2017, pp. 278–283.

[5] I. E. Sutherland, "Micropipelines," *Communications of the ACM*, vol. 32, no. 6, pp. 720–738, June 1989.

[6] G. Gimenez, A. Cherkaoui, G. Cogniard, and L. Fesquet, "Static Timing Analysis of Asynchronous Bundled-Data Circuits," in *24th International Symposium on Asynchronous Circuits and Systems*. IEEE, May 2018, pp. 110–118.

[7] K. S. Stevens, Y. Xu, and V. Vij, "Characterization of Asynchronous Templates for Integration into Clocked CAD Flows," in *15th International Symposium on Asynchronous Circuits and Systems*. IEEE, May 2009, pp. 151–161.

[8] F. Burns, D. Shang, A. Koelmans, and A. Yakovlev, "An Asynchronous Synthesis Toolset using Verilog," in *Design, Automation and Test in Europe (DATE)*, vol. 1, Feb 2004, pp. 724–725.

[9] C. J. Myers, T. G. Rokicki, and T. H.-Y. Ming, "Automatic Synthesis of Gate-Level Timed Circuits with Choice," in *16th Conference on Advanced Research in VLSI*, March 1995, pp. 42–58.

[10] R. Kol, R. Ginosar, and G. Samuel, "Statechart Methodology for the Design, Validation, and Synthesis of Large Scale Asynchronous Systems," in *2nd International Symposium on Advanced Research in Asynchronous Circuits and Systems*, March 1996, pp. 164–174.

[11] M.-D. Shieh, J.-M. Horng, M.-H. Sheu, and Y.-C. Hsu, "A CAD System for Automatic Synthesis of Generalized Asynchronous Circuits," in *International Symposium on Circuits and Systems (ISCAS)*, vol. 4. IEEE, May 1996, pp. 818–821.

[12] K. S. Stevens, S. Rotem, S. M. Burns, J. Cortadella, R. Ginosar, M. Kishinevsky, and M. Roncken, "CAD Directions for High Performance Asynchronous Circuits," in *Proceedings of the Digital Automation Conference (DAC99)*. IEEE, June 1999, pp. 116–121.

[13] W. Belluomini and C. J. Myers, "Efficient Timing Analysis Algorithms for Timed State Space Exploration," in *3rd International Symposium on Advanced Research in Asynchronous Circuits and Systems*, April 1997, pp. 88–100.

[14] M. Bibiluka, M. T. Moreira, and N. L. V. Calzans, "A Bundled-Data Asynchronous Circuit Synthesis Flow Using a Commercial EDA Framework," in *Euromicro Conference on Digital System Design*, Aug 2015, pp. 79–86.

[15] G. Miorandi, M. Balboni, S. M. Nowick, and D. Bertozzi, "Accurate Assessment of Bundled-Data Asynchronous NoCs Enabled by a Predictable and Efficient Hierarchical Synthesis Flow," in *23rd International Symposium on Asynchronous Circuits and Systems*. IEEE, May 2017, pp. 10–17.

[16] Y. Thonnart, E. Beigné, and P. Vivet, "A Pseudo-Synchronous Implementation Flow for WCHB QDI Asynchronous Circuits," in *International Symposium on Asynchronous Circuits and Systems*. IEEE, May 2012, pp. 73–80.

[17] E. Quist, P. Beerel, and K. S. Stevens, "Enhanced SDC Support for Relative Timing Designs," in *Digital Automation Conference*. IEEE/ACM, July 2009, user Track Poster.

978-1-7281-5410-7/20 $31.00 © 2020 IEEE

Mining Hyperproperties from Behavioral Traces

Mayank Rawat   Sujit Kumar Muduli   Pramod Subramanyan*

Indian Institute of Technology, Kanpur

{*mayankr, smuduli, spramod*}@cse.iitk.ac.in

This paper is dedicated to the loving memory of Pramod Subramanyan* (1984 - 2020),
our academic advisor who guided and motivated us for this work.

Abstract—**Many important specifications of hardware and software systems, such as secure information flow and determinism are expressible only as *hyperproperties*. In contrast to the well-studied class of trace properties, which specify sets of valid runs (aka traces) of a system, hyperproperties can specify relations that must hold between the traces of a system. While hyperproperties have many applications, primarily in security verification, coming up with hyperproperties for SoC validation is challenging. In this paper, we work toward addressing this challenge by introducing a framework for mining hyperproperties from execution traces of SoC designs. We introduce novel algorithms based on coverage-guided fuzzing that enable the generation of good input traces for the hyperproperty miner. We also present novel optimistic and pessimistic semantics for Hyper Linear Temporal Logic (HyperLTL) that enable principled evaluation of HyperLTL formulas over finite traces. Finally, we propose algorithms for scalably evaluating non-trivial satisfaction of candidate hyperproperties on sets of traces. Experiments on a small but realistic SoC design show the framework is effective in identifying useful hyperproperties.**

I. INTRODUCTION

An important methodology for the verification and validation of modern systems-on-chip (SoC) platforms has been assertion-based verification (ABV). ABV enables SoC designers to declaratively specify properties that the design must satisfy. These properties can be verified/validated using formal techniques such as bounded and unbounded model checking, semi-formal techniques such as concolic execution as well as simulation-based validation. Commonly used property specification languages include SystemVerilog Assertions (SVA) [4] and the Property Specification Language (PSL) [12]. These specification languages are based on the underlying formalism of linear temporal logic (LTL) [19].[1] Unfortunately, several important classes of requirements, including information flow properties like confidentiality and integrity *cannot* be specified in formalisms based on LTL/CTL [18]. This exacerbates the critical challenge of SoC security verification.

Recent efforts, both academic [6, 11, 24] and commercial [1, 23], are tackling the security verification problem by introducing novel specification languages that can express information flow assertions, based on the theory of hyperproperties [7]. However, two important bottlenecks in using these specification languages are: (i) coming up with a meaningful set of security-related assertions, and (ii) keeping this set of assertions updated as the design evolves. In the

context of traditional assertions in LTL-based formalisms like SVA/PSL, analogous problems are mitigated via *assertion mining*: techniques for algorithmically identifying likely assertions satisfied by the design by the examination of behavioral simulation traces [8–10, 13–15, 17, 25, 26, 29]. However, existing methods cannot mine hyperproperties and extending them to hyperproperties poses unique challenges (discussed later in this section). In this paper, we address this gap in the literature by proposing novel techniques for mining hyperproperties from behavioral traces of SoC designs.

A. Hyperproperties and their Applications

LTL-based formalisms can only specify *trace properties* [2]. Roughly speaking, a trace property is a set of good traces and a system satisfies this trace property if the set of traces of the system is a subset of this set of good traces.

1) Beyond trace properties: Consider the property of determinism. Determinism requires that the output of a module be a deterministic function of only its input(s). Assume we have a hardware module whose input is the 32-bit signal x and the output is a 16-bit signal y that is produced one cycle after the input is given. We would like to show that y is a function of only x. If we are given a trace where the input is $x = 1$ and the output is $y = 2$, can we determine whether the module behaved deterministically or not in this trace? We cannot! A counterexample to determinism requires two executions or two traces with the same inputs and different outputs. So, if we had two traces, both with the same input $x = 1$ but the outputs being $y = 2$ in one trace and $y = 3$ in the other, this pair of traces would be a counterexample to determinism.

The above property can be expressed in the logic HyperLTL, introduced by Clarkson et al. [6] as follows: $\forall \pi_1. \forall \pi_2.\ \mathbf{G}\left((x_{\pi_1} = x_{\pi_2}) \rightarrow \mathbf{X}\left(y_{\pi_1} = y_{\pi_2}\right)\right)$. We will provide a more detailed introduction to HyperLTL in § III-B; we informally describe its meaning here. This property is satisfied by a system if for every pair of traces of the system, named here π_1 and π_2, if the value of x in those traces is equal at some cycle i, then the value of the variable y is also equal at cycle $i + 1$. This property is a relation over traces and not a set of traces. It is an example of a *hyperproperty* [7].

2) Information Flow Hyperproperties: A particularly important class of hyperproperties are secure information flow properties which state that information must not flow from a specified source to a specified destination [1, 11, 23, 24].

---

[1]PSL includes an optional extension based on computation tree logic [5].

978-1-7281-5410-7/20 $31.00 © 2020 IEEE

Information flow properties can capture both confidentiality (e.g. secret key must flow to output) and integrity (e.g. firmware input must not influence register).

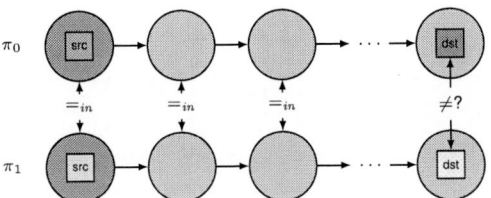

Fig. 1: Information flow property.

To check whether information can flow from a source component to a destination component, it is sufficient to check the following. Consider two almost-identical executions (i.e. traces) of system state where all registers except those of the source component have the same values in the initial state. The two source components have arbitrarily different initial values. Now suppose we ask a model checker whether these two different initial states can result in different values of the registers in the destination component when given the same input in both traces. If the answer is in the affirmative, that means information can flow from the source to the destination component. If the answer is negative, that means that values at the source are effectively indistinguishable at the destination, so no information flow exists. This is also a hyperproperty that corresponds to a symmetric binary relation over traces. It can be expressed in HyperLTL as $\forall \pi_1. \forall \pi_2. (\sigma_{\pi_1} = \sigma_{\pi_2} \wedge \mathbf{G} (\iota_{\pi_1} = \iota_{\pi_2})) \rightarrow \mathbf{G} (dst_{\pi_1} = dst_{\pi_2})$, where σ refers to all the variables in the design except for src. This is illustrated in Figure 1. Like determinism, secure information flow is not a trace property.

B. Challenges in Mining Hyperproperties

As we see above, some important specifications of SoCs are only expressible as hyperproperties. Therefore, it would be desirable to have assertion mining techniques for hyperproperties. However, in comparison to traditional trace properties, there are three challenges in mining hyperproperties.

Hyperproperties are almost always implications; i.e. formulas like $\varphi \rightarrow \psi$ where both φ and ψ are relations between states. If traces for behavioral mining are generated randomly, we may end up with no traces that satisfy φ, so formulas of the form $\varphi \rightarrow \psi$ are satisfied vacuously. Such vacuously satisfied formulas are unlikely to be valid or interesting. This is the first problem in hyperproperty mining: *generating good traces in order to drive the property miner towards likely properties.*

We collect traces of the system by simulating it and capturing how values of a set of interesting variables evolve. Traces collected in such a manner must necessarily be of finite length. However, a tuple of traces satisfying a hyperproperty is defined mathematically in terms of infinite-length traces. As we will demonstrate in § III-B1, the traditional semantics for satisfaction poses problems in determining satisfaction over finite traces. Therefore, defining finite trace satisfaction in a principled manner is another important challenge.

The final challenge is in efficient evaluation of the hyperproperties. Since hyperproperties are k-ary relations over traces, the time taken to check that n traces with maximum length L satisfy such a relation is $O(L \times n^k)$. This can quickly blow-up for long traces or for large sets of traces and efficient techniques are required to ensure that we test only promising hyperproperties, and reject invalid hyperproperties early.

C. An Overview of HYPERMINER

In this paper, we address each of the above challenges and introduce an new framework called HYPERMINER for mining hyperproperties from behavioral traces of SoC designs. Figure 2 shows an overview of the HYPERMINER framework. The framework consists of two main parts: (a) the tracer, and (b) the miner. The tracer is responsible for generating a set of traces of SoC execution capturing how the values of the variables evolve over time. This set of traces is fed to the miner which uses a formula generator and satisfaction tester to shortlist hyperproperties satisfied by these traces.[2]

D. Contributions of This Paper

This paper makes the following novel contributions.

- We introduce a framework for mining temporal hyperproperties from behavioral traces of SoC designs. To the best of our knowledge, it is the first framework that mines temporal hyperproperties, and the first to mine any kind of hyperproperty in the context of hardware designs.
- We introduce a novel coverage-guided algorithm that generates traces for hyperproperty mining. Our algorithm modifies coverage-guided fuzzing with three novelties: (i) test sketches instead of "seed inputs", (ii) a test mutator to converts sketches into concrete tests, and (iii) novel coverage metrics to help generate meaningful traces.
- Building on the HyperLTL specification language, we introduce novel optimistic and pessimistic semantics for satisfaction of formulas over finite traces. We show the need for these semantics when finding satisfying and violating traces of a candidate hyperproperty.

We evaluate the HYPERMINER framework on a small but realistic SoC consisting of two μ-controller cores, accelerators for AES, SHA, RSA along with an MMU and shared memory. The framework identifies many tens of likely hyperproperties in various modules of the SoC design.

II. TRACE GENERATION IN HYPERMINER

This section describes the tracer component of HYPERMINER, shown in Figure 2(a). Our tracer is based on coverage-guided fuzzing [20, 27] with the following novel modifications: (i) introducing *test sketches* instead of seed inputs, (ii) introducing *test mutators* to concretize sketches, and (iii) system-level *coverage metrics*. A test sketch is essentially a test with "holes". The "filling of holes" is done by the *test mutator*, which takes in a random bitstream from the fuzzer

---

[2]Currently, HYPERMINER only outputs likely hyperproperties. However, it is a straightforward matter to hook-up the output of HYPERMINER to hyperproperty model checking tool like MCHyper [11] to check their validity.

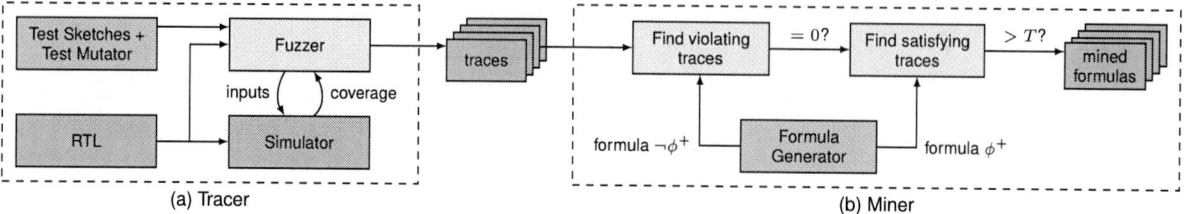

(a) Tracer   (b) Miner

Fig. 2: Overview of the HYPERMINER framework. The color-coding is as follows. Yellow boxes show new reusable components developed as part of this paper. Blue boxes show existing components from SoC designs. Violet components are automatically generated. Cyan boxes show where design-specific inputs are required.

and produces a concrete test without holes. This concrete test is executed using the register-transfer level (RTL) simulator. During execution, the fuzzer measures one or more *coverage* metrics. The coverage metric is an estimate of which parts of the design were exercised by the test. Tests that increase coverage are prioritized for additional mutation.

A. Test Sketches and Test Mutator

We used two kinds of test sketches in our tracing.

a) NOP Sketch: This is a firmware-based test sketch where a firmware program (e.g. one that performs AES encryption) has a sequence of NOP instructions inserted into it. The test mutator concretizes this sketch by replacing the NOPs with an arbitrary sequence of instructions.

b) FSMWriter Sketch: This sketch has finite state machine connected to the SoC interconnect that repeatedly generates write-traffic to a sequence of memory locations (some of which refer to the on-chip memory-mapped I/O). The mutator concretizes this sketch by programming the sequence of address/data accesses made by the FSM.

The NOP Sketch guides the fuzzer towards executing new types of instructions in the microcontroller while the FSMWriter sketch generates new memory and memory-mapped I/O (MMIO) accesses.

B. Coverage Metrics

The coverage metric is used by the fuzzer to evaluate test quality. A good metric is extremely important for effective fuzzing. The default coverage metric used by traditional software fuzzers like `afl-fuzz` and `libFuzzer` is based on basic block coverage, and therefore inapplicable for hardware designs. The hardware fuzzer RFUZZ [16] uses mux-coverage: a measure of how many times select signals were toggled for the multiplexers (muxes) in the design. We found that this metric also did not work well; the fuzzer found very few new paths (i.e. inputs that increased coverage). This is likely because with a good test most muxes in the design are already toggled, so finding a new input that toggles some previously unexercised mux via random mutation is unlikely. Our insight is that system-level coverage metrics that correspond to software-visible events are likely to provide better feedback to the fuzzer. Therefore, we used the following novel coverage metrics in this work.

a) Instruction Bigrams: the number of unique 2-tuples of consecutive instructions executed by the primary μ-controller.

b) Memory Access Bigrams: estimates the number of unique memory access bigrams at the interface to the shared memory space in the SoC (note this includes MMIO).

These metrics were helpful in guiding the fuzzer towards tests that either execute new instructions or make new memory accesses. Results show that this helped the tracer generate traces with more interesting events for mining.

C. Putting it all together

In typical usage, the tracer is run until a certain number of traces are generated. These traces are examined by the miner (shown in Figure 2b) to determine likely hyperproperties that hold on the collected traces. Note both the sketches/mutator and the coverage metric need to work in concert in order to generate good traces. If we have the right sketch but do not have the appropriate coverage metrics, the fuzzer will be unable to distinguish good mutations from bad ones. This will cause it potentially waste a lot of time exploring redundant or meaningless paths. Similarly, a good coverage metric with a poor mutator is also useless because there will not be an easy way for the fuzzer to generate inputs that maximize the coverage metric. Finally, the use of a mutator that modifies only a part of the test, as opposed to one that randomizes the entire test is extremely important for hyperproperty mining. This ensures that most traces are related to one or more other traces – they have the same or similar values for various SoC state variables – thus preventing vacuous satisfaction of the antecedents in conjectured hyperproperties.

III. HYPERPROPERTY MINING

In this section, we first introduce the hyperproperty specification language, its syntax and semantics. We then describe the architecture of the miner component in HYPERMINER.

A. Preliminaries

A transition system M is defined as the tuple $\langle S, S_0, R, L \rangle$ where S is the set of states of the transition system, $S_0 \subseteq S$ is the set of initial states, $R \in S \times S$ is the transition relation. Our definition of the labelling function $L : S^k \to 2^{AP}$ is a little non-standard. L maps k-tuples of states to a set of the atomic propositions AP. If $k = 1$, this corresponds to the standard definition where each state has zero or more atomic propositions associated with it. However, if $k = 2$, then every *pair of states* has zero or more atomic propositions associated with it. The k-tuple labels encode relations between traces.

A trace of the system M is a (finite or infinite) sequence of states $\pi = s_0 s_1 s_2 \ldots s_i \ldots$ such that $s_0 \in S_0$ and for all $i \geq 0$, $(s_i, s_{i+1}) \in R$. We will use the notation π^i to denote the i-th element of a trace. In the above example, $\pi^0 = s_0$, $\pi^2 = s_2$, $\pi^i = s_i$, etc. For a finite-length trace π, its length is given by $len(\pi)$; $len(\pi)$ is undefined for an infinite trace. We will write $\pi[i, \infty]$ for the suffix of the trace π starting from element i. If $len(\pi) = N$, then $len(\pi[i, \infty]) \leq N - i$. The set of all traces of the system M is denoted by Φ_M.

B. HyperLTL

$$\psi \quad ::= \forall \pi.\ \psi \mid \exists \pi.\ \psi \mid \varphi$$
$$\varphi \quad ::= \mathsf{AP}_{\pi_1, \ldots, \pi_k} \mid \neg \varphi \mid \varphi \wedge \varphi \mid \varphi\, \mathbf{U}^{\pm}\, \varphi \mid \mathbf{X}^{\pm}\, \varphi$$
$$\pm \quad ::= + \mid -$$

Fig. 3: HyperLTL Syntax

Figure 3 shows the grammar for the HyperLTL-variant that is considered in this paper. Formulas are required to be in *prenex form* – all quantifiers must appear in the beginning of the formula. We do not support quantifier alternation. For simplicity of presentation, we assume the specification formula is universally quantified. Extending the ideas to existentially-quantified formulas is straightforward. Atomic propositions (APs) apply to k-tuples of traces and the specific traces involved are denoted by the subscript of the AP. The other logical and temporal operators are standard except for the introduction of optimistic (superscript $^+$) and pessimistic (superscript $^-$) versions of each temporal operator.

Given the operators in Figure 3, we can define the usual abbreviations: $\varphi \vee \psi \equiv \neg(\neg \varphi \wedge \neg \psi)$, $\mathbf{F}^{\pm}\, \psi \equiv \text{true}\, \mathbf{U}^{\pm}\, \psi$, etc. Note polarity (optimism/pessimism) is preserved in the above identities. However, negation reverses polarity of optimism/pessimism. For example, $\mathbf{G}^{\pm}\, \psi \equiv \neg \mathbf{F}^{\mp}\, \neg \psi$.

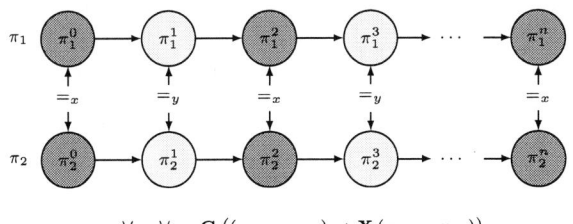

$$\forall \pi_1. \forall \pi_2.\ \mathbf{G}\left((x_{\pi_1} = x_{\pi_2}) \to \mathbf{X}(y_{\pi_1} = y_{\pi_2})\right)$$

Fig. 4: Example traces for the above property.

1) Need for Optimistic and Pessimistic Variants: To see why optimistic and pessimistic variants are required, consider the formula $\forall \pi_1. \forall \pi_2.\ \mathbf{G}\left((x_{\pi_1} = x_{\pi_2}) \to \mathbf{X}(y_{\pi_1} = y_{\pi_2})\right)$ described in § I-A1. The notation $v_{\pi_1} = v_{\pi_2}$ is syntactic sugar for an atomic proposition that holds for all pairs of states in

which the valuation of the variable v are equal. Recall the hyperproperty means that y is a deterministic function of x with a delay of one cycle. Figure 4 shows two traces of length $n + 1$ that appear to satisfy the above formula, but with a catch. The AP $x_{\pi_1} = x_{\pi_2}$ holds at steps 0, 2 and n of the two traces while the AP $y_{\pi_1} = y_{\pi_2}$ holds at steps 1 and 3 of the traces. In the final states of the traces, at step n, we have $x_{\pi_1} = x_{\pi_2}$. Since the trace has been truncated at this point, we cannot determine the value of y at step $n + 1$. Therefore, we cannot determine satisfaction with traditional (Hyper-)LTL. The optimistic operator $\mathbf{G}^+$ allows us to conclude that the formula is indeed satisfied in such scenarios if some extension to the trace exists that will satisfy the formula.

One may wonder why we need pessimistic variants in addition to the optimistic variants. Consider the negation of the above formula: $\exists \pi_1. \exists \pi_2.\ \mathbf{F}\left((x_{\pi_1} = x_{\pi_2}) \wedge \mathbf{X} \neg(y_{\pi_1} = y_{\pi_2})\right)$. If we use optimistic operators, then the same pair of traces in Figure 4 satisfies the negated formula as well! This contradiction must be avoided. *Therefore, we use optimistic operators when searching for satisfying traces and pessimistic variants when searching for counterexamples/violations.*

2) Formal Satisfaction Semantics: As in standard Hyper-LTL [11] without the optimistic/pessimistic variants, the validity judgement of a property ψ by system $M = \langle S, S_0, R, L \rangle$ is defined with respect to a trace assignment $\Pi : \mathcal{V} \to \Phi_M$. Here, $\mathcal{V}$ is a trace variable; recall that Φ_M is the set of traces of the system M. The partial function Π is a mapping from trace variables to traces. We use the notation $\Pi[\pi \mapsto \rho]$ to refer to a trace assignment that is identical to Π except that the variable π maps to trace ρ. We write $\Pi \models_M \psi$ if the system M satisfies the property ψ under the trace assignment Π. We use the notation $\Pi[i, \infty]$ as an abbreviation for the new trace assignment obtained by taking the suffix starting from index i of every trace in Π: $\Pi'(\pi) = \Pi(\pi)[i, \infty]$ for every trace $\pi \in dom(\Pi)$, where $dom(\Pi)$ is the domain of Π. We write $\Pi \not\models_M \psi$ when $\Pi \models_M \psi$ is not satisfied. We write $len(\Pi)$ to refer to the length of the minimum length trace in Π: $len(\Pi) = \min\{len(\pi) \mid \pi \in dom(\Pi)\}$. Given the above definitions, satisfaction semantics are shown in Figure 5. We say that system M satisfies the property ψ, denoted by $M \models \psi$ if $\Pi_\emptyset \models_M \psi$ for the empty trace assignment $\Pi_\emptyset$.

Lemma 1. *Equivalence of optimistic/pessimistic variants: The optimistic and pessimistic variants of the operators coincide in satisfaction for infinite-length traces.*

The proof is by induction on the structure of the formula and relies on the fact that the additional satisfaction introduced by the optimistic variants only applies to finite-length traces. (Recall $len(\pi)$ is undefined for infinite-length traces.)

C. Overview of the Miner in HYPERMINER

We now describe the miner component in Figure 2. It consists of a formula generator which outputs candidate HyperLTL formulas. These are checked on the set of collected traces to: (i) ensure that no violations exist, and (ii) the number of satisfying traces is more than the threshold T.

$$\Pi \models_M \forall \pi.\ \psi \quad \text{iff for all } \rho \in \Phi_M : \Pi[\pi \mapsto \rho] \models_M \psi$$

$$\Pi \models_M \exists \pi.\ \psi \quad \text{iff exists } \rho \in \Phi_M : \Pi[\pi \mapsto \rho] \models_M \psi$$

$$\Pi \models_M \psi\, \mathbf{U}^-\, \varphi \quad \text{iff there exists } i \geq 0 : \Pi[i, \infty] \models_M \varphi$$
$$\text{and for all } 0 \leq j < i : \Pi[j, \infty] \models_M \psi$$

$$\Pi \models_M \psi\, \mathbf{U}^+\, \varphi \quad \text{iff } \Pi \models_M \psi\, \mathbf{U}^-\, \varphi,\ \text{or}$$
$$\text{for all } 0 \leq j < len(\Pi) : \Pi[j, \infty] \models_M \psi$$

$$\Pi \models_M \mathbf{X}^-\, \psi \quad \text{iff } \Pi[1, \infty] \models \psi$$

$$\Pi \models_M \mathbf{X}^+\, \psi \quad \text{iff } \Pi \models \mathbf{X}^-\, \psi \text{ or } len(\Pi[1, \infty]) = 0$$

$$\Pi \models_M a_{\pi_1, \dots \pi_k} \quad \text{iff } a \in L(\Pi(\pi_1)^0, \dots, \Pi(\pi_k)^0)$$

$$\Pi \models_M \neg \psi \quad \text{iff } \Pi \not\models_M \psi$$

$$\Pi \models_M \psi \wedge \varphi \quad \text{iff } \Pi \models_M \psi \text{ and } \Pi \models_M \varphi$$

Fig. 5: Satisfaction semantics for our HyperLTL-variant.

1) Formula Generator: The formula generator produces candidate formulas for further examination. Our current implementation uses template-based enumeration for generating these candidate formulas. We use two templates. The first is of the form $\forall \pi_1.\forall \pi_2.\ \mathbf{G}\,(u_{\pi_1} = u_{\pi_2}) \rightarrow \mathbf{G}\,(v_{\pi_1} = v_{\pi_2})$ where u and v are variables in the design. This is a form of observational determinism, a secure information flow property [28]. The second is a template of the form $\forall \pi_1.\forall \pi_2.\ \mathbf{G}\,(u_{\pi_1} = u_{\pi_2} \rightarrow v_{\pi_1} = v_{\pi_2})$. As before, u and v are variables in the design. This captures a deterministic dependency between two variables in the design. We also enumerated random formulas including the temporal operators $\mathbf{G}$ and $\mathbf{X}$ and variable equalities. We note that extension of this component to support additional templates and/or static/dynamic analyses as proposed in past work (e.g. [9, 17, 25]) is straightforward.

2) Finding Violating and Satisfying Traces: The formula generator outputs a formula in plain HyperLTL without optimistic or pessimistic variants of the operators. The negation of the optimistic variant is checked against the trace set to determine if there are any counterexamples for the formula. If there are none, then the optimistic variant of the original formula is checked to determine if there are more than T traces that non-trivially satisfy it. If so, the formula is output.

Example: Suppose the formula generator outputs $\mathbf{G}\,(x_{\pi_1} = x_{\pi_2}) \rightarrow \mathbf{G}\,(y_{\pi_1} = y_{\pi_2})$. The optimistic version of this formula is: $\mathbf{G}^+\,(x_{\pi_1} = x_{\pi_2}) \rightarrow \mathbf{G}^+\,(y_{\pi_1} = y_{\pi_2})$. This is equivalent to $\mathbf{F}^-\,\neg(x_{\pi_1} = x_{\pi_2}) \vee \mathbf{G}^+\,(y_{\pi_1} = y_{\pi_2})$, and its negation is $\mathbf{G}^+\,(x_{\pi_1} = x_{\pi_2}) \wedge \mathbf{F}^-\,\neg(y_{\pi_1} = y_{\pi_2})$. This negation is checked on the mined traces to determine whether a counterexample exists. If none is found, we count satisfying traces exist for the optimistic variant. If the count is more than the threshold, the formula is shortlisted for output.

IV. EXPERIMENTAL EVALUATION

This section presents our experimental evaluation of HYPERMINER. We present details of the methodology, an overview of the test SoC design and the experimental results.

A. Implementation and Methodology

We implemented the tracer using `afl-fuzz` [27] and Verilator [22] for RTL simulation. Formula generation was implemented in Python while property satisfaction testing was implemented in C++. In keeping with the focus of VLSI-SoC 2020, the HYPERMINER and SoC platform have been open-sourced at https://github.com/c0demag/HyperMiner_SourceCode.

1) SoC Overview: We evaluated HYPERMINER by examining various modules in a small SoC design consisting of two microcontroller cores and accelerators for AES, SHA, modular exponentiation, a memory management unit (MMU), shared memory and I/O devices. The accelerators are controlled using memory-mapped I/O. The SoC RTL description consists of about 16000 lines of Verilog. The firmware used for the test sketches is about 1000 lines of C code.

2) Evaluation Methodology: We collected system-level traces for this SoC and then tried to mine hyperproperties within individual modules. We also provide a comparison with Bach [21], a tool for inferring (non-temporal) hyperproperties in a functional programming language. We report the number of hyperproperties found, the time taken to check satisfaction/violation and briefly discuss some mined properties.

B. Results

Table I summarizes the result of our evaluation. We show a comparison between random mutations and the coverage-guided mutation strategy for generating traces described in § II. We show results for the two hyperproperty templates used in the formula generator as well as random formula enumeration (§ III-C1). For each experiment, we report the number of mined likely hyperproperties (#hp) and the time taken for mining these hyperproperties in seconds.

TABLE I: Summary of Assertion Mining Results.

Module	Random Mutations				Coverage-guided Mutations			
	Template		Random		Template		Random	
	#hp	time	#hp	time	#hp	time	#hp	time
μC	239	561	33	318	**310**	558	**41**	96
AES	39	43	12	52	**42**	115	**33**	91
SHA	108	83	37	74	**152**	281	**60**	102
MMU	**178**	59	**57**	59	174	196	25	26

In three out of four cases, the coverage-guided tracing strategy works better than random mutations. In the fourth case, the difference for template-based formula enumeration is minimal (174 vs 178) properties. This supports our claim that coverage-guided tracing is effective. The specific templates chosen by us produce many more likely hyperproperties than random formula enumeration – this is an expected result.

The time taken to evaluate the traces is only a few minutes. While we do not report detailed results due to a lack of space, we compared runtime with Bach [21] and found that our custom-built hyperproperty evaluator was about $200\times$ faster than the Datalog-based approach proposed in that paper.

978-1-7281-5410-7/20 $31.00 © 2020 IEEE

Interesting Assertions Mined: : We provide a few examples of assertions mined by the tool. In the μcontroller, the tool identified the assertion $\mathbf{G}\,(decoderop_{\pi_1} = decoderop_{\pi_2} \rightarrow decoderrdsel_{\pi_1} = decoderrdsel_{\pi_2})$. This property states that the decoder's read-select signal is determined by the current operation in the decoder. Note the hyperproperty does not say *how* the read-select signal is determined by the decoder operation; this allows it to remain valid even if the encoding of decoder op is changed. Similar assertions capturing the relation between input data length and the byte iterator registers were identified in the AES and SHA modules.

V. RELATED WORK

GoldMine [15, 25] mines assertions automatically using static analysis and decision trees and uses a formal verification engine to check their validity. A-Team [8] introduces a methodology to mine temporal assertions of the form $\mathbf{G}\,(\varphi \rightarrow \psi)$ by combining coverage analysis with data mining. Danese et al. [9] is an earlier effort that proposes a similar methodology but with multiple templates and using static and dynamic techniques. Ghasempouri [14] present a way to measure the relevance or "interestingness" of a specification by the use of contingency tables and support metrics. This provides a way to order/rank mined assertions. Malburg et al. [17] propose an approach to mine temporal properties by building dynamic dependency graphs and deducing the properties via graph traversal. In comparison to our work, none of these efforts can mine hyperproperties. Further, these ideas – the use of static analysis, decision trees, dependency graphs, ranking assertions, etc. are orthogonal to our approach and can potentially be incorporated in the formula generator component of HYPERMINER to further improve its performance.

A noteworthy effort that does mine hyperproperties of functions, in the context of a pure functional language as opposed to hardware or SoC designs, is Bach, by Smith et al. [21]. Unlike our work, Bach cannot mine temporal hyperproperties. Further, as shown in our evaluation, extending it support temporal properties results in extremely poor performance.

Three-valued semantics for LTL proposed by Bauer et al. [3] is related to our notion of optimistic and pessimistic operators. However, their semantics was developed for monitoring and yields indeterminate for the corner cases similar to Figure 4. An indeterminate result is not useful for mining, as it does not tell us whether to shortlist a property.

VI. CONCLUSION

In this paper, we introduced a framework for mining *hyperproperties* from execution traces of SoC designs. Our framework had two components: a tracer that is based on coverage-guided trace generation and a miner that evaluates the collected traces to find likely hyperproperties in those traces. An important theoretical contribution of this paper are the novel optimistic and pessimistic semantics for Hyper Linear Temporal Logic (HyperLTL) that enable principled evaluation of HyperLTL formulas over finite traces. Experiments showed that the framework was effective in finding interesting hyperproperties in an SoC design.

REFERENCES

[1] JasperGold: Security Path Verification App. https://www.cadence.com/en_US/home/tools/system-design-and-verification/formal-and-static-verification/jasper-gold-verification-platform/security-path-verification-app.html?CMP=SVG_JasGApp_IntDgn, 2020.

[2] Bowen Alpern and Fred B. Schneider. Defining liveness. *Information Processing Letters*, 21(4):181 – 185, 1985.

[3] Andreas Bauer, Martin Leucker, and Christian Schallhart. Monitoring of real-time properties. In *Foundations of Software Technology and Theoretical Computer Science*, pages 260–272. Springer, 2006.

[4] Doron Bustan, Dmitry Korchemny, Erik Seligman, and Jin Yang. SystemVerilog Assertions: Past, present, and future SVA standardization Experience. *IEEE Design & Test of Computers*, 29(2):23–31, 2012.

[5] Edmund M. Clarke and E. Allen Emerson. Design and synthesis of synchronization skeletons using branching time temporal logic. In *Workshop on Logic of Programs*, pages 52–71. Springer, 1981.

[6] Michael R. Clarkson, Bernd Finkbeiner, Masoud Koleini, Kristopher K Micinski, Markus N Rabe, and César Sánchez. Temporal logics for hyperproperties. In *Principles of Security and Trust*, pages 265–284. Springer, 2014.

[7] Michael R. Clarkson and Fred B. Schneider. Hyperproperties. *Journal of Computer Security*, 18(6):1157–1210, 2010.

[8] Alessandro Danese, Nicolò Dalla Riva, and Graziano Pravadelli. A-team: Automatic template-based assertion miner. In *Design Automation Conference*, pages 1–6. IEEE, 2017.

[9] Alessandro Danese, Tara Ghasempouri, and Graziano Pravadelli. Automatic extraction of assertions from execution traces of behavioural models. In *Design, Automation & Test in Europe*, pages 67–72. IEEE, 2015.

[10] Calvin Deutschbein and Cynthia Sturton. Mining security critical linear temporal logic specifications for processors. In *Workshop on Microprocessor and SOC Test and Verification (MTV)*, pages 18–23. IEEE, 2018.

[11] Bernd Finkbeiner, Markus N Rabe, and César Sánchez. Algorithms for model checking HyperLTL and HyperCTL*. In *Computer Aided Verification*, pages 30–48. Springer, 2015.

[12] Harry Foster, Erisch Marschner, and Yaron Wolfsthal. IEEE 1850 PSL: The Next Generation. In *Design and Verification Conference and Exhibition*, 2005.

[13] Tara Ghasempouri, Jan Malburg, Alessandro Danese, Graziano Pravadelli, Goerschwin Fey, and Jaan Raik. Engineering of an effective automatic dynamic assertion mining platform. In *Very Large Scale Integration (VLSI-SoC)*, pages 111–116. IEEE, 2019.

[14] Tara Ghasempouri and Graziano Pravadelli. On the estimation of assertion interestingness. In *Very Large Scale Integration (VLSI-SoC)*, pages 325–330. IEEE, 2015.

[15] Samuel Hertz, David Sheridan, and Shobha Vasudevan. Mining Hardware Assertions with Guidance from Static Analysis. *IEEE Transactions on Computer-Aided Design of Integrated Circuits and Systems*, 32(6):952–965, 2013.

[16] Kevin Laeufer, Jack Koenig, Donggyu Kim, Jonathan Bachrach, and Koushik Sen. RFUZZ: coverage-directed fuzz testing of RTL on FPGAs. In *International Conference on Computer-Aided Design*, pages 1–8. IEEE, 2018.

[17] Jan Malburg, Tino Flenker, and Görschwin Fey. Property mining using dynamic dependency graphs. In *Asia and South Pacific Design Automation Conference (ASP-DAC)*, pages 244–250. IEEE, 2017.

[18] John Mclean. Proving Noninterference and Functional Correctness Using Traces. *Journal of Computer Security*, 1:37–58, 1992.

[19] A. Pnueli. The Temporal Logic of Programs. In *Foundations of Computer Science*, pages 46–57. IEEE, 1977.

[20] Kostya Serebryany. libFuzzer – a library for coverage-guided fuzz testing. 2015.

[21] Calvin Smith, Gabriel Ferns, and Aws Albarghouthi. Discovering Relational Specifications. In *Foundations of Software Engineering*, pages 616–626, 2017.

[22] Wilson Snyder. Verilator and systemperl. In *North American SystemC Users' Group, Design Automation Conference*, 2004.

[23] P. Subramanyan and D. Arora. Formal Verification of Taint-Propagation Security Properties in a Commercial SoC Design. In *Design, Automation & Test in Europe*, 2014.

[24] P. Subramanyan, S. Malik, H. Khattri, A. Maiti, and J. Fung. Verifying Information Flow Properties of Firmware using Symbolic Execution. In *Design Automation & Test in Europe*, 2016.

[25] Shobha Vasudevan, David Sheridan, Sanjay Patel, David Tcheng, Bill Tuohy, and Daniel Johnson. Goldmine: Automatic assertion generation using data mining and static analysis. In *Design, Automation & Test in Europe*, pages 626–629. IEEE, 2010.

[26] Chenguang Wang, Yici Cai, Qiang Zhou, and Haoyi Wang. ASAX: Automatic security assertion extraction for detecting Hardware Trojans. In *Asia and South Pacific Design Automation Conference*, pages 84–89. IEEE, 2018.

[27] Michal Zalewski. Technical whitepaper for afl-fuzz, 2014.

[28] Steve Zdancewic and Andrew C Myers. Observational determinism for concurrent program security. In *Proceedings of the 16th IEEE Computer Security Foundations Workshop*, pages 29–43. IEEE, 2003.

[29] Tong Zhang, Daniel Saab, and Jacob A Abraham. Automatic assertion generation for simulation, formal verification and emulation. In *IEEE Computer Society Annual Symposium on VLSI*, pages 471–476. IEEE, 2017.

978-1-7281-5410-7/20 $31.00 © 2020 IEEE

A Hybrid Cache HW/SW Stack for Optimizing Neural Network Runtime, Power and Endurance

William Andrew Simon*, Alexandre Levisse*, Marina Zapater[†]* and David Atienza*
*Embedded Systems Laboratory (ESL), Swiss Federal Institute of Technology Lausanne (EPFL)
[†]University of Applied Sciences Western Switzerland (HEIG-VD / HES-SO)
Email: {william.simon, alexandre.levisse, marina.zapater, david.atienza}@epfl.ch

Abstract—Hybrid caches consisting of both SRAM and emerging Non-Volatile Random Access Memory (eNVRAM) bitcells increase cache capacity and reduce power consumption by taking advantage of eNVRAM's small area footprint and low leakage energy. However, they also inherit eNVRAM's drawbacks, including long write latency and limited endurance. To mitigate these drawbacks, many works propose heuristic strategies to allocate memory blocks into SRAM or eNVRAM arrays at runtime based on block content or access pattern. In contrast, this work presents a HW/SW Stack for Hybrid Caches (SHyCache), consisting of a hybrid cache architecture and supporting programming model, reminiscent of those that enable GP-GPU acceleration, in which application variables can be allocated explicitly to the eNVRAM cache, eliminating the need for heuristics and reducing cache access time, power consumption, and area overhead while maintaining maximal cache utilization efficiency and ease of programming. SHyCache improves performance for applications such as neural networks, which contain large numbers of invariant weight values with high read/write access ratios that can be explicitly allocated to the eNVRAM array. We simulate SHyCache on the gem5-X architectural simulator and demonstrate its utility by benchmarking a range of cache hierarchy variations using three neural networks, namely, Inception v4, ResNet-50, and SqueezeNet 1.0. We demonstrate a design space that can be exploited to optimize performance, power consumption, or endurance, depending on the expected use case of the architecture, while demonstrating maximum performance gains of 1.7/1.4/1.3x and power consumption reductions of 5.1/5.2/5.4x, for Inception/ResNet/SqueezeNet, respectively.

Index Terms—eNVRAM, STT-MRAM, hybrid caches, neural networks, low-power systems

I. INTRODUCTION

In recent years, Neural Networks (NNs) have gained popularity for performing a variety of tasks such as image recognition [1], object detection [2], and natural language processing [3]. In an effort to enable NNs on as many devices as possible, many optimizations to reduce NN memory and compute overhead have been proposed, such as quantization, pruning, and custom layers [4]–[6]. Even so, the memory footprint of "small" NNs often still measure in the order of MBs [7]; therefore, memory enhancements that exploit the invariant nature of these weights can continue to improve NN performance on area restricted devices. In this regard, Figure 1 displays the read access to write access (read/write) ratios of the memory blocks that account for 98% of total inference-time

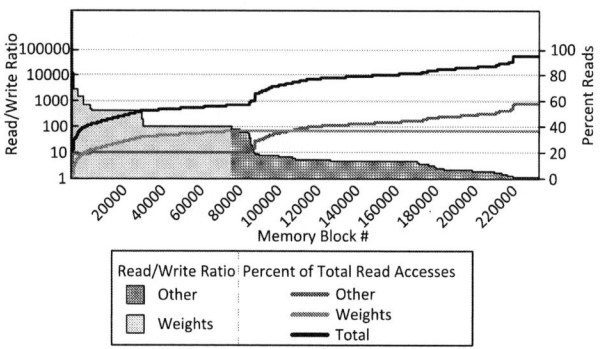

Fig. 1: Read/write access ratios in relation to total read accesses. Weight accesses account for nearly 40% of all reads.

read accesses for the SqueezeNet neural net [6]. As can be seen, the memory blocks with the highest read/write ratio contain weight values, as these blocks are only written during line fills from lower memory levels during inference, never from the processor. Weight values also account for almost 40% of all memory accesses performed at runtime. It can be inferred that improving processor read access to these weight values will result in overall application performance gain.

In this context, Hybrid Caches (HCs), consisting of SRAM and emerging Non-Volatile Random Access Memories (eN-VRAMs), can be used to accelerate NNs. eNVRAM's low area footprint and leakage energy enable more efficient execution of memory intense algorithms by increasing cache capacity with little area overhead, while simultaneously reducing power consumption. However, eNVRAMs also incur a high write energy cost and have limited endurance. It is therefore necessary to optimize write strategies to avoid unnecessary writes. Many works have proposed heuristical, predictive placement strategies. In contrast, a deterministic cache allocation strategy enables the utilization of eNVRAM allocated variables to choose which values are written to eNVRAM and avoid unnecessary transfer between SRAM and eNVRAM, thus providing maximum cache usage efficiency. In the case of NNs, invariant weight values are an excellent candidate for eNVRAM storage.

To this end, we present a HW/SW Stack for Hybrid Caches (SHyCache), consisting of a HC architecture and deterministic

978-1-7281-5410-7/20 $31.00 © 2020 IEEE

cache allocation strategy, supported via a programming model reminiscent of those utilized to enable GP-GPU computation, illustrated in Figure 2-a. SHyCache enables precise control over data placement within the cache, and is compatible with heuristical hybrid cache strategies. Then, we explore the HC design space by considering various eNVRAM/SRAM HC ratios, and benchmark SHyCache in the gem5-X [8] architectural simulator, on a range of NNs of varying computational complexity and memory footprint.

The contributions of this paper are as follows:

- We introduce SHyCache, an HC architecture with a deterministic allocation strategy allowing for precise data allocation within an HC. SHyCache's allocation strategy is compatible with other hybrid cache allocation strategies.
- We develop a programming model with a C++ support library allowing easy integration of SHyCache support into any existing application.
- We implement SHyCache in the gem5-X architectural simulator and explore the HC design space to optimize for performance, power, and endurance, demonstrating performance gains of 1.7/1.4/1.3x and power consumption reductions of 5.1/5.2/5.4x for the Inception v4, ResNet-50, and SqueezeNet 1.0 NNs, respectively.

The rest of this paper is organized as follows. Section II explores related state-of-the-art work. Section III details SHyCache's HC architecture. Section IV details SHyCache's programming model and support library and discusses tandem implementation with other allocation strategies. Section V details our benchmarking methodology, while Section VI discusses results. Finally, Section VII concludes this work.

II. RELATED WORK

A. Resistive Random Access Memory

Emerging nonvolatile memories, including phase change [9], resistive [10] and spin-torque transfer [11] memories, have gained popularity in recent years thanks to their small size, up to 4x smaller than 6T SRAM cells [12], and low leakage energy resulting from their nonvolatility. However, eNVRAM also suffers from long/high-energy write operations, and low endurance due to the underlying physics of the technology. In order to efficiently utilize eNVRAM within an architecture, eNVRAM-specific optimizations must be implemented to magnify their advantages while mitigating or masking drawbacks.

B. Hybrid Cache Design and Allocation Strategies

One implementation of eNVRAM within the memory hierarchy involves placement alongside standard SRAM cache arrays, creating a Hybrid Cache (HC) hierarchy, as illustrated in Figure 2-b. This architecture increases cache capacity while also reducing power consumption [13]. However, HCs also inherit eNVRAM's disadvantages as described above. Further, a naive HC implementation may magnify these disadvantages, as the frequency of cache writes, and therefore cache lifetime, is highly variant depending on the application [14], as well as reducing performance even while not in use due to slower access time. Many works have therefore proposed memory

management strategies [15] for allocating blocks in either SRAM or eNVRAM depending on a variety of factors. The majority of these strategies are heuristic [14], [16], [17] or compiler based [18], [19]. In contrast, this work presents an application driven allocation strategy which obviates the need for heuristics and takes advantage of cases in which an application's data is constant, such as neural networks.

C. Neural Networks

Neural networks are a class of applications that accept inputs in various forms such as images, text, or audio, process them through the use of consecutive compute layers, and return an output, for example, the class of the input. Each hidden layer consists of one or more "neurons" of various function. The two most widely used neuron layers are the fully connected and convolutional layer. Both layers perform multiply-and-accumulate operations between the outputs of the previous layer and an array of previously trained weights. These layers require a massive number of weight values; the classical Alexnet neural network utilizes 3.78M weights (144MB for floating point weights) in its first fully connected layer [1]. Convolutional layers reduce memory footprint by using small (ex. 3x3) weight kernels that are convolved with the layer input. While convolutional layers greatly reduce the NN's memory footprint, they are generally still large in an absolute sense; for example, the SE-ResNeXt-50 NN achieves the highest Top-1 and Top-5% accuracy on the ImageNet-1k database at a low operational complexity, yet still contains over 10MB of weights [7]. Managing such large quantities of weights is imperative for efficient NN execution.

III. HYBRID CACHE ARCHITECTURAL DESIGN

SHyCache's hybrid cache consists of arrays of two memory types, one being standard 6T SRAM based memory and the other a flavor of eNVRAM, as illustrated in Figure 2-b. Each bitcell array is indexed by a separate tag array. The combined area of the tag array memory macros is equivalent to a single tag array of an equivalently sized monolithic cache memory, plus overhead for tag array periphery. As our data placement strategy is deterministic, as described in Section IV, only one data/tag array needs be accessed per read/write, reducing power consumption in comparison to heuristic strategies that must check both arrays for the data as its location is not known beforehand. In regards to cache access latency, it is important to note that, as only either the SRAM or eNVRAM is accessed, SHyCache's allocation strategy does not impact access latency of programs not utilizing the eNVRAM, i.e. the system kernel, and thus does not impact standard system performance. This is not necessarily the case if other heuristic or compiler-based allocation strategies are implemented alongside SHyCache's allocation strategy, as discussed in Section IV-C.

As illustrated in Figure 2-b, we consider HC configurations at both the L1 and L2 levels. We utilize an inclusive cache policy for reasons explained in Section VI. The L1 cache utilizes parallel tag/data access to reduce access time, while the L2 uses sequential tag/data access to reduce power consumption.

978-1-7281-5410-7/20 $31.00 © 2020 IEEE

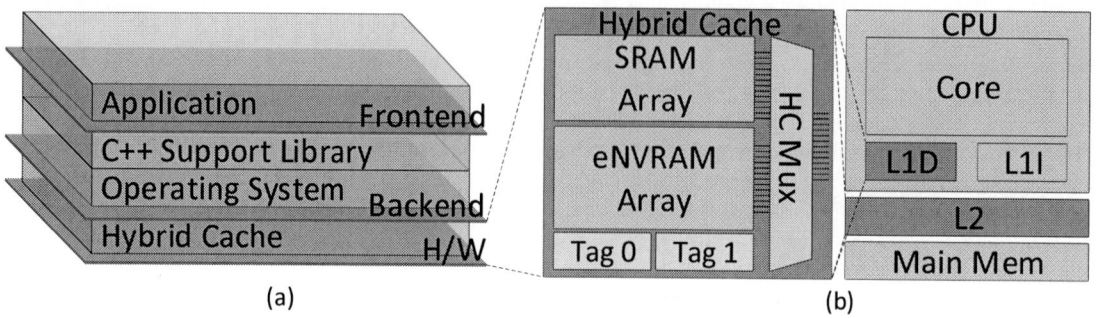

Fig. 2: SHyCache is a HW/SW stack (a) that enables efficient use of a hybrid cache (b).

```
using namespace SHyCache;
void loadWeights(string weightsFile, size_t len) s{
    // Declare var to be stored to eNVRAM portion of
    // cache. Allocation handled by helper library.
    float32_nv *weightsPtr = new float32_nv[len];
    // Open file containing pre-calculated weights.
    ifstream wIn(weightsFile);
    // Store weights to previously allocated memory.
    wIn.read((char *)weightsPtr,len);
    //...Perform inference...
    // Clean up
    delete weightsPtr;
}
```

Listing 1: Allocating the hybrid cache is done by allocating the variable pointer within the memory mapped region reserved for eNVRAM.

IV. INTEGRATING SHYCACHE'S PROGRAMMING MODEL INTO NEURAL NETWORK FRAMEWORKS

Several characteristics of NN weights enable NNs to be accelerated by HCs. The first is that, as previously mentioned, most NNs that achieve >80% Top-5% accuracy utilize large amounts (in the order of MBs) of weights. Second, weight values are calculated at training time and not modified during inference. Finally, fully connected and convolutional layers result in spatially local data accesses. These characteristics make eNVRAM suitable for storing NN weights. High eNVRAM bitcell density allows more weights to be stored without the need for eviction, while the long write latency of eNVRAM is mitigated by the read-only nature of weights.

A. Enabling HC allocation at the Operating System Level

Most previous HC works utilize heuristic strategies to allocate data either in the SRAM or eNVRAM bitcell arrays depending on various factors. In contrast, because the location and value of NN weight values are deterministic, no heuristic strategy is necessary for weight allocation in SHyCache. This is accomplished at the system level by reserving a portion of memory at operating system startup that can be mapped by an application in the same manner that peripherals can be mapped and accessed by user applications. When variables allocated to the memory range reserved for eNVRAM caching are fetched into the cache hierarchy, an address predecoder analyzes the MSBs of the incoming address. Addresses within the reserved

memory region will be automatically cached in eNVRAM array when accessed. Such a strategy does not require any compiler modification and minimal application modification. Architectural modifications will depend on the nature of the architectures virtual-physical memory address translation. If the reserved memory is virtual, when address translation occurs the processor can tag the memory access with a bit to indicate if it is a standard or eNVRAM memory access before passing the access to the cache hierarchy. If the reserved memory is physical, or there is no virtual-physical translation, for example in embedded systems that use tightly coupled memory [20], the type of memory access will be attained as a byproduct of the address decoding that occurs during cache access, hence, no modification to the processor architecture is necessary.

B. Enabling HC allocation at the Application Level

At the application level, the programmer utilizes SHyCache's C++ data types to instantiate variables that will be allocated to the eNVRAM, as seen in the example function in Listing 1. The support library then facilitates the allocation of variables to the eNVRAM memory region without further programmer intervention by allocating the variables to the memory mapped region described in Section IV-A. Current NN frameworks such as Tensorflow, Caffe, and the ARM Compute Library perform several preprocessing stages upon weights before storing them in their final tensor, after which this tensor is not modified during inference. Framework extension to support HCs consists therefore of redirecting the output of the final preprocessing stage to store weight values in a tensor stored in the eNVRAM cache, resulting in no extra data movement overhead. In this work, we extend the ARM Compute Library, a neural network framework optimized for ARM processors [21], with SHyCache's C++ support library, however such extensions could be applied to any of the aforementioned frameworks to enable HC support. It should be noted that the use of a support library obviates the need for any language compiler modifications, simplifying the deployment process.

C. Co-Implementation with Other Hybrid Cache Allocation Strategies

One advantage of SHyCache is that it does not preclude the use of other heuristical [14] or compiler-driven [18] HC

TABLE I: Simulator Parameters

Processor	2GHz, 4 stage pipeline, ARMv8 ISA in-order core, 7 entry LSQ
NEON Co-processor	128 bit registers 16 parallel 8-bit operations
L1-I Cache	32kB, 4-way, 2 cycle access
L1-D Cache	32/0kB SRAM, 0/128kB STT-MRAM 4-way, 2 cycle access
L2 Cache	1024/0kB SRAM, 0/4096kB STT-MRAM mostly-inclusive, 16-way, 20 cycle access
STT-MRAM Write Time	50ns [11]
Memory	DDR3 2133MHz, 4GB

TABLE II: Neural Network Benchmark Parameters

Benchmark	# Parameters	Weight Memory Footprint (MB)
Inception v4	41.1M	156.8
ResNet-50	23.5M	89.6
SqueezeNet v1.0	1.25M	4.76

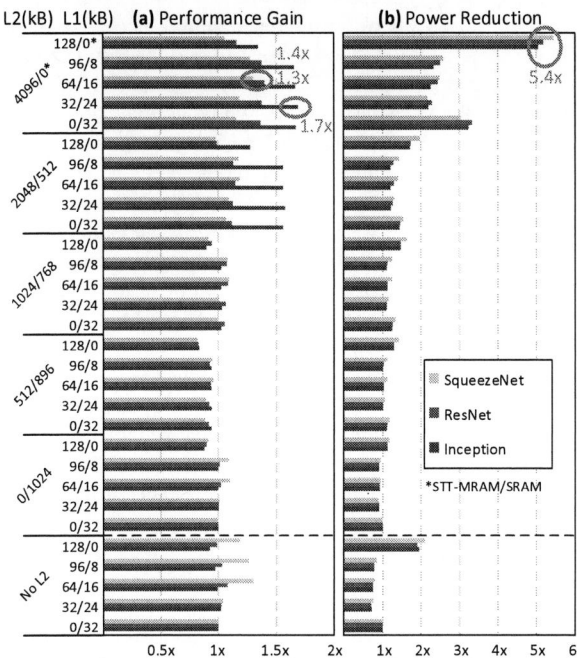

Fig. 3: Performance gain (a) and power reduction (b) across the L1/L2 design space. HC label format is STT-MRAM/SRAM Capacity (kB.)

allocation strategies. Such strategies can be implemented in tandem by excluding the memory region utilized by SHyCache from the data migration scheme. Even a heuristical allocation strategy with oracle prediction abilities would benefit from SHyCache, as, in order to maintain fast access times, the tag array (and data array in the case of simultaneous tag/data access) of both the SRAM and eNVRAM portions of the HC cache must be accessed simultaneously, as the location of the data is unknown prior to access. On the other hand, SHyCache determines the location of the data at compile time, and the address decoding process routes data access to only the portion of cache in which the data is located, reducing power consumption.

V. EXPERIMENTAL SETUP

To assess SHyCache's application level performance, we extend the gem5-X architectural simulator [8] [22] to support HC caches. We then simulate three NNs of differing computational complexity and memory footprint, and extract performance, power, and endurance trends across a range of HC geometries.

A. gem5-X Simulator Parameters and Hybrid Cache Access Latency Simulation

We emulate an ARMv8 A53 in-order core by calibrating gem5-X with the simulation parameters illustrated in Table I, and simulating an Ubuntu 18.04 LTS software environment. CPU and interconnect power statistics are extracted via the McPAT power estimation framework [23]. SRAM timing and power values are extracted from an implemented subarray in 28nm using TSMC's high performance technology PDK [24]. We draw eNVRAM power values from literature, considering STT-MRAM [11] for this work, however the allocation strategy is technology independent. In order to illustrate SHyCache's performance and power trends, we extract performance and power statistics across multiple HC hierarchies, in addition to SRAM-only baseline simulations. Hybrid cache geometries are defined by assuming a 4x area ratio between SRAM and eNVRAM bitcell arrays [12], and then sweeping eNVRAM capacity between 0-128kB and 0-4096kB for the L1/L2 caches, respectively, while maintaining an equivalent area footprint.

In order to accurately simulate HC access, SRAM and STT-MRAM access latency is defined in cycles, as documented in Table I. This access latency represents the time to access a cache block through the decoding logic and H-tree, and is pipelined

in this implementation, allowing consecutive cache accesses to overlap without blocking. Additionally, STT-MRAM write latency includes an additional write time measured in *ns*, representing the time taken to write a line of data to STT-MRAM. During this time, the subarrays being written to cannot be accessed; therefore, this time is not pipelined and subsequent accesses to busy subarrays are blocked. As SHyCache is deterministic in that only the SRAM or eNVRAM portions of the memory need to be accessed for any given cache block, this added latency is not present in standard SRAM accesses, and hence does not impact system performance in cases where the eNVRAM is not accessed.

B. Neural Network Benchmarks

In order to benchmark SHyCache, we utilize three modern neural networks of differing sizes, namely, Inception v4 [25], ResNet-50 [26] and SqueezeNet v1.0 [6], whose parameters are outlined in Table II. We choose these networks to benchmark SHyCache under a wide range of network complexities and memory footprints. All weights and inputs are in floating point, and input batch sizes are set to one. We use the ARM Compute Library (ACL) [21] as our software framework. ACL is a graph dataflow framework, specially designed to optimally utilize the ARM NEON SIMD co-processor to accelerate neural networks.

VI. Experimental Results and Analysis

Our experimental results reveal trends in relation to runtime performance, power consumption, and eNVRAM endurance.

A. Performance Results

To observe SHyCache's impact on NN runtime, summarized in Figure 3-a, we perform inference with a batch size of one for the three neural networks, normalizing the results to pure SRAM cache hierarchies. The No-L2 portion is normalized to a pure SRAM cache of 32kB, while all other portions are normalized to a 32/1024kB L1/L2 pure SRAM cache hierarchy.

As can be seen, runtime acceleration varies widely across cache geometries. On one hand, if we consider solely the L1 cache in Figure 3-a, we measure performance gains of up to 1.31/1.09/1.03x for Inception/ResNet/SqueezeNet, respectively, as we increase the STT/SRAM HC ratio up to 64kB/16kB. However, increasing the size of the STT-MRAM array past this point degrades performance, as less SRAM cache space is allocated for the remainder of the application. Generally, a 128kB pure STT-MRAM L1 cache results in a steep decrease in performance as all memory accesses, including those to memory with low read/write ratios, are relegated to STT-MRAM. It should also be noted that performance gain attributable to L1 STT-MRAM decreases as computational complexity and memory footprint increases, as the tiny L1 cache becomes insignificant in comparison to the size of the weights.

On the other hand, if we consider the L2 we find a very different trend. Increasing the HC ratio consistently improves performance for all NN benchmarks, up to a pure eNVRAM array of 4096kB. The larger cache size results in fewer weight evictions, and the mostly-inclusive cache policy mitigates the effects of constant L1 evictions. Additionally, having such a large ratio between L2 and L1 STT-MRAM capacity (64 in the case of a 64kB L1 and 4096kB L2), reduces the negative effects of data repetition that results from an inclusive cache policy. Overall, we achieve maximum possible performance gains of 1.7/1.4/1.3x for Inception/ResNet/SqueezeNet, respectively, when normalized against pure SRAM L1/L2 cache hierarchies.

B. Power Results

Next, we consider SHyCache's implications on power consumption. Figure 3-b summarizes the results of the HC design space, from which two trends can be drawn. First, regardless of the L2 cache, a spike in power reduction is seen at a 128kB pure SRAM cache. This is because in a pure STT-MRAM cache the power-hungry SRAM bitcell array is replaced with a low leakage STT-MRAM bitcell array. A similar, more pronounced power reduction occurs when replacing the L2 SRAM array entirely with STT-MRAM. Figure 4 provides an in-depth breakdown of the power consumption of pure SRAM, pure L1 STT-MRAM, and pure L1/L2 STT-MRAM cache hierarchies. As can be seen, while L1/L2 STT-MRAM write power is substantial, eliminating the energy-leaking SRAM caches provides an excellent reduction in power consumption. STT-MRAM read energy is on par with SRAM read energy, and is too low to be visible in Figure 4. Overall, we see

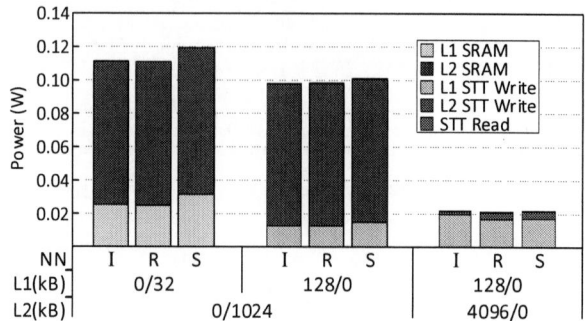

Fig. 4: Power consumption of all-SRAM, SRAM+eNVRAM, and all-eNVRAM caches for Inception (I), ResNet (R) and SqueezeNet (S) NNs.

a maximum possible power reduction of 5.1/5.2/5.4x, for Inception/ResNet/SqueezeNet, respectively.

C. Endurance Results

Lastly, we analyze the number of bitflips that occur within the STT-MRAM array at different cache geometries. eNVRAM life expectancy is tied to its endurance with respect to bitcell value flips, or bitflips. This is measured by counting every $1\rightarrow0/0\rightarrow1$ flip during writes to the STT-MRAM arrays. As eNVRAM technologies have significantly lower endurance compared to CMOS-based memories, it is imperative to consider bitflip frequency of any architecture utilizing eNVRAM.

Figure 5-a illustrates the STT-MRAM bitflip count at all L1 HC geometries with no L2. Consistent with the performance results and reasoning presented in Section VI-A, bitflip count drops for 64 and 96kB STT-MRAM caches, before increasing again for pure STT-MRAM caches, with the bitflip reduction more pronounced in the smaller SqueezeNet NN.

Meanwhile, Figure 5-b presents STT-MRAM bitflip count for all L2 HC geometries with a pure SRAM L1 cache. The first point of note is that the geometry with the highest bitflip count is not a pure STT-MRAM cache, but in fact an HC of 512/896kB. This is consistent with the performance drop seen across all NNs at this geometry in Section VI-A, and is a result of cache thrashing due to the small cache size in relation to the number of weights. Bitflip count then drops as the HC ratio increases and less cache blocks are evicted. Finally, at a pure STT-MRAM cache, the bitflip count for the smaller SqueezeNet NN spikes, as the whole application utilizes STT-MRAM. ResNet and Inception's larger weight footprints dilute this effect, as they gain more from keeping weights in-cache.

In this paper, we consider only overall bitflip count, not flip counts for individual bits. We do observe a drop in average flip per bit as cache capacity increases; however, this metric does not account for uneven intra-word flips skewed toward the LSBs. Many works have explored various eNVRAM wear reducing and leveling optimizations to alleviate this skew. These optimizations are out of this paper's scope, however, and have not been applied in this work; hence, the numbers demonstrated here are worst case values, with room for future optimization.

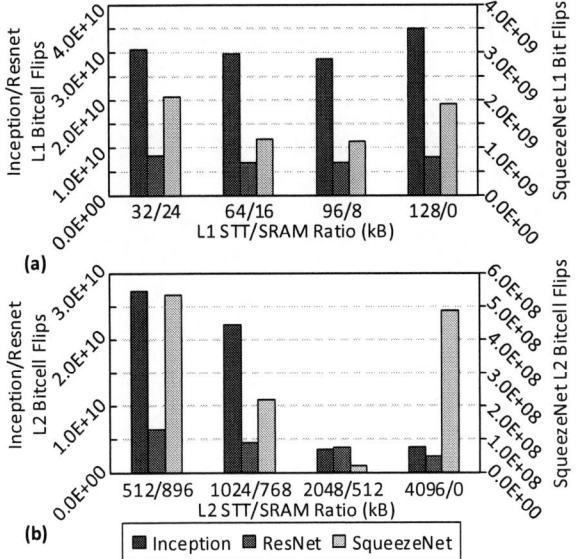

Fig. 5: STT-MRAM bitcell flips across varying L1 (a) and L2 (b) cache sizes.

D. Optimizing HCs for Performance, Power, or Endurance

As seen in Sections VI-A-C, proper selection of HC geometry for the L1 and L2 caches depends on the system's expected use case. Different geometries optimize either performance, power, or endurance. For example, performance is maximized with a 64/16kB L1 HC cache and a pure STT-MRAM 4096kB L2 cache. However, such a configuration may have a poor endurance when small NNs are the target application. In terms of power, a pure STT-MRAM L1 and L2 provides significant power reduction; however, endurance suffers greatly from such a configuration. From an endurance perspective, the highly active L1 cache accounts for nearly half of all bitflips during inference; a good trade-off between performance, power, and endurance, therefore, may be a pure SRAM L1 with a 2048/512kB L2 HC cache. This architecture provides performance and power improvements of 1.6/1.1/1.1x and 1.5/1.5/1.5x, respectively, while incurring the lowest bitflip count of any architecture.

VII. CONCLUSION

In this work, we presented SHyCache, a hybrid cache with a deterministic allocation strategy and supporting programming model designed to improve NN runtime while reducing power consumption. SHyCache enables NN frameworks to explicitly allocate weight values to the eNVRAM cache, eliminating data transitions between SRAM and eNVRAM arrays and providing maximal cache efficiency. In this work, we explained how SHyCache can be implemented at the system and application level and in tandem with other HC allocation strategies, we have developed a C++ support library allowing implementation in current applications, and we benchmarked SHyCache on three neural network applications of varying computational complexity and memory footprint. Our experimental results have demonstrated a maximum performance gains of

1.7/1.4/1.3x and power consumption reductions of 5.1/5.2/5.4x, for our Inception/ResNet/SqueezeNet benchmarks, respectively. Finally, we have considered the implications of our results for optimizing an architecture based on expected use case, and propose a middle-ground solution that provides optimal trade-off between performance, power, and endurance.

ACKNOWLEDGMENTS

This work has been partially supported by EC H2020 RECIPE project (GA No. 801137), EC H2020 WiPLASH project (GA No. 863337), ERC Consolidator Grant COM-PUSAPIEN (GA No. 725657), and by the Swiss NSF ML-Edge Project (GA No. 200020_182009).

REFERENCES

[1] A. Krizhevsky, I. Sutskever, and G. Hinton, "Imagenet classification with deep convolutional neural networks," *Commun. ACM*, 2017.

[2] S. Ren, K. He *et al.*, "Faster R-CNN: Towards real-time object detection with region proposal networks," in *NIPS 28*, 2015.

[3] R. Collobert and J. Weston, "A unified architecture for natural language processing: Deep neural networks with multitask learning," *ICML*, 2008.

[4] A. Zhou, A. Yao *et al.*, "Incremental network quantization: Towards lossless cnns with low-precision weights," *CoRR*, 2017.

[5] S. Han, H. Mao, and W. J. Dally, "Deep compression: Compressing deep neural networks with pruning, trained quantization and huffman coding," *CoRR*, 2015.

[6] F. N. Iandola, M. W. Moskewicz *et al.*, "Squeezenet: Alexnet-level accuracy with 50x fewer parameters and <0.5mb model size," *CoRR*, 2016.

[7] S. Bianco, R. Cadene *et al.*, "Benchmark analysis of representative deep neural network architectures," *IEEE Access*, vol. 6, 2018.

[8] Y. M. Qureshi, W. A. Simon *et al.*, "Gem5-x: A gem5-based system level simulation framework to optimize many-core platforms," *HPC*, 2019.

[9] G. W. Burr, M. J. Brightsky *et al.*, "Recent progress in phase-change memory technology," *JETCAS*, vol. 6, no. 2, pp. 146–162, June 2016.

[10] R. Fackenthal, M. Kitagawa *et al.*, "19.7 a 16gb reram with 200mb/s write and 1gb/s read in 27nm technology," in *ISSCC*, Feb 2014.

[11] Q. Dong, Z. Wang *et al.*, "A 1mb 28nm stt-mram with 2.8ns read access time at 1.2v vdd using single-cap offset-cancelled sense amplifier and in-situ self-write-termination," in *ISSCC*, Feb 2018.

[12] L. Wei, J. G. Alzate *et al.*, "13.3 a 7mb stt-mram in 22ffl finfet technology with 4ns read sensing time at 0.9v using write-verify-write scheme and offset-cancellation sensing technique," in *ISSCC*, Feb 2019.

[13] J. Li, C. J. Xue, and Yinlong Xu, "Stt-ram based energy-efficiency hybrid cache for cmps," in *VLSI-SOC 19*, Oct 2011.

[14] Y. Li, Y. Chen, and A. K. Jones, "A software approach for combating asymmetries of non-volatile memories," in *ISLPED*, 2012.

[15] D. Atienza, J. M. Mendias *et al.*, "Systematic dynamic memory management design methodology for reduced memory footprint," *ACM TODAES*, Apr. 2006.

[16] J. Ahn, S. Yoo, and K. Choi, "Prediction hybrid cache: An energy-efficient stt-ram cache architecture," *TC*, March 2016.

[17] Z. Wang, D. A. Jiménez *et al.*, "Adaptive placement and migration policy for an stt-ram-based hybrid cache," in *HPCA 20*, Feb 2014.

[18] Y.-T. Chen, J. Cong *et al.*, "Static and dynamic co-optimizations for blocks mapping in hybrid caches," in *ISLPED*, 2012, p. 237–242.

[19] Q. Li, J. Li *et al.*, "Compiler-assisted stt-ram-based hybrid cache for energy efficient embedded systems," *TVLSI*, vol. 22, no. 8, Aug 2014.

[20] F. Conti, D. Rossi *et al.*, "Energy-efficient vision on the pulp platform for ultra-low power parallel computing," in *SiPS*, Oct 2014, pp. 1–6.

[21] (2018). [Online]. Available: https://developer.arm.com/technologies/compute-library

[22] (2019). [Online]. Available: https://github.com/esl-epfl/gem5-X

[23] S. L. Xi, H. Jacobson *et al.*, "Quantifying sources of error in mcpat and potential impacts on architectural studies," in *HPCA 21*, 02 2015.

[24] W. A. Simon, Y. M. Qureshi *et al.*, "An in-cache computing architecture for edge devices," *TC*, 2020.

[25] C. Szegedy, S. Ioffe *et al.*, "Inception-v4, inception-resnet and the impact of residual connections on learning," *AAAI 31*, 2017.

[26] K. He, X. Zhang *et al.*, "Deep residual learning for image recognition," in *CVPR*, June 2016.

An ULP Self-Supplied Brain Interface Circuit

Amin Aghighi, Massood Tabib-Azar, and Armin Tajalli

Electrical & Computer Engineering Department, University of Utah, Salt Lake City

Email: armin.tajalli@utah.edu

Abstract—An Ultra-Low-Power (ULP) brain interface (BI) circuit, including signal processing front-end (SPFE) and wireless data transmitter (TX), supplied from the incident sensor power, is presented. While a rectifier/voltage multiplier (RVM) provides the required power for the wireless TX circuitry, some of the SPFE circuits are directly supplied from the sensor. This approach helps to reduce the overall required power and simplifies the design of the power management system. The proposed BI circuit is designed in 0.18 μm CMOS technology. Isolated Schottky diodes that are available in the process are used to design an RVM for the transmitter block. The SPFE circuitry samples the input, filters the noise and launches the transmitter in case of detecting a sufficient number of target events. The SPFE unit consumes 500 pA from a 400 mV supply, which is provided by the RVM circuit. Data transmission circuitry, which utilizes the same supply, consumes 205 pA leakage power, and 2.125 μA average active current during data transmission mode.

Index Terms—Brain interface circuit, ultra-low-power, near-zero-power sensors, energy harvesting, self-supplied interface circuit.

I. INTRODUCTION

The ever-increasing demand for efficient and ULP implantable devices requires reliable smart interface circuits that guarantee robust performance and extensive lifetime. Battery-powered implantable devices such as neural recording and brain interface circuits suffer from supply and performance degradation over time. Besides, having a re-surgery to replace the battery is very risky and undesirable for the patient even after a few years. These issues are not limited only to implantable biomedical applications. The advent of *"Internet of Things"* (IoT), implementation of wireless sensor nodes (WSN), and radio frequency identification (RFID) technology necessitate very efficient and robust power management systems to harvest the incident energy and guarantee acceptable life-time [1]-[4].

This article presents a front-end circuit designed as part of an implantable brain interface (BI) system, illustrated in Fig. 1. This circuit detects the neuron signals, amplifies them, and start counting the number of occurrences of the neuron spikes. When the number of spikes reaches a certain number, then the transmitter circuit will be activated to produce some high-frequency pulses and communicate the event to the outside body. An analog counter (integrator) has been employed to simplify the design and avoid adding digital circuits. As the device is implanted inside the body, a very efficient energy harvesting mechanism is required to power the signal processing, and also data transmission parts of the circuit. Fig. 1 also demonstrates the fabricated micro-sensor [5].

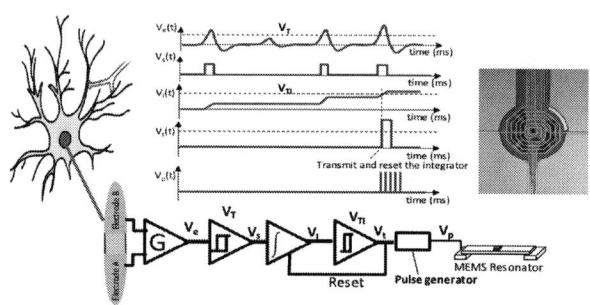

Fig. 1. Conceptual block diagram of the proposed brain interface (BI) structure, including the mico-fabricated device.

In energy harvesting-based platforms, systems working with AC input signals, which require an AC-DC rectifier front-end stage [6], suffer from an inherently lower energy efficiency than those of working with DC input signals. Moreover, input amplitude and frequency affect the RVM performance [7]. While smaller input amplitudes reduce the achievable output voltage at a fixed number of stages, lower input frequency hinders achieving acceptable efficiencies in two ways. First, it is very hard to design a practical integrated matching network between the input source and RVM at a lower frequency. This intermediary matching network is widely used in high-frequency applications such as RFID to boost the total available input power [6], [8]. Second, low input frequency increases output series impedance of the RVM, which hampers drawing the desired output current. Poor efficiency of different RVM typologies at low input power range and difficulties aroused by low-frequency operation make it interesting to propose system-level techniques to reduce the energy amount that should be provided by the RVM circuit.

An ULP BI circuit has been proposed here, in which the SPFE blocks are directly supplied from sensor output to reduce the power that is drawn from the RVM circuit. In the case of detecting the desired event, the SPFE output triggers the TX circuitry which is powered by the RVM circuit.

The rest of this paper is organized as follows: While the BI system is described in Section II, the transistor level building blocks are discussed in Section III. Section IV presents simulation results of the BI system and finally, the paper is concluded in Section V.

978-1-7281-5410-7/20 $31.00 © 2020 IEEE

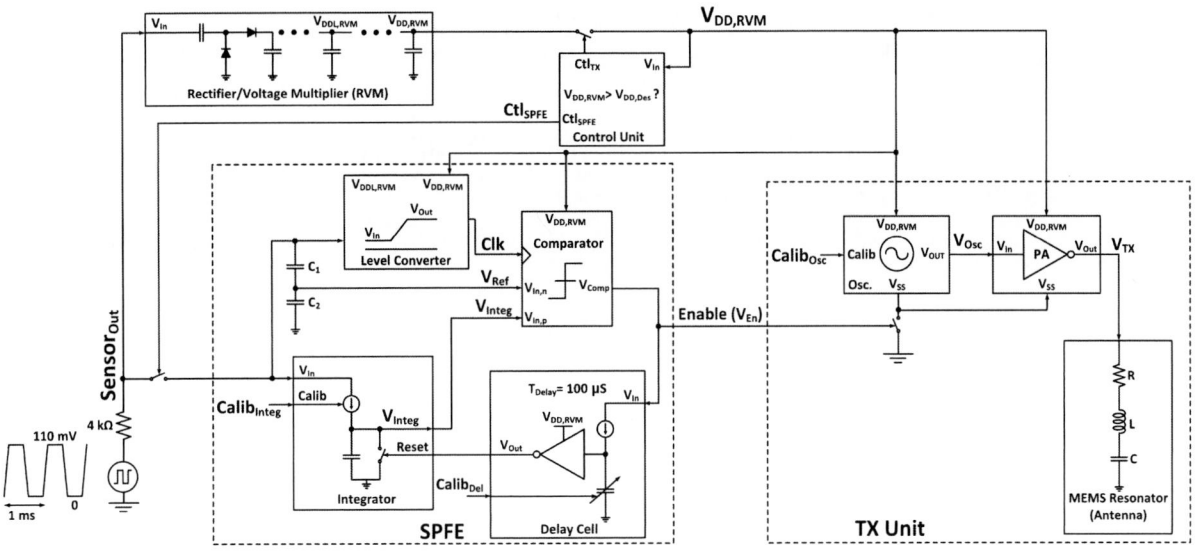

Fig. 2. Architecture of the proposed BI system, including self-supplied and RVM-supplied sections.

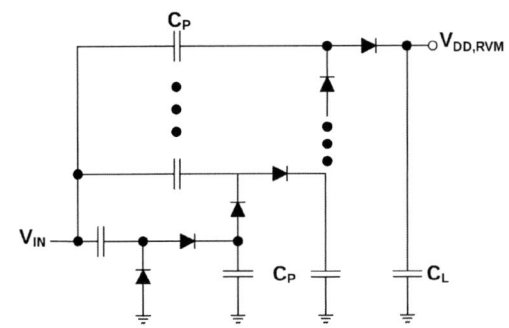

Fig. 3. Dickson rectifier/voltage multiplier (RVM) used in the BI system.

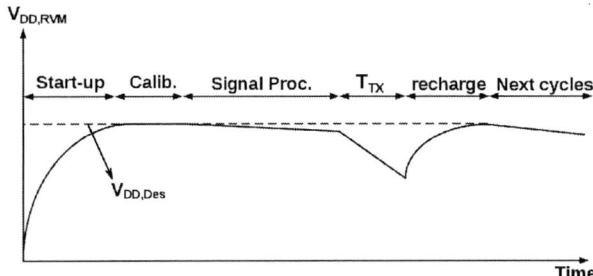

Fig. 4. Timing diagram of the RVM system, showing $V_{DD,RVM}$ in different modes of operation.

II. SYSTEM DESCRIPTION

Fig. 2 illustrates the high-level block diagram of the proposed BI system. A piezoelectric sensor, injected inside the brain, will be used to detect neuron activities [9]. The piezoelectric sensor exhibits an equivalent resistance of $R_T = 4k\Omega$, while the detected signal is periodic with an amplitude and frequency of about 110-mV$_{PP}$ and 1-KHz, respectively. Shown in Fig. 2, the sensor output is fed into the RVM block (Fig. 3) in order to generate a robust supply for the wireless TX unit. Moreover, the sensor output directly supplies the SPFE unit, which performs some signal processing operations.

An integrator circuit is directly connected to the sensor output to collect information by accumulating charge on a storage capacitance. This integrator which is working as an input sampler, filters the input noise and provides the required data for the rest of the SPFE unit, simultaneously. As soon as the integrator output reaches a target voltage of V_{Ref}, a comparator triggers the transmitter circuitry for a period of

T_{TX}. During this period, the TX is activated and transfers data. After the transmission period, the control unit resets the integrator and shuts down the transmitter. The reference level, V_{Ref}, is chosen based on observing a certain number of pulses at the input of the BI circuit, which is typically happening at $\Delta T_{Target} = 50ms$.

As illustrated in Fig. 4, the RVM block needs a one-time start-up in order to reach the stable output voltage of around 400 mV, i.e.: $V_{DD,Des} \approx 400$ mV. As the load on the RVM block is very low during signal processing and counting operation, there is a very little drop in its output during this mode of operation. Each time that transmitter is activated, the output of RVM will drop gradually, which will be recovered in the time period between two transmissions, as shown in Fig. 4.

A conventional Power-On-Reset (POR) circuit is employed as the control unit for the timing management of the system at the startup. The sensing mode starts when $V_{DD,RVM}$ reaches the desired value, $V_{DD,Des}$. Thanks to the ULP operation of

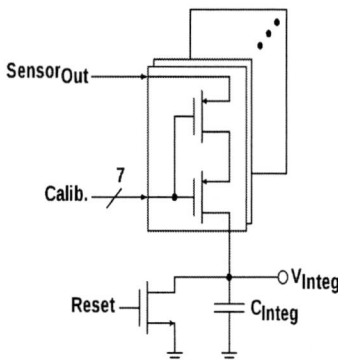

Fig. 5. The proposed integrator, constructed of an array, to be calibrated using 7 control bits.

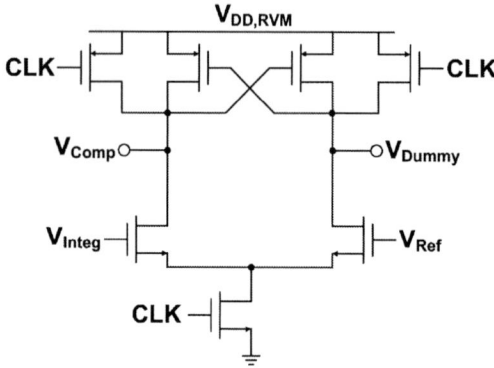

Fig. 6. StrongArm comparator used to detect the target event.

the SPFE unit, $V_{DD,RVM}$ drop during this mode is small enough ($\approx$ 5mV) not to affect the SPFE performance. In data transmission mode, turning on the TX unit leads to a larger drop on $V_{DD,RVM}$ ($\approx$ 40mV). The control unit senses this voltage and after a period of T_{TX}, produces a standby command to rest of the system in order for the RVM block to recharge its output voltage for the next cycle of operation. the transmission period, $T_{TX} = 100$ μs, is chosen such that the drop is yet small and does not affect the amount of the transmitted energy.

III. BLOCK DESCRIPTION

A. Charge-Pump Rectifier/Voltage Multiplier (RVM)

A conventional Dickson charge pump multiplier [7], in which its first stage is included in Fig. 2, is used to implement the RVM stage. Available isolated Schottky diodes that provide a large enough ratio between their forward and leakage current are selected in this work as the rectifying device. The output voltage, $V_{DD,RVM}$, of an N stage Dickson voltage multiplier for a desired load current of I_{Load} can be found through [7]:

$$V_{Out} = N \times (V_a - V_d)(\frac{C_P}{C_P + C_{Par}}) - \frac{I_{Load} \times N}{C_P \times f} \quad (1)$$

where V_a is half of the input peak-to-peak voltage, V_d is the voltage drop across each diode, C_P is the pumping capacitance in each stage, C_{Par} models the total parasitic capacitance at each node, and f represents the input signal frequency. The second term in (1) models the output series impedance of the RVM, which is inversely proportional to f and hampers drawing an arbitrarily large amount of current from the output node. Consequently, large capacitance values are needed to draw even a very small amount of current in low input amplitudes and frequencies. In this design, a 6-stage voltage multiplier (N=12) with $C_P = 50$ pF to reach $V_{DD,RVM} = 400$ mV, and $V_{DDL,RVM} = 265$ mV. While all of the C_P capacitors are integrated on-chip, an off-chip load cap of 10 nF is used to keep the supply drop within an acceptable range in the TX mode.

B. Signal Processing Front-End (SPFE)

Fig. 5 demonstrates the proposed zero-power integrator. A storage capacitance, C_{Integ}, is being charged through an array of current sources, each unit branch consists of two series PMOS transistors that are operating in the sub-threshold region. Hence:

$$I_{Integ} = N \times I_S \times e^{\frac{V_{SG} - |V_{th}|}{nV_t}} \cdot (1 - e^{\frac{-V_{SD}}{V_t}}) \quad (2)$$

where, N is the number of unit current sources that are turned on, I_S is the sub-threshold saturation current of the PMOS device, n is the sub-threshold slope factor, V_{th}, V_t, V_{SG} and V_{SD} are the threshold, Thermal, source-gate and source-drain voltages, respectively. Also:

$$I_{Integ} = C_{Integ} \frac{V_{Integ}}{\Delta T_{target}} \quad (3)$$

where, V_{Integ} is the voltage across C_{Integ} and $\Delta T_{target} = 50$ ms. Since only ΔT_{target} is known, we should reasonably select the other three parameters. $V_{Integ,max}$ is equal to the input amplitude, 110 mV. However, choosing values close to the maximum voltage reduces V_{SD} towards zero, and consequently pushes the target event closer to the nonlinear region of the I_{Integ} and V_{Integ} curve due to the last term in (2). This non-linearity adversely affects the calibration of the integrator over different process corners. Hence, the desired V_{Integ} (V_{Ref}) is chosen to be 50 mV. Obviously, a lower I_{Integ} is chosen, the smaller C_{Integ} is required. Therefore, the bulk terminals of the PMOS devices are connected to $V_{DD,RVM}$, in order to increase their threshold voltage. Moreover, longer channel devices are utilized to reduce V_{th} variation over different corners. Thus, I_{Intrg} is selected to be 100 pA. Based on (3), and having the other three parameters known, C_{Integ} is equal to 100 pF. The integrator current is calibrated over different process corners through a set of 7-bit control signal.

A conventional strongArm circuit, shown in Fig. 6, compares V_{Integ} with a reference voltage, V_{Ref}, to detect the target event. In order to further simplify the SPFE design and reduce the power consumption, both V_{Ref} and comparator clock signals are generated directly using sensor output. V_{Ref} is

978-1-7281-5410-7/20 $31.00 © 2020 IEEE

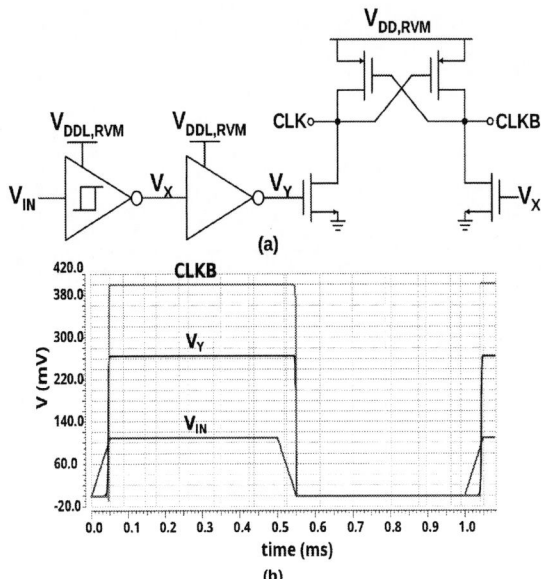

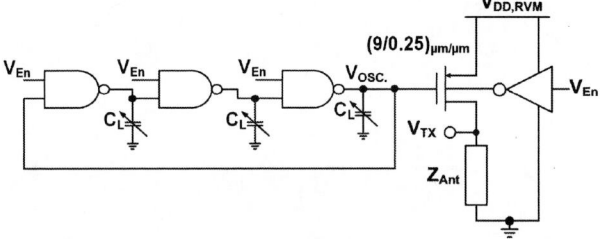

Fig. 9. A 3-stage gated ring oscillator, and the proposed leakage controlled PA.

Fig. 7. (a) Proposed level converter circuit, and (b) its input/output waveform signals.

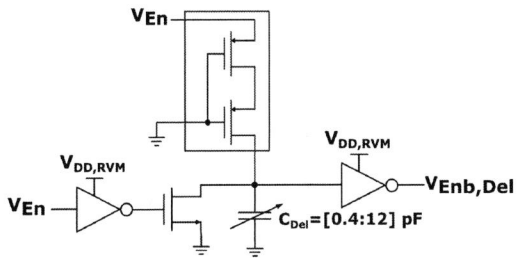

Fig. 8. The proposed micro-second delay generation circuit.

produced by using a capacitive divider connected to the sensor output. As shown in Fig. 2, $C_1 = 1$ pF, and $C_2 = 1.2$ pF to have a $V_{Ref} = 50$ mV. For accurate detection, the comparator clock signal is needed to be synchronized with the sensor output frequency and phase. Hence, an ultra-low-power multi-step level converter is proposed to generate the clock signal directly from the sensor output. Fig. 7 represents the level converter schematic and its input/output waveform. A multi-step level conversion approach is adopted to not only maintain the functionality over different process corners, but also to reduce the leakage current of the pull-up devices when they are supposed to be off. In the first step, the sensor output pulses (110 mV_{PP}) is converted to 265 mV_{PP} through two inverter stages with hysteresis supplied from $V_{DDL,RVM}$. Then, a circuit with a weak positive feedback converts the signal to a rail-to-rail clock, between the ground and $V_{DD,RVM} \approx 400$ mV. $V_{DDL,RVM}$ which is an intermediate stage in the RVM, is selected such that to minimize the leakage current in the off-state.

Fig. 8 shows the proposed delay cell circuit. The idea is

based on charging a storage capacitor until it reaches the threshold voltage of the following inverter stage. As the enable signal (comparator's output) rises, the unit PMOS current source starts to charge the capacitance and as soon as it reaches the threshold voltage of the inverter, the output signal goes to zero. Once the enable signal falls, the storage capacitance is reset to the ground potential, while the output signal rises consequently. The load capacitance, C_{Del} is calibrated through a 5-bit control circuit to compensate for the variations of integrator current and the inverter threshold voltage over PVT.

C. TX Circuit

Fig. 9 shows the TX circuitry which is composed of a gated ring oscillator and an output power amplifier (PA) that drives a MEMS resonator load, acting as an antenna. A 3-stage gated ring oscillator with NAND delay gates is used to simplify turning on and off the transmitter circuit using V_{En} signal. The delay of the NAND gates is calibrated through a capacitive bank array to keep frequency variation within $\pm 3\%$ of the target carrier frequency over PVT.

The PA consists of a properly-sized PMOS transistor in order to drive the output load, while the carrier frequency is produced by the preceding oscillator. Given the large size of the PA PMOS device, the leakage current of this device can excessively increase in fast corners, causing extra drop over $V_{DD,RVM}$. To mitigate this issue, a small size PMOS device is selected to implement the PA. To improve the current driving during the TX mode, an inverter is utilized to control the bulk potential of this device and reduce its threshold voltage during the transmission period. Shown in Fig. 9, this inverter connects the bulk terminal of the PMOS driver to $V_{DD,RVM}$ (in the sensing mode), and to ground (in the TX mode). Using this technique, the threshold voltage of the PMOS device reduces from $|V_{th}| = 550$ mV to $|V_{th}| = 380$ mV. Therefore, while the PMOS leakage is kept small in the sensing mode, its driving strength is kept high enough in the TX mode.

IV. RESULTS AND DISCUSSION

The SPFE loop which is the heart of the BI system, is simulated over different process corners (37 °C) for functionality verification. Fig. 10 depicts the integrator output voltage over different corners. In addition to the voltage value at which the event is detected, V_{Ref}, the time that it takes for

978-1-7281-5410-7/20 $31.00 © 2020 IEEE

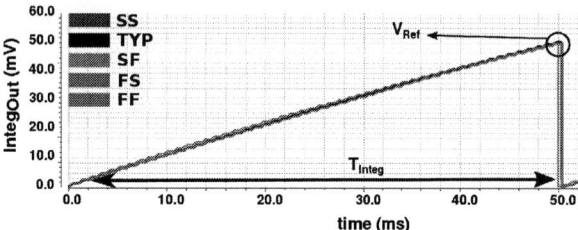

Fig. 10. Simulated integrator output voltage over different process corners.

TABLE I
CURRENT DRAWN FROM $V_{DD,RVM}$ BY EACH BLOCK IN THE PROPOSED
BI SYSTEM.

Block	Sensing Mode	Transmission Mode
Integrator	0	0
Comparator	250 pA	≈ 0
Level converter	200 pA	≈ 0
Other blocks in SPFE loop	40 pA	≈ 0
Oscillator	80 pA	125 nA
Output PA	125 pA	2.0 μA

the integrator to reach this voltage, T_{Integ}, needs to be very well controlled as well. Based on Fig. 10, the maximum variation over $T_{Integ} = 50$ ms for different corners remain below ± 0.7 ms.

Fig. 11 shows the functionality simulation result of the entire BI system depicted in Fig. 1. As discussed before, $V_{DD,RVM}$ is very stable during the sensing mode. Detecting the target event by comparator triggers the transmitter circuitry, which causes a noticeable drop in $V_{DD,RVM}$ due to current drawn by the PA. The voltage drop during the TX mode causes a gradual drop during the TX mode, shown in Fig. 11.

Table I provides a detailed report on the consumption of each block from $V_{DD,RVM}$ in the proposed BI system. The ULP SPFE consumes less than 500 pA in sensing mode which makes it a promising solution for ultra-low-power WSN or implantable sensors. In addition, the entire TX consumes an average current of 2.125 μA in the transmission mode.

V. CONCLUSION

An ultra-low-power brain interface (ULP BI) circuit is presented, which significantly simplifies the power management scheme by performing most of the signal processing tasks with near-zero power consumption. The proposed system exploits the incident sensor power for both signal processing and data transmission to reduce the required energy from a charge-pump based voltage multiplier. While signal processing front-end circuit consumes less than 500 pA current during the sensing mode, the transmitter circuit consumes an average current of 2.125 μA in the data transmission mode.

REFERENCES

[1] T. Shimamura et al., "MEMS-switch-based power management with zero-power voltage monitoring for energy accumulation architecture on dust-size wireless sensor nodes," *Symp. VLSI Circ.*, pp. 276-277, 2011.

[2] T. Ruan, et all., "Energy-aware approaches for energy harvesting powered wireless sensor nodes," *IEEE Sens. J.*, vol. 17, no. 7, pp. 2165-2173, 2017.

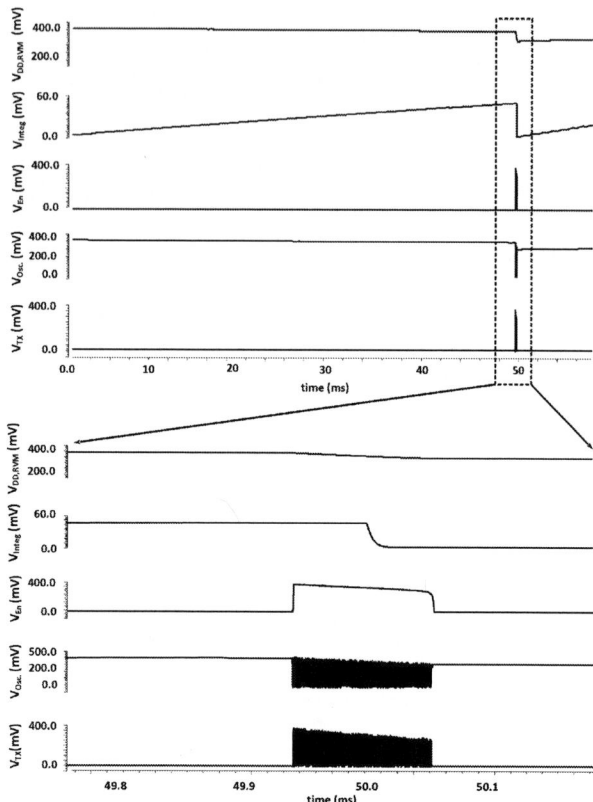

Fig. 11. Functionality simulation of the BI system.

[3] S. Bose, et all., "Fully-integrated 57 mV cold start of a thermoelectric energy harvester using a cross-coupled complementary charge pump," *2018 IEEE Cust. Integr. Circ. Conf. (CICC)* pp. 1-4, 2018.

[4] E. Rahiminejad, et all., "A Power-Efficient Signal-Specific ADC for Sensor-Interface Applications," *IEEE Transactions on Circuits and Systems II: Express Briefs,* vol. 64, no. 9, pp. 1032-1036, Sept. 2017.

[5] M. Tabib-Azar, "Nervous System Interface Device," US Patent 20160128589.

[6] P. Saffari, et. all., "An RF-powered wireless temperature sensor for harsh environment monitoring with non-intermittent operation," *IEEE Trans. Circ. Syst. I*, vol. 65, no. 5, pp. 1-14, 2017.

[7] J. F. Dickson, "On-chip high-voltage generation in NMOS integrated circuits using an improved voltage multiplier technique," *IEEE J. Solid-State Circ.,* vol. 11, no. 3, pp. 374-378, June 1976.

[8] U. Karthaus and M. Fischer, "Fully integrated passive uhf rfid transponder IC with 16.7-μW minimum rf input power," *IEEE J. Solid-State Circ.,* vol. 38, no. 10, pp. 1602-1608, 2003.

[9] K. Sinha and M Tabib-Azar, "Remote power transfer using magneto-electric devices," *J. Phys.: Conf. Ser.*, 2015.

Energy and Area Efficient Mixed-Mode MCMC MIMO Detector

Amin Aghighi, Behrouz Farhang-Boroujeny, and Armin Tajalli

Electrical and Computer Engineering Department, University of Utah, Salt Lake City, USA

Abstract—**A hybrid analog/digital signal processor has been proposed to implement energy-efficient multi-input-multi-output (MIMO) detectors. A sub-optimum MIMO detector based on Markov Chain Monte Carlo (MCMC) algorithm for a 4×4 MIMO system is presented. The main part of the required signal processing occurs in analog domain to reduce power consumption. The outputs of the proposed analog processor are converted to digital using a low-resolution analog-to-digital converter (ADC) in order to close the loop in digital domain. The proposed 4×4 MCMC MIMO detector is designed in a conventional 45 nm CMOS technology, that consumes 29.3 mW from 1.0 V supply. A throughput of 235.3 Mbps is achieved while operating at 1.0 GHz clock frequency. The design occupies 0.11 mm^2 silicon area.**

Index Terms—**Optimum detector, Markov Chain Monte Carlo (MCMC), VLSI MIMO, Mixed-mode MIMO, Mixed-mode circuits.**

I. INTRODUCTION

Modern wireless communications use the Multi-Input Multi-Output (MIMO) approach to improve data throughput at a lower cost. Moreover, the ever-growing number of users makes MIMO systems even more desirable [1]. Since MIMO systems use the same frequency band for transmitting parallel data streams, data transfer bandwidth improves with the number of transmit antennas [2]. Therefore, receiver-joint detection is crucial for exploiting the full capacity of the system. Although the optimum detectors can harness the full channel capacity, their complexity increases exponentially with the number of transmit antennas [3]. As a result, improving the performance of sub-optimum detectors is a demanding research topic [4]–[7]. Few implementations for sub-optimum detectors operating based on the Markov-Chain Monte Carlo (MCMC) algorithm are reported in [2], [8], [9]. Although this detector can achieve full channel capacity, the existing implementations, which are mainly based on Digital Signal Processors (DSP), result in fairly complex and power-hungry circuits. This paper targets lowering power consumption and increasing throughput of the MCMC detectors by moving high-speed and energy-hungry operations from DSP to analog/mixed-mode domain. A set of system-level simulations are carried out to show the performance of the proposed analog/mixed-mode approach. Several analog building blocks are proposed to implement target signal processing schemes in a more energy-efficient way. The power and area cost of these blocks are

calculated through simulations to have a good cost estimation of the proposed detector.

The rest of this paper is organized as follows: Section II provides a brief overview on MCMC detectors, and describes the high level implementation of the proposed MCMC detector. System level simulation results are presented in Section III. Section IV demonstrates circuit-level implementation, and Section V provides comparison between he proposed MCMC detector and the state-of-the-art.

II. SYSTEM LEVEL MCMC DETECTOR

A. Conventional MCMC Detectors

Considering a flat fading channel. The inputs and outputs for each channel use can be described by:

$$y = A \cdot d + n \qquad (1)$$

where, $\mathbf{y}$ is the received data vector, $\mathbf{d}$ is the transmitted data bits, $\mathbf{n}$ is the channel additive noise vector, and $\mathbf{A}$ is the channel gain matrix. The MCMC detector is based on an iterative approach. While the detection algorithm is thoroughly described in [2], a brief concept review is presented here. The MCMC detector takes a random set of initial bits, b_0, by a Gibbs sampler [10], and calculates the error function based on:

$$e_0 = y_0 - A \cdot b_0 \qquad (2)$$

where each element of $\mathbf{e_0}$ represents the initial error associated with each channel. In other words:

$$e_{0,i} = y_{0,i} - (A_{1,i} \cdot b_{0,1} + A_{2,i} \cdot b_{0,2} + \cdots + A_{N,i} \cdot b_{0,N}) \quad (3)$$

where, i is the desired channel index and N is equal to the number of transmit antennas. When all of the $\mathbf{e_0}$ elements are found, one of the bits in b_0 will be flipped in each iteration, and the error vector, $\mathbf{e}$, will be recalculated. The new error vector in each iteration can be readjusted from the previous value:

$$e_k = e_{k-1} + A_{:,m} \cdot (2b_{(k-1),m}) \qquad (4)$$

where, k is the index for the iteration steps, m is the index for the bit which is flipped, and $\mathbf{A_{:,m}}$ represents the m-th column of $\mathbf{A}$. In order to compare the error functions between every two consecutive iterations, and decide about the m-th bit, the summation of squared components of $\mathbf{e}$ is calculated and defined as:

$$E = e_1^2 + e_2^2 + \cdots + e_N^2 \qquad (5)$$

978-1-7281-5410-7/20 $31.00 © 2020 IEEE

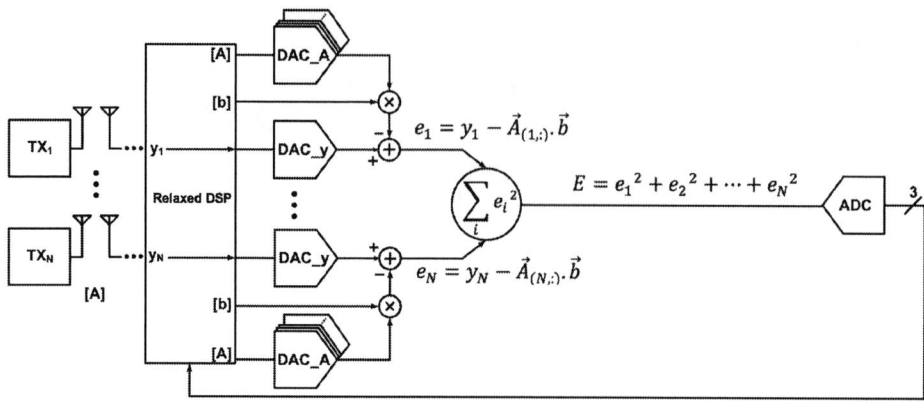

Fig. 1. The proposed MCMC detector block diagram.

Based on the E value at each iteration step, it will be decided to keep either "+1" or "-1" for the m-th bit. When the decision has been made for all the bits in b_0, the MCMC detector repeats this operation for N_{gs} times for different initial conditions. The performance (accuracy) of the MCMC detector improves when T parallel Gibbs samplers run with different initial random set of bits [2]. As including randomness to the decision process reduces the stalling problem, the MCMC detector introduced in [2] utilizes a random variable, v, and considers both of E and v for making decisions.

Usually, MCMC detectors employ Forward Error Correction (FEC), in order to reduce the Bit Error Rate (BER). In this work, we will analyze only the raw BER, before applying FEC.

B. Proposed MCMC Detector Circuit

Fig. 1 shows the proposed MCMC detector, in which most of the speed-limited operations are moved into the analog domain. In order to calculate each component of the error vector, e, a set of digital-to-analog converters (DACs) are employed to convert the digitized received data, y, and error terms, e, to the analog domain. The DACs are implemented based on very low-power and low complexity circuits. Bit-wise multiplication is implemented in the analog domain to multiply DAC_A by b_0, and then simply subtract the result from DAC_y to produce the new error vector, e. As the summation occurs in current domain, indeed no extra hardware is required. There are N parallel operators in total needed to produce all the components of the e. A square generator produces the e_i^2 for each channel separately and adds them together to estimate the updated value for E. Finally, a low-resolution analog-to-digital converter (ADC) closes the loop by digitizing E value, leaving the final step of decision, i.e. determining b_0, to digital circuits. As will be shown later, the entire operation explained above can be implemented very simple analog circuits and switches, occupying a small area and occurs in only one clock cycle.

Considering that it takes one cycle to generate each component of e, having N transmit antennas and based on (2), it takes one clock cycle to produce N components of e_0 by the array

of DAC_A, and DAC_y, which are working in parallel. Since all of the building blocks in Fig. 1 are synchronized, and working at the same frequency, producing and digitizing E_0 also occurs at the same clock rate. Hence, it takes one clock cycle to fulfill all the operations shown in Fig. 1. Based on (3), another clock cycle is required to calculate e_k, and decide about k-th bit of $k \geq 1$. Therefore, while making decision for the first bit takes two clock cycles, the next bits require only one clock cycle to be determined. Hence, it takes $(N + 1)$ clock cycles in order to determine the polarity of all the bits in b.

III. SYSTEM LEVEL SIMULATION RESULTS

High-level simulations have been carried out to determine the performance of the circuit shown in Fig. 1, and determine the required specifications for the key building blocks in this architecture. Some design parameters, such as the number of parallel Gibbs samplers, T, the number of iterations for each Gibbs sampler, N_{gs}, and the required resolution for each data converters will be determined base don this study. For simplicity, it is assumed that each antenna transmits bits over only one carrier, i.e. there is no sub-carrier. The number of transmit antennas is set to N. Also, in order to have enough number of bits for calculating BER, the whole MCMC detector iterates for $N_{MC} = 500$ times, which results in $N_{MC} \times T \times N_{gs} \times N$ total number of bits. Here, T and N_{gs} are the most critical design parameters that determine the cost and the performance of the detector. This work uses the same T and N_{gs} values that have been employed in [2].

Fig. 2 shows Bit-Error-Rate (BER) versus signal-to-noise ratio (SNR) for the received signal, while T and N_{gs} have been swept. As it is expected, BER improves by N_{gs} and T. However, improvement for $N_{gs} = T > 10$ is marginal. It is reported in [2] that $N_{gs} = T = 6$ to 8 is a good compromise.

In order to determine the right resolution for each of the data converters (DAC_A, DAC_y and ADC), a two-step verification has been implemented. Initially, the resolution of each data converter has gradually reduced, while all the other data converters is considered to be infinite. This approach prevents the other data converters to contribute to the total quantization

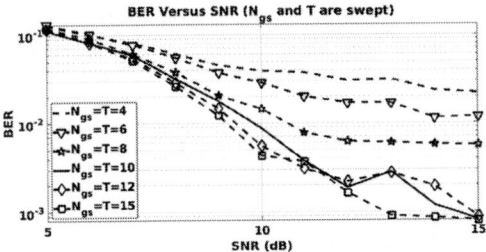

Fig. 2. BER versus SNR while sweeping N_{gs} and T.

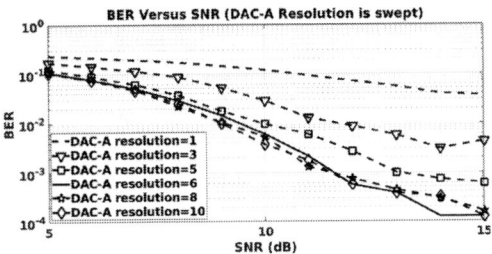

Fig. 3. BER versus SNR while sweeping DAC_A resolution.

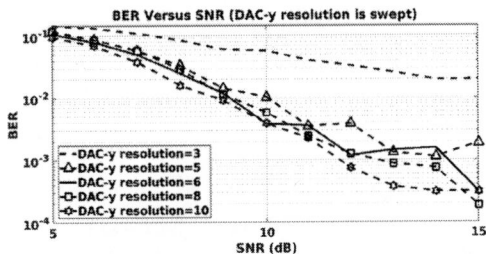

Fig. 4. BER versus SNR while sweeping only DAC_y resolution.

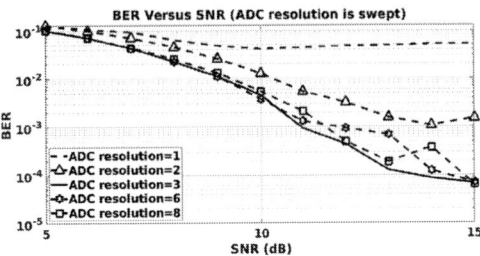

Fig. 5. BER versus SNR while sweeping the resolution of ADC (only).

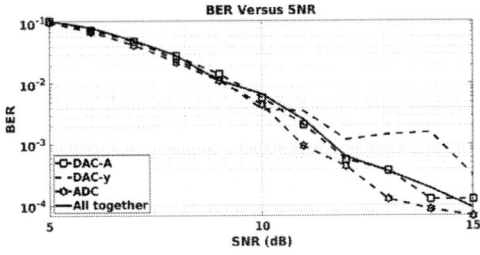

Fig. 6. BER versus SNR when including all the data converters.

noise, and affect the system performance. In the second step, the entire system is simulated while all the data converters have been employed with their limited resolution. Fig. 3 to Fig. 5 represent the MCMC performance, while the resolution for DAC_A, DAC_y and ADC have been swept, respectively. As it is shown in Fig. 3, increasing DAC_A resolution higher than 6 bits, does not improve the MCMC performance. Therefore, the resolution of DAC_A needs to be 6b or more, in order not to lose much performance. When a resolution of DAC_A is more than 6b, the system needs to be simulated for a much longer time in order to produce a precise BER estimation. As can be seen in Fig. 3, the results for resolutions higher than 6b is not very accurate. Based on the results shown in Fig. 4, and Fig. 5, one can conclude that the minimum resolution for DAC_y and the ADC are 6b and 3b, respectively.

Fig. 6 represents the second step of our analysis, in which the resolution of all the data converters have been limited to the values discussed above. Based on these results, the performance of the proposed MCMC detector, which is utilizing a realistic model for the data converters, is very close to those reported in Figs. 3 to 5.

IV. CIRCUIT LEVEL SIMULATIONS

This Section demonstrates circuit-level implementations and simulation results for the proposed detector shown in Fig. 1.

Fig. 7 shows the merged DAC_A, DAC_y, and the multiplier for producing the error vector, e. Current mode logic (CML) based circuits have been used to simplify the design, and also make it possible to linearly operate at very high frequencies with a low level of consumption and complexity [11]–[14]. While DAC_A and multiplier are shown in the right-hand side of the schematic (Fig. 7), the programmable differential pair

at the left-hand side represents DAC_y. Current-mode DAC architecture has been used to simplify the multiplication and summation operations. Depending on sign of **b** values, the output of DAC_A is multiplied by "+1" or "-1". The outputs of the N parallel DACs are shorted properly to implement a summer, as required in (3).

The consumption of the DACs depends on the resolution as well as their speed of operation. In order to make sure that the DACs can operate properly at the desired clock frequency, the time constant at the output node should be chosen carefully:

$$R_L < 1/(2\pi C_{L,DAC} \times 5 f_{bit}) \qquad (6)$$

where $C_{L,DAC}$ is the DAC load capacitance, and f_{bit} is the input bit frequency. Here, a factor of five is considered to assure settling with an error of less than 1%. The maximum DAC output swing will be achieved when all of the binary-weighted current sources are turned on in all of the $N + 1$ parallel DACs on each channel. Since there are a total of $2^6 - 1$ copies of the unit current source for each DAC, the

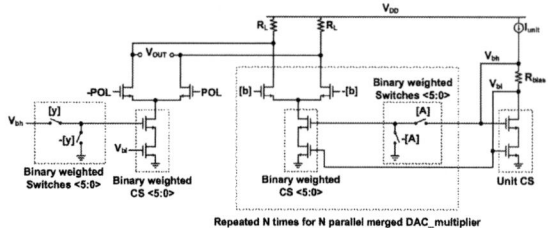

Fig. 7. DAC$_A$ merged with multiplier and DAC$_y$ for producing e function.

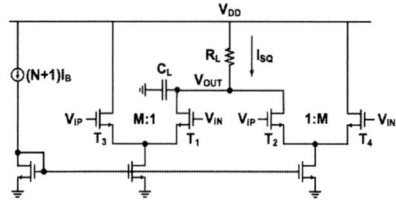

Fig. 8. Squarer circuit schematic.

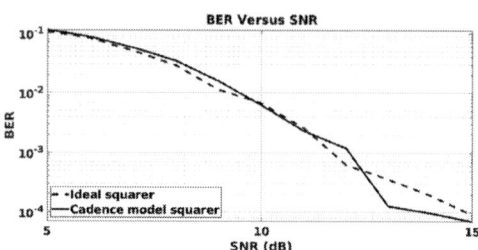

Fig. 9. Input-output transfer characteristics of the simulated squarer circuit.

Fig. 10. System level simulation results with ideal and transistor level squarer circuit.

minimum required unit current, I_{unit}, is equal to:

$$I_{unit} = \frac{V_{swing,sq}}{(2^6 - 1) \times (N + 1) \times R_{L,DAC}} \quad (7)$$

where, $V_{Swing,sq}$ is the squarer required input swing, and R_L is the load resistance determined by (6). Fig. 8 represents the circuit level implementation of the squarer block. Given that all of the devices work in the saturation region, and assuming ideal long channel device characteristics, it can be shown that [15]:

$$I_{SQ} = I_{D1} + I_{D2} = kV_{id}^2 + 2I_B \quad (8)$$

where, V_{id} is the input differential voltage, $k = (W/L)\mu_n C_{ox}$ and (W/L) is the aspect ratio of T1 and T2. There should be N parallel squarer blocks in an N×N MIMO system, which their outputs are combined together to produce E. Hence, N squarer output nodes have been shorted together to produce $I_{SQ,total}$:

$$I_{SQ,tot} = I_{SQ,1} + I_{SQ,2} + \ldots I_{SQ,N} \quad (9)$$

and E:

$$E = e_1^2 + e_2^2 + \cdots e_N^2 = I_{SQ,tot} \times R_L \quad (10)$$

While having large M as the ratio between transistors T1/T2 to T3/T4 increases the squaring accuracy [15], a ratio of M=4 provides enough accuracy for our system. In order to prove this, the system is simulated while replacing the ideal squarer by the transistor level circuit. Fig. 9 shows the input-output transfer characteristics of the squarer circuit with an ideal squarer, whose gain is 4 [V/V]. The offset introduced by the last term in (8) can be removed by comparing the output of the target circuit with a reference (replica) circuit. As can be seen in Fig. 9, the maximum error is limited to about 25%.

Fig. 10 compares the expected system performance with a system that uses transistor-level squarer circuit. Here, a realistic model for all of the other building blocks including data converters has been included. Although we are using a replica circuit to eliminate the offset of the squarer circuit, system-level simulations show that the absolute value of the offset does not influence the system performance. Based on Fig. 10, the system performance with a transistor-level model of the squarer circuit is consistent whit that of an ideal squarer.

A careful design procedure needs to be employed to minimize the energy consumption of the circuit shown in Fig. 7, while the dynamic range is maintained. Based on this design procedure, the size of the load resistance, R_L, and the bias current of the DAC unit cells can be determined.

The 3b ADC block in Fig. 1 has been implemented using a conventional flash ADC structure, that employs StrongArm based comparator topology, shown in Fig. 11. Since PVT variations affect the output swing of the squarer, two replicas of the squarer are provided to determine the maximum and minimum reference levels for the ADC. While one of the replicas mimics the squarer when input swing is minimum, i.e. $V_{id} = 0$, the other replica produces the expected squarer output swing with $V_{id} = V_{id,max} = 180$ mV. These two voltages are then utilized to generate different reference levels for comparators using a resistor ladder.

V. COMPARISON AND DISCUSSION

This Section provides a high-level comparison between the proposed MCMC detector and the state-of-the-art, especially [2], as well as other MIMO detector implementations [16]–[18].

Table I reports the detailed occupied area and power consumption of the proposed MCMC detector while considering biasing circuits and other auxiliary blocks. Pessimistic

978-1-7281-5410-7/20 $31.00 © 2020 IEEE

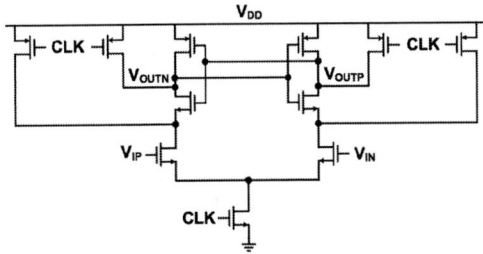

Fig. 11. StrongARM comparator that is used in the ADC implementation.

TABLE I
AREA AND POWER CONSUMPTION OF THE PROPOSED MCMC DETECTOR

	DAC	Squarer	ADC	DSP	1 Gibbs iteration
Power, mW	3.87	1.6	0.84	1.0	7.3
Area, mm^2	0.012	0.0026	0.011	0.002	0.0276
Power Share, %	53.2	21.8	11.4	13.6	100
Area Share, %	43.4	9.6	39.8	7.2	100

parasitic capacitance estimation is also included to account for routing and layout considerations. Based on the results in table I, the DAC arrays have the biggest contribution in the total power consumption. In terms of area, DACs and the ADC are the most dominant contributors. The entire system of Fig. 1 consumes 7.3 mW, while occupying a core area of 0.0276 mm^2.

The MCMC presented in [2] is implemented on an FPGA, which makes it hard to have a detailed comparison with our proposed implementation. However, they have synthesized their proposed detector in a 130-nm VLSI IBM process. Therefore, we used their synthesized simulation results to compare it with that of our proposed mixed-mode MCMC detector. Table II compares the performance of the proposed MCMC detector with the synthesized version reported in [2]. For a fair comparison, similar system parameters have been selected (e.g. throughput per Gibbs sampler, which is defined as $f_{clk}/(N_{gs}T)$). It is assumed that $T = N_{gs} = 4$ for both cases. In addition, the area efficiency is defined to be the ratio of the occupied area of each Gibbs sampler to the throughput per Gibbs sampler. Proper scaling factors have been employed to convert power and area between the two technologies. Based on Table II, our proposed MCMC detector area efficiency outweigh that of [2] by a factor of about 2.66. Unfortunately, there is no power consumption reported for the implementation in [2].

Table III provides a performance comparison between the proposed mixed-mode MCMC detector and some of the state-of-the-art MIMO detectors with VLSI implementations. The throughput of the proposed MCMC detector is calculated based on:

$$Throughput = \frac{N_{ant} \times f_{clk}}{N_{Cycles}} \qquad (11)$$

where, N_{ant} is the number of antennas, N, f_{clk} is the clock frequency, and N_{Cycles} is the total number of cycles which is

TABLE II
PERFORMANCE COMPARISON WITH SYNTHESIZED VERSION IN [2]

	This Work, 45 nm	[2], 0.13 μm
Power, mW	7.3	N/A
Area, mm^2	0.0276	0.37
Clock Freq, MHz	1000	620
Throughput per Gibbs Sampler, Mbps	62.5	38.75
Area Efficiency mm^2/(μm × Mbps)	0.21	0.56

TABLE III
PERFORMANCE COMPARISON WITH STATE-OF-THE-ART MIMO DETECTORS

	[18]	[17]	[16]	This Work
SD Algorithm	Relaxed K-best	MMSE-PIC	LMMSE	MCMC
Technology	0.13μm	90nm	65nm	45nm
Supply, (V)	1.3	1.2	1.2	1.0
Clock Freq, (MHz)	270	568	400	1000
Throughput, (Mbps)	8.57	757	600	235.3
Area, (mm^2)	2.38	1.5	1.4	0.11
Power, (mW)	94	189.1	266	29.44
Area Efficiency (mm^2/(Mbps× μm^2))	16.4	0.245	0.552	0.23
Power Efficiency*, (pJ/b)	2920.6	104.08	255.77	125.11
Simulation (S)/ Measurement (M)	M	M	M	S

*Normalized to 45 nm technology

equal to:

$$N_{Cycles} = (N_{gs} \times N) + 1 \qquad (12)$$

Also, power efficiency (PE) for each reference is normalized to 45 nm technology as calculated by:

$$PE = Power \times \frac{1(V)}{Supp.(V)} \times \frac{45(nm)}{Tech.(nm)} \times \frac{1}{Throughput} \qquad (13)$$

VI. CONCLUSION

An analog/mixed-mode approach for designing MCMC MIMO detectors is presented. The proposed system relaxes some of the complexities in the design of conventional digital detectors, especially by moving some high-speed operations to analog domain. While the proposed system consumes 29.3 mW, the proposed detector operates at 1 GHz clock frequency. Achieving a throughput of 235.3 Mbps, the circuit occupies 0.11 mm^2 Silicon area (estimated). The proposed approach can be applied for implementing similar processing systems in which speed and energy efficiency is the concern.

REFERENCES

[1] G. J. Foschini, "Layered space-time architecture for wireless communication in a fading environment when using multi-element antennas," *Bell labs technical journal*, vol. 1, no. 2, pp. 41–59, 1996.

[2] S. A. Laraway and B. Farhang-Boroujeny, "Implementation of a markov chain monte carlo based multiuser/mimo detector," *IEEE Transactions on Circuits and Systems I: Regular Papers*, vol. 56, no. 1, pp. 246–255, 2008.

[3] S. Verdu, "Minimum probability of error for asynchronous gaussian multiple-access channels," *IEEE Trans. on Information Theory*, vol. 32, no. 1, pp. 85–96, 1986.

[4] H. Zhu, Z. Shi, and B. Farhang-Boroujeny, "Mimo detection using markov chain monte carlo techniques for near-capacity performance," in *Proceedings.(ICASSP'05). IEEE Int. Conference on Acoustics, Speech, and Signal Processing, 2005.*, vol. 3.   IEEE, 2005, pp. iii–1017.

[5] B. Farhang-Boroujeny, H. Zhu, and Z. Shi, "Markov chain monte carlo algorithms for cdma and mimo communication systems," *IEEE transactions on Signal Processing*, vol. 54, no. 5, pp. 1896–1909, 2006.

[6] J. C. Hedstrom, C. H. Yuen, R.-R. Chen, and B. Farhang-Boroujeny, "Achieving near map performance with an excited markov chain monte carlo mimo detector," *IEEE Transactions on Wireless Communications*, vol. 16, no. 12, pp. 7718–7732, 2017.

[7] H. El Gamal and A. R. Hammons, "A new approach to layered space-time coding and signal processing," *IEEE Transactions on Information Theory*, vol. 47, no. 6, pp. 2321–2334, 2001.

[8] D. Auras, U. Deidersen, R. Leupers, and G. Ascheid, "A parallel mcmc-based mimo detector: Vlsi design and algorithm," in *IFIP/IEEE Int. Conf. on Very Large Scale Integration-System on a Chip*.   Springer, 2014, pp. 149–169.

[9] U. Deidersen, D. Auras, and G. Ascheid, "A parallel vlsi architecture for markov chain monte carlo based mimo detection," in *Proceedings of the 23rd ACM Int. Conf. on Great Lakes Symp. on VLSI*, 2013, pp. 167–172.

[10] S. D. Tribble, "Markov chain monte carlo algorithms using completely uniformly distributed driving sequences," Ph.D. dissertation, Stanford University, 2007.

[11] A. Aghighi, J. Atkinson, N. Bybee, S. Anderson, M. Crane, A. Bailey, R. Morell, A. Hassanin, and A. Tajalli, "Cmos amplifier design based on extended gm/id methodology," *2019 17th IEEE Int. New Circuits and Systems Conference (NEWCAS)*, pp. 1–4, 2019.

[12] J. Atkinson, A. Aghighi, S. Anderson, A. Bailey, M. Crane, and A. Tajalli, "Multi-stage current-steering amplifier design based on extended gm/i d methodology," *2019 IEEE 62nd Int. Midwest Symposium on Circuits and Systems (MWSCAS)*, pp. 129–132, 2019.

[13] S. Shahsavari and M. Saberi, "A highly linear 8-bit m–2m digital-to-analog converter for neurostimulators," *IEEE Transactions on Circuits and Systems II: Express Briefs*, vol. 67, no. 6, pp. 989–993, 2020.

[14] A. Aghighi, A. H. Alameh, M. Taherzadeh-Sani, and F. Nabki, "A 10-gb/s low-power low-voltage ctle using gate and bulk driven transistors," *IEEE International Conference on Electronics, Circuits and Systems (ICECS)*, pp. 217–220, 2016.

[15] A. Gerosa, S. Soldà, A. Bevilacqua, D. Vogrig, and A. Neviani, "An energy-detector for noncoherent impulse-radio uwb receivers," *IEEE Transactions on Circuits and Systems I: Regular Papers*, vol. 56, no. 5, pp. 1030–1040, 2009.

[16] X. Chen, A. Minwegen, S. B. Hussain, A. Chattopadhyay, G. Ascheid, and R. Leupers, "Flexible, efficient multimode mimo detection by using reconfigurable asip," *IEEE Transactions on Very Large Scale Integration (VLSI) Systems*, vol. 23, no. 10, pp. 2173–2186, 2014.

[17] C. Studer, S. Fateh, and D. Seethaler, "Asic implementation of soft-input soft-output mimo detection using mmse parallel interference cancellation," *IEEE Journal of Solid-State Circuits*, vol. 46, no. 7, pp. 1754–1765, 2011.

[18] S. Chen, T. Zhang, and Y. Xin, "Relaxed k-best mimo signal detector design and vlsi implementation," *IEEE Transactions on Very Large Scale Integration (VLSI) Systems*, vol. 15, no. 3, pp. 328–337, 2007.

MIST: monitor generation from informal specifications for firmware verification

Samuele Germiniani*, Moreno Bragaglio[†] and Graziano Pravadelli[§]

University of Verona, department of computer science

Email: *samuele.germiniani@univr.it, [†]moreno.bragaglio@univr.it, [§]graziano.pravadelli@univr.it

Abstract—This paper presents MIST, an all-in-one tool capable of generating a complete environment to verify C/C++ firmwares starting from informal specifications. Given a set of specifications written in natural language, the tool guides the user in translating each specification into an XML formal description, capturing a temporal behavior that must hold in the design. Our XML format guarantees the same expressiveness of linear temporal logic, but it is designed to be used by designers that are not familiar with formal methods. Once each behavior is formalized, MIST automatically generates the corresponding test-bench and checker to stimulate and verify the design. In order to guide the verification process, MIST employs a clustering procedure that classifies the internal states of the firmware. Such classification aims at finding an effective ordering to check the expected behaviors and to advise for possible specification holes. MIST has been fully integrated into the IAR System Embedded Workbench. Its effectiveness and efficiency have been evaluated to formalize and check a complex test-plan for an industrial firmware.

Index Terms—assertion, verification, testing, simulation, checker, PSL, LTL, specification

The research has been partially supported by the project "Dipartimenti di Eccellenza 2018 -2022" funded by the Italian Ministry of Education, Universities and Research (MIUR); and with the collaboration of IDEA S.p.a.

I. INTRODUCTION

In the last few decades, verification has become one of the most crucial aspects of developing embedded systems. Thoroughly verifying the correctness of a design often leads to identifying bugs and specification holes far earlier in the deployment process, exempting the developing company from wasting resources in costly maintenance.

However, our experience suggests that many companies have to cut down the verification process due to lack of time, tools and specialized engineers. That is even more critical in case of firmware verification, which requires exceptional consideration to deal also with the underlying hardware.

Indeed, one of the main problems that prevent an effective and efficient firmware verification process is the incapability of formalizing the initial design specification, which is generally written in extremely long and ambiguous natural-language descriptions. Such descriptions risk to be differently interpreted by designers and verification engineers, as well as by the project's customers themselves, thus leading to the misalignment between the initial specification and the final implementation. Besides, the lack of formalization prevents the engineer from exploiting automatic tools for verification, with the consequent adoption of ineffective and inefficient (semi-)manual approaches. In particular, without a well-defined specification, it becomes impractical to define any formal or semi-formal verification strategy. Generally, those strategies require to describe the expected behaviors in terms of logic assertions unambiguously, and, in case of semi-formal approaches, to define a set of test-benches to stimulate the design under verification.

To fill in the gap, we present MIST: an all-in-one tool capable of generating a complete environment to verify C/C++ firmware starting from informal specifications. The tool provides a user-friendly interface to allow designers and their customers, which are not familiar with temporal logic, to formalize the initial specifications into a set of non-ambiguous temporal behaviors. From those, MIST generates a verification environment composed of monitors (checkers) and test-benches to verify the correctness of the firmware implementation automatically. Then, in order to guide the verification process, MIST employs a clustering procedure that classifies the internal states of the firmware. Such classification aims at finding an effective ordering to check the expected behaviors and to advise for possible specification holes.

The verification environment has been fully integrated with the popular IAR Embedded Workbench toolchain [1]. It has been evaluated for verifying the correctness of an already released industrial firmware, allowing the discovery of bugs that were never detected previously.

The rest of the paper is organized as follows. Section II summarizes the state of the art. Section III overviews the methodology. Sections IV, V, VI, VII explain in detail the methodology implemented in MIST. Section VIII reports the experimental results. Finally, in IX we draw our conclusions.

II. RELATED WORK

Formalization of specifications is the process of translating requirements of a design into logic properties that can be used to verify its correctness automatically. Usually, the procedure consists of two main steps. Firstly, the verification engineer has to disambiguate the informal specifications written in natural language. Secondly, a formal specification language must be adopted to formalize the specifications into logical formulas that will be used to verify the design. During the past decades, numerous approaches have been developed to perform verification with the above paradigm. [2] introduces an automated collaborative requirements engineering tool, called TestMEReq, to promote effective communication and collaboration between client stakeholders and requirements engineers for better requirements validation. [3] proposes an approach to verify firmware security properties using symbolic execution. The paper introduces a property specification language for information flow properties, which intuitively captures the requirements of confidentiality and integrity. In [4], the authors review plenty of existing approaches for

978-1-7281-5410-7/20 $31.00 © 2020 IEEE

requirements formalization that apply to the generation of discrete-time temporal behaviors.

To perform a meaningful simulation, verification engineers often have to provide good test-benches to the employed simulator. These test-benches are required to stimulate the design to verify all its functionalities, to maximize its statement/branch coverage, and if possible, to discover hidden bugs. In [5], the authors propose a self-tuning approach to guide the generation of constrained random test-benches using a sat solver. It employs a greedy search strategy to obtain a high-uniform distribution of stimuli. [6] presents KLEE, a symbolic simulation tool capable of automatically generating tests that achieve high coverage for C/C++ programs. In [7], the authors introduce a purely SAT-based semi-formal approach for generating multiple heterogeneous test-cases for a propositional formula.

III. METHODOLOGY

As shown in Fig. 1, the proposed tool is composed of four main steps executed sequentially. The input of MIST is a set of informal specifications written in natural language. The output is a collection of files that need to be added to a target simulator to perform the verification of the design.

(1) Formalization of specifications: The first step consists of translating the informal requirements into logic formulas. Initially, the user has to reinterpret the specifications into a set of cause/effect propositions, which naturally translate to logic implications $a \rightarrow c$. The user must fill in an XML scheme containing the implications, where each antecedent/consequent pair (a, c) is still written in natural language. After that, (a, c) pairs are formalized into formulas predicating on inputs/outputs and internal variables of the design under verification (DUV). To do so, the user uses an intuitive language of our craft to easily model complex temporal behaviors.

(2) Checker synthesis: In the second step, the tool parses the formalized specifications from the XML schema and generates a checker for each formula. Firstly, each formula is translated into a Büchi automaton. Secondly, a C/C++ representation of a corresponding checker is obtained from the automaton.

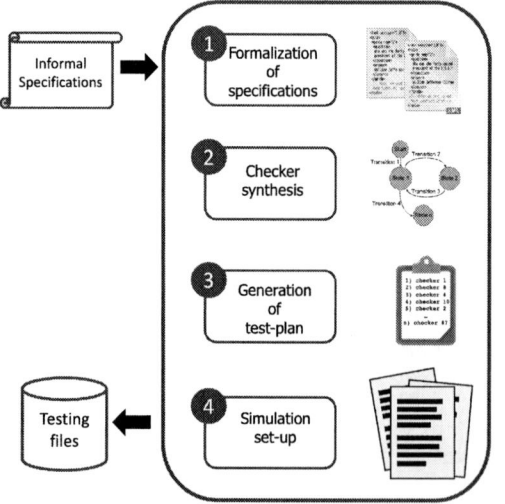

Fig. 1. Execution flow of MIST.

(3) Generation of test-plan: The third step of the methodology aims at finding an effective verification order for the given specifications. Each behavior must be verified when the firmware reaches a specific memory state that we call "precondition state", otherwise the verification would be vacuous. In this state, the behavior can be verified by providing the proper stimuli. During the verification of a behavior, the firmware changes to a new memory state that we call "postcondition state". Considering these assumptions, we identify a sorted list of behaviors that would connect each "postcondition state" to the "precondition state" of the following behavior in the list to guarantee an effective verification process.

(4) Simulation set-up: In the last step, the tool generates all the files necessary to set-up the verification environment. This phase handles the architecture-dependent features of the employed simulator, such as time flow, interrupts and breakpoints. The output files can be described as follows:

- A set C/C++ source files implementing the checkers;
- A set of test-benches to stimulate the design;
- An orchestration file to verify each behavior in the optimal "pre/postcondition" order;

Details related to the four steps implemented by MIST are reported hereafter.

IV. FORMALIZATION OF SPECIFICATIONS

In the first step of the methodology, informal specifications are formalized into logic formulas. This procedure consists in completing two XML files containing the formalized specifications at two different levels of abstraction. The first XML file contains high-level descriptions of firmware behaviors. It describes specifications in a semi-formal language mixing propositional implications with natural language. This partial formalization helps the engineer disambiguate the informal specifications during the interaction with the customer. After that, the engineer completes the second XML file by adding low-level details to the semi-formal behaviors. To clarify the whole procedure, we refer to the formalization of the following example of specification:

"Firmware is in standard mode. If switches A and B are pressed for at least 2000 ms, then the firmware enters in comfort mode and sends an acknowledgment as output"

Firstly, the user has to interpret the specification and translate it into a cause/effect behavior, which is represented by a high-level XML file as follows.

```
1  <assertion id=66>
2      <precondition>
3      Firmware is in standard mode
4      </precondition>
5      <postcondition>
6      Firmware is in comfort mode
7      </postcondition>
8      <antecedent>
9      Switches A and B are pressed for at least 2000 ms
10     </antecedent>
11     <consequent>
12     The firmware enters in comfort mode and sends an
       acknowledgment as output
13     </consequent>
14 </assertion>
```

As depicted in the example, the high-level XML file consists of 5 tags:

- <assertion> contains the *id* attribute to uniquely identify the behavior;

- <antecedent> contains the antecedent part of the informal specification;
- <consequent> contains the consequent part of the informal specification;
- <precondition> contains the memory state the firmware must reach before checking the antecedent;
- <postcondition> contains the memory state reached by the firmware after the consequent has been successfully verified.

When the high-level XML file is completed, the user fills in the low-level XML file by adding unambiguous details to formalize the behaviors. To help non-expert in formal logic and temporal methods during the formalization process, we defined a new language whose grammar is showed below.

```
1 assertion : antecedent -> consequent | precondition |
     postcondition
2 precondition : proposition
3 postcondition : proposition
4 antecedent : next_fragment
5 consequent : next_fragment
6 next_fragment : fragment | fragment; next_fragment
7 fragment : proposition [min, max, times, delay, forced,
     until]
8 proposition : c_boolean_expression
```

Through this language, the user can formalize the specifications in forms of implications, where each antecedent/consequent is an ordered list of *fragments*. Each fragment contains a proposition p and a set of attributes specifying the temporal behavior of p. A proposition is a C/C++ boolean expression. From a temporal perspective, the verification of a consequent starts in the same instant in which the antecedent becomes true, and each fragment is evaluated one instant after the evaluation of the previous fragment completes. For example, in the implication $a \rightarrow c$, where a contains the sequence of fragments $[f_1; f_2; f_3]$ and c contains $[f_4; f_5]$: if f_1 holds in the interval $[t_0, t_n]$, f_2 evaluation starts at time t_{n+1}; on the contrary, if f_3 holds in the interval $[t_k, t_l]$, f_4 evaluation starts at t_l, since t_3 belongs to the antecedent while t_4 to the consequent. A fragment represents then a sequence of boolean events, similar to a PSL SERE.

Given a fragment f with a set of attributes $[min, max, times, delay, until]$ containing a proposition p, the semantics of the evaluation of f at time t_0 can be described as follows:

- **min** $= n$ with $n > 0$: f is true if p holds from t_0 to t_{n-1}
- **max** $= n$ with $n > 0$: f is true if p is false before t_n
- **times** $= m$ with $m > 0$ and **max** $= n$ with $n > 0$: f is true at time $t_k <= t_n$ if p holds for m instants. If attribute *times* is set, then *max* must be set, while *min* and *until* are ignored
- **delay** $= n$ with $n > 0$: f is true at time t_{n-1}
- **until** $= q$ where q is a proposition, and **max** $= n$ with $n > 0$: f is true if q holds at time t_f with $t_0 \leq t_f \leq t_{n-1}$ and p holds from time t_0 to t_{f-1}. If attribute *until* is set then *max* must be set, while *min* and *times* are ignored

In addition, the attribute **forced** can be specified for a fragment f to guide the test-bench generator during the DUV simulation. If $forced = n$ with $n > 0$, MIST calls a SAT solver to generate a model for the proposition p that returns an assignment $var_i = val_i$ for each variable var_i included in p. If f is evaluated at time t_0, then each var_i is forced to value val_i in the interval $[t_0, t_{n-1}]$.

To exemplify the use of the proposed language, we report hereafter the low-level XML resulting from the formalization of the behavior previously used as a running example.

```
1  <assertion id=66>
2    <precondition>
3        mode == 0
4    </precondition>
5    <postcondition>
6        mode == 1
7    </postcondition>
8    <antecedent>
9        <fragment forced=2000 delay=2000>
10           P0 == 0  && P4 == 16 && P12 == 4
11       </fragment>
12   </antecedent>
13   <consequent>
14       <fragment min=1>
15           mode == 1 && P16 == 1
16       </fragment>
17       <fragment min=1>
18           (P16 >> 1) == 1
19       </fragment>
20   </consequent>
21 </assertion>
```

The precondition (postcondition) is represented as a proposition identifying a concrete memory state that must be reached before (after) the verification of the behavior. The antecedent contains a single fragment that, according to the described semantics, identifies the following behavior: the antecedent becomes true after 2000 instants while proposition $P0 == 0 \,\&\&\, P4 == 16 \,\&\&\, P12 == 4$ is forced to be true for 2000 instants. In this scenario, the antecedent is then used only to provide the design with the proper stimuli. The consequent contains two fragments. In the first fragment the proposition $mode == 1 \,\&\&\, P16 == 1$ must be true for one instant. In the following instant, the second fragment is evaluated, and the proposition $(P16 >> 1) == 1$ must be true. From a temporal perspective, the antecedent is evaluated from time t_0 to t_{1999} while the consequent is evaluated from t_{1999} to t_{2000}.

V. CHECKER SYNTHESIS

In the second step of the methodology, MIST parses the formalized specifications in the low-level XML files and generates a C/C++ checker for each implication. The process works in three main sub-steps. Firstly, the tool translates each XML assertions to a PSL formula. Secondly, each PSL formulas is used to generate its equivalent Büchi automaton. Finally, the Büchi automaton is translated to C/C++.

We treat each implication as two independent formulas, one for the antecedent and one for the consequent. This separation is necessary to pinpoint scenarios where the implication is vacuously true. If we considered the implication as a whole, a true evaluation could either mean that the consequent was true or the antecedent was false, we want to distinguish both cases to better warn the user. To convert an XML assertion to PSL, each sequence of *fragments* is treated as a PSL SERE. For example, the consequent of the specification used in Section IV translates to the following PSL formula $\{mode == 1 \,\&\&\, P16 == 1; (P16 >> 1) == 1\}$. Since the PSL syntax does not allow the use of many C operators such as the bit shift operator ($<<$), we execute an intermediate step to provide support to all C operators that can be used to form a boolean expression. In this step, the tool substitutes each fragment's proposition with a placeholder boolean variable representing the proposition. For example, the above formula would be translated to $\{ph1; ph2\}$ where $ph1$ is the placeholder for $mode == 1 \,\&\&\, P16 == 1$ and $ph2$ is the placeholder for $(P16 >> 1) == 1$; Once

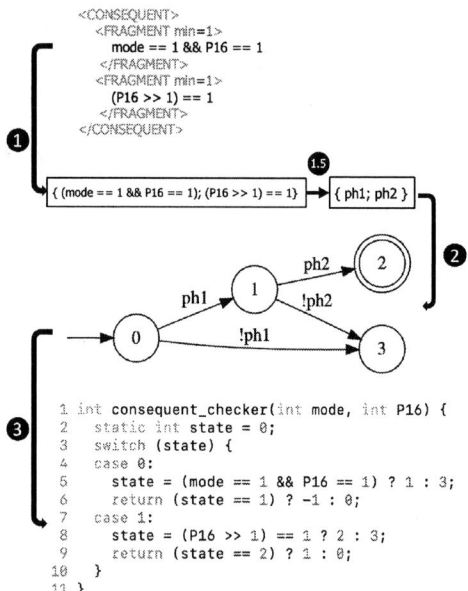

Fig. 2. Example of checker synthesis.

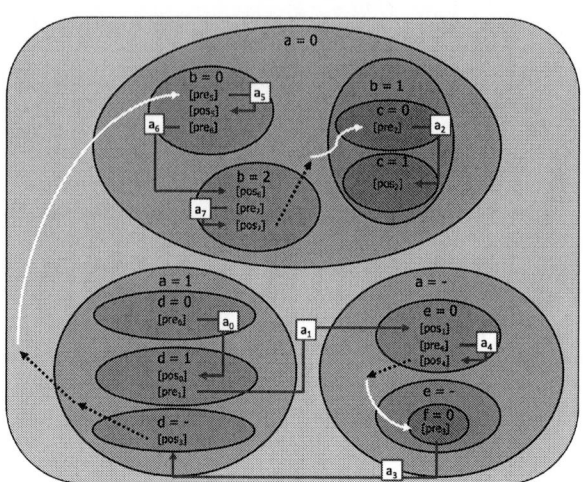

Fig. 3. Example of test-plan generation

the translations above are completed, we generate a Büchi automaton for each formula. To do so, we use spotLTL [8], an external library capable of generating automata from LTL/PSL formulas. Finally, the resulting automaton is visited to generate a C/C++ implementation of the corresponding checker.

Fig. 2 shows an example to clarify the process. In steps (1) and (1.5), the fragment is converted to PSL, and its proposition is substituted with placeholders according to the aforementioned procedure. In step (2), the LTL formula is given as input to spotLTL to generate the depicted Büchi automaton. Before synthesizing the C/C++ checker, each placeholder is substituted back to its original proposition. In Fig. 2, placeholder $ph1$ and $ph2$ are substituted back to $mode == 1 \,\&\&\, P16 == 1$ and $(P16 >> 1) == 1$. In step (3), the automaton is visited starting from the first state. For each state, the tool generates a *case* of a C *switch*, for each edge the tool generates the next-state function in each case. Note that the accepting (rejecting) state is optimized away. For example, the generated checker contains a *case* in which *state* is equal to 0. In this case, if the condition $mode == 1 \,\&\&\, P16 == 1$ is satisfied then *state* is changed to 1, otherwise it is changed to 3. In this scenario, states 2 and 3 are respectively the accepting and rejecting states where the checker returns 1 (true) and 0 (false). In all other states, the checker returns -1 (unknown).

VI. TEST PLAN GENERATION

In the third step of the methodology, the low-level XML file is used to generate an effective verification order. Such an order is intended for generating test-benches that make the firmware evolving in the right memory state before the verification of a checker is started. Otherwise, the checker may pass vacuously or fail due to a wrong precondition state reached by the firmware when the checker is executed.

This step consists of two main procedures. Firstly, all assertions formalized in the low-level XML file are divided into subsets through a clustering procedure. Secondly, each

subset is treated as a node of a multilevel graph, and a verification order is defined by generating a path that connects all nodes. Such a path is then traversed to generated effective test-benches.

To generate the verification order, we consider precondition and post-condition tags of each assertion. Each precondition/post-condition consists of a propositional formula in the form $\langle variable_1 \rangle == \langle constant_1 \rangle \,\&\, \langle variable_2 \rangle == \langle constant_2 \rangle \,\&\, ... \,\&\, \langle variable_n \rangle == \langle constant_n \rangle$ that represents a concrete memory configuration. To simplify the exposition, we will use the term "memory state" while referring to a precondition/post-condition.

In the clustering phase, the goal is to divide the set of all memory states into subsets. We will refer to the example depicted in Fig. 3 to clarify the procedure. The clustering process starts by considering the whole set of memory states, and then it is recursively repeated for each generated subset until no set can be further divided. The process counts the occurrences of each variable in all memory states in the current set; the variable with the highest count is used to perform the split. In the example, the most frequent variable in the whole set is a. The current set is split into as many sub-sets as the number of different assignments of the most frequent variable. Also, we add an optional sub-set containing all memory states that do not include the most frequent variable (do not care sub-set). In the example, the whole set is divided into three clusters, two clusters for $a = 0$ and $a = 1$ and one don't care cluster $a = -$. The same process is repeated until all sub-sets contain only memory states with equivalent assignments. In the example, the cluster identified by $a = 0$ and $b = 0$ contains three equivalent memory states $[pre_5], [pos_5], [pre_6]$ that have the same assignments $[a = 0 \,\&\, b = 0]$. This heuristic approach is intuitively justified by the assumption that the most frequent variables represent better the whole state; therefore, it is reasonable to make them represent wider clusters than those represented by less frequent variables. The clustering procedure aims at making all similar memory states "close" to each other.

In the second part of the approach, each sub-state is used to infer an effective verification order. Starting from the precondition of an assertion chosen by the user, the tool finds a path that covers all the memory states. To move from one memory state to the next, the procedure applies the following rules:

R1: Checking an assertion i in memory state $[pre_i]$ moves the process to $[pos_i]$ (solid red arrow);

R2: If the process can not find any other unused precondition in the current state cluster, it must jump to its upper cluster and continue the search (dotted black arrow);

R3: After a jump, the process searches for the first unused precondition $[pre_j]$ in the current cluster. If it finds one, it continues the process from that state (rounded white arrow).

To clarify the procedure, we explain the process by considering the example of Fig. 3. In this example, the user chooses to start with assertion a_0; therefore, the starting state is $[pre_0]$. By applying rule R1, assertion a_0 is added to the test-plan, and the execution moves to state $[pos_0]$. In the destination cluster, we find an unused precondition $[pre_1]$. We apply again rule R1, assertion a_1 is added to the test-plan, and the execution is moved to pos_1. We repeat the process for assertion a_4, and we reach the state pos_4. In this case, no more preconditions are available in the current cluster; therefore, the execution must apply rule R2 and jump to the upper cluster identified by $a = -$. By applying rule R3, the process finds an unused precondition pre_3 and continues from there. Again, we add assertion a_3 to the test-plan, and we move the execution to pos_3. We apply rule R2 as no other preconditions can be found in the current cluster, and we reach cluster $a = 1$. We must apply rule R2 again for the same reason and jump to the upper cluster. The procedure continues as described above until all assertions are added to the test-plan. The resulting test-plan is $[a_0, a_1, a_4, a_3, a_5, a_6, a_7, a_2]$.

The reason why this approach produces an effective verification order resides in the assumption that the DUV was developed by following a coherent logic flow. In other words, the generated ordering tries to mimic the behavior of a human that manually tests the DUV. Indeed, to check the correctness of a design, the human starts from the initial state and provides a sequence of stimuli to the DUV. Each sequence of stimuli moves the DUV from one configuration to the next in a coherent flow, such that the ending configuration represents the starting precondition for effectively checking the next behavior in a cause-effect cascade fashion. On the opposite, when moving randomly from one behavior to an unrelated one, the completeness of the verification suffers from the absence of a coherent test-plan; in fact, many behaviors are unlikely to be verified as the DUV never reaches the correct configuration that satisfies the precondition of any behavior. To solve this issue, one could force the correct DUV configuration before testing each behavior. However, the user would have to annotate each formalized behavior with the required memory state to force. This process can be extremely time-consuming and error-prone. In the worst case, errors in this procedure lead to a vacuous verification; the random stimuli are unable to fire the precondition and the antecedent of the target assertion. In this situation, if no feedback on the vacuous passing is

provided to the verification engineer, the DUV is released without proper verification.

Through our clustering approach, the specifications are verified in the order intended by the designer, thus reducing the necessity of forcing the memory state that represents the precondition of the target behavior, since the DUV gets naturally brought to the proper state. Note that the ideal case, where all behaviors described by the initial specification perfectly connect to form a coherent path, requires the user to completely formalize the specifications such that all assertions belong to a unique cluster. This requirement could be extremely tedious to achieve manually and could be unfeasible for most large-scale designs. For this reason, each time we identify a hole in the specification, such that the postcondition of an assertion does not connect with the precondition of any other assertion, our heuristic approach jumps to a similar close state and warn the verification engineer. To be clear, in case of fully connected specifications, our approach uses only rule R1. Each time rules R2 and R3 are used, we are approximating.

After generating the test-plan, MIST informs the user of the *completeness* of the given set of behaviors by comparing the total number of assertions with the number of times rule R2 was applied to continue the clustering process. The completeness index is calculated with the following formula: $(1 - exceeded\_maxR2\_applications / tot\_assertions)$. Where $exceeded\_maxR2\_applications$ represents the number of times the process has to violate the maximum number of consecutive applications of rule R2. Intuitively, the resulting completeness is an index describing how much the set of behaviors is likely to cover all functionalities of the DUV without holes. Each time a missing link is found, the completeness is reduced.

Table I shows the completeness for the running example. The first row of the table shows the completeness when no approximation is allowed, or in other words, when the process should not use rule R2 to continue. In the example, rule R2 is used 3 times non-consecutively; therefore, the resulting completeness is $(1 - 8/3) = 0.625$. In the example, the second (third) row shows the completeness reachable by allowing the consecutive application of rule R2 at most once (twice).

The user can exploit this information to improve the set of formalized behaviors such that rule R2 is applied as less as possible while achieving high completeness.

VII. SIMULATION SETUP

In the last step of the methodology, the verification environment is set up. This phase handles the architecture-dependent features of the target simulator. For now, MIST is capable of generating a verification environment for the IARsystem workbench, which is an industrial compiler and debugger toolchain for ARM-based platforms. In particular, we exploit the provided breakpoint system to evaluate the checkers and to handle the time flow. Since our checkers provide support for temporal behaviors, we need a way to sample the time flow. To

TABLE I
COMPLETENESS ANALYSIS FOR EXAMPLE IN FIG. 3

max applications of rule R2	completeness
0 times	62.5%
1 times	87.5%
2 times	100 %

accomplish that, we provide a debugging variable sim_time that can be used by the user to simulate the advancement of time in the DUV. To capture this event in the debugger, we place a breakpoint on that variable to recognize *write* operations. Each time sim_time is incremented, the simulated time advances by one instant producing a re-evaluation of the active checker. Usually, the best way to use sim_time is to place it in a timed interrupt that keeps increasing it at a constant rate. Furthermore, we use breakpoints to inject stimuli in the ports and variables of the fragments using the *forced* attribute. Following the above paradigms, MIST generates the files to perform the verification of the DUV using IARsystem. The generated files consist of an entry point to set up the verification environment, utility functions to handle the time events, the orchestration file that executes each checker according to the methodology described in Section 3, and a set of files containing the checkers. To execute the verification with IARSystem, the user must provide the entry point file generated by MIST; this file will include all other files when needed.

VIII. EXPERIMENTAL RESULTS

The experimental results have been carried out on a 2.9 GHz Intel Core i7 processor equipped with 16 GB of RAM and running Windows 10. We evaluated the effectiveness of our tool to verify an industrial firmware composed of over 10000 lines of C code. The firmware implements the controller of a boiler plant. The user can interact with the firmware through an HMI (Human-machine interface). Moreover, the firmware is connected to several external devices providing inputs/outputs such as thermostats, boilers, clocks and an internet gateway. The internal time flow is handled using timed interrupts.

Finally, the firmware runs on an RL78 microcontroller, allowing communications with the external devices through Modbus and I2C protocols.

In particular, we put emphasis on the timing results of the complete verification process, from the formalization of specifications to the simulation of the behaviors. Starting from the informal specification of the firmware, we formalized 100 behaviors. On average, each behavior takes 30 seconds to be formalized into the high-level XML format. The formalization of the low-level XML format depends significantly on the skill of the verification engineer and his/her knowledge of the underlying implementation details. After some practice, we were capable of formalizing a behavior in less than three minutes. Overall, we formalized all 100 behaviors in less than 6 hours. After that, MIST generated the testing files and produced an effective test-plan in less than 10 seconds. Finally, we set-up the verification environment in the simulator (IAR System Workbanch). The simulation took less than 40 minutes to verify non-vacuously each behavior and to produce a report of the verification.

The employment of our methodology to an industrial legacy firmware discovered numerous bugs related to an inaccurate sampling of time. One notable example concerns the usage of switches in the HMI. Many specifications implied that some switches needed to be pressed for a certain amount of time to active a functionality. However, during simulation, the correct behavior did not occur even when providing the correct stimuli. Using MIST for the verification of such a firmware was considerably helpful in identifying a temporal

TABLE II
COMPLETENESS ANALYSIS FOR THE CONSIDERED CASE STUDY.

max applications of rule R2	completeness
0	45.5%
1	72.73%
2	79.22%
3	81.82 %
4	97.73%
5	100%

TABLE III
COMPLETENESS ANALYSIS OF CASE STUDY AFTER THE IMPROVEMENTS.

max applications of rule 2	completeness
0	48.5%
1	75.73%
2	88.2%
3	100 %

inconsistency of Modbus and I2C protocols that caused a delay in its execution.

Furthermore, the generation of the test-plan for 100 behaviors suggested a remarkable incompleteness in the firmware specifications. In table II we can observe the completeness estimations produced for the case study by considering the approach proposed in Section VI. We used those statistics to improve the completeness of the specifications by adjusting the behaviors underlining the highest incompleteness and by adding 10 behaviors to cover some specifications holes. After completing this procedure, we achieved new completeness estimations reported in Table III. To achieve 100% completeness with the new specifications, we needed to apply rule R2 only 3 times, while with the initial specifications, it was used 5 times.

IX. CONCLUSIONS

In this paper, we presented MIST, an all-in-one tool capable of generating a complete environment to verify C/C++ firmwares starting from informal specifications. MIST reduces the verification effort by providing a user-friendly interface to formalize specifications into assertions and to generate the verification environment automatically. Furthermore, MIST employs a clustering procedure to generate an effective test-plan that reduces potential mistakes while formalizing the specifications.

REFERENCES

[1] [Online]. Available: https://www.iar.com/iar-embedded-workbench
[2] N. A. Moketar, M. Kamalrudin, S. Sidek, M. Robinson, and J. Grundy, "An automated collaborative requirements engineering tool for better validation of requirements," in *2016 31st IEEE/ACM International Conference on Automated Software Engineering (ASE)*, 2016, pp. 864–869.
[3] P. Subramanyan, S. Malik, H. Khattri, A. Maiti, and J. Fung, "Architecture of a tool for automated testing the worst-case execution time of real-time embedded systems firmware," in *2016 Design, Automation & Test in Europe Conference & Exhibition (DATE)*, 2016, pp. 337–342.
[4] I. Buzhinsky, "Formalization of natural language requirements into temporal logics: a survey," in *2019 IEEE 17th International Conference on Industrial Informatics (INDIN)*, 2019, pp. 400–406.
[5] Y. Zhao, J. Bian, S. Deng, and Z. Kong, "Random stimulus generation with self-tuning," in *13th International Conference on Computer Supported Cooperative Work in Design*, 2009, pp. 62–65.
[6] C. Cadar, D. Dunbar, D. R. Engler *et al.*, "Klee: Unassisted and automatic generation of high-coverage tests for complex systems programs." in *OSDI*, vol. 8, 2008, pp. 209–224.
[7] S. Agbaria, D. Carmi, O. Cohen, D. Korchemny, M. Lifshits, and A. Nadel, "Sat-based semiformal verification of hardware," in *Formal Methods in Computer Aided Design*, 2010, pp. 25–32.
[8] A. Duret-Lutz, A. Lewkowicz, A. Fauchille, T. Michaud, E. Renault, and L. Xu, "Spot 2.0 — a framework for ltl and ω - automata manipulation," 10 2016, pp. 122–129.

Breaking ACORN at Bitstream Level

Michail Moraitis, Elena Dubrova and Kalle Ngo
Royal Institute of Technology (KTH), Stockholm, Sweden
{micmor,dubrova,kngo}@kth.se

Abstract—Assuring the security of the Internet of Things (IoT) is much more challenging than assuring the security of centralized environments, like the cloud. A reason for this is that IoT devices are often deployed in domains that are remotely managed and monitored. Thus, they cannot be protected from physical attacks as reliably as data centers. Up till now, implementations of many established, standardized algorithms including AES and SNOW 3G have been broken by physical attacks. In this paper, we show that even the most recently designed algorithms are also vulnerable. We attack an SRAM-based FPGA implementation of ACORN v3 stream cipher, a finalist of CAESAR cryptographic competition for authenticated encryption. By modifying the content of several look-up tables directly in the bitstream, we inject faults which reduce the non-linear feedback function of ACORN to a linear one. As a result, it becomes possible to extract the full key from $2^{15.34}$ bits of faulty keystream by an algebraic attack using $2^{35.46}$ operations. Our results, once again confirm the necessity to rethink the way cryptographic algorithms are implemented in FPGAs.

Index Terms—FPGA, bitstream modification, fault attack, algebraic attack, ACORN, stream cipher, IoT security.

I. INTRODUCTION

A rapid growth in the Internet of Things (IoT) applications is expected in the coming years [1]. Household appliances, meters, sensors, and vehicles will be accessible and controlled via local networks or the Internet to provide new services appealing to users.

However, assuring IoT security is much more challenging than assuring security of centralized environments, like the cloud. While IoT inherits old problems such as weak zero-day vulnerabilities and a lack of updates, it also creates new ones.

On one hand, the attack surface of future IoT with billions of connected devices will be enormous. On the other hand, IoT relies on many different types of devices, including resource-constrained sensors and actuators which may not have enough storage, computing and energy resources for implementing a strong cryptographic protection. The security of a network is only as strong as its weakest link. A compromised sensor can potentially be used as an entry point for cyberattacks on other devices connected to the network, or the network itself [2]. Moreover, IoT devices are often deployed in domains that are remotely managed and monitored. Thus, their physical security cannot be guaranteed as reliably as the physical security of data centers.

Some believe that physical security becomes less important if all the data that are processed and stored within a device are encrypted and secure access is assured [3]. However, an attacker with physical access to a device implementing an encryption algorithm may be able to extract the encryption key and decrypt the data. Up till now, implementations of many established, standardized algorithms including Advanced Encryption Standard (AES) and SNOW 3G have been broken by physical attacks, e.g. [4], [5]. At the time these algorithms were designed physical attacks were not an issue. However, in this paper, we show that even the most recently designed algorithms are vulnerable. We attack an FPGA implementation of ACORN stream cipher, a finalist of CAESAR cryptographic competition for authenticated encryption (2014-2018) [6].

Previous Work. Due to its importance, ACORN has been actively cryptanalyzed before. In [7], leakage of ACORN is evaluated using t-test, and in [8] using cube testers and d-monomial test. In [9], a differential power analysis attack on ACORN is described. In [10], an EM-based side-channel attack on ACORN is presented. A SAT-based cryptanalysis of ACORN v1 and v2 is made in [11]. A state recovery attack on ACORN v1 and ACORN v2 with 2^{120} complexity is described in [12]. In [13], the key is recovered from ACORN by re-using the nonce several times to encrypt the same chosen plaintext. In [14], a cube attack on the reduced versions of ACORN v1 and v2 is presented. In [15], a state recovery attack on ACORN v2 with 2^{40} complexity is presented under a chosen-plaintext attack model which assumes that, for a given key, one can find a message m whose corresponding polynomial is equal to the non-linear feedback function f of ACORN. As a result, $m \oplus f = 0$ and f is canceled from the state update. However, it is not clear how m can be found if the key is unknown.

Several faults attacks on ACORN has also been presented. In [16], a differential fault attack on ACORN v3 which requires 9 bit flips to be injected into the initial state to recover the state with the complexity $2^{25.40}$ is described. Zhang et al. [17] presented another differential fault attack on ACORN v3 in which a bit of the initial state is flipped at random. Then, the fault is located and the resulting equations are solved. With k faults, the initial state can be recovered with time complexity $c \cdot 2^{146.5-3.52k}$, where c is the complexity of solving linear equations and $26 < k < 43$. In [18], an attack on ACORN v1 and v2 is described in which a random bit of the fifth LFSR of ACORN is stuck to the constant-1 value during the encryption. For certain faulty bit positions, the attack complexity is $2^{55.85}$.

Our Contribution. In this paper, we present a new fault attack on an SRAM-based FPGA implementation of ACORN v3 in which the content of several LUTs is modified directly in the bitstream to reduce the non-linear feedback function of ACORN to a linear one. As a result, it becomes possible

978-1-7281-5410-7/20 $31.00 © 2020 IEEE

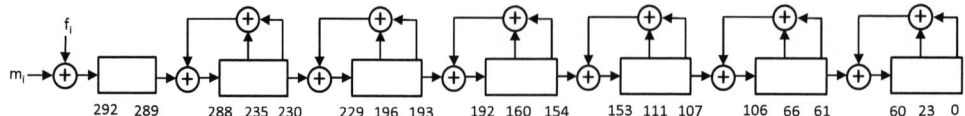

Fig. 1. Block diagram of ACORN v3.

to express the state update function of ACORN in terms of the linear equations depending on the key and unfold the function to the initial encryption state without its size blowing-up exponentially, as in the case of the fault-free ACORN. It also becomes possible to recover the initial encryption state from $2^{15.34}$ keystream bits with $2^{35.46}$ operations by applying known linearization methods. Once the state is recovered, the key is obtained by solving the system of linear equations representing the unfolded update function, which takes $2^{16.63}$ operations. We demonstrate the attack on a Xilinx Artix-7 FPGA and discuss potential countermeasures.

II. DESIGN DESCRIPTION OF ACORN

ACORN is a bit-oriented authenticated stream cipher [19]. It is constructed from six LFSRs which are composed into a 293-bit register, as shown in Fig. 1. It uses several functions: a function to generate the keystream bit, a function to compute the feedback bit, and 6 functions to update the state.

The keystream generation function, ks, is defined by:

$$ks(s) = s_{12} \oplus s_{154} \oplus maj(s_{235}, s_{61}, s_{193}) \oplus ch(s_{230}, s_{111}, s_{66})$$
$$(1)$$

where s_j is the jth bit of the state $s = (s_0, \ldots, s_{292})$ while maj and ch are the *majority* and the *choice* operations:

$$maj(x, y, z) = xy \oplus xz \oplus yz$$
$$ch(x, y, z) = xy \oplus \overline{x}z,$$

where "$\oplus$" is the Boolean XOR and $\overline{x}$ is for the Boolean complement of x, $\overline{x} = 1 \oplus x$.

The feedback function, f, is given by:

$$f(s, ca, cb) = s_0 \oplus \overline{s}_{107} \oplus maj(s_{244}, s_{23}, s_{160}) \oplus ca \cdot s_{196} \oplus cb \cdot ks,$$
$$(2)$$

where "$\cdot$" is the Boolean AND, and ca and cb are the control bits. The ca bit is used to separate the processing of associated data, the processing of plaintext, and the generation of authentication tag. The cb bit is used to let the keystream bit to affect a feedback bit during initialization, processing of associated data, and the tag generation.

The state is updated in 4 steps. First, the LFSRs are updated:

$$\begin{aligned} s_{289} &= s_{289} \oplus s_{235} \oplus s_{230} \\ s_{230} &= s_{230} \oplus s_{196} \oplus s_{193} \\ s_{193} &= s_{193} \oplus s_{160} \oplus s_{154} \\ s_{154} &= s_{154} \oplus s_{111} \oplus s_{107} \\ s_{107} &= s_{107} \oplus s_{66} \oplus s_{61} \\ s_{61} &= s_{61} \oplus s_{23} \oplus s_0 \end{aligned} \qquad (3)$$

Second, the keystream bit is computed as (1). Third, the feedback bit is generated as (2). Finally, all but the input

register bits are updated by a shift as $s_j^+ = s_{j+1}$, for $j \in \{0, \ldots, 291\}$ and the input bit is updated as $s_{292}^+ = f(s) \oplus m_i$, where m is the data bit at the step i and s_j^+ denotes the value of s_j at the next step.

The initialization of ACORN is done as follows. Initially, the 293-bit register is set to the all-0 state. Then, the cipher is run for 1792 steps during which a 128-bit key K and a 128-bit IV are loaded into the register bit-by-bit through the input m_i as shown below:

$$\begin{aligned} m_i &= K_i & \text{for } i = 0 \text{ to } 127; \\ m_i &= IV_{i \bmod 128} & \text{for } i = 128 \text{ to } 255; \\ m_i &= \overline{K}_{i \bmod 128} & \text{for } i = 256; \\ m_i &= K_{i \bmod 128} & \text{for } i = 257 \text{ to } 1792. \end{aligned} \qquad (4)$$

After the initialization, the associated data is used to update the state. Even when there is no associated data, the cipher is run for 256 steps. Otherwise, the cipher is run for $256 + adlen$ steps, where $adlen$ is the length of the associated data. Then, the encryption starts.

At each step of the encryption, one plaintext bit p_i is used as the data bit $m_i = p_i$ to update the state. The ciphertext bit c_i is obtained by XORing the plaintext bit p_i with the keystream bit computed according to (1).

After processing all the plaintext, the authentication tag is generated. We omit the description since it is not related to the presented work.

III. ATTACK: THE THEORETICAL PART

The attack is performed as follows[1]:

1) Inject faults which modify the updating function of the input state bit from $s_{292}^+ = f(s) \oplus m_i$ to $s_{292}^+ = m_i$.
2) Encrypt some r-bit plaintext while the fault is active.
3) Analyze the resulting r-bit faulty keystream as described in Section III-A and III-B to recover the key.

The fault is required to be active for at least $2048 + adlen + r$ steps since the initialization takes 1792 steps and the associated data processing takes $256 + adlen$ steps. Assuming $adlen = 0$, ACORN starts the encryption after 2048 steps.

The analysis consists of two parts: (1) deriving the value of the initial encryption state from the keystream bits, and (2) deriving the key from the initial encryption state.

A. Deriving the key form the initial state

If the input state bit is updated as $s_{292}^+ = m_i$ for $i = \{0, 1, \ldots, 2047\}$, its value does not affect the next state

---

[1]An early version of the theoretical part of the attack is described in the technical report [20] (unpublished).

of ACORN during the initialization and the associated data processing. In this case, the state is updated by the linear function L defined by the six LFSRs and the data bit m_i, namely

$$
\begin{aligned}
s_{292}^+(m_i) &= m_i \\
s_{288}^+(s) &= s_{289} \oplus s_{235} \oplus s_{230} \\
s_{229}^+(s) &= s_{230} \oplus s_{196} \oplus s_{193} \\
s_{192}^+(s) &= s_{193} \oplus s_{160} \oplus s_{154} \\
s_{153}^+(s) &= s_{154} \oplus s_{111} \oplus s_{107} \\
s_{106}^+(s) &= s_{107} \oplus s_{66} \oplus s_{61} \\
s_{60}^+(s) &= s_{61} \oplus s_{23} \oplus s_0 \\
s_j^+ &= s_{j+1}, \text{ for all other } j \in \{0, \dots, 291\}
\end{aligned}
\tag{5}
$$

Since m_i is a function of the key K and IV, see (4), the initial encryption state is defined by $L^{2048}(K, IV)$. It can be expressed by a system of 293 linear equations depending on 128 unknown key bits and 128 known IV bits.

It is known that a linear system with k variables can be solved by the Gaussian elimination in time k^ω, where $\omega \leq 2.376$ is the exponent of the Gaussian reduction [21]. So, in our case, finding the solution takes at most $128^{2.376} = 2^{16.63}$ operations.

B. Deriving the initial state from the keystream

Given the initial encryption state $s = (s_0, \dots, s_{292})$, ACORN with the fault $f_i = 0$ for $i = \{2048, \dots, 2047 + r\}$, generates the keystream bits for the encryption of the plaintext $p_0, \dots, p_{r-1}$ as follows

$$
\begin{aligned}
b_0 &= ks(s, p_0) \\
b_1 &= ks(L(s, p_0, p_1)) \\
b_2 &= ks(L^2(s, p_0, p_1, p_2)) \\
&\dots \\
b_{r-1} &= ks(L^{r-1}(s, p_0, p_1, p_2, \dots, p_{r-1}))
\end{aligned}
\tag{6}
$$

where L is the linear state updating function defined by (5) with $m_j = p_j$, for $j = \{0, 1, \dots, r-1\}$.

In equation (1), the keystream generation function $ks(s)$ is expressed in terms of 8 variables. However, since each keystream bit is computed *after* the state bits $s_{289}, s_{230}, s_{193}, s_{154}, s_{107}, s_{61}$ are updated as shown in (3), the function $ks(s)$ actually depends on 13 bits of the state. After substituting s_{61}, s_{154}, s_{193} and s_{230} by their corresponding expressions from (3) and expanding, we get the following algebraic normal form for $ks(s)$:

$$
\begin{aligned}
ks(s) = \; & s_{12} \oplus s_{66} \oplus s_{107} \oplus s_{111} \oplus s_{154} \oplus s_{235}s_{61} \\
\oplus \; & s_{235}s_{23} \oplus s_{235}s_0 \oplus s_{61}s_{193} \oplus s_{61}s_{160} \oplus s_{61}s_{154} \\
\oplus \; & s_{23}s_{193} \oplus s_{23}s_{160} \oplus s_{23}s_{154} \oplus s_0s_{193} \oplus s_0s_{160} \\
\oplus \; & s_{235}s_{154} \oplus s_{230}s_{111} \oplus s_{196}s_{111} \oplus s_{193}s_{111} \oplus s_{230}s_{66} \\
\oplus \; & s_0s_{154} \oplus s_{235}s_{193} \oplus s_{235}s_{160} \oplus s_{196}s_{66} \oplus s_{193}s_{66}
\end{aligned}
$$

The function $ks(s)$ is non-linear, however, there are known linearization methods, e.g [22], [23], which can find the solution to the system of the non-linear equations (6) given $r \geq \binom{n}{d}$ keystream bits and within r^ω computations, where n is the state size and d is the algebraic degree of the output function. The linearization is done by introducing a new

variable for each monomial and solving the resulting system of linear equations by the Gaussian elimination. Since the number of monomials of degree d is $\binom{n}{d}$ such a method is efficient for a small d.

In our case, the algebraic degree of $ks(s)$ is only two. The four state bits $s_{289}, \dots, s_{293}$ can be ignored since they just delay the input plaintext bits by 4 clock cycles. The remaining 289 bits of the state s can be recovered using $r = \binom{289}{2} = 2^{15.34}$ keystream bits with $2^{15.34 \times 2.376} = 2^{35.46}$ operations.

It is worth mentioning that there are other methods, such as XL algorithm or Gröbner basis algorithms [22], [24], [25], that can find a solution of an overdefined system using less than $\binom{n}{d}$ keystream bits. However, the complexity of the attack increases substantially.

IV. ATTACK: THE PRACTICAL PART

In this section, we describe how we injected the fault into ACORN by direct bitstream modification. In the experiments, we used the VHDL implementation of ACORN v3 with the degree of parallelization 8 from [26]. The attack was mounted on a bitstream synthesized for the Xilinx Artix-7 (XC7A35T-CPG236-1) FPGA.

A. Attack Model

Our attack model is based on the following assumptions[2]:
1) The attacker has limited physical access to the victim FPGA device implementing the encryption algorithm.
2) The encryption key K is stored on chip/board.

Nowadays the supply chain of electronic products involve several parties (e.g logistics providers, distributors, retailers etc). All these parties at one point have physical access to the products. Furthermore, during its lifetime a device may need to be returned for maintenance or repair. Thus the first assumption is very realistic. As for the second assumption, it seems to be a common way to store keys for FPGA implementations [27], [28], [29].

The attacker's ultimate goal is to recover the key K. To achieve that, he first extracts the bitstream from the FPGA (e.g by using a probe while it is being loaded from the Flash memory to the FPGA during configuration). If the bitstream is encrypted, the attacker has several options to overcome it. Techniques such as side-channel analysis [30], [31], [32], optical probing [33] or thermal laser stimulation [34] can be used to extract the bitstream encryption key and decrypt the bitstream. The recently discovered Starbleed [35] vulnerability can also be exploited to recover the decrypted bistream in Xilinx 7 series FPGAs. After that the recovery of the authentication key is possible since it is stored as plaintext in the bitstream.

The next steps are to modify the bitstream, encrypt it if necesary and load it to the victim FPGA to generate the exploitable keystream. From that, k is extracted. The original bitstream is then loaded back in tho the FPGA and the compromised device is returned to its legitimate user.

---

[2]Those assumptions are commonly made in bitstream modification attacks [4], [5].

TABLE I
POSSIBLE k-LUTs CONTAINING A FAULT INJECTION POINT.

$$l_1 = \dot{1} \oplus x_1 \oplus x_2 \oplus x_3 \oplus \dot{x}_4\dot{x}_5$$
$$l_2 = \dot{1} \oplus x_1 \oplus x_2 \oplus x_3 \oplus \dot{x}_4\dot{x}_5 \oplus \dot{x}_4\dot{x}_6$$
$$l_3 = \dot{1} \oplus x_1 \oplus x_2 \oplus x_3 \oplus \dot{x}_4\dot{x}_5 \oplus \dot{x}_6\dot{x}_1$$
$$l_4 = \dot{1} \oplus x_1 \oplus x_2 \oplus x_3 \oplus x_4 \oplus \dot{x}_5\dot{x}_6$$
$$l_5 = \dot{1} \oplus x_1 \oplus x_2 \oplus x_3 \oplus x_4 \oplus \dot{x}_1\dot{x}_5$$
$$l_6 = \dot{1} \oplus x_1 \oplus x_2 \oplus x_3 \oplus x_4 \oplus \dot{x}_5\dot{x}_6 \oplus \dot{x}_5\dot{x}_1$$
$$l_7 = \dot{1} \oplus x_1 \oplus x_2 \oplus x_3 \oplus x_4 \oplus \dot{x}_3\dot{x}_4 \oplus \dot{x}_5\dot{x}_6$$
$$l_8 = \dot{1} \oplus x_1 \oplus x_2 \oplus x_3 \oplus x_4 \oplus x_5 \oplus \dot{x}_6\dot{x}_1$$
$$l_9 = \dot{1} \oplus x_1 \oplus x_2 \oplus x_3 \oplus x_4 \oplus x_5 \oplus \dot{x}_1\dot{x}_2$$
$$l_{10} = \dot{1} \oplus x_1 \oplus x_2 \oplus x_3 \oplus x_4 \oplus x_5 \oplus \dot{x}_1\dot{x}_2 \oplus \dot{x}_1\dot{x}_3$$
$$l_{11} = \dot{1} \oplus x_1 \oplus x_2 \oplus x_3 \oplus x_4 \oplus x_5 \oplus \dot{x}_1\dot{x}_2 \oplus \dot{x}_3\dot{x}_4$$
$$l_{12} = \dot{1} \oplus x_1 \oplus x_2 \oplus x_3 \oplus x_4 \oplus x_5 \oplus \dot{x}_1\dot{x}_2 \oplus \dot{x}_6\dot{x}_3$$

B. Analyzing LUT Network

Let $\mathcal{B}$ be the bitstream under attack. Our target is to find in $\mathcal{B}$ the set of LUTs whose content should be modified to change $s_{292}^+ = f(s) \oplus m_i$ to $s_{292}^+ = m_i$. The implementation of ACORN used in our experiments has the degree of parallelization 8. This means that 8 most significant bits of the state, $s_{292}, s_{291}, \ldots, s_{285}$, are updated at each clock cycle. Hence there are 8 fault injection points corresponding to $f[1], f[2], \ldots, f[8]$, where $f[i]$ denotes the bit of the feedback function $f(s)$ updating the state bit s_{293-i}, for $i \in \{1, 2, \ldots, 8\}$. Note that the degree of parallelization 8 makes the presented attack more challenging compared to the attack on an implementation with a degree of parallelization one.

First, we extract a Boolean network $\mathcal{N} = (V, \mathcal{E})$ representing the combinatorial part of ACORN and analyze possible ways the fault injection points can be covered by k-LUTs in $\mathcal{N}$. The Xilinx 7 series FPGAs use 6-input dual-output LUTs, so $k = 6$. A 6-input dual-output LUT can implement either a single Boolean function of up to 6 independent variables or two Boolean functions of up to 5 dependent variables [36]. Table I summarizes potential candidates. The notation $\dot{x}$ represents both, a variable x and its complement $\overline{x}$. Similarly, $\dot{1}$ stands for both, the constant-1 and the constant-0. Note that a Boolean expression containing m symbols with "dot" on the top encodes 2^m different Boolean expressions. Thus, Table I represents 264 different candidates.

C. Finding LUTs in Bitstream

Next, we search the bitstream for all functions in Table I using the FINDLUT() algorithm presented in [5]. FINDLUT() takes as input a k-variable Boolean function l and a bitstream $\mathcal{B}$ and returns a set of candidates into LUTs implementing l in $\mathcal{B}$, together with the location of each LUT in $\mathcal{B}$. FINDLUT() treats all Boolean functions in the same P equivalence class as one function[3]. Therefore, it considers all possible permutations of variables.

[3]Two Boolean functions belong to the same P equivalence class, if they can be transformed to each other through a permutation of inputs [37].

From this step, we get $N = 30$ candidate LUTs that possibly contain fault injection points.

D. Identifying Inputs

To find which of the LUT-candidates contain the fault injection points, we need to identify which variables in the expressions of Table I are related to the feedback function f and which represent the message bit m_i. We solve this problem using the *key independent bitstream exploration* technique introduced in [5]. To make the bitstream $\mathcal{B}$ key independent, we find in $\mathcal{B}$ the LUTs which load K and IV into the LFSRs and modify their content to load the all-0 state instead. Obviously, if ACORN is initialized to the all-0 state, the resulting keystream does not depend on the secret key K any longer and can be calculated from the ACORN's algorithm. Note that, due to the presence of the complemented term $\overline{s}_{107}$ in the expression (2), the LFSRs are capable of escaping from the all-0 initial state. The knowledge of the keystream allows us to find target variables in the expressions in Table I.

At the first iteration, we use a software implementation of ACORN which is modified to start from the all-0 state to encrypt some plaintext P in presence of the fault $\beta : s_{293-k} = 0$ for each $k \in \{1, 2, \ldots, 8\}$, one-by-one. The resulting 8 faulty ciphertexts C_k^β are recorded. Then, for each of the N LUT-candidates l_i, $i \in \{1, 2, \ldots, N\}$, we change the content of l_i in $\mathcal{B}$ to $l_i^\beta = 0$. We upload the resulting faulty bitstream $\mathcal{B}^\beta$ into the FPGA, compute the ciphertext C^β and compare it to the ciphertexts C_k^β, for all $k \in \{1, 2, \ldots, 8\}$. If $C^\beta = C_k^\beta$, then the output of LUT l_i is the input of the state bit s_{293-k}. Let $J_i \subset \{1, 2, \ldots, 6\}$ denote the set of indexes of single-variable terms of l_i. From Table I we can see that the size of J_i is $|J_i| \leq 5$ for all functions.

At the second iteration, we once again use the software implementation of ACORN which is modified to start from the all-0 state to encrypt some plaintext P in presence of the fault $\alpha : f[k] = 0$, for each $k \in \{1, 2, \ldots, 8\}$, and record the faulty ciphertexts C_k^α. Then, for each s_{293-k}, $k \in \{1, 2, 3, 4\}$, and each single-variable term x_j, $j \in J_i$, in the expression of the LUT l_i providing the input to s_{293-k}, we change the content of l_i in $\mathcal{B}$ to $l_i^\alpha = x_j$ (i.e. we guess that x_j corresponds the message bit). Then, we load the faulty bitstream $\mathcal{B}^\alpha$ into the FPGA, compute the ciphertext C^α and compare it to the ciphertexts C_k^α, for $k \in \{1, 2, 3, 4\}$. If $C^\alpha = C_k^\alpha$, then x_j is the input of l_i corresponding to the message bit.

For each s_{293-k}, this procedure takes $O(c)$ steps, where $c = \max |J_i|$ and $k \in \{1, 2, 3, 4\}$. The first 4 rows of Table II summarize the resulting fault injection points. The variables $A_1, A_2, \ldots, A_6$ correspond to the actual LUT inputs. The last column of Table II shows the relation with Table I.

The procedure for finding fault injection points for the state bits s_{293-k}, $k \in \{5, 6, 7, 8\}$, is more complex because, in this case, not only the message bit, but also the state bits s_{235} and s_{230} which update the left-most LFSR in Fig. 1 should remain in the "faulty" expression of l_i^α, where l_i is the LUT providing the input to s_{293-k}. Thus, the number of possibilities is larger, including, for each s_{293-k}:

TABLE II
LUTs COVERING FAULT INJECTION POINTS.

State bit	Original LUT function	With fault injected	Table I
s_{292}	$l = 1 \oplus A_1 \oplus A_3 \oplus A_6 \oplus \overline{A_4}\overline{A_5} \oplus A_2\overline{A_4}$	$l^\alpha = A_3$	l_2
s_{291}	$l = 1 \oplus A_1 \oplus A_4 \oplus A_6 \oplus \overline{A_2}\overline{A_5} \oplus A_3\overline{A_5}$	$l^\alpha = A_4$	l_2
s_{290}	$l = 1 \oplus A_1 \oplus A_5 \oplus A_6 \oplus \overline{A_2}\overline{A_4} \oplus \overline{A_2}\overline{A_3}$	$l^\alpha = A_5$	l_2
s_{289}	$l = 1 \oplus A_1 \oplus A_2 \oplus A_3 \oplus A_4 \oplus A_5 \oplus A_2 A_4 \oplus A_2 A_5$	$l^\alpha = A_4$	l_{10}
s_{288-1}	$l = 1 \oplus A_1 \oplus A_2 \oplus A_3 \oplus \overline{A_5}\overline{A_6}$	$l^\alpha = A_2 \oplus A_3$	l_1
s_{288-2}	$l = A_2 \oplus A_3 \oplus A_4 \oplus A_5 \oplus A_6 \oplus \overline{A_2}\overline{A_4}$	$l^\alpha = A_3 \oplus A_5$	l_9
s_{287-1}	$l = A_1 \oplus A_2 \oplus A_3 \oplus A_4 \oplus A_5 \oplus A_4\overline{A_6}$	$l^\alpha = A_1 \oplus A_2$	l_8
s_{287-2}	$l = 1 \oplus A_1 \oplus A_2 \oplus A_4 \oplus A_5 \oplus A_6 \oplus \overline{A_2}A_4 \oplus A_3\overline{A_6}$	$l^\alpha = A_1 \oplus A_5$	l_{12}
s_{286-1}	$l = 1 \oplus A_1 \oplus A_2 \oplus A_3 \oplus A_5 \oplus A_6 \oplus \overline{A_4}A_5$	$l^\alpha = A_1 \oplus A_3 \oplus A_6$	l_8
s_{286-2}	$l = 1 \oplus A_1 \oplus A_3 \oplus A_4 \oplus A_5 \oplus A_6 \oplus \overline{A_1}A_5 \oplus A_2\overline{A_4}$	$l^\alpha = A_6$	l_{12}
s_{285-1}	$l = A_2 \oplus A_4 \oplus A_5 \oplus \overline{A_3}\overline{A_6}$	$l^\alpha = A_4 \oplus A_5$	l_1
s_{285-2}	$l = A_1 \oplus A_2 \oplus A_3 \oplus A_4 \oplus A_5 \oplus \overline{A_2}\overline{A_5} \oplus A_1\overline{A_5}$	$l^\alpha = A_3$	l_{10}
s_{285-3}	$l = A_3 \oplus A_5 \oplus A_6 \oplus A_2\overline{A_4} \oplus \overline{A_1}A_3$	$l^\alpha = A_5 \oplus A_6$	l_3

1) $l_i^\alpha = x_{j_1} \oplus x_{j_2} \oplus x_{j_3}$, $j_1, j_2, j_3 \in J_i$, $O\binom{c}{3}$ choices.
2) $l_i^\alpha = x_{j_1} \oplus x_{j_2}$, $l_r^\alpha = x_{j_3} \oplus x_{j_4}$, $j_1, j_2 \in J_i$, $j_3, j_4 \in J_r$, $O\left(\binom{c}{2}^2 N'\right)$ choices.
3) $l_i^\alpha = x_{j_1} \oplus x_{j_2} \oplus x_{j_3}$, $l_r^\alpha = x_{j_4}$, $j_1, j_2, j_3 \in J_i$, $j_4 \in J_r$, $O\left(c\binom{c}{3}N'\right)$ choices.

where l_r is any LUT from the remaining set of $N' = N - 8$ candidates. Using this procedure, the fault injection points shown in the rows 5-10 of Table II are found.

The state bit s_{285} is the most difficult case. To find its fault injection points, we need to search among pairs of the remaining $N'' = N - 11$ candidates l_{r_1}, l_{r_2}, including

1) $l_i^\alpha = x_{j_1} \oplus x_{j_2}$, $l_{r_1}^\alpha = x_{j_3} \oplus x_{j_4}$, $l_{r_2}^\alpha = x_{j_5}$, $j_1, j_2 \in J_i$, $j_3, j_4 \in J_{r_1}$, $j_5 \in J_{r_2}$, $O\left(c\binom{c}{2}^2 ((N'')^2)\right)$ choices.
2) $l_i^\alpha = x_{j_1} \oplus x_{j_2} \oplus x_{j_3}$, $l_{r_1}^\alpha = x_{j_4}$, $l_{r_2}^\alpha = x_{j_5}$, $j_1, j_2, j_3 \in J_i$, $j_4 \in J_{r_1}$, $j_5 \in J_{r_2}$, $O\left(c^2\binom{c}{3}((N'')^2)\right)$ choices.

where l_i is the LUT providing the input to s_{285}. The successful combination is shown in the last three rows of Table II.

E. Key extraction

To extract the key, the content of 13 LUTs in Table II is modified as shown in column 3. The resulting faulty bitstream $\mathcal{B}^\alpha$ is loaded into the FPGA and used to generate $2^{15.34}$ bits of the faulty ciphertext C^α for some IV and plaintext P. Then, the keystream $C^\alpha \oplus P$ is analyzed to recover the key K as described in Section III.

To verify the key, a software implementation of ACORN is used to encrypt P using the same K and IV. The resulting ciphertext is compared with the reference ciphertext generated by the FPGA loaded with the original bitstream $\mathcal{B}$. If ciphertexts are the same, the key is correct.

V. COUNTERMEASURES

In this section, we discuss how the presented attack can be mitigated.

The attack is successful because the LUTs containing the fault injection points have a distinct structure (see Table II). This happens because FPGA technology mappers usually "pack" in a k-LUT as many nodes as possible, in order to minimize the depth and size of the resulting k-LUT network [38]. Contrary, if the function $s_{292}^+ = f(s) \oplus m_i$ is covered by an individual LUT implementing a 2-input XOR, finding such a LUT becomes more difficult because cryptographic algorithms typically contain many XORs.

The idea to constrain the technology mapping to generate k-LUT networks with smaller LUTs has been exploited in several countermeasures, e.g. [4], [39]. The *trivial cut*[4] technique proposed in [39] seems to be particularly suitable for ACORN. Given a Boolean network $\mathcal{N} = (V, \mathcal{E})$ implementing the cryptographic algorithm and a set of nodes $V_t \subseteq V$ with the function f_v to be protected, the trivial cut technique constrains the technology mapping as follows:

1) Cover each $v \in V_t$ by the trivial cut $C_v = \{v\}$.
2) Cover r additional nodes $u \in V - V_t$ with the same function as the nodes $v \in V_t$ by trivial cuts $C_u = \{u\}$.

It is shown in [39] that, for any functionally equivalent k-LUT network for $\mathcal{N}$ satisfying (1) and (2),

$$\epsilon \geq \left(\frac{e(m+r)}{m}\right)^m \qquad (7)$$

operations are required to find all nodes of V_t by an exhaustive search, where $m = |V_t|$.

Since the implementation of ACORN in our experiments has the degree of parallelization 8, the set V_t consists for eight 2-input XOR gates, e.g. $m = 8$. Let us estimate how many additional 2-input XORs have to be covered by trivial cuts to assure the security bound $\epsilon \geq 2^{128}$. To satisfy the condition $\left(\frac{e(8+r)}{8}\right)^8 \geq 2^{128}$, we need $r = 2^{17.6}$ XORs, which is much larger than the total number of 2-input XORs in the ACORN design. Note, however, that if the ACORN implementation has the degree of parallelization 32, then to satisfy $\left(\frac{e(32+r)}{32}\right)^{32} \geq 2^{128}$ we need only $r = 157$ XORs.

---

[4]A set of nodes $C \subseteq V$ is a *cut* of a node v if any path from any primary input in the cone of influence of v to v passes through at least one node in C [38]. Node v itself is a trivial cut.

We can conclude that, for ACORN, a higher degree of parallelization helps achieve stronger protection with lower performance penalty. A constrained k-LUT network is likely to have a larger depth than an unconstrained, depth-optimal k-LUT network. If less nodes are to be covered by trivial cuts, it might be possible to minimize performance penalty by choosing the nodes which are at non-critical paths.

VI. CONCLUSION

In this paper, we demonstrated an attack which can recover the full key from an SRAM-based FPGA implementation of ACORN stream cipher. By modifying the content of 13 LUTs directly in the bitstream, we inject faults which reduce the non-linear feedback function of ACORN to a linear one. As a result, it becomes possible to extract the key from the faulty keystream by an algebraic attack.

Besides demonstrating the attack, we also analyzed potential countermeasures and argued that ACORN's implementations with a higher degree of parallelization might be easier to defend. Our results are expected to enrich the understanding of the threats which FPGA implementations of cryptographic algorithms are facing and contribute towards protecting them more effectively.

VII. ACKNOWLEDGEMENT

This work was supported in part by the research grant 2018-04482 from the Swedish Research Council.

REFERENCES

[1] Ericsson, "More that 50 billions connected devices," 2012. www.ericsson.com/ res/docs/whitepapers/wp-50-billions.pdf.

[2] I. Rouf et al., "Security and privacy vulnerabilities of in-car wireless networks: A tire pressure monitoring system case study," in *Proc. of the 19th USENIX Conf. on Security*, pp. 21–21, 2010.

[3] J. Edwards, "Edge computing security dos and don'ts." Network Computing, November 2018. https://www.networkcomputing.com/network-security/edge-computing-security-dos-and-donts.

[4] P. Swierczynski et al., "FPGA trojans through detecting and weakening of cryptographic primitives," *IEEE Trans. on Computer-Aided Design of Integrated Circuits and Systems*, vol. 34, pp. 1236–1249, Aug 2015.

[5] M. Moraitis and E. Dubrova, "Bitstream modification attack on SNOW 3G," in *Proc. of the 2015 Design, Automation & Test in Europe Conf. & Exhibition (DATE'20)*, 2020.

[6] CESAR, "Cryptographic competition for authenticated encryption: Security, applicability, and robustness," 2018. https://competitions.cr.yp.to.

[7] W. Diehl et al., "Comparison of cost of protection against differential power analysis of selected authenticated ciphers," in *IEEE Int. Symp. on Hardware Oriented Security and Trust*, pp. 147–152, April 2018.

[8] A. Ghafari and H. Hu, "A new chosen IV statistical distinguishing framework to attack symmetric ciphers, and its application to ACORN-v3 and Grain-128a," *Journal of Ambient Intelligence and Humanized Computing*, pp. 1–6, 2018.

[9] A. Adomnicai et al., "Masking the lightweight authenticated ciphers ACORN and Ascon in software," *IACR Cryptology ePrint Archive*, vol. 2018, p. 708, 2018.

[10] A. Adomnicai et al., "Practical algebraic side-channel attacks against ACORN," in *Proc. of Int. Conf. on Information Security and Cryptology (ICISC'2018)*, (Seoul, Korea), 2018.

[11] F. Lafitte et al., "SAT-based cryptanalysis of ACORN," *IACR Cryptology ePrint Archive*, vol. 2016, p. 521, 2016.

[12] D. Dalai and D. Roy, "A state recovery attack on ACORN-v1 and ACORN-v2," in *Proc. of Int. Conf. on Network and System Security*, pp. 332–345, Springer, 2003.

[13] C. Chaigneau et al., "Full key-recovery on ACORN in nonce-reuse and decryption-misuse settings." Posted on the Crypto-competition mailing list, 2015.

[14] Md Iftekhar Salam et al., "Investigating cube attacks on the authenticated encryption stream cipher ACORN." Cryptology ePrint Archive, Report 2016/743, 2016. https://eprint.iacr.org/2016/743.pdf.

[15] D. Roy and S. Mukhopadhyay, "Some results on ACORN." Cryptology ePrint Archive, Report 2016/1132, 2016. https://eprint.iacr.org/2016/1132.pdf.

[16] A. Siddhanti et al., "Differential fault attack on Grain v1, ACORN v3 and Lizard," in *Proc. of 7th Int Conf. on Security, Privacy and Applied Cryptorgaphy Engeneering*, pp. 247–263, Springer, 2017.

[17] X. Zhang et al., "Fault attack on ACORN v3." Cryptology ePrint Archive, Report 2017/855, 2017. http://eprint.iacr.org/2017/855.pdf.

[18] P. Dey et al., "Full key recovery of ACORN with a single fault," *Journal of Information Security and Applications*, vol. 29, pp. 57–64, 2016.

[19] H. Wu, "ACORN: A lightweight authenticated cipher," 2016. https://competitions.cr.yp.to/round3/acornv3.pdf.

[20] E. Dubrova, "Breaking ACORN with a single fault." Cryptology ePrint Archive, Report 2019/697, 2019. https://eprint.iacr.org/2019/697.

[21] D. Coppersmith and S. Winograd, "Matrix multiplication via arithmetic progression," *J. Symboic Computation*, vol. 9, pp. 251–280, 1990.

[22] A. Shamir et al., "Efficient algorithms for solving overdefined systems of multivariate polynomial equations," in *Proc. of Eurocrypt'2000, LNCS 2612*, pp. 141–157, Springer, 2000.

[23] N. Courtois and W. Meier, "Algebraic attacks on stream ciphers with linear feedback," in *Proc. of Eurocrypt'2003, LNCS*, pp. 345–359.

[24] N. Courtois and J. Patarin, "About the XL algorithm over GF(2)," in *Proc. of Cryptographers' Track RSA'2003, LNCS 2612*, pp. 141–157, Springer, 2003.

[25] N. Courtois, "Higher order correlation attacks, XL algorithm and cryptanalysis of Toyocrypt," in *Proc. of ICISC'2002, LNCS 1807*, pp. 392–407, Springer, 2000.

[26] ATHENA, "Authomated tool for hardware evaluation." https://cryptography.gmu.edu/athena/index.php?id=CAESAR_source_codes.

[27] S. Drimer, "Volatile FPGA design security – a survey," 2007. http://citeseerx.ist.psu.edu/viewdoc/summary?doi=10. 1.1.105.3354.

[28] J. Vliegen et al., "A single-chip solution for the secure remote configuration of FPGAs using bitstream compression," in *2013 Int. Conf. on Reconfigurable Computing and FPGAs (ReConFig)*, pp. 1–6, Dec 2013.

[29] M. Majzoobi et al., "Fpga time-bounded unclonable authentication," in *Int. Workshop on Information Hiding*, pp. 1–16, Springer, 2010.

[30] A. Moradi et al., "Side-channel attacks on the bitstream encryption mechanism of Altera Stratix II: facilitating black-box analysis using software reverse-engineering," in *Proc. of the ACM/SIGDA Int. Symp. on FPGAs*, pp. 91–100, 2013.

[31] A. Moradi et al., "On the vulnerability of FPGA bitstream encryption against power analysis attacks: extracting keys from Xilinx Virtex-II FPGAs," in *Proc. of the 18th ACM Conf. on Computer and Communications Security*, pp. 111–124, 2011.

[32] A. Moradi and T. Schneider, "Improved side-channel analysis attacks on Xilinx bitstream encryption of 5, 6, and 7 series," in *Int. Workshop on Constructive Side-Channel Analysis and Secure Design*, pp. 71–87, Springer, 2016.

[33] S. Tajik et al., "On the power of optical contactless probing: Attacking bitstream encryption of FPGAs," in *Proc. of the 2017 ACM SIGSAC Conf. on Computer and Communications Security*, pp. 1661–1674, 2017.

[34] H. Lohrke et al., "Key extraction using thermal laser stimulation," *IACR Transactions on Cryptographic Hardware and Embedded Systems*, pp. 573–595, 2018.

[35] M. Ender et al., "The unpatchable silicon: A full break of the bitstream encryption of Xilinx 7-series FPGAs," in *29th USENIX Security Symposium (USENIX Security 20)*, (Boston, MA), 2020.

[36] Xilinx, "7 series FPGAs Configurable Logic Block User Guide (UG474 v1.8)," Sept. 27, 2016.

[37] S. Hurst et al., *Spectral Techniques in Digital Logic*. Academic Press, 1985.

[38] M. Teslenko and E. Dubrova, "Hermes: LUT FPGA technology mapping algorithm for area minimization with optimum depth," in *IEEE/ACM Int. Conf. on Computer Aided Design, 2004. ICCAD-2004.*, pp. 748–751, Nov 2004.

[39] M. Moraitis and E. Dubrova, "Bitstream modification attack on SNOW 3G." Cryptology ePrint Archive, Report 2020/038, 2020. https://eprint.iacr.org/2020/038.

978-1-7281-5410-7/20 $31.00 © 2020 IEEE

Breaking Barriers: Maximizing Array Utilization for Compute In-Memory Fabrics

Brian Crafton[†1], Samuel Spetalnick[†1], Gauthaman Murali[1],
Tushar Krishna[1], Sung-Kyu Lim[1], and Arijit Raychowdhury[1]

[1]Georgia Institute of Technology, Atlanta, GA
[1]School of Electrical and Computer Engineering
brian.crafton@gatech.edu, arijit.raychowdhury@ece.gatech.edu

Abstract—**Compute in-memory (CIM) is a promising technique that minimizes data transport, the primary performance bottleneck and energy cost of most data intensive applications. This has found wide-spread adoption in accelerating neural networks for machine learning applications. Utilizing a crossbar architecture with emerging non-volatile memories (eNVM) such as dense resistive random access memory (RRAM) or phase change random access memory (PCRAM), various forms of neural networks can be implemented to greatly reduce power and increase on chip memory capacity. However, compute in-memory faces its own limitations at both the circuit and the device levels. Although compute in-memory using the crossbar architecture can greatly reduce data transport, the rigid nature of these large fixed weight matrices forfeits the flexibility of traditional CMOS and SRAM based designs. In this work, we explore the different synchronization barriers that occur from the CIM constraints. Furthermore, we propose a new allocation algorithm and data flow based on input data distributions to maximize utilization and performance for compute-in memory based designs. We demonstrate a 7.47× performance improvement over a naive allocation method for CIM accelerators on ResNet18.**

I. INTRODUCTION

Modern computing systems are heavily dependent on the capacity and access time of expensive memory banks due to the ever increasing performance gap between main memory and logic. Furthermore, the cost of moving data has become more expensive than operating on it [1], and thus not only has the memory become the fundamental bottleneck of computing, but both reading and transporting the data has become more expensive than the operation we seek to perform. Popularization of data intensive applications like machine learning and artificial intelligence have further exacerbated this problem. To address these issues, new architectures based on traditional CMOS attempt to minimize the transport of data by optimizing for data reuse [1] and adopting constraints inspired by the brain [2]. While these techniques yield strong results, they still face the fundamental technological limitations of CMOS.

Fortunately a new class of embedded non-volatile memory (eNVM) is positioned to minimize data transport by performing compute in-memory. In-memory computing seeks to perform matrix multiplication ($\vec{y} = W\vec{x}$) in a crossbar structure using Ohm's law and the non-volatile conductance

† These authors contributed equally

state provided by the non-volatile memory. Using this technique, each weight of the matrix (W_{ij}) is programmed as a conductance to a cell and each value of the vector ($\vec{x}_i$) is converted to voltage and applied to the rows of the memory crossbar. By Ohm's law, the current through each cell is proportional to the product of the programmed conductance (W_{ij}) and applied voltage ($\vec{x}_i$). By Kirchhoff's current law (KCL), the resulting currents summed along the columns of the crossbar are proportional to the product of the matrix and vector, ($\vec{y}$). Under this procedure, the only data transport required for matrix multiplication is the feature vector ($\vec{x}$) from memory and result ($\vec{y}$) to memory. Therefore, in-memory computing eliminates the majority of data transfer and thus energy cost of data intensive operations.

Although compute in-memory using the crossbar architecture can greatly reduce data transport, the rigid nature of these large fixed weight matrices forfeits the flexibility of traditional CMOS and SRAM based designs. Given that eNVM has high density and unfortunately high write energy compared to traditional SRAM, CIM-based inference-only designs avoid writing to the eNVM cells once programmed. While this is advantageous for data transport and energy efficiency, it means each CIM processing element (PE) can only perform operations it has the weights for. This implies that if there is an unbalanced workload where some PEs operations take longer than others, we cannot simply re-allocate these operations to other PEs. Therefore, we must use synchronization barriers for all PEs so distributed matrix multiplication completes before another is started. In contrast, every CMOS and SRAM based PE are computationally identical and can perform any operation in the DNN graph.

Therefore a fundamental problem in CIM based designs is array utilization, the percent of time an array is in use. Recent large scale CIM designs [3], use weight duplication and layer pipelining techniques to maximize performance. We describe these techniques in detail in Section II. While impressive performance is achieved, these techniques only perform well when the workloads are deterministic. Circuit level techniques like zero-skipping greatly increase performance, but create non-deterministic workloads that compromise array utilization. In this work we identify and profile these new challenges using a simple simulator framework. We then propose a novel algorithm, which makes use of input statistics to find optimal

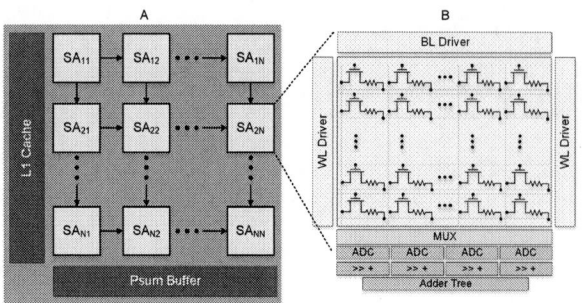

Fig. 1. Typical compute in-memory PE (processing engine) and sub-array (SA) architecture. (A) NxN sub-array PE with L1 cache and psum buffer. In this work N is 8. (B) Typical sub-array design with dual word line drivers, ADCs, shift and add units, and an adder tree.

array allocation policies to maximize utilization and *break* synchronization *barriers*. Furthermore, we introduce a new data flow that generalizes CIM arrays to maximize their utilization. We run our experiments on ImageNet using ResNet18 and CIFAR10 using VGG11. Although we apply our techniques to deep learning, we claim that the techniques we propose can be extended to any compute in-memory application. We note that a combination of these strategies yield 7.47× improvement in performance over a baseline naive array allocation.

II. BACKGROUND AND MOTIVATION

Compute in-memory systems use binary or multi-level cells as weights to perform matrix multiplication in memory. In this work we will focus our attention to binary cells given the current state of the art in eNVM [4] already struggles with variance thus making multi-level cells even more difficult to utilize. However, the same techniques demonstrated in this work can easily be applied to multi-level cells as well. Given binary cells, we must use 8 adjacent cells to form a single 8-bit weight, like those shown in the columns of Figure 1. The 8-bit vector inputs to this array are shifted in 1 bit at a time, and the resulting binary product collected at the ADCs is shifted left by the same amount the inputs are shifted right. In this way, each array is able to perform an 8-bit matrix multiplication.

There are two common techniques for performing compute in memory. The first technique, we call *baseline*, is simply reading as many rows as the ADC precision allows (e.g. for a 3-bit ADC, we read 8 rows simultaneously). The next technique is commonly called zero skipping [5], where only rows with '1's are read. This technique exploits sparsity in the input features or activations (for neural networks). Zero skipping performs faster than the baseline technique because for most cases it will process more total rows per cycle. In Figure 2, we provide an example case for zero-skipping where 8 total rows are read using a 2-bit ADC. Baseline (2A) requires 2 cycles since it targets four consecutive rows at a time. Zero-skipping (2B) is able to finish all 8 rows in a single cycle because we only consider the '1's in the input vector. There are few reasons not to perform zero skipping, unless there is limited input data bandwidth or the eNVM has high variance and accumulated too many errors.

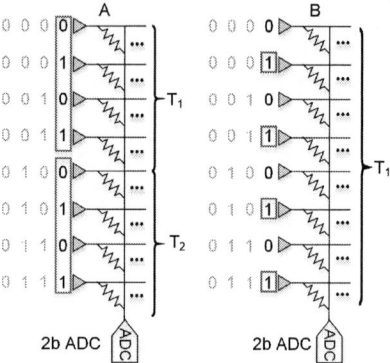

Fig. 2. Simplified breakdown of ADC reads in baseline and zero-skipping with 2-bit ADC precision. (A) Baseline targets four consecutive rows at a time since the 2-bit ADCs are capable of distinguishing 4 states. (B) Zero skipping targets the next 4 rows where the word line is enabled. This way we can read more rows and not overflow our ADC.

By encapsulating the array, ADCs, and shift and add logic, a matrix multiplication engine can be created. Using these arrays as building blocks, prior work has implemented CNNs (Convolutional Neural Networks) where a group of arrays implement a larger matrix multiplication. Despite performing more complex operations, the core operations of CNNs are converted into matrix multiplication. In Figure 1 we illustrate this idea, showing how a group of arrays is tiled together to form a PE. In Figure 3 we further depict how these arrays can be pieced together to form a larger matrix. In this example, both input feature maps and filters are vectorized with the filters forming the columns of a matrix. The vectorized feature maps are input to the crossbar to perform matrix multiplication, where the results are output feature maps for this layer in a CNN.

Given the high density of these PEs, hundreds or thousands of them can be tiled in the same area used by modern ICs. Although similar in concept, CIM-based DNN accelerators have numerous differences from traditional CMOS based designs that introduce challenges in maximizing performance. First off, a CIM-based PE has fixed weights that cannot be reprogrammed due to the high energy cost of writing eNVM. Traditional CMOS based PEs are generalized compute units that can operate on any input data, since they do not contain fixed weights. Thus a fundamental issue in CIM-based accelerators is array utilization. Several works have addressed this issue introducing ideas such as weight duplication and layer pipelining.

Weight duplication [3] is used to maximize throughput in large scale CIM accelerators where the amount of on-chip memory exceeds the number of weights in the model. In [6], 24,960 arrays are used for a total on-chip memory capacity of nearly 104 MB (2b cells), while only using an area of $250mm^2$. Using this enormous on-chip memory capacity, they not only fit ResNet [7] but duplicate shallow layers up to 32×. When weights are duplicated, the input data is divided equally amongst each duplicate array so they can process in parallel.

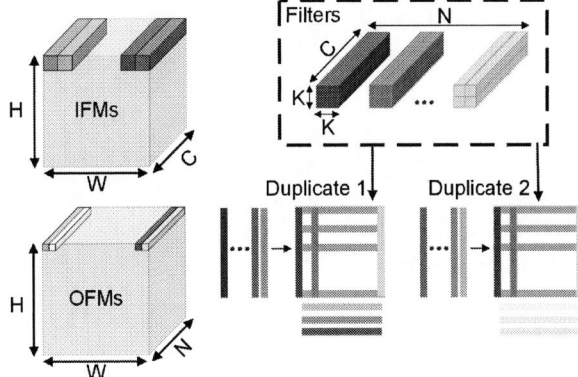

Fig. 3. Convolutional layer mapped to a CIM array. Both input features maps (IFM) and filters are vectorized with the filters forming the columns of a matrix. The vectorized feature maps applied to the crossbar to perform matrix multiplication, where the results are output feature maps (OFMs).

We illustrate this idea for a convolutional layer in Figure 3. The input patches from the input feature maps (IFMs) are divided into groups based on the number of duplicates, and then mapped to each duplicate.

Layer pipelining [3] is used to maximize throughput in eNVM CIM accelerator, where arrays are not re-programmed due to large amounts of on-chip memory and high write energy. At the same time, most modern neural networks contain 20 or more layers that must be processed sequentially. Given that most designs use 128×128 arrays, it becomes infeasible to partition arrays such that they can be used for each layer without being re-programmed. This implies that the majority of PEs would sit idle waiting for their layer to be processed. To solve this problem, images are pipelined through the network to keep all arrays utilized. Although this compromises single example latency, it maintains maximum throughput.

III. BLOCK-WISE ARRAY ALLOCATION

In the previous section, we discussed several techniques that are used in CIM accelerators to increase throughput, but each introduces it's own synchronization barrier that limits array level utilization. In this work, we identify two of these barriers and propose our solution to mitigate this problem. The two techniques that create these barriers are weight duplication and layer pipelining. In previous work these barriers were not a problem because array performance was deterministic. When zero-skipping is introduced, it instigates these barriers because it introduces non-deterministic computation time for each array. Zero skipping will only improve the performance of a CIM accelerator because it simply means each array will perform equal to or faster than the baseline algorithm. However, since the number of ones in the input vector of the CIM operation follows a random distribution, the amount of time to finish a dot product is non-deterministic. This means that several arrays performing a part of a larger matrix multiplication need to be synchronized to the slowest preforming array. As the size of the operation (and number of arrays) increases, the

more stalls occur. In the following section, we explore the implications of zero skipping at the architectural level.

A. Identifying Synchronization Barriers

The non-determinism introduced by zero-skipping induces the need for synchronization barriers. A synchronization barrier is required when a group arrays are processing a distributed workload and finish at different times, but must be synchronized before starting another task. The first barrier occurs at the layer level and is a result of using layer pipelining. When the arrays are distributed to each layer, we attempt to divide them evenly so that all layers finish at the same time. If any layer is consistently performing faster than other layers, it will have to stall because layers downstream will not be able to buffer its outputs. Previous work [6] have allocated arrays to layers based on the number of duplicates required such that all layers in the pipeline complete their workload at the same time, and thus sustain full utilization.

This allocation method works under the assumption that all arrays perform at the same rate and we can choose the number of arrays on chip. However, as [5] points out neither of these assumptions will hold in a realistic design. Prior works [3], [6] assume 128 cells can be read at once using 5 and 8 bit ADCs. Although feasible in theory, we note that such a design will yield very high error given that the state of the art devices have 5% device-to-device variance [4], [8], and thus at most 8 rows (3-bit) can be read at once. Such a design also yields very poor memory density since large (5-8 bit) ADCs occupy over $10 \times$ the area of eNVM. Instead columns must be processed in batches using zero-skipping, where current summation is used for 8 rows and then intermediate results are stored and accumulated using existing digital logic in the array.

When zero skipping is used, each array performs at a non-deterministic speed that follows the distribution of input data it receives. In Figure 4, we plot the average time for an array to perform a 128×16 matrix multiplication versus the percentage of '1's in all the 8-bit input features for the 20 convolutional layers in ResNet18. To compute the percentage of '1's for a layer, we average the 8 bits in all 8-bit input features together. For example, a 1000-entry 8-bit input vector contains 8000 bits and to determine the percentage of '1's, we average over 8000 bits to compute this percentage. From Figure 4, we infer a linear relationship between the percentage of '1's in the input features to a layer, and the expected number of cycles to perform the matrix multiplication.

Naturally, we can use this information to better allocate duplicates to each layer in our design. We approach this problem by quantifying the total number of multiply-and-accumulate (MAC) operations in each layer, and the average number of MAC operations per cycle an array can perform. In prior works, performance per array is constant since each array takes the same number of cycles to perform a matrix multiplication. Therefore, arrays are allocated to each layer based only on the total MACs per layer. When zero-skipping is introduced and performance per array is not constant, this allocation method fails to allocate evenly. To achieve equal

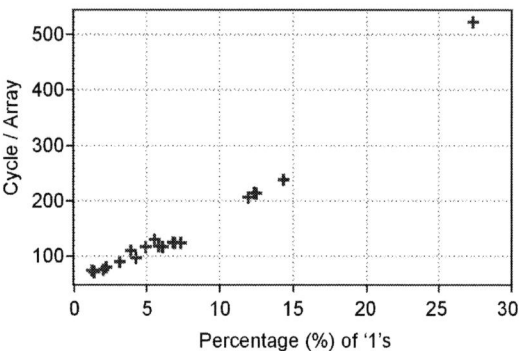

Fig. 4. Cycles per array versus the percentage of '1's in all 8-bit input features. Each point represents the average percentage for one of the 20 layers in ResNet18.

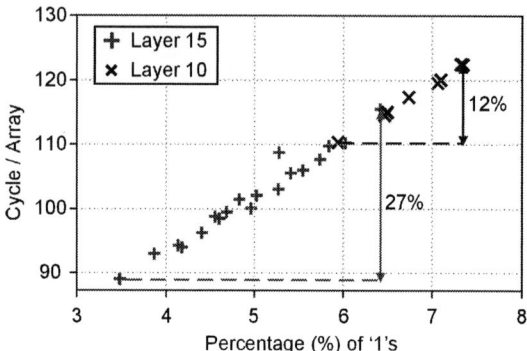

Fig. 6. Cycles per array versus the percentage of '1's in all 8-bit input features. The blue crosses represent the average percentage for 1 of the 18 blocks in layer 15 of ResNet18. The black ×s represent 1 of the 9 blocks in layer 10.

utilization, we can instead allocate arrays to each layer based on the expected number of cycles it will take to finish without any duplicate arrays. We can compute the expected number of cycles it will take a layer to finish by dividing the total MACs in a layer by the average performance of each array in the layer. We call this allocation method *performance-based* allocation, whereas allocation that assumes all arrays perform evenly is *weight-based* allocation.

While this technique ensures that all our layers will be equally utilized, it does not ensure that the arrays inside each layer will be equally utilized. Each layer in our DNN (convolution or fully connected) is implemented as a matrix consisting of eNVM arrays. We visualize this idea in Figure 5, where a $3 \times 3 \times 128 \times 128$ filter is mapped to 72 arrays arranged in a 9×8 grid. In each of the 9 rows, all 8 arrays share the same input data and, consequently, the same word lines. This implies that all 8 arrays will operate at the same speed and form our minimal deterministic compute unit that we call a *block*. Because the 9 different rows do not share the same input vectors, they will operate at different speeds. If some arrays receive fewer '1's than other arrays, they will sit idle waiting for arrays that receive more '1's to finish. In Figure 6, we plot the average cycle time of the arrays in each block of layers 10 and 15 (ResNet18) versus the percent of '1's they receive. Layer 10 is a $3 \times 3 \times 128 \times 128$ filter (Figure 5) that contains 9 different blocks, and Layer 15 is a $3 \times 3 \times 256 \times 256$ filter that contains 18 different blocks. Just as before, we observe a linear

relationship between cycle time and the percentage of '1's. Since layer 15 contains more blocks, it is more susceptible to longer delays because the expected slowest block's cycle time increases with the number of arrays. In this figure, we observe a 12% and 27% difference in cycle time for layers 10 and 15, which motivates a better allocation technique to prevent significant idle time.

B. Optimizing Array Allocation

Finding the optimal allocation policy for blocks is more difficult. We cannot add redundant blocks to the same layer, because each layer only uses each weight once per operation. Instead, we adopt a new grouping strategy for arrays: rather than duplicating layers of arrays, we duplicate blocks of arrays. To find the optimal array allocation policy, we propose a linear time ($O(N)$ complexity) solution. This is especially important for larger networks like ResNet18, where there are 247 blocks and finding an optimal solution could be quite difficult.

With this new grouping strategy, we can allocate using the same technique as before. First we gather an approximation of the average MAC per cycle for each block of arrays. We can do this two ways. The first option, is running a cycle accurate simulator on some example data to get a very accurate approximation. The second option is to profile the distribution of '1's in the activations gathered from a large set of examples run on a GPU. Once we have an approximation for the MAC per cycle of each block, we can compute the expected number of cycles each block will take to perform it's partial dot product. Once we have cycle approximations for each block, we begin allocating arrays to each block. While we have free (not allocated) arrays, we loop through and allocate arrays to the block with the highest expected latency. Once we run out of arrays or the number of arrays left over is not enough to allocate to the slowest block we have found the optimal allocation. We call this allocation method *block-wise*, whereas allocation based on the layer is *layer-wise*.

C. Block-wise Data Flow

To make use of our new allocation policy, a new data flow strategy is required. Since arrays from the same layer are not

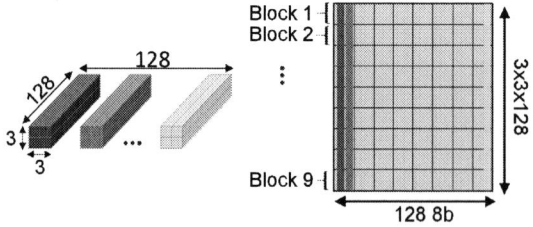

Fig. 5. The $3 \times 3 \times 128 \times 128$ filter used in layer 10 from ResNet18 converted into a matrix with annotated blocks. This filter requires 72 128×128 arrays to store in a 9×8 grid.

978-1-7281-5410-7/20 $31.00 © 2020 IEEE

grouped together, we treat these blocks as generalized compute units rather than being bound to a specific duplicate. Therefore, we no longer stall for the slowest block in a layer, but rather just send work to the next available block. This means that the same blocks will no longer be working together on the same input data, and thus will not be part of the same gather and accumulate procedure. As a result, a new routing and scheduling policy is required because blocks will not always send their partial sums to the same accumulator for every input feature map. To implement this idea, we include output feature destination addresses in the packet containing data when sending input features to each block. Upon completing a partial dot product, a block sends their computed partial sums to the designated accumulator and requests additional work from the memory controller.

IV. CIM-BASED ARCHITECTURE

Although our allocation policy will work for any general CIM based accelerator, we adopted a similar architecture to previous work [3], [6]. Our basic processing element (PE) contains 64 128×128 arrays. We choose 64 arrays because it provides each block with sufficient network bandwidth and SRAM capacity, while maintaining good SRAM density and low interconnect overhead. Our input data, weights, and activations are all 8 bits. Each array has 1 3-bit ADC for every 8 columns where a single column is pitch-matched with a comparator. We choose 3-bit because state of the art devices [4] have 5% variance and 3-bits is the maximum precision that can be read with no error. We shift one bit from each of the 128 inputs in one at a time which takes 8 cycles. In the best case scenario, we perform all 128 rows at the same time. In the worst case scenario, it takes 16 cycles since we enable every single row. Therefore, each array takes anywhere from 64 to 1024 cycles and performs a 128×16 dot product. In all designs we consider, we use use the same 64 array PE and simply increase the count per design.

The activation inputs to the RRAM sub-arrays are stored in on-chip SRAM, while the input images are read in from external DRAM. Matrix multiplication is performed by the

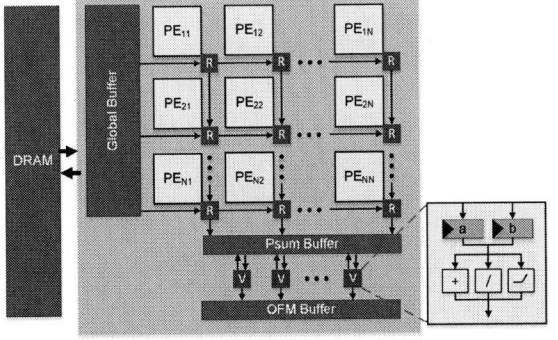

Fig. 7. Block-wise network architecture with 1 router (R) per PE. All input features are routed from the global buffer to PEs. All partial sums are routed from PE to vector unit (V), and vector unit to output feature buffer.

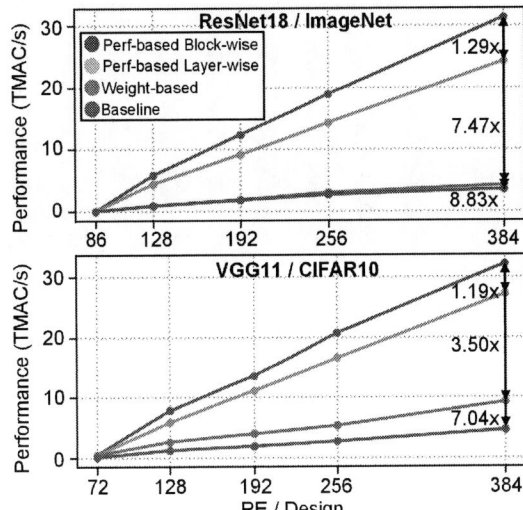

Fig. 8. Inference performance for ResNet18 and VGG11 by algorithm and design size assuming 100MHz clock. For ResNet18, block-wise allocation sustains a 8.83×, 7.47×, and 1.29× speedup over baseline (no zero-skipping), weight-based, and performance-based layer-wise allocation. For VGG11, block-wise allocation sustains a 7.04×, 3.50×, and 1.19× speedup.

PEs, while custom vector units are used to perform vector-wise accumulation, bias addition, quantization, and relu. We use a $N \times N$ mesh network for communication between PEs, memory, and vector units shown in Figure 7. Since blocks vary in size and no block contains 64 sub-arrays, we have to partition the PE to contain several blocks. This configuration implies that the different blocks share the same virtualized input and output ports. As discussed in Section III, input and output vectors are packetized to include destination information. Each block in the PE is given an id that is used to route packets to and from. Upon completing a partial dot product, a block sends its partial sum to vector units where they are accumulated and activation functions and quantization is applied.

V. RESULTS

To benchmark block-wise allocation, we compare with several other techniques: weight-based allocation, performance-based layer-wise allocation, and the baseline algorithm which does not use zero-skipping. We empirically evaluate performance and array utilization for the three techniques on ImageNet using ResNet18 and CIFAR10 using VGG11. We run these techniques in a custom simulation framework designed to evaluate performance and power of compute in-memory using standard CMOS and RRAM models from [9]. In this work we focus on performance evaluations, however higher array utilization will result in less leakage power and improved energy efficiency.

Our simulator performs cycle-accurate implementations of convolutional and fully connected layers. It is based in Python, but runs array level operations in C for faster evaluation. We model components in the design in object oriented fashion,

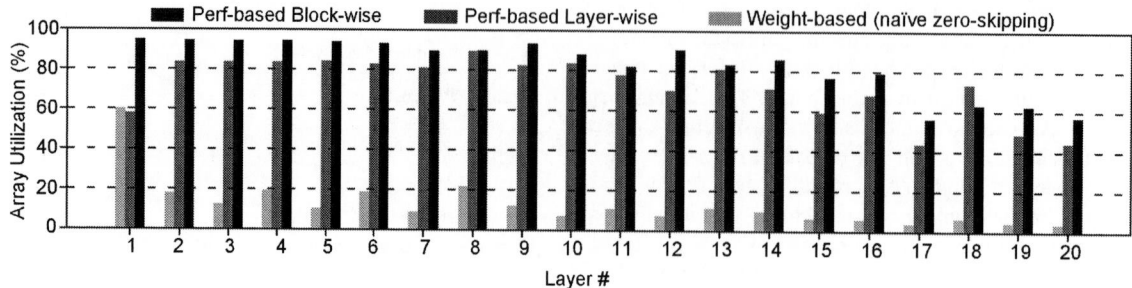

Fig. 9. Array utilization by layer for ResNet18 on ImageNet. Baseline not shown because zero skipping is not used.

iterating through all components in all PEs each cycle. We embed performance counters in our ADC and sub-array objects to track metrics like stalls so we can calculate utilization. As input, the simulator takes the network weights, input images, PE level configuration, and chip-level configuration. The PE-level configuration includes details like the precision of eac ADC and size of the sub-array. The chip-level configuration contains the number of PEs and details about array allocation and mapping. As output, the simulator produces a table with all desired performance counters and all intermediate layer activations that are verified against a TensorFlow implementation for correctness.

To show how our algorithm scales by the size of the design, we have evaluated the different allocation algorithms on several different designs with increasing numbers of PEs. In Figure 8, we plot performance versus the number of PEs in the design for both ResNet18 and VGG11. For ResNet18, we begin at 86 PEs since this contains the minimum number of arrays (5472) required to store ResNet18. At 86 PEs, all algorithms yield the same result since no duplication can be done and weights are simply allocated to store ResNet18. From there, we begin increasing the design size by $\frac{1}{2}$ powers of 2. Block-wise allocation performs the best achieving 29% improvement over layerwise-allocation and $7.47\times$ improvement over both weight-based and baseline (not zero-skipping) algorithms. We follow the same procedure for VGG11, however we observe that block-wise allocation yields less performance advantage. This is because VGG11 has roughly half the layers that ResNet18 has. It is more difficult to allocate evenly amongst a deeper network and therefore, block-wise allocation yields better results on deeper networks.

To better understand why we get these large performance improvements, it is useful to analyze array utilization. In Figure 9, we visualize layer-wise utilization of the 20 convolutional layers from ResNet18 using the different techniques. It is clear that block-wise allocation sustains the highest array utilization across nearly all layers in the network, easily outperforming the other techniques. Weight-based allocation performs very poorly because of the very different speeds of each layer and block we showed in Figures 4 and 6.

VI. CONCLUSION

In this paper we demonstrate the efficacy of a new technique and data flow to improve array utilization in CIM accelerators.

Given that the write energy of eNVM is high, CIM arrays contain fixed weights unlike CMOS PEs which can perform any operation in a DNN. Thus array utilization becomes a key challenge since only some arrays can perform particular operations. By profiling input statistics and relaxing our data flow, we can allocate arrays to maximize utilization and as a result, performance. The proposed allocation algorithm and data flow performs $7.47\times$ better than naive allocation and a layer-wise dataflow.

VII. ACKNOWLEDGEMENT

This work was funded by the U.S. Department of Defense's Multidisciplinary University Research Initiatives (MURI) Program under grant number FOA: N00014-16-R-FO05 and the Semiconductor Research Corporation under the Center for Brain Inspired Computing (C-BRIC) and Qualcomm.

REFERENCES

[1] Y.-H. Chen, T. Krishna, J. S. Emer, and V. Sze, "Eyeriss: An energy-efficient reconfigurable accelerator for deep convolutional neural networks," *IEEE Journal of Solid-State Circuits*, vol. 52, no. 1, pp. 127–138, 2017.

[2] M. Davies, N. Srinivasa, T.-H. Lin, G. Chinya, Y. Cao, S. H. Choday, G. Dimou, P. Joshi, N. Imam, S. Jain, *et al.*, "Loihi: A neuromorphic manycore processor with on-chip learning," *IEEE Micro*, vol. 38, no. 1, pp. 82–99, 2018.

[3] A. Shafiee, A. Nag, N. Muralimanohar, R. Balasubramanian, J. P. Strachan, M. Hu, R. S. Williams, and V. Srikumar, "Isaac: A convolutional neural network accelerator with in-situ analog arithmetic in crossbars," *ACM SIGARCH Computer Architecture News*, vol. 44, no. 3, pp. 14–26, 2016.

[4] J. Wu, Y. Chen, W. Khwa, S. Yu, T. Wang, J. Tseng, Y. Chih, and C. H. Diaz, "A 40nm low-power logic compatible phase change memory technology," in *2018 IEEE International Electron Devices Meeting (IEDM)*, pp. 27–6, IEEE, 2018.

[5] T.-H. Yang, H.-Y. Cheng, C.-L. Yang, I.-C. Tseng, H.-W. Hu, H.-S. Chang, and H.-P. Li, "Sparse reram engine: joint exploration of activation and weight sparsity in compressed neural networks," in *Proceedings of the 46th International Symposium on Computer Architecture*, pp. 236–249, 2019.

[6] X. Peng, R. Liu, and S. Yu, "Optimizing weight mapping and data flow for convolutional neural networks on processing-in-memory architectures," *IEEE Transactions on Circuits and Systems I: Regular Papers*, 2019.

[7] K. He, X. Zhang, S. Ren, and J. Sun, "Deep residual learning for image recognition," in *Proceedings of the IEEE conference on computer vision and pattern recognition*, pp. 770–778, 2016.

[8] B. Crafton, S. Spetalnick, and A. Raychowdhury, "Counting cards: Exploiting weight and variance distributions for robust compute in-memory," *arXiv preprint arXiv:2006.03117*, 2020.

[9] P.-Y. Chen, X. Peng, and S. Yu, "Neurosim: A circuit-level macro model for benchmarking neuro-inspired architectures in online learning," *IEEE Transactions on Computer-Aided Design of Integrated Circuits and Systems*, vol. 37, no. 12, pp. 3067–3080, 2018.

SANSCrypt: A Sporadic-Authentication-Based Sequential Logic Encryption Scheme

Yinghua Hu, Kaixin Yang, Shahin Nazarian, and Pierluigi Nuzzo

Department of Electrical and Computer Engineering
University of Southern California, Los Angeles, CA, USA
{yinghuah, kaixinya, shahin.nazarian, nuzzo}@usc.edu

Abstract—We propose SANSCrypt, a novel sequential logic encryption scheme to protect integrated circuits against reverse engineering. Previous sequential encryption methods focus on modifying the circuit state machine such that the correct functionality can be accessed by applying the correct key sequence only once. Considering the risk associated with one-time authentication, SANSCrypt adopts a new temporal dimension to logic encryption, by requiring the user to sporadically perform multiple authentications according to a protocol based on pseudo-random number generation. Analysis and validation results on a set of benchmark circuits show that SANSCrypt offers a substantial output corruptibility if the key sequences are applied incorrectly. Moreover, it exhibits an exponential resilience to existing attacks, including SAT-based attacks, while maintaining a reasonably low overhead.

I. INTRODUCTION

The design process of modern VLSI systems often relies on a supply chain where several services, such as verification, fabrication, and testing, are outsourced to third-party companies. If these companies gain access to a sufficient amount of critical design information, they can potentially reverse engineer the design. One possible consequence of reverse engineering is Hardware Trojan (HT) insertion, which can be destructive for many applications. HTs can either disrupt the normal circuit operation [1] or provide the attacker with access to critical data or software running on the chip [2].

Countermeasures such as logic encryption [3]–[6], integrated circuit (IC) camouflaging [7], watermarking [8], and split manufacturing [9] have been developed over the past decades to prevent IC reverse engineering. Among these, logic encryption has received significant attention as a promising, low-overhead countermeasure. Logic encryption modifies the circuit in a way such that a user can only access the correct circuit functionality after providing a correct key sequence. Otherwise, the circuit function remains hidden, and the output different from the correct one.

Various logic encryption techniques [3]–[6] and potential attacks [10]–[12] have appeared in the literature, as well as methods to systematically evaluate them [13], [14]. A category of techniques [3]–[5] is designed to modify and protect the combinational logic portions of the chip and can be extended to sequential circuits by assuming that the scan chains are not accessible by the attacker, e.g., due to scan chain encryption and obfuscation [15]–[17]. Another category of techniques, namely, sequential logic encryption [6], [18], [19], targets, instead, the state transitions of the original finite state machine (FSM). Sequential logic encryption introduces additional states

and transitions in the original FSM, essentially partitioning the state space into two sets. After being powered on or reset, the FSM enters the *encrypted mode*, exhibiting an incorrect output behavior. The FSM transitions, instead, to the *functional mode*, providing the correct functionality, upon receiving a sequence of key patterns.

A set of attacks have been reported against sequential encryption schemes, aiming to retrieve the correct key sequence or circuit function. Shamsi *et al.* [20] adapted the Boolean satisfiability (SAT)-based attack [10], traditionally targeted to combinational logic encryption, by leveraging methods from bounded model checking to unroll the sequential circuit. Recently, an attack based on automatic test pattern generation (ATPG) [21] uses concepts from excitation and propagation of stuck-at faults to search the key sequence among the input vectors generated by ATPG. When the attackers have some knowledge of the topology of the encrypted FSM, then they can extract and analyze the state transition graph and bypass the encrypted mode [22]. Overall, the continuous advances in FSM extraction and analysis tools tend to challenge any of the existing sequential encryption schemes and call for approaches that can significantly increase their robustness.

This paper proposes *SANSCrypt*, a Sporadic-Authentication-based Sequential Logic Encryption (SANSCrypt) scheme, which raises the attack difficulty via a *multiple-authentication protocol*, whose decryption relies on *retrieving a set of key sequences as well as the time at which the sequences should be applied*. Our contributions can be summarized as follows:

- A robust, multi-authentication based sequential logic encryption method that for the first time, to the best of our knowledge, systematically incorporates the robustness of multi-factor authentication (MFA) [23] in the context of hardware obfuscation.
- An architecture for sporadic re-authentication where key sequences must be applied at multiple random times, determined by a random number generator, to access the correct circuit functionality.
- Security analysis and empirical validation of SANSCrypt on a set of ISCAS'89 benchmark circuits [24], showing exponential resilience against existing attacks, including sequential SAT-based attacks, and reasonably low overhead.

Analysis and validation results show that SANSCrypt can significantly enhance the resilience of sequential logic encryption under different attack assumptions.

978-1-7281-5410-7/20 $31.00 © 2020 IEEE

II. BACKGROUND AND RELATED WORK

Among the existing sequential logic encryption techniques, HARPOON [6] defines two modes of operation. When powered on, the circuit is in the encrypted mode and exhibits an incorrect functionality. The user must apply a sequence of input patterns during the first few clock cycles to enter the functional mode, in which the correct functionality is recovered. However, the encrypted mode and functional mode FSMs are connected by only one transition (edge), which can be exploited by an attacker to perform FSM extraction and analysis, and bypass the encrypted mode [22].

Interlocking [18] sequential encryption modifies the circuit FSM such that multiple paths are available between the states of the encrypted and the ones of the functional FSMs, making it harder for the attacker to detect the only correct transition between the two modes. However, in both HARPOON and Interlocking encryption, once the circuit enters the functional mode, it remains there until reset.

Dynamic State-Deflection [25] requires, instead, an additional key input verification step while in the functional mode. If the additional key input is incorrect, the FSM transitions to a black-hole state cluster which can no longer be left. However, because the additional key input is fixed over time, the scheme becomes more vulnerable to sequential SAT-based attacks [20].

Finally, instead of corrupting the circuit function immediately after reset, DESENC [19] counts the occurrence of a specific but rare event in the circuit. Once the counter reaches a threshold, the circuit enters the encryption mode. This scheme is more resilient to sequential SAT-based attacks [26] because it requires unrolling the circuit FSM a large number of times to find the key. However, the initial transparency window may still expose critical portions of the circuit functionality.

III. SANSCRYPT

We introduce design and implementation details for SANSCrypt, starting with the underlying threat model.

A. Threat Model

SANSCrypt assumes a threat model that is consistent with the previous literature on sequential logic encryption [6], [20], [22]. The goal of the attack is to access the correct circuit functionality, by either reconstructing the deobfuscated circuit or finding the correct key sequence. To achieve this goal, the attacker can leverage one or more of the following resources: (i) the encrypted netlist; (ii) a working circuit providing correct input-output pairs; (iii) knowledge of the encryption technique. In addition, we assume that the attacker has no access to the scan chain and cannot directly observe or change the state of the circuit.

B. Authentication Protocol

As shown in Fig. 1a, existing logic encryption techniques are mostly based on a single-authentication protocol, requiring users to be authenticated only once before using the correct circuit function. After authentication, the circuit remains functional unless it is powered off or reset. To attack the circuit,

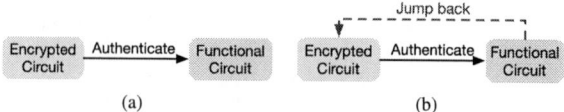

Fig. 1. Conventional (a) and proposed (b) authentication protocols for logic encryption.

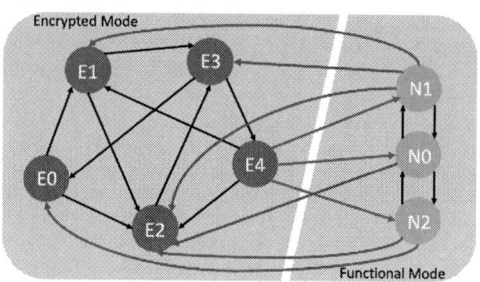

Fig. 2. State transition diagram of SANSCrypt.

it is then sufficient to discover the correct key sequence that must be applied in the initial state.

We adopt, instead, the authentication protocol in Fig. 1b, where the functional circuit can "jump" back to the encrypted mode from the functional mode. Once the back-jumping occurs, another round of authentication is required to resume the normal operation. The back-jumping can be triggered multiple times and involve a different key sequence for each re-authentication step. The hardness of attacking this protocol stems from both the increased number of the key sequences to be produced and the uncertainty on the time at which each sequence should be applied. A new temporal dimension adds to the decryption procedure, which poses a significantly higher threshold to the attackers.

C. Overview of the Encryption Scheme

SANSCrypt is a sequential logic encryption scheme which supports random back-jumping, as represented in Fig. 2. When the circuit is powered or reset, the circuit falls into the reset state $E0$ of the encrypted mode. To transition to the initial (or reset) state $N0$ of the functional mode, the user must apply at startup the correct key sequence to the primary input ports.

Once in the functional mode, the circuit can deliberately, but randomly, jump back, as denoted by the blue edges in Fig. 2, to a state s_{bj} in the encrypted mode, called *back-jumping state*, after a designated number of clock cycles t_{bj}, called *back-jumping period*. The user needs to apply another key sequence to resume normal operations, as shown by the red arrows. Both the back-jumping state s_{bj} and the back-jumping period t_{bj} are determined by a pseudo-random number generator (PRNG) embedded in the circuit. Therefore, when and where the back-jumping operation happens is unpredictable unless the attacker is able to break the PRNG or find its seed. The schematic of SANSCrypt is shown in Fig. 3 and consists of two additional blocks, a back-jumping module and an encryption finite state machine (ENC-FSM), besides the original circuit. We discuss each of these blocks in the following subsections.

978-1-7281-5410-7/20 $31.00 © 2020 IEEE

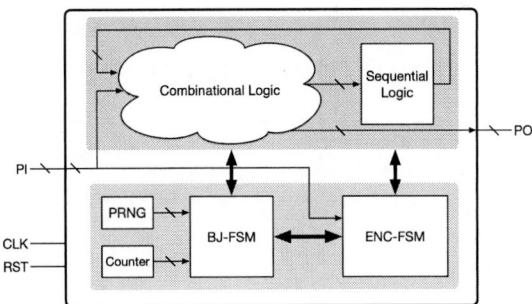

Fig. 3. Schematic view of SANSCrypt.

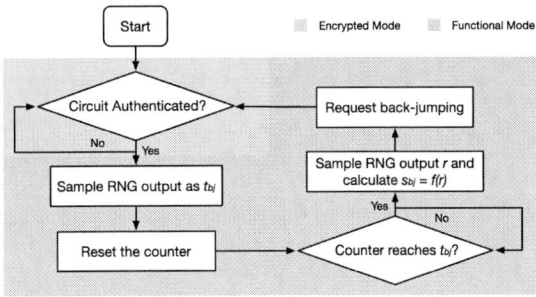

Fig. 4. Flowchart of BJ-FSM.

D. Back-Jumping Module

The back-jumping module consists of an n-bit *PRNG*, an n-bit *Counter*, and a *Back-Jumping Finite State Machine* (BJ-FSM) which sends back-jumping commands to the rest of the circuit. As summarized in the flowchart in Fig. 4, when the circuit is in the encrypted mode, BJ-FSM checks whether the authentication has occurred. If this is the case, BJ-FSM stores the current PRNG output as the back-jumping period t_{bj} and initializes the counter.

The counter increments its output at each clock cycle until it reaches t_{bj}. This event triggers BJ-FSM to sample again the current PRNG output r, which is generally different from t_{bj}, and use it to determine the back-jumping state $s_{bj} = f(r)$. For example, if s_{bj} is an l-bit binary number, BJ-FSM can arbitrarily select l bits from r and assign the value to s_{bj}. If the first l bits of r are selected, we have $f(r) = r[0 : l - 1]$. At the same time, BJ-FSM sends a back-jumping request to the other blocks of the circuit and returns to its initial state, where it keeps checking the authentication status of the circuit. On receiving the back-jumping request, the circuit jumps back to state s_{bj} in the encrypted mode and will stay there unless re-authentication is performed. Any PRNG architecture can be selected in this scheme, based on the design budget and the desired security level. For example, linear PRNGs, such as Linear Feedback Shift Registers (LFSRs), provide higher speed and lower area overhead but tend to be more vulnerable than cipher algorithm-based PRNGs, such as AES, which are, however, more expensive.

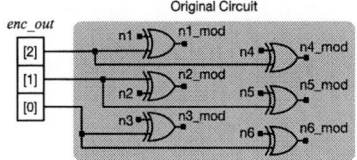

Fig. 5. *enc_out* controls the original circuit via XOR gates.

TABLE I
TRUTH TABLE FOR A 3-BIT *enc_out* ARRAY

State	E0	E1	E2	E3	E4	Auth
enc_out[0]	0	1	1	1	1	0
enc_out[1]	1	0	1	1	0	0
enc_out[2]	1	1	1	0	0	0

E. Encryption Finite State Machine (ENC-FSM)

The Encryption Finite State Machine (ENC-FSM) determines whether the user's key sequence is correct and, if it is not correct, takes actions to hide the functionality of the original circuit. The input of the ENC-FSM can be provided via the primary input ports, without the need to create extra input ports for authentication. The output enc_out of ENC-FSM, which is n bit long, together with a set of nodes in the original circuit netlist, can be provided as an input to a set of XOR gates, to corrupt the circuit function as in combinational logic encryption [3]. For example, in Fig. 5, a 3-bit array enc_out is connected to six nodes in the original circuit via XOR gates. In this paper, XOR gates are inserted at randomly selected nodes. However, any other combinational logic encryption technique is also applicable. As a design parameter, we denote by *node coverage* the ratio between the number of inserted XOR gates and the total number of combinational logic gates in the circuit.

Only one state of ENC-FSM, termed $auth$, is used in the functional mode. In state $auth$, all bits in enc_out are set to zero and the original circuit functionality is activated. In the other states, the value of enc_out changes based on the state, but at least one of its bits is set to one to guarantee that the final output is incorrect. A sample truth table for a 3-bit enc_out array is shown in Table I. Even if the circuit is in the encrypted mode, enc_out changes its value based on the state of the encryption FSM. Such an approach makes it difficult for signal analysis attacks, aiming to locate signals with low switching activity in the encrypted mode, to find enc_out and bypass ENC-FSM. After a valid authentication, the circuit resumes its normal operation. Additional registers are, therefore, required in the ENC-FSM to store the circuit state before back-jumping so that it can be resumed after authentication.

IV. PERFORMANCE ANALYSIS

We analyze SANSCrypt's resilience against existing attacks and estimate its overhead.

A. Brute-Force Attack

Let us suppose that the number of primary input bits used as key inputs is i and each re-authentication procedure requires

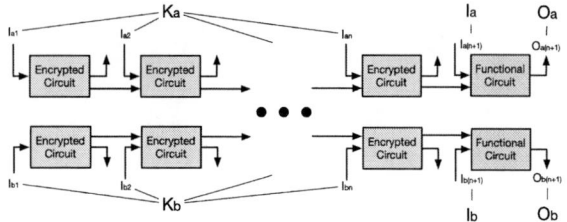

Fig. 6. An unrolled encrypted circuit which requires n clock cycles to find the key sequence.

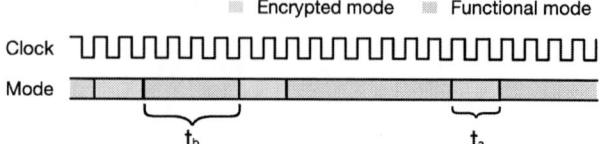

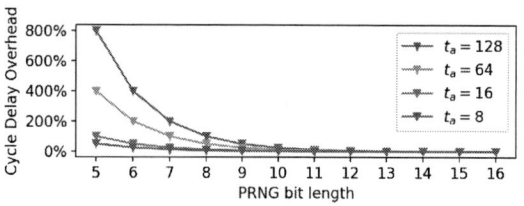

Fig. 7. Circuit mode switching for an authenticated user.

Fig. 8. Average cycle delay as a function of PRNG bit length when the key sequence cycle length t_a is 8, 16, 64, and 128.

c clock cycles to apply the key sequence. If the attacker has no preference in selecting the key sequence, then she would have, on average, $(2^{i \cdot c} + 1)/2 \approx 2^{i \cdot c - 1}$ attempts for each re-authentication procedure, which amounts to the same brute-force attack complexity of HARPOON. However, because the correct key sequence of each re-authentication procedure depends on the PRNG output, the number N_{prng} of possible values of the PRNG output will also contribute to the attack effort. If each PRNG output corresponds to a unique key sequence which is independent from other key sequences, the average attack effort will be $N_{prng} \cdot 2^{i \cdot c - 1}$. For a 10-bit PRNG, $i = 32$, and $c = 8$, the average attack effort will reach 5.93×10^{79}.

B. Sequential SAT-Based Attack

A SAT-based attack can be carried out on sequential encryption by unrolling the sequential portions of the circuit [20]. This attack can be remarkably successful especially when the correct key is the same at each time (clock cycle) and the key input ports are different from the primary input ports. Similarly to HARPOON, SANSCrypt is resilient to this SAT-based attack variant, since the correct keys are generally not the same at different clock cycles.

We therefore analyze the resilience of SANSCrypt via a modified version of the sequential SAT-based attack [22] that is appropriate for schemes such as HARPOON and SANSCrypt, as shown in Fig. 6. Let us first assume that the encryption scheme requires n clock cycles after reset to enter the functional mode. Then, the attacker can start the attack by unrolling the circuit $(n+1)$ times. The first n copies of the circuit receive the keys at their primary input ports (K_a and K_b), while the primary input and output ports of the $(n+1)^{th}$ circuit replica can be used to read the circuit input and output signals after n cycles. If the SAT-based attack fails to find the correct key with $(n + 1)$ circuit replicas, as in Fig. 6, the circuit will be unrolled one more time (see, e.g., [20]).

The attack above would be still ineffective on SANSCrypt, since it can retrieve the first key sequence but would fail to discover when the next back-jumping occurs and what would be the next key sequence. Even if the attacker knows when the next back-jumping occurs, the above SAT-based attack will fail due to the large number of circuit replicas needed to find all the key sequences, as empirically observed in Section V.

C. FSM Extraction and Structural Analysis

As discussed in Section II, a possible shortcoming of certain sequential encryption schemes is the clear boundary between the encrypted mode and the functional mode FSMs. As shown in Fig. 3, SANSCrypt addresses this issue by designing more than one transition between the two FSMs.

An attacker may also try to locate and isolate the output of ENC-FSM by looking for low signal switching activities when the circuit is in the encrypted mode. SANSCrypt addresses this risk by expanding the output of ENC-FSM from one bit to an array. The value of each bit changes frequently based on the state of the encrypted mode FSM, which makes it difficult for attackers to find the output of ENC-FSM based on signal switching activities.

D. Cycle Delay Analysis

Due to multiple back-jumping and authentication operations in SANSCrypt, additional clock cycles will be required in which no other operation can be executed. Suppose that authentication requires t_a clock cycles and the circuit stays in the functional mode for t_b clock cycles before the next back-jumping occurs, as shown in Fig. 7. The cycle delay overhead can be computed as the ratio $O_{cd} = t_a/t_b$.

Specifically, for an n-bit PRNG, the average t_b is equal to the average output value, i.e., 2^{n-1}. To illustrate how the cycle delay overhead is influenced by this encryption, Fig. 8 shows the relation between average cycle delay overhead and PRNG bit length. The clock cycles (t_a) required for (re-)authentication are set as 8, 16, 64, and 128. When the PRNG bit length is small, the average cycle delay increases significantly with the increase of t_a. However, the cycle delay can be reduced by increasing the PRNG bit length. For example, the average cycle delay overhead becomes negligible for all the four cases when the PRNG bit length is 11 or larger. A key manager, available to the trusted user, will be in charge of automatically applying the key sequences from a tamper-proof memory at the right time, as computed from a hard-coded replica of the PRNG.

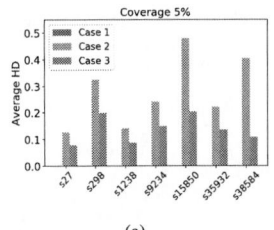 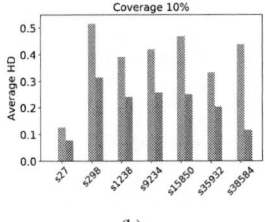 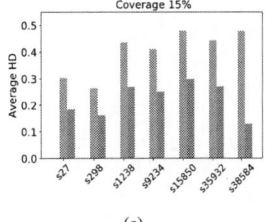 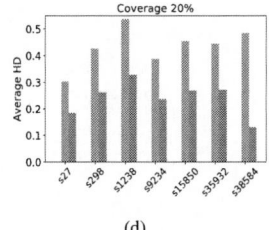

Fig. 9. The average HD for different node coverage: (a) 5%, (b) 10%, (c) 15%, and (d) 20%.

TABLE II
SYNTHESIS RESULT OF AREA, POWER, DELAY

Circuit	s27				s298				s1238				s9234			
Node Coverage	5%	10%	15%	20%	5%	10%	15%	20%	5%	10%	15%	20%	5%	10%	15%	20%
Area [%]	1418.5	1418.5	1403.2	1403.2	413.0	427.3	425.2	453.8	144.8	165.7	176.0	189.2	114.6	131.7	144.5	160.1
Power [%]	1627.7	1627.7	1627.5	1627.5	385.7	390.6	389.9	402.8	217.8	232.1	235.0	249.8	179.8	197.5	188.0	190.6
Delay [%]	0.0	0.0	1.4	1.4	0.0	0.0	0.0	0.5	0.0	0.0	0.0	5.8	0.0	0.0	0.9	3.6
Circuit	s15850				s35932				s38584				Average (s27 and s298 excluded)			
Node Coverage	5%	10%	15%	20%	5%	10%	15%	20%	5%	10%	15%	20%	5%	10%	15%	20%
Area [%]	92.9	112.1	120.1	133.9	116.3	129.5	139.4	151.6	133.5	140.9	158.7	165.6	120.4	136.0	147.8	160.1
Power [%]	127.4	142.3	153.2	163.0	98.4	101.9	101.2	103.0	123.9	128.8	142.0	140.3	149.5	160.5	163.9	169.4
Delay [%]	-0.3	0.0	0.1	0.6	-0.4	0.0	4.3	5.3	0.6	2.0	0.4	4.9	0.0	0.4	1.1	4.0

TABLE III
OVERVIEW OF THE SELECTED BENCHMARK CIRCUITS

Circuit	s27	s298	s1238	s9234	s15850	s35932	s38584
Input	4	3	14	36	77	35	38
Output	1	6	14	39	150	320	304
DFF	3	14	18	211	534	1728	1426
Gate	10	119	508	5597	9772	16065	19253

TABLE IV
SAT-BASED ATTACK RUNTIME FOR FINDING THE FIRST 7 KEY SEQUENCES

Key Seq. Index	1 (HARPOON)	2	3	4	5	6	7
Runtime [s]	4	123	229	1941	1301	2202	25571

V. SIMULATION RESULTS

We first evaluate the effectiveness of SANSCrypt on seven ISCAS'89 sequential benchmark circuits with different sizes, as summarized in Table III. All the experiments are executed on a Linux server with 48 2.1-GHz processor cores and 500-GB memory. We implement our technique on the selected circuits with different configurations and use a 45-nm NangateOpenCellLibrary [27] to synthesize the encrypted netlists for area optimization under a critical-path delay constraint that targets the same performance as in the non-encrypted versions. For the purpose of illustration, we realize the PRNG using Linear Feedback Shift Registers (LFSR) with different sizes, ranging from 5 to 15 bits. An LFSR provides an area-efficient implementation and has often been used in other logic encryption schemes in the literature [9], [28]. We choose a random 8-cycle-long key sequence as the correct key, and select 5%, 10%, 15%, and 20% as node coverage levels. Finally, we use the Hamming distance (HD) between the correct and the corrupted output values as a metric for the output corruptibility. If the HD is 0.5, the effort spent to identify the incorrect bits is maximum.

We run functional simulations on all the encrypted circuits with the correct key sequences (case 1) and without the correct sequences (case 2), by applying 1000 random input vectors.

We then compare the circuit output with the golden output from the original netlist and calculate the HD between the two. Moreover, we demonstrate the additional robustness of SANSCrypt by simulating a scenario (case 3) in which the attacker assumes that the encryption is based on a single-authentication protocol and provides only the first correct key sequence upon reset. Fig. 9a-d show the average HD in these three cases. For all the circuits, the average HD is zero only in case 1, when all the correct key sequences are applied at the right clock cycles. Otherwise, in case 2 (orange) and case 3 (green), we observe a significant increase in the average HD. The average HD in case 3 is always smaller than that of case 2 because, in case 3, the correct functionality is recovered for a short period of time, after which the circuit jumps back to the encrypted mode. The longer the overall runtime, the smaller will be the impact of the transparency window in which the circuit exhibits the correct functionality.

We then apply the sequential SAT-based attack in Section IV to circuit s1238 with 5-bit LFSR and 20% node coverage, under a stronger attack model, in which the attacker knows when to apply the correct key sequences. Table IV shows the runtime to find the first set of 7 key sequences. The runtime remains exponential in the number of key sequences, which makes sequential SAT-based attacks impractical for large designs.

Finally, Table II reports the synthesized area, power, and delay overhead due to the implementation of our technique.

978-1-7281-5410-7/20 $31.00 © 2020 IEEE

TABLE V
ADP OVERHEAD RESULTS FOR PARTIAL ENCRYPTION

Encrypted registers/Total registers	100%	50%	25%	10%	5%	2.5%	1%
Area [%]	133.5	71.6	49.1	33.4	27.8	23.5	22.4
Power [%]	123.9	40.2	9.6	-12.8	-20.5	-22.1	-25.0
Delay [%]	0.6	1.8	2.1	4.2	5.4	3.9	4.6

In more than 70% of the circuits the delay overhead is less than 1%, and exceeds the required clock cycle by at most 5.8%. Except for *s27* and *s298*, characterized by a small gate count, all the other circuits show average area and power overhead of 141.1% and 160.8%, respectively, which is expected due to the additional number of registers required in ENC-FSM to guarantee that the correct state is entered upon re-authentication. However, because critical modules in large SoCs may only account for a small portion of the area, this overhead becomes affordable under partial obfuscation. For example, we encrypted a portion of state registers in *s38584*, the largest ISCAS'89 benchmark, using SANSCrypt. We then randomly inserted additional XOR gates to achieve the same HD as in the case of full encryption. Table V reports the overhead results after synthesis, when the ratio between the encrypted state registers and the total number of state registers decreases from 100% to 1%. Encrypting 10% of the registers will only cost 33.4% of the area while incurring negative power overhead and 4.2% delay overhead.

VI. CONCLUSION

We proposed SANSCrypt, a robust sequential logic encryption technique relying on a sporadic authentication protocol, in which re-authentications are carried out at pseudo-randomly selected time slots to significantly increase the attack effort. Future work includes optimizing the implementation to further reduce the overhead and hide any structural traces that may expose the correct key sequence. Further, we plan to investigate key manager architectures to guarantee reliable timing and operation in real-time applications.

ACKNOWLEDGMENT

This work was partially sponsored by the Air Force Research Laboratory (AFRL) and the Defense Advanced Research Projects Agency (DARPA) under agreement number FA8560-18-1-7817.

REFERENCES

[1] R. Karri, J. Rajendran, K. Rosenfeld, and M. Tehranipoor, "Trustworthy hardware: Identifying and classifying hardware trojans," *Computer*, vol. 43, no. 10, pp. 39–46, 2010.

[2] M. Tehranipoor and F. Koushanfar, "A survey of hardware trojan taxonomy and detection," *IEEE Design & Test of Computers*, vol. 27, no. 1, pp. 10–25, 2010.

[3] J. Rajendran, H. Zhang, C. Zhang, G. S. Rose, Y. Pino, O. Sinanoglu, and R. Karri, "Fault analysis-based logic encryption," *IEEE Trans. Computers*, vol. 64, no. 2, pp. 410–424, 2013.

[4] M. Yasin, A. Sengupta, M. T. Nabeel, M. Ashraf, J. J. Rajendran, and O. Sinanoglu, "Provably-secure logic locking: From theory to practice," in *Proc. SIGSAC Conf. Computer and Communications Security*, pp. 1601–1618, 2017.

[5] M. Yasin, B. Mazumdar, J. J. Rajendran, and O. Sinanoglu, "SARLock: SAT attack resistant logic locking," in *IEEE Int. Symp. Hardware Oriented Security and Trust (HOST)*, pp. 236–241, 2016.

[6] R. S. Chakraborty and S. Bhunia, "HARPOON: an obfuscation-based SoC design methodology for hardware protection," *IEEE Trans. Computer-Aided Design of Integrated Circuits and Systems*, vol. 28, no. 10, pp. 1493–1502, 2009.

[7] M. Yasin, B. Mazumdar, O. Sinanoglu, and J. Rajendran, "CamoPerturb: Secure IC camouflaging for minterm protection," in *2016 IEEE/ACM Int. Conf. Computer-Aided Design (ICCAD)*, pp. 1–8, 2016.

[8] E. Charbon, "Hierarchical watermarking in IC design," in *IEEE Proc. Custom Integrated Circuits Conf.*, pp. 295–298, 1998.

[9] K. Xiao, D. Forte, and M. M. Tehranipoor, "Efficient and secure split manufacturing via obfuscated built-in self-authentication," in *IEEE Int. Symp. Hardware Oriented Security and Trust (HOST)*, pp. 14–19, 2015.

[10] P. Subramanyan, S. Ray, and S. Malik, "Evaluating the security of logic encryption algorithms," in *IEEE Int. Symp. Hardware Oriented Security and Trust (HOST)*, pp. 137–143, 2015.

[11] P. Chakraborty, J. Cruz, and S. Bhunia, "SURF: Joint structural functional attack on logic locking," in *IEEE Int. Symp. Hardware Oriented Security and Trust (HOST)*, pp. 181–190, 2019.

[12] Y. Shen, Y. Li, S. Kong, A. Rezaei, and H. Zhou, "SigAttack: New high-level sat-based attack on logic encryptions," in *Design, Automation and Test in Europe Conference and Exhibition (DATE)*, pp. 940–943, 2019.

[13] V. V. Menon, G. Kolhe, A. Schmidt, J. Monson, M. French, Y. Hu, P. A. Beerel, and P. Nuzzo, "System-level framework for logic obfuscation with quantified metrics for evaluation," in *Secure Development Conf. (SecDev)*, pp. 89–100, 2019.

[14] Y. Hu, V. V. Menon, A. Schmidt, J. Monson, M. French, and P. Nuzzo, "Security-driven metrics and models for efficient evaluation of logic encryption schemes," in *ACM-IEEE Int. Conf. Formal Methods and Models for System Design (MEMOCODE)*, pp. 1–5, 2019.

[15] G. Sengar, D. Mukhopadhyay, and D. R. Chowdhury, "Secured flipped scan-chain model for crypto-architecture," *IEEE Trans. Computer-Aided Design of Integrated Circuits and Systems*, vol. 26, no. 11, pp. 2080–2084, 2007.

[16] S. Paul, R. S. Chakraborty, and S. Bhunia, "Vim-scan: A low overhead scan design approach for protection of secret key in scan-based secure chips," in *IEEE VLSI Test Symp. (VTS)*, pp. 455–460, 2007.

[17] X. Wang, D. Zhang, M. He, D. Su, and M. Tehranipoor, "Secure scan and test using obfuscation throughout supply chain," *IEEE Trans. Computer-Aided Design of Integrated Circuits and Systems*, vol. 37, no. 9, pp. 1867–1880, 2017.

[18] A. R. Desai, M. S. Hsiao, C. Wang, L. Nazhandali, and S. Hall, "Interlocking obfuscation for anti-tamper hardware," in *Proc. Cyber Security and Information Intelligence Research Workshop*, pp. 1–4, 2013.

[19] Y. Kasarabada, S. R. T. Raman, and R. Vemuri, "Deep state encryption for sequential logic circuits," in *IEEE Computer Society Annual Symp. VLSI (ISVLSI)*, pp. 338–343, 2019.

[20] K. Shamsi, M. Li, D. Z. Pan, and Y. Jin, "KC2: Key-condition crunching for fast sequential circuit deobfuscation," in *Design, Automation and Test in Europe Conference and Exhibition (DATE)*, pp. 534–539, 2019.

[21] D. Duvalsaint, Z. Liu, A. Ravikumar, and R. Blanton, "Characterization of locked sequential circuits via ATPG," in *IEEE Int. Test Conf. in Asia (ITC-Asia)*, pp. 97–102, 2019.

[22] T. Meade, Z. Zhao, S. Zhang, D. Pan, and Y. Jin, "Revisit sequential logic obfuscation: Attacks and defenses," in *IEEE Int. Symp. Circuits and Systems (ISCAS)*, pp. 1–4, 2017.

[23] A. Bhargav-Spantzel, A. C. Squicciarini, S. Modi, M. Young, E. Bertino, and S. J. Elliott, "Privacy preserving multi-factor authentication with biometrics," *Journal of Computer Security*, vol. 15, no. 5, pp. 529–560, 2007.

[24] F. Brglez, D. Bryan, and K. Kozminski, "Combinational profiles of sequential benchmark circuits," in *IEEE Int. Symp. Circuits and Systems (ISCAS)*, pp. 1929–1934, 1989.

[25] J. Dofe and Q. Yu, "Novel dynamic state-deflection method for gate-level design obfuscation," *IEEE Trans. Computer-Aided Design of Integrated Circuits and Systems*, vol. 37, pp. 273–285, Feb. 2018.

[26] Y. Kasarabada, S. Chen, and R. Vemuri, "On SAT-based attacks on encrypted sequential logic circuits," in *Int. Symp. Quality Electronic Design (ISQED)*, pp. 204–211, 2019.

[27] Silvaco, "45nm open cell library," 2019.

[28] M. S. Rahman, A. Nahiyan, S. Amir, F. Rahman, F. Farahmandi, D. Forte, and M. Tehranipoor, "Dynamically obfuscated scan chain to resist oracle-guided attacks on logic locked design," *IACR Cryptol. ePrint Arch.*, vol. 2019, p. 946, 2019.

Basic Block Encoding Based Run-time CFI Check for Embedded Software

Love Kumar Sah, Srivarsha Polnati, Sheikh Ariful Islam and Srinivas Katkoori
Department of Computer Science and Engineering
University of South Florida
Tampa, FL 33620
Email: {lsah, spolnati, sheikhariful, katkoori}@usf.edu

Abstract—**Modern control flow attacks circumvent existing defense mechanisms to transfer the program control to attacker chosen malicious code in the program, leaving application vulnerable to attack. Advanced attacks such as Return-Oriented Programming (ROP) attack and its variants, transfer program execution to gadgets (code-snippet that ends with return instruction). The code space to generate gadgets is large and attacks using these gadgets are Turing-complete. One big challenge to harden the program against ROP attack is to confine gadget selection to a limited locations, thus leaving the attacker to search entire code space according to payload criteria. In this paper, we present a novel approach to label the nodes of the Control-Flow Graph (CFG) of a program such that labels of the nodes on a valid control flow edge satisfy a Hamming distance property. The newly encoded CFG enables detection of illegal control flow transitions during the runtime in the processor pipeline. Experimentally, we have demonstrated that the proposed Control Flow Integrity (CFI) implementation is effective against control-flow hijacking and the technique can reduce the search space of the ROP gadgets upto 99.28%. We have also validated our technique on seven applications from MiBench and the proposed labeling mechanism incurs no instruction count overhead while, on average, it increases instruction width to a maximum of 12.13%.**

I. INTRODUCTION

Embedded software uses C and its variant languages vulnerable to different attacks. Control-flow hijacking has been one of the most severe cyber threats for over 40 years. An attacker can directly compromise the program behavior and perform malicious activity because of an exploitable vulnerability such as a buffer overflow where specially crafted input exceeds the buffer size. Low-level programming languages (e.g., C) lack memory protection techniques and have been vulnerable to several memory attacks (stack smashing [1], return-into-libc, Return-Oriented Programming [2], etc.).

Numerous defenses have been proposed to defend against control-flow hijacking. Techniques such as stack canary, which places all buffers at the top of the stack frame is effective against buffer overflow attack. It ensures that the attacker has to corrupt canary value before overwriting any other variable space in the program [3]. However, the canary has been shown to vulnerable that leaks its secret value. For example, unsafe pointer arithmetic in yaSSL allows an attacker to bypass canary words and access critical state directly [4].

Bound Checking [5], [6] extracts the real-time memory space information of variables and restricts its use based on the information. This defense mechanism suffers from high performance degradation during monitoring of memory utilisation. Similarly, Address Space Layout Randomization (ASLR) has been proposed [7] which rearranges the address space of key memory sections of a process in a random fashion. This address space mostly includes the base address of executable, stack, heap, and libraries. On a 32-bit computing platform, ASLR is constrained by the number of available bits (usually 16) for randomization which makes brute-force attack feasible. On the contrary, in a 64-bit platform, the number of random bits is high, thus making brute-force attack infeasible. However, ASLR can still be circumvented for a special type of vulnerability (e.g., format string) which leaks information about the layout of address space. Furthermore, modern CPUs and operating systems include NX-bit (no-execute) to protect the memory page. To bypass these protection mechanisms, a new vulnerability, ROP, has evolved that allows an attacker to circumvent non-executable page protections without injecting any code. Gadget finding tool [8] explores the entire executable code space to extract the gadgets composed of different instruction sequences. In [2], authors show that GNU libc have enough gadgets to make Turing-complete attack.

CFI is one of the promising techniques, which, in theory, can protect the program from control-flow hijacking [9]. In general, CFI is a program-specific path-level access control model. To construct the CFI policy, we first analyze the control-flow graph (CFG) of the host program to extract all legal target(s) for each valid control transfer. During runtime, the inlined code checks whether the control flow target belongs to a set of white-listed ones. With this checking mechanism, CFI guarantees that the execution path of the program strictly follows an edge in its original CFG. Traditional CFI uses pre-assigned label at exit and entry of each basic block (BB). These labels are chosen randomly and one should perform a manual search for each successful control transfer. In doing so, we insert an additional instruction for each control transfer. However, this mandatory verification step can be bypassed by jumping to the next instruction in the line. For example, we could jump to second instruction instead of jumping to the first instruction in a BB. Instruction(s) skipping during verification forms the basis of the ROP attack to subvert CFI protection.

978-1-7281-5410-7/20 $31.00 © 2020 IEEE

In this paper, we propose a novel approach to enforce fine-grained CFI verification using special labeling mechanism for basic blocks of CFG. We use relative Hamming coded information as a label for BBs adjacent to a node. With a Hamming code comparator embedded within the pipeline architecture, we perform verification of control transfer between the BBs. We used Hamming code as the label in order to make label random yet related via chosen Hamming distance. INTEL PIN tool analysis over seven applications from MiBench testsuite results reduction of ROP gadgets by 99.28%. We validate fine grained CFI enforcement on each application using 8-bit label with Hamming distance of 5 over SimpleScalar toolset. Our approach annotates instructions with label thus incur no additional instruction overhead in application while, on average, it increases instruction width to a maximum of 12.13%.

The rest of the paper is organized as follows. In Section II, we discuss the ROP attack and existing defense mechanisms. In Section III, we present our approach consisting static analysis of BB for Hamming code generation followed by annotation and detection using Hamming code. We provide detailed experimental results in Section IV. Section V concludes the paper.

II. BACKGROUND AND RELATED WORK

Return-oriented Programming attack is an advanced buffer overflow attack. With the wide adoption of data execution prevention techniques classical code injection attack are difficult to stage. This opens the opportunity to build the attack using code already in the executable. ROP attack is return-to-libc attack where an attacker chains multiple gadgets (set of instructions ending at return) together. Gadget is searched from the entire executable code section. Once the chain of gadgets is formed, these gadgets are executed in sequence by indirect jumps. In Figure 1, sample ROP attack sequence is shown. Three gadget each ending with ret instruction, is chosen to stage an attack. Stack return address is overflown with gadget addresses sequence. After each gadget execution, control returns to stack and proceeds to the next gadget address along with data (if any) to be used to change the register content.

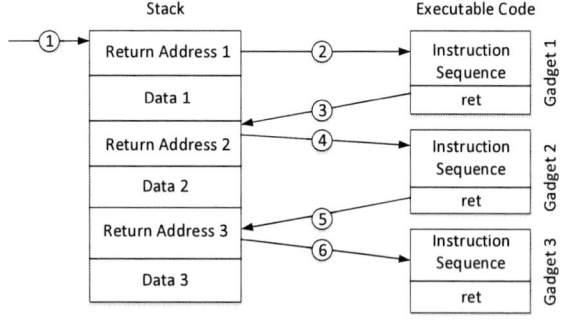

Figure 1: ROP gadget chaining

The ROP attack [2] presented initially handpicked sets of gadgets from libc. Automated tool [8] extracts all the available gadgets in the executable. With full knowledge of the gadget, a complex attack can be staged. Several invariants of the ROP attack have evolved such as JOP [10], COOP [11] which follow similar principles. Recently, several defense systems have been proposed to detect and prevent the attack. Separate shadow stacks have been utilized to store the return address in ROPdefender [12]. Gadget smashing [13] randomizes the memory component of process to harden the address guessing of the gadgets, resulting incorrect instruction to execute. Gfree [14] uses code rewriting to remove all unaligned instruction used to link the gadgets.

CFI enforcement [9] can effectively protect against code reuse attack. CFI implementation can be categorized into two broad categories: fine-grained and coarse-grained. A fine-grained approach has a set of target addresses for every indirect valid control transfer in a program. CFLocking [15], Forward-edge-CFI [16], RockJIT [17], MCFI [18], code pointer integrity [19], CCFI [20], uCFI [21] implement fine-grained CFI via complex source code analysis. In contrast, a coarse-grained CFI technique restricts control transfer into fixed equivalence classes. In several cases, the number of such classes is restricted to three. CFIMon [22], BinCFI [23], and kBouncer [24] implement coarse-grained approach. This approach to not covering all the valid target sets often fails to detect attacks during invalid control transfer.

We also classify CFI policy according to software or hardware support. During software-based CFI enforcement [25], pre-inserted IDs are checked for each target of call, return, and indirect jump instruction. Additional instructions are inserted to perform such comparisons and they incur performance overhead of about 21% [9]. CCFIR [26] uses limited IDs to reduce high-performance cost by redirecting indirect jump instructions. Hardware-assisted CFI enriches Instruction Set Architecture (ISA) with new instruction(s) [27], [28] to check valid targets during call, return, and jump. CCFI [20] uses encryption-based authentication to protect the control transfer and program-related information (e.g., frame pointer, return address, vtable pointers, and exception handler). PUF-based encryption of return address [29] is proposed to enforce the CFI check. However, an additional key is required to decode such encoding. PathArmor [30] resolves backward edges using a hardware-supported context-sensitive static analysis over the inter-procedural CFG. Similarly, kBouncer [24] and ROPecker [31] rely on Intel's Last Branch Record (LBR) to efficiently implement branch tracing using heuristic CFI policies which can be easily circumvented. CFIMon [22] relies on the significantly slower Intel BTS, yielding high detection latency. Unlike previous approaches, our technique (a) does not add any new instruction for fine-grained CFI enforcement; and (b) relies on Hamming coded information for gadget confinement and embeds a detector in pipeline architecture to valid transfer.

978-1-7281-5410-7/20 $31.00 © 2020 IEEE

III. PROPOSED APPROACH

We present the CFI implementation framework in Figure 2. It consists of three steps: 1) static analysis of the program CFG; 2) label determination and instruction annotation; and 3) run-time CFI verification. We analyze the program to extract the basic block information and the highest degree of a node in the program. Based on the number of BB(s) and their edge connectivity, we generate Hamming coded label for each BB in the CFG. During run time, the embedded CFI checker module in the pipelined processor will verify the control flow.

A. Static Analysis for Basic Block

A basic block is a sequential code sequence with a single entry and exit point. Control instructions such as jump, ret, branch might result in non-sequential control flow. Static analysis for such instruction in object code determines the number of BBs available in the program.

A sample C program is shown in Figure 3a. This program snippet consists of different control statements (if, for) and a function call (incr). The C program is compiled to generate object code shown in Figure 3b. Here, control instructions such as branch, jump, function call, and ret divert the sequential execution. Such control instruction ends the current basic block. The new BB starts at (1) target instruction of the current control instruction and/or (2) next immediate instruction following such instruction (based on the type of instruction). For example, bgtz instruction at address 4002b0 ends current BB1. Next BB starts at address 4002e0 which is target instruction of bgtz and at 4002b8 which is the next immediate instruction after bgtz. Static analysis of object code gives BB counts, starting address of each BB, and its degree (number of BBs connected to it).

B. Generation of Hamming Code Label

Once we perform static analysis of the object code, we obtain list of control transfer addresses (source and destination). Using these as edge information, we create a graph G. Limited number of unique fixed bit labels can be obtained for a particular hamming distance. For example, maximum of 56

Algorithm 1: Relative Hamming Codes for BB

Input: G= Graph; start_v= Starting Node
Output: Unique Hamming Coded Graph
Let Q be a queue
Label start_v as Visited and label_assigned
Q.enqueue(start_v)

while *!(Q.empty)* **do**
  v := Q.dequeue()
  if *v is the goal* **then**
    | return v
  end
  Hamming_label(v, G.adjacentEdges(v))
  for *w in G.adjacentEdges(v)* **do**
    if *w.visit = NULL* **then**
      w.visit = True
      w.parent := v
      Q.enqueue(w)
    end
  end
end

---

unique 8-bit labels can be produce with hamming distance of 5. Thus, the maximum node degree in the CFG determines the minimum hamming distance required for the n-bit label. Where as the total number of labels determines the number of bits required for a label.

We modify general breadth first search (BFS) algorithm with one extra variable; i.e., label_assigned for each graph node. As shown in Algorithm 1, for each node visited in the graph, we call a function Hamming_label() along with its adjacent nodes where labels are calculated (with hamming distance k) to assign adjacent node yet to be labeled.

Figure 4 illustrates Hamming_label() where BB1 has 3 incoming edges and 1 outgoing edge. Hence, the total degree of BB1 is 4. We encode each adjacent node with equidistant Hamming code for the node BB1 only. Nodes 2, 3, 4, and 5 do not share this rule between them if they are not connected.

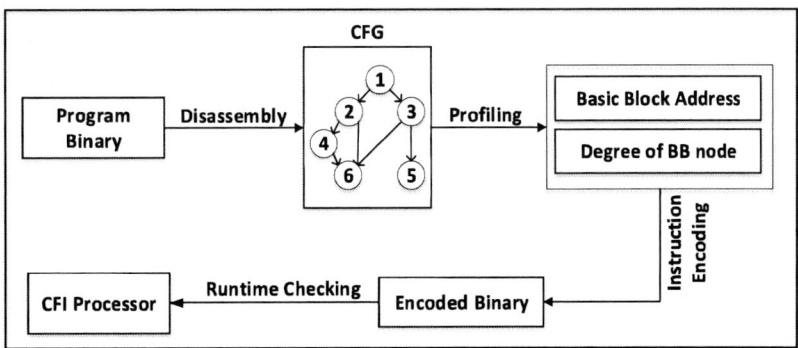

Figure 2: Proposed CFG labeling based CFI implementation framework

978-1-7281-5410-7/20 $31.00 © 2020 IEEE   137

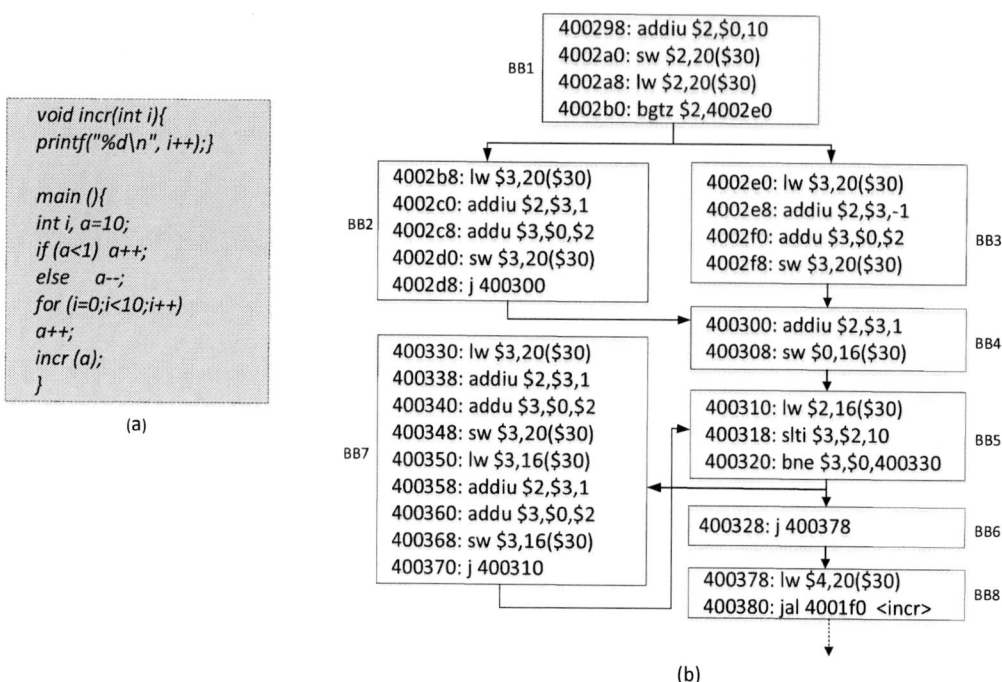

(a)

(b)

Figure 3: Basic block diagram of sample C code: (a) C code; (b) control flow graph representation.

Label information (00001, 00010, 00100, and 01000) from node 2 to 4 follows equal Hamming distance for label of BB1 (00000). During verification, only 4 adjacent nodes to/from BB1 are acceptable, which restrict the control flow within connected nodes.

C. Gadget Selection Restriction

Our approach follows relative encoding of adjacent nodes with respect to a node. All the valid adjacent nodes to a node will pass the CFI label check. Thus these are the only available BBs where attacker is confined for the gadget selection. With drastic reduction of BBs to select for gadgets, attacker will have very hard time finding all the components necessary for Turing-complete attack.

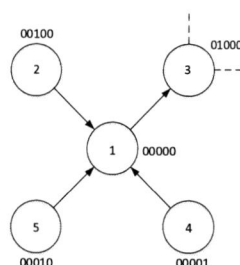

Figure 4: Adjacent basic blocks

D. Instruction Annotation

We compile the host program using gcc in SimpleScalar toolset which generates an object code and later compiled to assembly instructions. Given the total number of branch and jump instructions, we divide the assembly code into different BBs, similar to Figure 3. Then, we assign different Hamming code to each BB following Algorithm 1. In our technique, we maintain a property that all the basic blocks in a legal path will have a Hamming distance of less than 5, while BBs in illegal paths will have a Hamming distance greater than or equal to 5. However, depending on the depth of the program and the number of BBs, the Hamming distance value can be changed. Next, we present the three different approaches in which encoding can be performed along with its effectiveness:

1) Annotating the first instruction of a basic block: We annotate the first instruction of each basic block assuming control transfer is restricted to first instruction of BB. This is similar to conventional CFI check at the entry of each BB. The checker module in Figure 6 only needs to operate on instruction that follows right after the control instruction in BB. However, the limitation of this annotation technique is that we cannot detect attacks if the control instruction points to instruction in the following BB other than the first instruction (ROP gadget). We introduce a register to carry the label of the current BB. During valid control flow, next BB label is compared with the current register value.

978-1-7281-5410-7/20 $31.00 © 2020 IEEE   138

Table I: BB analysis of MiBench programs on PIN

Benchmark program	Instruction	Control instruction %	# of BB	Highest Degree of nodes	ROP-Gadget Visibility Reduction %
Automotive/basicmath	62323543	8.69	1264	26	97.94
Automotive/qsort	4584457	15.2	2568	32	98.75
Network/dijkstra	5482245	14.04	2214	42	98.1
Office/ispell	1894412	19.82	4625	33	99.28
Security/sha	55874246	3.14	625	13	97.92
Telecom/adpcm	45896714	15.76	1857	22	98.81
Telecom/adpcm	38854247	16.23	2356	29	98.76

2) Annotating all instructions in a basic block: In this case, we annotate all the instructions of a basic block with the hamming code set for that particular block. With all the instructions annotated, We overcome the limitation of previous case where control flow was restricted to start of BB only. ROP gadget attack can also be detected in this case where we have both labels available for comparison.

3) Software approach: For an architecture where we have no extra bits left in opcode for annotation, we use extra instructions to set the label for the BB. Verification can be done by inserting additional comparison instruction (e.g., `cmp`) during compilation.

In first two approaches for annotation, we do not add any new instructions. For the third approach, there is an instruction overhead which depends on the percentage of number of branch and jump instructions to the total number of instructions in a program (up to 25.4% according to our test).

E. Simplescalar Tool and Instruction Annotation

To validate our technique, we used SimpleScalar toolset. It is highly parameterized simulator and provides us the capability to update the opcode for encoding additional information. In our case, we perform instruction annotation using unused bits from 31 to 16. A snippet of annotated instructions is shown in Figure 5. We do not annotate the first instruction while the remaining 3 instructions are annotated with Hamming distance of 4.

```
. . . . . . . . . .
400238: 06 00 00 00    bne $3, $0, 400268
400240: a2 00 0f 00    lui $4, 4096
. . . . . . . . . .
400268: a2 00 f0 00    lui $4, 5020
. . . . . . . . . .
400288: 42 00 ff 00    addu $2, $0, $0
```

Figure 5: Annotated Instructions

F. Detection of CFI in Pipeline

Detection of control flow violation in hardware is shown in Figure 6. In pipeline architecture, different instructions propagate through stages of pipeline. During control instruction execution, immediate successor instruction in the pipeline is compared using hamming code detector (XOR operation + 1's counter) to validate the labels. Failed comparison results in an interrupt.

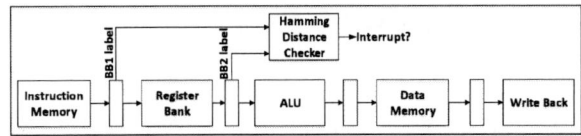

Figure 6: Runtime control flow violation detection based on Hamming distance check in a 5-stage pipelined processor

IV. EXPERIMENTAL RESULTS

We perform our experiment in two stages. First, we analyze seven different applications from the MiBench suite to find the number of BBs. We have used the Intel PIN tool [32] to trace the `branch`, `jump`, and `ret` instructions' source and destination address. With each BB edge information, We determine the highest degree of the node in a program. This information helps to determine the ROP gadget selection space which is ratio of total BB-count and highest Node-degree. Table I reports the BBs count and node degree for each program. For the case of Office/ispell benchmark, We observe maximum 99.28% restriction in BB use. In general, CFG with high number of BBs and lesser node degree produce more restrictive gadget space.

For second part of experimentation, we modified SimpleScalar simulation toolset [33] for instruction annotation (section III-E) and CFI enforcement. We annotated 128 BBs in each program listed in Table II with 8 bit label. We modified these programs to introduce buffer overflow attack to change control flow to different BBs. We modified the pipeline architecture of the Sim-outorder simulator to introduce a module to compare labels between two consecutive instructions. Table II shows the instruction count and conditional instructions for each benchmark program. After each control flow diversion, our simulator was able to compare the label to verify the diversion. Even we used 8 unused bits of simulator, for a 64-bit architecture it accounts 12.13% increment in instruction width.

Table II: MiBench benchmark programs on SimpleScalar

MiBench Program	Instruction Count	Control Instruction (%)	Attack Detected
Automotive/basicmath	180303082	8.0	Yes
Automotive/qsort	13753371	14.1	Yes
Network/dijkstra	17172622	13.5	Yes
Office/ispell	2014412	19.4	Yes
Security/sha	187622738	2.8	Yes
Telecom/timing	142690142	15.2	Yes
Telecom/rawaudio	127627413	25.4	Yes

978-1-7281-5410-7/20 $31.00 © 2020 IEEE

V. CONCLUSIONS AND FUTURE WORK

In this paper, we presented a new way to label the adjacent nodes relative to the node of a CFG. We perform instruction encoding using unused bits in the opcode and the technique can efficiently determine the Hamming distance between instructions' label to identify legal or illegal control transfer. This approach helps in enforcing CFI verification from different attacks. In particular, return-oriented programming attack is highly restricted because of less gadget available to craft an attack. As an extension of our current approach, we would like to enhance our labelling mechanism to detect transfer within adjacent BB.

REFERENCES

[1] A. One. Smashing the stack for fun and profit. *Phrack*, 7(49), November 1996.

[2] H. Shacham. The geometry of innocent flesh on the bone: Return-into-libc without function calls (on the x86). In *Proceedings of the 14th ACM Conference on Computer and Communications Security*, CCS '07, page 552–561, New York, NY, USA, 2007.

[3] C. Cowan, C. Pu, DD.ave Maier, H. Hintony, J. Walpole, P. Bakke, S. Beattie, A. Grier, P. Wagle, and Q. Zhang. Stackguard: Automatic adaptive detection and prevention of buffer-overflow attacks. In *Proceedings of the 7th Conference on USENIX Security Symposium - Volume 7*, SSYM'98, page 5, USA, 1998. USENIX Association.

[4] CyaSSL: Embedded SSL library WolfSSL. http://www.wolfssl.com/yaSSL/Home.html. Accessed: 2020-05-10.

[5] L. K. Sah, S. A. Islam, and S. Katkoori. An efficient hardware-oriented runtime approach for stack-based software buffer overflow attacks. In *2018 Asian Hardware Oriented Security and Trust Symposium (Asian-HOST)*, pages 1–6, 2018.

[6] L. K. Sah, S. Ariful Islam, and S. Katkoori. Variable record table: A run-time solution for mitigating buffer overflow attack. In *2019 IEEE 62nd International Midwest Symposium on Circuits and Systems (MWSCAS)*, pages 239–242, 2019.

[7] E. G. Barrantes, D. H. Ackley, S. Forrest, T. S. Palmer, D. Stefanovic, and D. Dai Zovi. Randomized instruction set emulation to disrupt binary code injection attacks. In *Proceedings of the 10th ACM Conference on Computer and Communications Security*, CCS '03, page 281–289, New York, NY, USA, 2003. Association for Computing Machinery.

[8] ROPgadget Tool. http://shell-storm.org/project/ROPgadget/. Accessed: 2020-05-10.

[9] M. Abadi, M. Budiu, U. Erlingsson, and J. Ligatti. Control-flow integrity. In *Proceedings of the 12th ACM Conference on Computer and Communications Security*, CCS '05, page 340–353, New York, NY, USA, 2005. Association for Computing Machinery.

[10] T. Bletsch, X. Jiang, V. Freeh, and Z. Liang. Jump-oriented programming: a new class of code-reuse attack. pages 30–40, 01 2011.

[11] F. Schuster, T. Tendyck, C. Liebchen, L. Davi, A. Sadeghi, and T. Holz. Counterfeit object-oriented programming: On the difficulty of preventing code reuse attacks in c++ applications. In *2015 IEEE Symposium on Security and Privacy*, pages 745–762, 2015.

[12] L. Davi, A. Sadeghi, and M. Winandy. Ropdefender: A detection tool to defend against return-oriented programming attacks. In *Proceedings of the 6th ACM Symposium on Information, Computer and Communications Security*, ASIACCS '11, page 40–51, New York, NY, USA, 2011. Association for Computing Machinery.

[13] V. Pappas, M. Polychronakis, and A. D. Keromytis. Smashing the gadgets: Hindering return-oriented programming using in-place code randomization. In *2012 IEEE Symposium on Security and Privacy*, pages 601–615, 2012.

[14] K. Onarlioglu, L. Bilge, A. Lanzi, D. Balzarotti, and E. Kirda. G-free: Defeating return-oriented programming through gadget-less binaries. In *Proceedings of the 26th Annual Computer Security Applications Conference*, ACSAC '10, page 49–58, New York, NY, USA, 2010. Association for Computing Machinery.

[15] T. Bletsch, X. Jiang, and V. Freeh. Mitigating code-reuse attacks with control-flow locking. In *Proceedings of the 27th Annual Computer Security Applications Conference*, ACSAC '11, page 353–362, New York, NY, USA, 2011. Association for Computing Machinery.

[16] C. Tice, T. Roeder, P. Collingbourne, S. Checkoway, U. Erlingsson, L. Lozano, and G. Pike. Enforcing forward-edge control-flow integrity in gcc llvm. In *Proceedings of the 23rd USENIX Conference on Security Symposium*, SEC'14, page 941–955, USA, 2014. USENIX Association.

[17] B. Niu and G. Tan. Rockjit: Securing just-in-time compilation using modular control-flow integrity. In *Proceedings of the 2014 ACM SIGSAC Conference on Computer and Communications Security*, CCS '14, page 1317–1328, New York, NY, USA, 2014. Association for Computing Machinery.

[18] B. Niu and G. Tan. Modular control-flow integrity. In *Proceedings of the 35th ACM SIGPLAN Conference on Programming Language Design and Implementation*, PLDI '14, page 577–587, New York, NY, USA, 2014. Association for Computing Machinery.

[19] V. Kuznetsov, L. Szekeres, G. Payer, M.and Candea, R. Sekar, and D. Song. Code-pointer integrity. In *Proceedings of the 11th USENIX Conference on Operating Systems Design and Implementation*, OSDI'14, page 147–163, USA, 2014. USENIX Association.

[20] A. J. Mashtizadeh, A. Bittau, D. Boneh, and D. Mazières. Ccfi: Cryptographically enforced control flow integrity. In *Proceedings of the 22nd ACM SIGSAC Conference on Computer and Communications Security*, CCS '15, page 941–951, New York, NY, USA, 2015. Association for Computing Machinery.

[21] B. Niu and G. Tan. Per-input control-flow integrity. In *Proceedings of the 22nd ACM SIGSAC Conference on Computer and Communications Security*, CCS '15, page 914–926, New York, NY, USA, 2015. Association for Computing Machinery.

[22] Y. Xia, Y. Liu, H. Chen, and B. Zang. CFImon: Detecting violation of control flow integrity using performance counters. In *IEEE/IFIP International Conference on Dependable Systems and Networks (DSN 2012)*, pages 1–12, 2012.

[23] M. Zhang and R. Sekar. Control flow and code integrity for cots binaries: An effective defense against real-world rop attacks. In *Proceedings of the 31st Annual Computer Security Applications Conference*, ACSAC 2015, page 91–100, New York, NY, USA, 2015. Association for Computing Machinery.

[24] V. Pappas, M. Polychronakis, and A. D. Keromytis. Transparent ROP exploit mitigation using indirect branch tracing. In *22nd USENIX Security Symposium (USENIX Security 13)*, pages 447–462, Washington, D.C., August 2013. USENIX Association.

[25] Z. Wang and X. Jiang. Hypersafe: A lightweight approach to provide lifetime hypervisor control-flow integrity. In *2010 IEEE Symposium on Security and Privacy*, pages 380–395, 2010.

[26] C. Zhang, T. Wei, Z. Chen, L. Duan, L. Szekeres, S. McCamant, D. Song, and W. Zou. Practical control flow integrity and randomization for binary executables. In *2013 IEEE Symposium on Security and Privacy*, pages 559–573, 2013.

[27] L. Davi, P. Koeberl, and A. Sadeghi. Hardware-assisted fine-grained control-flow integrity: Towards efficient protection of embedded systems against software exploitation. In *Proceedings of the 51st Annual Design Automation Conference*, DAC '14, page 1–6, New York, NY, USA, 2014. Association for Computing Machinery.

[28] D. Arora, S. Ravi, A. Raghunathan, and N. K. Jha. Hardware-assisted run-time monitoring for secure program execution on embedded processors. *IEEE Transactions on Very Large Scale Integration (VLSI) Systems*, 14(12):1295–1308, 2006.

[29] P. Qiu, Y. Lyu, D. Zhai, D. Wang, J. Zhang, X. Wang, and G. Qu. Physical unclonable functions-based linear encryption against code reuse attacks. In *2016 53nd ACM/EDAC/IEEE Design Automation Conference (DAC)*, pages 1–6, 2016.

[30] V. van der Veen, D. Andriesse, E. Göktaş, B. Gras, L. Sambuc, A. Slowinska, H. Bos, and C. Giuffrida. Practical Context-Sensitive CFI. In *Proceedings of the 22nd Conference on Computer and Communications Security (CCS'15)*, October 2015.

[31] Y. Cheng, Z. Zhou, M. Yu, D. Xuhua, and R. Deng. ROPecker: A generic and practical approach for defending against rop attacks. 01 2014.

[32] C. Luk, R. Cohn, R. Muth, H. Patil, A. Klauser, G. Lowney, S. Wallace, V. J. Reddi, and K. Hazelwood. Pin: Building customized program analysis tools with dynamic instrumentation. *SIGPLAN Not.*, 40(6):190–200, June 2005.

[33] T. Austin, E. Larson, and D. Ernst. SimpleScalar: an infrastructure for computer system modeling. *Computer*, 35(2):59–67, Feb 2002.

An Open-source Framework for Autonomous SoC Design with Analog Block Generation

Tutu Ajayi[*], Sumanth Kamineni[†], Yaswanth K Cherivirala[*], Morteza Fayazi[*], Kyumin Kwon[*],
Mehdi Saligane[*], Shourya Gupta[†], Chien-Hen Chen[†], Dennis Sylvester[*], David Blaauw[*],
Ronald Dreslinski Jr[*], Benton Calhoun[†], David D. Wentzloff[*]

* University of Michigan, Ann Arbor, MI
† University of Virginia, Charlottesville, VA

Abstract—**We present the world's first autonomous mixed-signal SoC framework, driven entirely by user constraints, along with a suite of automated generators for analog blocks. The process-agnostic framework takes high-level user intent as inputs to generate optimized and fully verified analog blocks using a cell-based design methodology.**

Our approach is highly scalable and silicon-proven by an SoC prototype which includes 2 PLLs, 3 LDOs, 1 SRAM, and 2 temperature sensors fully integrated with a processor in a 65nm CMOS process. The physical design of all blocks, including analog, is achieved using optimized synthesis and APR flows in commercially available tools. The framework is portable across different processes and requires no-human-in-the-loop, dramatically accelerating design time.

Index Terms—**analog synthesis, analog generator, SoC generator**

I. INTRODUCTION

There is an ever-growing need for automation in analog circuit design, validation, and integration to meet modern-day SoC requirements. Time-to-market constraints have become tighter, design complexity has increased and more functional blocks (in number and variety) are being integrated into SoCs. These challenges often translate to increased manual engineering efforts and non-recurring engineering (NRE) costs. We present FASoC, an open-source[1] framework for Fully-Autonomous SoC design. Coupled with a suite of analog generators, FASoC can generate complete mixed-signal system-on-chip (SoC) designs from user specifications. The framework leverages differentiating techniques to automatically synthesize correct-by-construction RTL descriptions for both analog and digital circuits, enabling a technology-agnostic, no-human-in-the-loop implementation flow.

Analog blocks like PLLs, LDOs, ADCs, and sensor interfaces are recast as structures composed largely of digital components while maintaining analog performance. They are then expressed as synthesizable Verilog blocks composed of digital standard cells and auxiliary cells (aux-cells). We employ novel techniques to automatically characterize aux-cells

---

[1]Source code for the framework and all generators developed as part of this work can be downloaded from https://fasoc.engin.umich.edu

and develop models required for generating bespoke analog blocks. The framework is portable across processes, EDA tools and scalable in terms of analog performance, layout, and other figures of merit.

The SoC generation tool translates user intent to low-level specifications required by the analog generators. To achieve full SoC integration, we leverage the IP-XACT [1] standard and added vendor extensions to capture additional meta-data from the generated blocks. This enables the composition of vast numbers of digital and analog components into a single correct-by-construction design. The fully composed SoC design is finally realized by running the Verilog through synthesis and automatic place-and-route (APR) tools to realize full design automation.

II. FRAMEWORK ARCHITECTURE

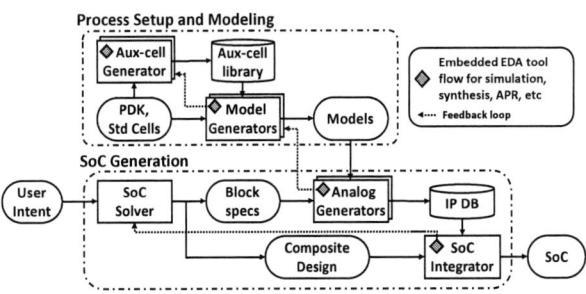

Fig. 1. FASoC Framework Overview

A high-level representation of the framework is shown in Fig. 1. The *Process setup and modeling* phase is performed once for the process design kit (PDK), and it involves the generation of the aux-cells and models for the generator. The *SoC generation* phase begins by translating high-level user-intent into analog specifications that satisfy the user constraints. The block generators are invoked as needed and the SoC integrator stitches the composed design and walks it through a synthesis and APR flow to create the final SoC layout. The FASoC framework is tightly integrated with analog generators for PLL, LDO, temperature sensor, and SRAM blocks. Section III describes the circuit architecture adopted by the different generators.

978-1-7281-5410-7/20 $31.00 © 2020 IEEE

A. Process Setup and Modeling

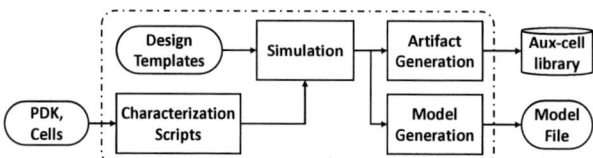

Fig. 2. Aux-cell and model file generation flow

FASoC employs a synthesizable cell-based approach for generating analog blocks, significantly cutting back on manual layout and verification efforts. Aux-cells are small analog circuits that buttress the standard cell library and provide specific analog functionality required by the generators. Each cell is no larger than a D flip-flop and can be placed on the standard cell rows. We simplify the creation of aux-cells by using a suite of design templates in tandem with PDK characterization scripts. The templates capture the aux-cell's precise circuit behavior without including any PDK-specific information. The characterization scripts operate on the PDK to derive technology-specific parameters required to set knobs within the templates. Example parameters extracted from the PDK include threshold voltage, metal parasitics, MOSFET behavior, and Fan-out of 4. The knobs set within the template include device type, transistor sizing, and other circuit design options. The results from aux-cell generation include the netlist, layout, timing library, and other files required to proceed with conventional synthesis and APR. At present, the layouts for the aux-cells are manually created, however, we are currently evaluating several layout automation tools [2]–[4] that are showing promising results. We find our template-based methodology for creating aux-cells enhances process-portability and significantly cuts down on design time. All of the generators presented in this work leverage 8 aux-cells that are depicted in Fig. 3.

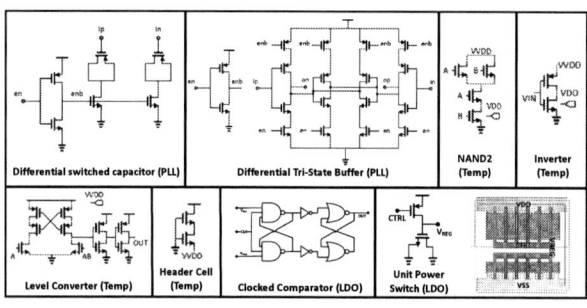

Fig. 3. Schematic for aux-cells used across PLL, LDO and temperature sensor generators

The analog generators use models to predict performance and select design parameters to create optimized block designs that satisfy the input specifications. The models are derived from the parameterized templates that incorporate the aux-cells. The models for each generator vary and are

developed from a combination of mathematical equations, machine learning, and design space exploration. The modeling exercise is also performed once per PDK and the results are saved into a model file. Sections III briefly describes the modeling approach adopted by each generator integrated into the framework.

B. SoC Generator

This stage begins with an iterative *SoC solver* to determine the optimal *composite design* description which is a combination of blocks, analog specifications, and connections. The strategy is guided by high-level user intent (i.e. target application and power/area budgets), available analog block generators, and a database of IPs. Analog generators are invoked as necessary to generate bespoke blocks required to satisfy the specifications within the composite design. The generator outputs include all artifacts required to push the block through standard synthesis and APR tools. The outputs are also cached in an *IP database*, allowing for block generation to be skipped if a matching entry already exists. Entries in the database can also be populated with 3rd party IPs such as processors and other peripherals.

We adopt the IP-XACT format to describe the composite design as well as the block designs stored in the database. We use an extended format [5] to capture additional analog data, simulation, and verification information.

The *SoC integrator* begins by stitching the composite design together and translating it to its structural Verilog equivalent that can be run through digital simulation tools. The structural Verilog, along with all required artifacts from the database, is then passed through the embedded tool flow to generate the final verified GDS. This same flow is pervasive across the framework and is also used by all generators (aux-cell, model, and analog). Tools within the flow cover all aspects of chip design including SPICE simulations, digital simulations, synthesis, APR, DRC, LVS, and extraction.

III. ANALOG GENERATOR ARCHITECTURE

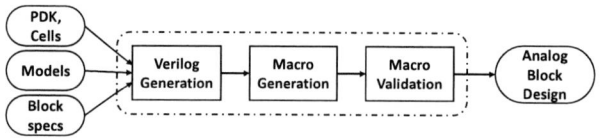

Fig. 4. Analog generator flow

Synthesizable analog blocks were introduced a few decades ago and have continued to evolve, closely matching the performance obtainable by full custom designs. Prior works have described techniques for synthesizing analog blocks for UWB transmitters [6], PLLs [7], DACs [8], and other types of analog blocks [9]–[11]. This approach lowers engineering design costs, increases robustness, eases portability across PDKs, and continues to show promise even at advanced process nodes [12]–[14]. The analog generators developed as part of this work can be likened to ASIC memory compilers

978-1-7281-5410-7/20 $31.00 © 2020 IEEE

that take in a specification file and produce results in industry-standard file formats, which can then be used in standard synthesis and APR tools. Unlike typical memory compilers, our generators are open-source, process agnostic, and share a scalable framework amenable to different types of blocks. The framework is modular and share a similar process as depicted in Fig. 4. The full generation process is broken down into three steps:

Verilog Generation: This step leverages models to produce a synthesizable Verilog description of the block that conforms to the input specifications. It also generates guidance information in a vendor-agnostic format. The guidance includes synthesis constraints, placement instructions, and other information that may be required by the synthesis and/or APR tool to generate blocks that achieve the desired performance. In addition, this step also reports early estimates on performance and the characteristics of the block to be created.

Macro Generation: The Verilog and guidance information is passed to a digital flow to create macros that can be embedded into larger SoC designs. The digital flow in this step performs synthesis, APR, DRC, and LVS verification. The digital flow includes an adapter to translate the guidance into vendor-specific commands used in synthesis and APR. The adapter abstraction allows us to (1) express additional design intent without exposing protected vendor-specific commands and (2) easily support multiple cad tools including open-source alternatives [15]–[17]

Macro Validation: The last step is a comprehensive verification and reporting of the generated block. The full circuit goes through parasitic extraction, SPICE simulations, requirement checks and other verification to culminate in a detailed datasheet report.

The generators can be invoked standalone, outside of the full SoC generator flow. To simplify the system integration, we adopt the AMBA™ APB protocol as the register interface to all blocks. The following sub-sections briefly describe the analog generators currently integrated into the FASoC framework.

A. PLL

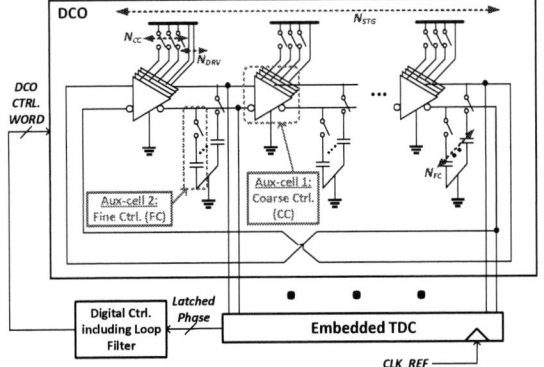

Fig. 5. DCO architecture indicating the aux-cells and designs parameters

The generated PLLs (Fig. 5) share the same base architecture as ADPLL [18]. The phase difference of the reference and output clocks are captured by the embedded time-to-digital converter (TDC), while the digital filter calculates the frequency control word for the digitally controlled oscillator (DCO). The input specification to the generator defines the nominal frequency range and in-band phase noise (PN). The PLL generator uses a physics-based mathematical model [19] for characterization. We first build a mathematical relationship between DCO design parameters (number of aux-cells and stages) and the required DCO specifications. Using simulation results from a parametric sweep, we then find the effective ratio of drive strength and capacitance for each aux-cell. This ratio enables us to predict frequency and power results (frequency range, frequency resolution, frequency gain factor, and power consumption) given a set of input design parameters.

B. LDO

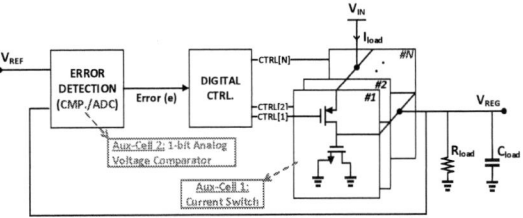

Fig. 6. LDO architecture indicating the aux-cells and design parameters derived from input specifications of V_{IN}, I_{load} and desired transients

The generated LDOs (Fig. 6) share the same base architecture as DLDO [20]. The LDO leverages an array of small power transistors that operate as switches for power management. Based on design requirements, the generator can swap the clocked comparator with a synthesizable stochastic flash ADC [21] to improve transient response. The input specifications to the LDO generator are the V_{IN} range, $I_{load,max}$ range, and the dropout voltage. The generator uses a poly-fit model of the load current ($I_{load,max}$) performance with respect to various combinations of aux-cell connections (connected in parallel and for different VDD inputs) in both ON and OFF states. We create the model by simulating various test circuits after parasitic extraction.

C. Temperature Sensor

The generated sensors (Fig. 7) share the same base architecture as [22]. The sensor relies on a temperature-sensitive ring oscillator and stacked zero-VT devices for better line sensitivity. The input specifications include the temperature range and optimization strategy, for either error or power. For a given temperature range, the models attempt to select the optimal circuit topology that minimizes error and/or performance. The generator relies on a predictive Bayesian neural network model to select design parameters that satisfy the input specifications.

978-1-7281-5410-7/20 $31.00 © 2020 IEEE

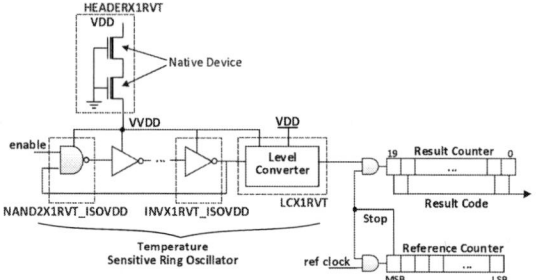

Fig. 7. Temperature sensor architecture indicating the aux-cells

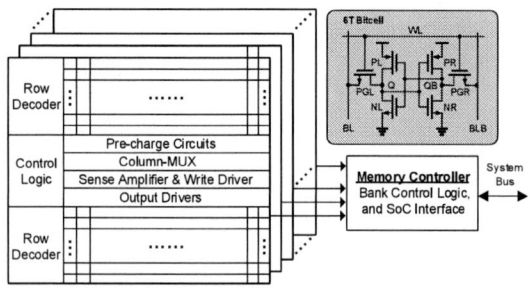

Fig. 8. SRAM architecture showing macros and bank strategy

D. SRAM

The compiled SRAMs (Fig. 8) follow a standard multi-bank memory architecture. Unlike other generators in the framework, the memory generator uses a combination of macros instead of aux-cells. The macros used include a 6T bitcell, a row decoder, column mux, wordline driver, sense amplifier, write driver, and a pre-charge circuit. The macros are stitched together, bottom-up, to form a bank. The user input specifications are capacity, word size, operating voltage, and operating frequency. The generator adopts a hierarchical meta compiler (HMC) [23] for technology characterization and a hierarchical memory model to determine the optimal row and column periphery. The model helps to select the SRAM architecture and the leaf-level components that best satisfy the user specifications while minimizing energy consumption and delay.

IV. EVALUATION

The framework has been fully validated in a 65nm process. Our evaluation begins with a focus on the individual generators. We present results that explore the design-space possible with each generator and demonstrate full adherence to the user input specification. We then present results from a prototype SoC created using this framework.

A. Analog Generation Results

Fig. 9 presents the results of several PLLs generated using different input specifications. It compares the input requirements against the simulated results after parasitic extraction. The results show that the generated frequency ranges cover that of the input requirements and with better phase noise

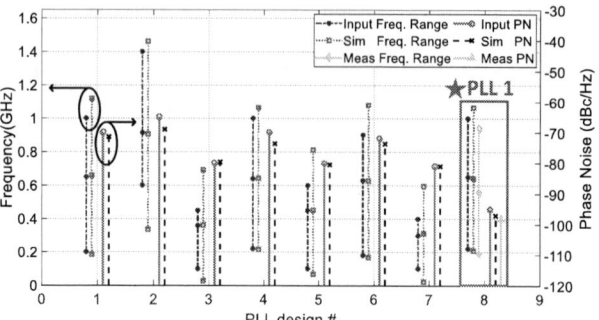

Fig. 9. Generated PLL designs for eight different input specifications. PLL1 is taped-out in the SoC prototype

levels. The highlighted PLL 8, corresponds to one of the PLLs integrated into the SoC prototype and also shows measured results that satisfy the given specifications

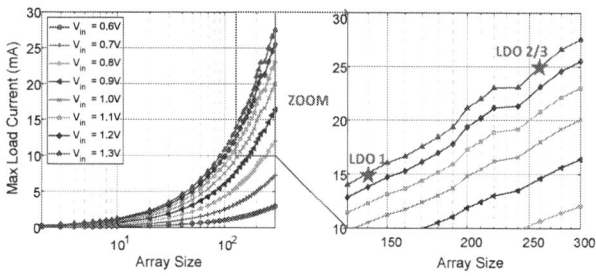

Fig. 10. $I_{load,max}$ vs. array size, for multiple LDO designs generated

Fig. 10 shows the spice simulation results of multiple LDO designs after parasitic extraction. The graph shows the maximum load current at different input voltages corresponding to the input parameter array size for a dropout voltage of 50mV. The highlighted measurements correspond to the input specification for blocks integrated into the SoC prototype with $V_{IN} = 1.3V$ and $V_{REG} = 1.2V$.

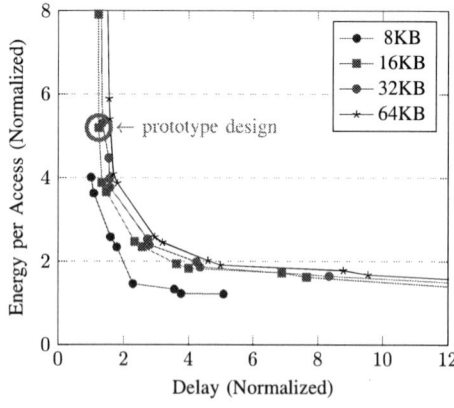

Fig. 11. Normalized energy and delay plots for various memory sizes while sweeping VDD.The results are normalized with respect to the 8KB memory.

978-1-7281-5410-7/20 $31.00 © 2020 IEEE

Fig. 11 presents the simulation results of various memory capacities across a broad range of architectural options and operating voltages (VDD). Each point on the curve corresponds to an energy-delay pair specific to an architecture (rows, columns, and banks) and VDD combination. The generator selects the Pareto-optimal design that satisfies the user requirements. The highlighted point on the 16KB curve corresponds to the memory block integrated into the SoC prototype.

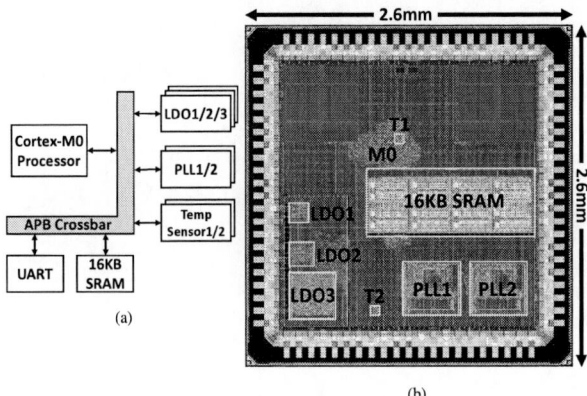

(a)

(b)

Fig. 13. Simplified block diagram (a) and annotated die photo (b) for the 65nm prototype SoC

TABLE I
PLL SIMULATION VS MEASUREMENT RESULTS

Output Specifications	PLL1		PLL2	
	Sim	Meas	Sim	Meas
Min Freq (MHz)	200	190	170	150
Max Freq (MHz)	1,060	920	1,080	930
F_{nom} (Mhz)	643	558	627	548
Power@F_{nom} (mW)	7.20	6.90	8.06	7.70
Area (μm^2)	167,639.04		167,639.04	

Table II shows the LDO $I_{load,max}$ measurements closely matching the input specification requirements. Compared to the comparator-based architecture (LDO1/2), the ADC based controller architecture (LDO3) achieves better transient performance with a 10x and 7x improvement in settling time and undershoot voltage respectively. The line and load regulation values are measured at V_{IN}=1.3V, V_{REF}=1.2V, and I_{load}=10mA. LDO3 load regulation is comparatively worse due to the high gain of the ADC based controller. As we operate at lower V_{REF} and I_{load} conditions, the line/load regulation degrades for all the LDOs because of the increase in relative switch strength.

The temperature sensor has an area of 2,620μm^2. A 2-pt calibration is performed at 0°C and 80°C. Measured results show a sensing range between -20°C and 100°C with an accuracy of ±4°C.

Fig. 14 summarizes the SRAM measured and simulated performance across the input operating voltage range of 0.8V to 1.2V. The SRAM peak performance is at 65MHz with the power consumption of 2.09mW at 1.2V, which exceeds the targeted frequency of 50MHz. The measured power for

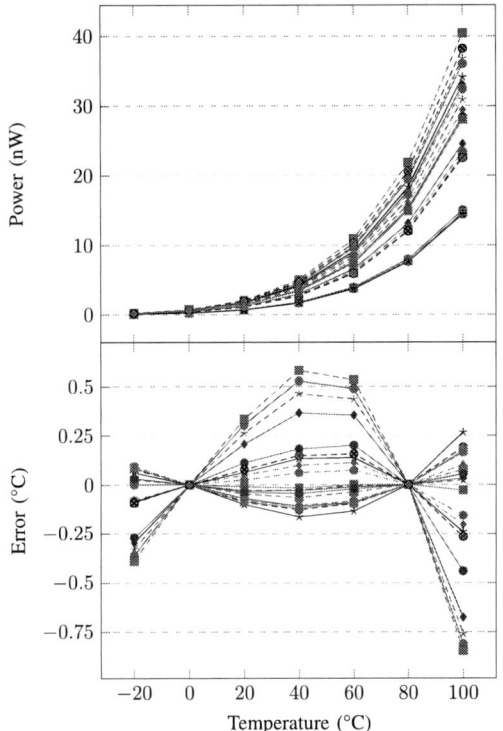

Fig. 12. Power and Error results against temperature for various temperature sensor designs (each fitted plot represents a unique design)

Fig. 12 shows the spice simulation results of multiple temperature sensor designs after parasitic extraction.

B. Prototype Chip Results

The prototype SoC design (Fig. 13) includes 2 PLLs, 3 LDOs, 1 16KB SRAM, and 2 temperature sensors fully integrated with an Arm® Cortex™-M0 in a 65nm CMOS process. Using off-chip connections, we were successfully able to power the entire SoC using one of the LDOs and clock it using the PLLs while monitoring the temperature of the chip.

Fig. 9 presents results for 8 PLL designs generated from different input specifications, including one from the prototype, and the results show output performances in-line with the input specifications. The measured frequency is 10% slower while the phase noise matches the simulation and specification requirement. Table I summarizes the results for all PLLs in the prototype.

TABLE II
LDO SIMULATION VS MEASUREMENT RESULTS @ 200MHz CONTROL CLOCK

Output Specifications	LDO1		LDO2		LDO3	
	Sim	Meas	Sim	Meas	Sim	Meas
Dropout Voltage (mV)	50	70	50	80	50	80
$I_{load,max}$ (mA)	15.00	15.38	25.00	24.84	25.00	23.72
Settling Time - Ts (µs)	1.1	1.8	2.1	2.9	0.12	0.19
Max Undershoot (V)	0.35	0.98	0.57	0.98	0.38	0.14
Max Current Eff. (%)	94.2	96.4	95.7	94.5	81.9	74.0
Load Regulation (mV/mA)	-	-1.00	-	-0.35	-	-3.6
Line Regulation (V/V)	-	0.180	-	0.004	-	0.950
Area (μm^2)	17,318.56		31,187.56		127,163.56	

978-1-7281-5410-7/20 $31.00 © 2020 IEEE

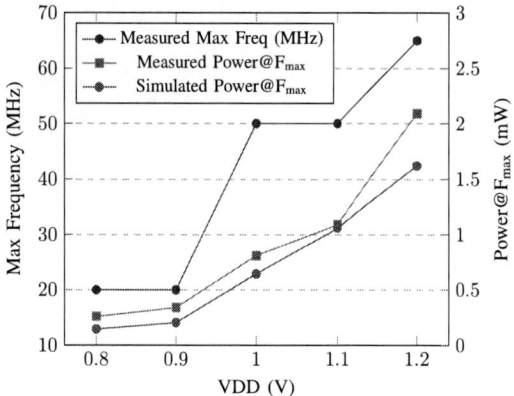

Fig. 14. Measured and simulated performance and power results of SRAM across VDD

the SRAM also include the leakage power of the processor and peripheral interface. The generated SRAM has an area of $0.68mm^2$ with the custom bitcell area occupying $0.4mm^2$.

V. CONCLUSION

We presented an autonomous framework that generates a completely integrated SoC design based on user input specifications. This framework is PDK agnostic and allows for faster turn-around times when building custom analog blocks and integrating them into larger SoC designs. The framework includes generators for PLL, LDO, temperature sensor and SRAM blocks. The framework can easily be extended to support more generators and different PDKs. We fabricated an SoC prototype in a 65nm process and presented silicon measurements to validate the framework's accuracy. Our work establishes a new milestone in creating a silicon compiler [24] that further reduces the complexity of realizing modern SoCs and cuts down on design time.

ACKNOWLEDGMENT

This material is based on research sponsored by Air Force Research Laboratory (AFRL) and Defense Advanced Research Projects Agency (DARPA) under agreement number FA8650-18-2-7844. The U.S. Government is authorized to reproduce and distribute reprints for Governmental purposes notwithstanding any copyright notation thereon.

REFERENCES

[1] Accellera, "IP-XACT - Accellera." https://www.accellera.org/downloads/standards/ip-xact. Last accessed 2020-05-03.

[2] C.-Y. Wu, H. Graeb, and J. Hu, "A pre-search assisted ilp approach to analog integrated circuit routing," in *2015 33rd IEEE International Conference on Computer Design (ICCD)*, pp. 244–250, IEEE, 2015.

[3] K. Kunal, M. Madhusudan, A. K. Sharma, W. Xu, S. M. Burns, R. Harjani, J. Hu, D. A. Kirkpatrick, and S. S. Sapatnekar, "ALIGN: Open-source analog layout automation from the ground up," in *Proceedings of the 56th Annual Design Automation Conference 2019*, pp. 1–4, 2019.

[4] B. Xu, K. Zhu, M. Liu, Y. Lin, S. Li, X. Tang, N. Sun, and D. Z. Pan, "MAGICAL: Toward fully automated analog ic layout leveraging human and machine intelligence," in *2019 IEEE/ACM International Conference on Computer-Aided Design (ICCAD)*, pp. 1–8, IEEE, 2019.

[5] R. Dreslinski, D. Wentzloff, M. Fayazi, K. Kwon, D. Blaauw, D. Sylvester, B. Calhoun, M. Coltella, and D. Urquhart, "Fully-autonomous soc synthesis using customizable cell-based synthesizable analog circuits," tech. rep., University of Michigan Ann Arbor United States, 2019.

[6] Y. Park and D. D. Wentzloff, "An all-digital 12pj/pulse 3.1–6.0 ghz ir-uwb transmitter in 65nm cmos," in *2010 IEEE International Conference on Ultra-Wideband*, vol. 1, pp. 1–4, IEEE, 2010.

[7] Y. Park and D. D. Wentzloff, "An all-digital pll synthesized from a digital standard cell library in 65nm cmos," in *2011 IEEE Custom Integrated Circuits Conference (CICC)*, pp. 1–4, IEEE, 2011.

[8] E. Ansari and D. D. Wentzloff, "A 5mw 250ms/s 12-bit synthesized digital to analog converter," in *Proceedings of the IEEE 2014 Custom Integrated Circuits Conference*, pp. 1–4, IEEE, 2014.

[9] S. Bang, A. Wang, B. Giridhar, D. Blaauw, and D. Sylvester, "A fully integrated successive-approximation switched-capacitor dc-dc converter with 31mv output voltage resolution," in *2013 IEEE International Solid-State Circuits Conference Digest of Technical Papers*, pp. 370–371, IEEE, 2013.

[10] W. Jung, S. Jeong, S. Oh, D. Sylvester, and D. Blaauw, "A 0.7 pf-to-10nf fully digital capacitance-to-digital converter using iterative delay-chain discharge," in *2015 IEEE International Solid-State Circuits Conference-(ISSCC) Digest of Technical Papers*, pp. 1–3, IEEE, 2015.

[11] M. Shim, S. Jeong, P. Myers, S. Bang, C. Kim, D. Sylvester, D. Blaauw, and W. Jung, "An oscillator collapse-based comparator with application in a 74.1 db sndr, 20ks/s 15b sar adc," in *2016 IEEE Symposium on VLSI Circuits (VLSI-Circuits)*, pp. 1–2, IEEE, 2016.

[12] S. Bang, W. Lim, C. Augustine, A. Malavasi, M. Khellah, J. Tschanz, and V. De, "A fully synthesizable distributed and scalable all-digital ldo in 10nm cmos," in *2020 IEEE International Solid-State Circuits Conference-(ISSCC)*, IEEE, 2020.

[13] S. Kundu, L. Chai, K. Chandrashekar, S. Pellerano, and B. Carlton, "A self-calibrated 1.2-to-3.8 ghz 0.0052mm² synthesized fractional-n mdll using a 2b time-period comparator in 22nm finfet cmos," in *2020 IEEE International Solid-State Circuits Conference-(ISSCC)*, pp. 276–278, IEEE, 2020.

[14] A. Rovinski, C. Zhao, K. Al-Hawaj, P. Gao, S. Xie, C. Torng, S. Davidson, A. Amarnath, L. Vega, B. Veluri, *et al.*, "A 1.4 ghz 695 giga risc-v inst/s 496-core manycore processor with mesh on-chip network and an all-digital synthesized pll in 16nm cmos," in *2019 Symposium on VLSI Circuits*, pp. C30–C31, IEEE, 2019.

[15] C. Wolf, "Yosys open synthesis suite." http://www.clifford.at/yosys/. Last accessed 2020-05-08.

[16] "Ngspice, the open source spice circuit simulator." http://ngspice.sourceforge.net/. Last accessed 2020-05-08.

[17] S. N. Laboratories, "Xyce parallel electronic simulator (xyce)." https://xyce.sandia.gov/. Last accessed 2020-05-08.

[18] D. M. Moore, T. Xanthopoulos, S. Meninger, and D. D. Wentzloff, "A 0.009 mm 2 wide-tuning range automatically placed-and-routed adpll in 14-nm finfet cmos," *IEEE Solid-State Circuits Letters*, vol. 1, no. 3, pp. 74–77, 2018.

[19] M. H. Perrott, M. D. Trott, and C. G. Sodini, "A modeling approach for Σ-Δ fractional-N frequency synthesizers allowing straightforward noise analysis," *IEEE Journal of Solid-State Circuits*, vol. 37, no. 8, pp. 1028–1038, 2002.

[20] Y. Okuma, K. Ishida, Y. Ryu, X. Zhang, P.-H. Chen, K. Watanabe, M. Takamiya, and T. Sakurai, "0.5-v input digital ldo with 98.7% current efficiency and 2.7-μa quiescent current in 65nm cmos," in *IEEE Custom Integrated Circuits Conference 2010*, pp. 1–4, IEEE, 2010.

[21] S. Weaver, B. Hershberg, P. Kurahashi, D. Knierim, and U.-K. Moon, "Stochastic flash analog-to-digital conversion," *IEEE Transactions on Circuits and Systems I: Regular Papers*, vol. 57, no. 11, pp. 2825–2833, 2010.

[22] M. Saligane, M. Khayatzadeh, Y. Zhang, S. Jeong, D. Blaauw, and D. Sylvester, "All-digital soc thermal sensor using on-chip high order temperature curvature correction," in *2015 IEEE Custom Integrated Circuits Conference (CICC)*, pp. 1–4, IEEE, 2015.

[23] S. Nalam, M. Bhargava, K. Mai, and B. H. Calhoun, "Virtual prototyper (vipro): An early design space exploration and optimization tool for sram designers," in *Design Automation Conference*, pp. 138–143, IEEE, 2010.

[24] D. Johannsen, "Bristle blocks: A silicon compiler," in *16th Design Automation Conference*, pp. 310–313, IEEE, 1979.

Ultra-Compact, Scalable, Energy-Efficient VO_2 Insulator-Metal-Transition Oxide Based Spiking Neurons for Liquid State Machines

Samiran Ganguly, Nikhil Shukla, Avik W. Ghosh

Charles L. Brown Dept. of Electrical and Computer Engineering, University of Virginia, Charlottesville, VA, USA
{sganguly, ns6pf, ag7rq} @virginia.edu

Abstract—**Spiking neural networks, inspired by biological neural systems, could process immense volumes of spatio-temporal data by representing them as spikes. Here, we propose to implement compact, scalable, energy-efficient spiking neurons based on the unique insulator-metal transition in Vanadium dioxide (VO_2) which interact through memristive synapses, to emulate a Liquid State Machine (LSM). Further, we demonstrate the implementation of this recurrent neural network as a temporal auto-encoder, and adaptive channel equalizer for application in neuromorphic signal processing. Our approach provides a pathway to reduce component count ($50 - 100X$) and improve energy efficiency ($> 50X$) over conventional CMOS based implementations.**

Index Terms—**Insulator-Metal-Transition, Spiking Neurons, Liquid State Machine, Reservoir Computing, Extended Kalman Filter**

I. Introduction

High-performance computing has historically developed around the von Neumann architecture and the Boolean computing paradigm, executed on Silicon Complementary Metal Oxide Semiconductor (CMOS) hardware. Software development model has for decades has worked around the CMOS fabric that has singularly dictated our choice of materials, devices, circuits and architecture to build our technology. Over the last decade however, Moore's law for hardware scaling has significantly slowed down with its end predicted soon, primarily due to the prohibitive energy cost of computing with increasingly smaller devices. At the same time, software development has taken off with Machine Learning and Artificial Intelligence dominating the roost. This has given rise to the notion of 'neuromorphic computing' which promises to radically transform this model of computing to match the ultimate computer in existence, the brain.

It is well known that the temporal dynamics of biological neurons encode immense volumes of processing power [1], [2] in extremely small footprint: $\sim 10^{12}$ neurons and $\sim 10^{15}$ synapses or neural interconnection in a volume of $\sim 1200cm^3$, and power consumption: $10 - 15W$. It has been argued that spike encoded information (spike interval/density modulation) can be energy-efficient as information is encoded in the dead times between sparse spikes [3]–[5], and their phase sequencing [6] and time multiplexing [7] in reservoir computing.

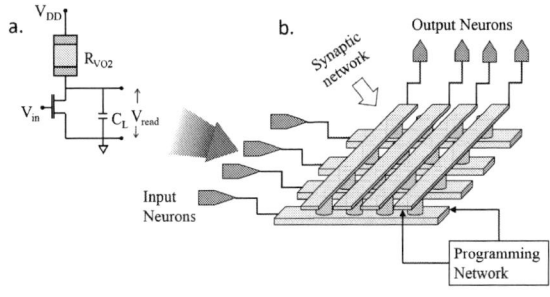

Fig. 1. a. The proposed hardware unit of the spiking neuron built from a 1T-1R pull-up pull-down configuration of a VO_2 layer and an n-channel MOSFET. b. Large scale neural networks can be fabricated with spiking neurons by coupling them with dense cross-bar arrays of programmable memristors forming a synaptic network between various layers of a neural network.

Spiking networks can have higher representation power and computational capacity compared to other neural networks [8]–[10], and are shown to be optimally entropy-efficient (maximal # of bits/Joule) [11], [12] in biological neurons. Commercial silicon spiking nets like IBM's TrueNorth chip [13]–[15] are seen to be $\sim 800X$ more energy efficient than silicon multiprocessors, although lacking integrated on-chip training. The true challenge with CMOS only hardware substrate is to find a physical mechanism to initiate, stabilize and terminate the spikes in a controllable way to leverage the advantages of scalable networks of spiking neurons, with small energy-area footprint, in particular for low size, weight and power (SWaP) applications.

In this work we propose to use Vanadium Dioxide's (VO_2) unique electrical volatile characteristics arising from the Insulator-Metal transition (IMT) property to build ultra-compact and highly energy-efficient hardware units that behaves as spiking neuron (fig. 1a) that can encode information in the spike intervals/frequency and when coupled with programmable memristive synapses, say in a cross-bar array arrangement (fig. 1b), can directly implement dense spiking recurrent neural networks in hardware. In this work we specifically illustrate the networks of these spiking neurons to build a Liquid State Machine (LSMs), the spiking version of a

978-1-7281-5410-7/20 $31.00 © 2020 IEEE

Reservoir Computer. We then show that these LSMs built from spiking neurons can perform non-linear signal processing tasks such as an adaptive channel equalizer, and learn to predict a complex time series running blind, i.e. without an external input. We then conclude by discussing the relative reduction in component count and power consumption provided by these hardware units over a purely CMOS implementations of such spiking neural networks.

II. VO_2 INSULATOR-METAL-TRANSITION: ULTRA-COMPACT SPIKING NEURON

A. Insulator-Metal-Transition, VO_2 Characteristics and Controllable Spiking Oscillator

VO_2 shows a sharp Insulator Metal Phase Transition (IMT) [16]–[19] when driven thermally, optically and electrically (fig. 2a). It has been shown that IMT at 340K shows a typical ON-OFF ratio $\sim 1-5 \times 10^3$ to a max of 10^5, with a switching speed $\sim 10ns$ electronically, $\sim 10fs$ optically, and measured switching reliability over 10^9 cycles [20]. The underlying mechanism is still in debate and seems to be a combination of purely electronic Mott transition (abrupt opening of a Coulomb gap) and a simultaneous structural change between monoclinic and tetragonal/rutile configurations [21]. The main feature of its output characteristics is a unique volatile hysteresis loop (fig. 2a,b) with two stable states (solid horizontal arms) and two unstable states (dashed vertical arms) which is distinct from pinched hysteresis memristor curves used as selectors (e.g. $Ag|HfO_2$), or magnetic hysteresis curves used in MRAM technology where both the stables states are accessible at $V \sim 0$ point of the characteristics.

This volatile hysteresis has its center shifted from $V = 0$ to half the transition voltages V_H, V_L which proves quite important for our application. This characteristics allows the device to move naturally from an oscillatory spiking to a non-spiking state and vice-versa, simply by adjusting a load line using an external transistor in the hardware unit as shown in the schematic fig. 2b. Fig. 3c,d show experimental data and simulations on the device whose characteristics are presented in fig. 2a, showing the non-spiking and spiking behavior and is obtained by a slight shift in the load line (determined by the resistance of the transistor).

When the transistor load line crosses the stable high resistance state section of the I-V curve, the IMT neuron remains in the resting state (solid blue line). However as gate voltage V_{in} increases the transistor load-line (solid orange line) now periodically crosses both the unstable (black dashed lines) arms, and the neuron fires and generates an oscillatory spike train, as can be seen in fig. 2e. Due to stochastic cycle-to-cycle variations in V_{IMT}, firing occurs with probability $p < 1$ leading to probabilistic spiking of the IMT neuron. Further, the spiking probability can be modulated using the gate voltage, V_{in}, which decides the fraction of the voltage range of V_T intersected by the transistor load line.

This unique volatile characteristic generating spike trains, along with in-built activation and deactivation (much like a real biological neuron) is naturally suited to provide ultra-compact

and efficient, controlled spiking behavior. Additionally, the built-in stochasticity of the phase transition in VO_2 naturally provides controllable noise in the device. It is well recognized that noise is an important feature in building robust neural network models by preventing over-fitting to training data and can provide a hardware mechanism to build a natural 'echo-state property' feature useful in recurrent neural networks like Reservoir Computing as we discuss later.

B. Spike Encoder and Spiking Stochastic Leaky-Integrate-and-Fire Neuron

We can build an efficient spike-interval/density modulator using the proposed VO_2 unit as shown in fig. 3a. The input signal is scaled appropriately using an impedance divider circuit to reflect the movement of the load-line between the stable high resistance state to the oscillatory state using the gate voltage applied on the input MOSFET. The simulation presented in fig. 3b shows the normalized output of the circuit measured across the load capacitor. A "high" signal, in this scheme, shifts the load-line to the oscillatory stage by reducing the transistor resistance and the unit keeps firing, till the input gate voltage reduces shifting the load-line back to the stable high R state, causing the spikes to end. Therefore, a simple $1R - 1T$ unit can generate an efficient and compact spike modulator based on the intrinsic electrical characteristics of the device itself.

The same unit can also be used to build a spiking neuron, where the input signal gets integrated over time in the gate capacitor and if it reaches a threshold, the neuron fires and generates spikes. Our proposed unit naturally embodies a spike generator based on the input signal at the input terminal. Instead of a smooth analog signal from some sensor (as in case of the spike encoder), if we provide spiking signals using a resistive-capacitive metallic interconnect to the input terminal, the gate capacitor can integrate the charge provided by these spikes naturally over a few characteristic RC time constants of the interconnect + gate capacitor and can move the unit from non-spiking to spiking state. Being a volatile device, this built-up charge will leak away. Therefore if the input spike train is stopped, the unit will move back to a non-spiking state. This illustrates the principle of the same spike encoder effectively working as a leaky-integrate-and-fire (LIF) neurons. Additionally, the inherent stochastic nature of the transition voltage V_H and V_L discussed previously, changes the spiking voltage/charge threshold creating inherent stochasticity in the activation/spike generation.

III. LIQUID STATE MACHINES FROM VO_2 SPIKING NEURONS

A. Dynamical Recurrent Neural Networks: Reservoir Computing

Reservoir Computing [22] is a biologically inspired model of computing where a set of non-linear units or neurons are assembled together in recurrent fashion, i.e. with bidirectional feedbacks between various neurons (fig. 4a). To this "reservoir" of neurons a temporal signal is imposed, giving rise

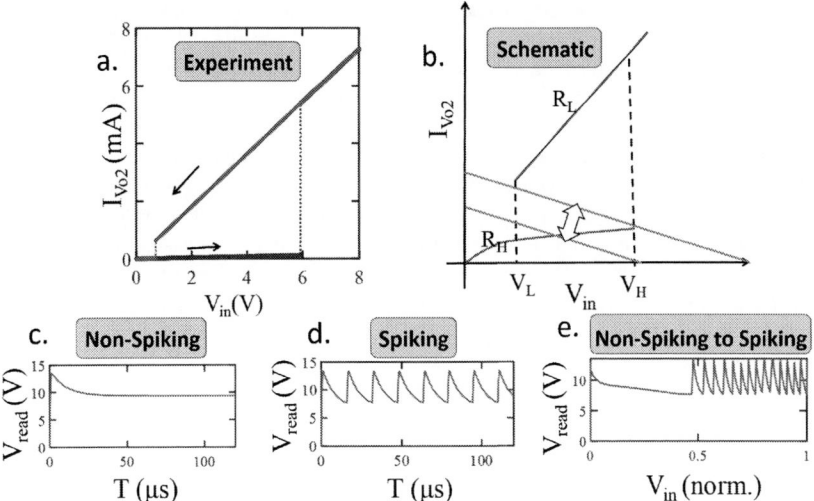

Fig. 2. a. Experimental characteristics of the a VO_2 film showing its shifted volatile hysteresis I-V curve. b. Schematic of the I-V curve to illustrate the utilization of a controllable load-line to go from non-spiking to spiking states. c. Non-spiking behavior of the neuron. d. Spiking behavior of the neuron. e. Movement from non-spiking to spiking behavior by moving the load-line using the gate input gate voltage to the device. It can be seen that spike frequency or spike density increases with the increasing input voltage.

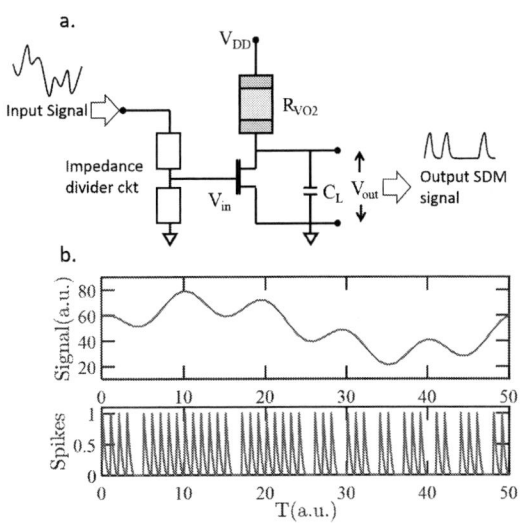

Fig. 3. a. Spike encoder using the proposed unit along with an impedance divider circuit at the input to appropriately scale the input voltage to control the movement of the load-line created by the transistor. b. Example of a spike interval/density encoding of an arbitrary input signal.

to collective excitations. An inference is made by harvesting or sampling these collective reservoir states using a readout layer which can be a linear or logistic readout, or even a deep network. In this model of computing, the reservoir is never trained - only the readout layer(s) is trained (Fig. 2b). This model of computing also allows attaching different readouts to the same reservoir to carry out different inferencing tasks, embodying the inherent parallelism of biological neural

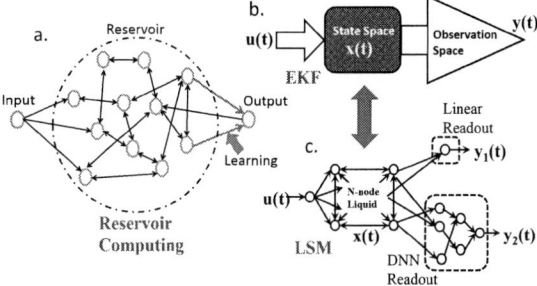

Fig. 4. a. A schematic of a reservoir computing. b. Comparison between the mathematical model of an Extended Kalman Filter (EKF) and a Liquid State Machine (LSM).

systems (fig. 4c). Reservoir computing is particularly suited to spatio-temporal inferencing tasks due to the recurrence, as the signature of an input signal provided to the reservoir continues to persist in the reservoir for a certain time period (also called the echo-state property) which allows the readout to harvest different time samples of the same input signal by reading from different nodes at the same time. Reservoir computing is presented in two different forms in the literature: if the neurons are logistic and the signal analog, it is called an Echo State Network (ESN) [23], whereas if the neurons are spiking and signals are encoded in spike domain, the model is called a Liquid State Machine (LSM) [24]–[26].

B. Reservoir Computing and Signal Processing

Kalman filters [27], [28] are a class of adaptive filters that can be used to learn and predict the states of a dynamical system from their histories and extract independent variables from dependent variables of state of such a system. A Kalman filter consists of a dynamical system composing a "state space", and an "observer system" that samples the state space to generate an estimate or an inference. Being a dynamical system, the state and therefore the observations are correlated in time (fig. 4b). In a regular Kalman filter, the state space is a linear function of the input signal and the observation is made by sampling the state of the dynamical system to yield the observed signal. Extended Kalman Filters (EKFs) extend conventional Kalman filters to nonlinear signal processing. EKFs are the de-facto standard in GPS tracking, navigation systems and nonlinear state estimation theory.

It can be seen that there are strong parallels between Reservoir computing and EKFs. The reservoir corresponds to the non-linear dynamical state space, while the readout corresponds to the observer (fig. 4b,c). ESNs have been presented in literature [29]–[31] as an approach to train EKFs composed of logistic neurons, whereas LSMs were developed as a plausible model of biological cognitive systems where extremely rich cognitive tasks can be carried out with rather small sets of neurons and limited localized learning. This obvious connection makes VO_2 based Liquid State Machines work as hardware embodiment of the mathematical model of an EKF to perform signal processing and control tasks. Compactness and energy efficiency is achieved by directly implementing the reservoir in the dynamics of the network of spiking neurons and programmable synapses built from memristive interconnects at the readout layer. Our proposed unit and their networks embed the paradigm of spiking neurons and EKF inherently in their physics and make them particularly suited for signal processing tasks as we show next.

IV. TEMPORAL PROCESSING TASKS

In this section we discuss two example temporal data/signal processing tasks using the LSMs by leveraging their connection to EKFs. These examples show that it is possible to implement highly compact and energy-efficient signal processors using small spiking neural networks built from VO_2 neurons.

A. Temporal Predictor

An autoencoder learns the underlying representation or generating function of a set of a data. When applied to a time series data it can be useful in predictive tasks such as navigation and control. In this mode the encoder predicts the next few steps of the sequence which can be used to carry out the navigation and control task. Best in class systems for such tasks are EKFs which can be used to predict the "next step" for a moving object given a history of the past and the current trajectory. Therefore, EKFs can work as temporal autoencoder, i.e. an encoder for time-series data. We implement a temporal auto-encoder using a VO_2 LSM based EKF, shown in fig. 5a. In this task we attempt to learn and reproduce this signal blindly, i.e. once it is learned we want to test if we can generate the "next steps" purely from the LSMs self generated output and how well this predicted signal matches with the original signal generator.

A periodic and subtly chaotic source is provided to the system as the input signal. The input signal is first converted into the spike domain and then provided to a 20-node LSM. A linear weighted sampler is used to readout the reservoir states and thereafter demodulated to generate the output signal. After harvesting a large number of reservoir states, we adjust the sampler weight using a linear regression technique as popular in the reservoir computing community.

During the testing phase, we disconnect the source and connect the readout of the LSM back to its own input, providing the input signal for the next time-step. Therefore, in this testing step the LSM tries to keep generating the source signal blindly looking at its own output, without any corrective feedback or knowledge of the testing signal, generated from the same source that was used during the training phase. Our simulations show that the LSM can do this predictive task reasonably well for 10-15 steps for the given function, however it starts to diverge as the number of steps increase without a corrective feedback (fig. 5b,c). This is to be expected due to the finite memory capacity of the LSM. Even excellent cognitive machines like humans cannot predict such trajectories indefinitely.

During our simulations, we found that extremely small reservoirs (# of nodes = 1-5) find it difficult to learn complex signals, however as the network size is increased the accuracy begins to improve as the state space of the LSM expands with the size of the network allowing it to remember more complex 'high frequency' components of the input signal.

B. MMSE Channel Equalizer

A communication channel can inherently inject all kinds of non-linear distortions, inter symbol interference, and noise in a signal. The task of equalization is to recover the underlying "clear" signal from the distorted signal. An equalizer can be constructed in multiple ways depending on how they are supposed to be used. We implement an EKF based MMSE (minimum mean square error) equalizer using VO_2 LSM which attempts to reconstruct a signal which has been distorted by a telecommunication media (fig. 5d).

In this task the LSM tries to generate output $y(t)$ that is as close to input $d(t)$ from the distorted signal $u(t)$ as possible. The original signal $d(t)$ passes through a medium which introduces inter-symbol interference, non-linear distortions and noise to generate the distorted signal $u(t)$. This distorted signal is then converted into the spike domain using VO_2 spike modulator, which is then provided to a 20 node LSM built from assembling together VO_2 neurons. Then we generate a weighted sample of the reservoir states, which is then demodulated to recover the original signal. During the training phase, we adjust the weighted sampling matrix using linear regression. After successful training, we test the equalizer which can now recover the original signal from the distorted one successfully as shown in fig. 5e. In our simulations we found that we could achieve a symbol recovery rate between $85-100\%$ depending on the degree of distortion and LSM network size.

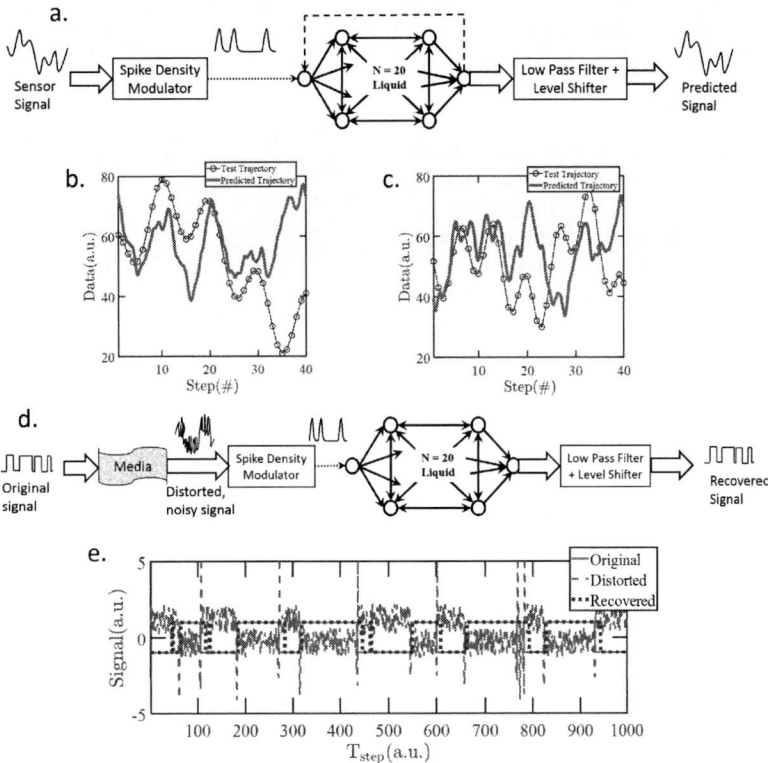

Fig. 5. a. Temporal Predictor setup using a VO_2 LSM. The input signal is spike encoded and provided to the reservoir and readout using a linear sampler and decoder built from low pass filter and level shifter. During the testing phase, the reservoir's own output is fed back to the input and the encoder is decoupled from reservoir. b. & c. Two example simulations of blind signal prediction task. The network reproduces the signal reasonably well for first 15-20 steps but diverges after that without corrective feedback. d. Channel Equalizer setup using a VO_2 LSM. The LSM attempts to reverse the effect of the distortion created by the media on an input signal. e. Original, distorted, and recovered signal obtained from the equalizer shown together.

V. VO_2 SPIKING NEURONS VS. BOOLEAN CMOS NEURONS: A COMPARATIVE ASSESSMENT

Spiking Neural Networks like LSMs can be emulated numerically on any Turing Machine, including a Boolean CMOS processor. However, the hardware primitives of such a computer do not match the algorithmic model of these networks. A network composed of the VO_2 based hardware unit directly implements a reservoir, where the collective dynamics of VO_2 switching itself gives rise to reservoir states being used for the computation. Therefore, the compactness, energy efficiency, and scalability that can be obtained from these units can never be matched, even in principle, by a Boolean emulator. Our proposed unit works in the analog domain as a controllable spike generator and is composed of just one transistor with VO_2 grown directly on the drain terminal so its area cost of just 1 transistor. The power dissipation and energy-delay product of this unit is comparable to a CMOS inverter due to high ON-OFF ratio, gain, and steep sub-threshold swing of the device and this has been previously demonstrated experimentally in a similar structure called the HyperFET [32]. Therefore, the energy cost of a single self-contained VO_2 spiking neuron is equivalent to the smallest

unit of computing in CMOS circuits. Additionally, since the signals are represented as spikes in time, it is an inherently analog mode of computation where a signal can be carried by just 1 wire, instead of multiple wires as required in a Boolean implementation, which further reduces the area and energy cost significantly.

As an illustrative example, consider the work referenced in [33] as an example of a liquid state machine implementation using a stochastic Boolean accelerator in a ring-like reservoir topology, demonstrating a similar temporal prediction task as this work, required ≈ 13 logical elements, ≈ 13 combinational functions, and ≈ 6 logic registers per neuron. While without knowing the exact detailed design of the FPGA used it is difficult to estimate the total component count, an efficient design presented in another work [34], demonstrating similar signal processing tasks as this work, required 4 4-LUT structures per neuron and a simple count will suggest this translates to 420 CMOS inverters or 840 MOSFETs. Therefore, it can be seen that *a Boolean implementation can easily require $\sim 500X$ more components over the presented unit to implement a spiking neural network.* Even assuming great leaps in designs of Boolean accelerators for neural networks in future, we can safely promise at least $50 - 100X$ reduction in component count. Considering that the dissipation in VO_2 neurons is similar to a CMOS inverter, we expect energy-efficiency to scale approximately linearly with the component count improvement, i.e. $\sim 500X$ in the best case scenario for VO_2 neurons or $\sim 50 - 100X$ in best case scenario for CMOS implementations. In [35] it was found that the VO_2 neurons had $\sim 30X$ decrease in power dissipation over an equivalent CMOS implementation in a classification task. This strongly supports our lower bound estimates for energy-efficiency improvement of VO_2 spiking neurons over CMOS implementations.

It should be noted that the estimates provided here are preliminary and cursory. One of the future directions will be extending these benchmarks to a wider set of implementations of LSMs and Kalman filters in FPGAs and ASICs, as well to other material systems, such as low energy-barrier magnets which have also been used to implement spiking neurons. The central goal of this paper is to demonstrate, using proof-of-concept simulations, that ultra-compact, energy-efficient LSMs can be fabricated directly from IMT spiking neurons, and elicit research and exploration interest in the same from the community.

REFERENCES

[1] D. V. Buonomano and M. M. Merzenich, "Temporal information transformed into a spatial code by a neural network with realistic properties," *Science*, vol. 267, no. 5200, pp. 1028–1030, 1995.

[2] W. Maass and H. Markram, "On the computational power of circuits of spiking neurons," *Journal of computer and system sciences*, vol. 69, no. 4, pp. 593–616, 2004.

[3] A. L. Hodgkin and A. F. Huxley, "A quantitative description of membrane current and its application to conduction and excitation in nerve," *The Journal of physiology*, vol. 117, no. 4, pp. 500–544, 1952.

[4] R. FitzHugh, "Impulses and physiological states in theoretical models of nerve membrane," *Biophysical journal*, vol. 1, no. 6, pp. 445–466, 1961.

[5] J. Nagumo, S. Arimoto, and S. Yoshizawa, "An active pulse transmission line simulating nerve axon," *Proceedings of the IRE*, vol. 50, no. 10, pp. 2061–2070, 1962.

[6] G. Buzsáki and A. Draguhn, "Neuronal oscillations in cortical networks," *science*, vol. 304, no. 5679, pp. 1926–1929, 2004.

[7] D. Marković, N. Leroux, M. Riou, F. Abreu Araujo, J. Torrejon, D. Querlioz, A. Fukushima, J. Yuasa, J. Trastoy, and P. Bortolotti, "Reservoir computing with the frequency, phase, and amplitude of spin-torque nano-oscillators," *Applied Physics Letters*, vol. 114, no. 1, p. 012409, 2019.

[8] W. Maass, "Networks of spiking neurons: the third generation of neural network models," *Neural networks*, vol. 10, no. 9, pp. 1659–1671, 1997.

[9] W. Maas, "Noisy spiking neurons with temporal coding have more computational power than sigmoidal neurons," *Advances in Neural Information Processing Systems*, vol. 9, pp. 211–217, 1997.

[10] R. Legenstein and W. Maass, "What makes a dynamical system computationally powerful," *New directions in statistical signal processing: From systems to brain*, pp. 127–154, 2007.

[11] P. Crotty and W. B. Levy, "Energy-efficient interspike interval codes," *Neurocomputing*, vol. 65, pp. 371–378, 2005.

[12] W. B. Levy and R. A. Baxter, "Energy efficient neural codes," *Neural computation*, vol. 8, no. 3, pp. 531–543, 1996.

[13] D. S. Modha, "Introducing a brain-inspired computer," *Published online at http://www. research. ibm. com/articles/brain-chip. shtml*, 2017.

[14] P. A. Merolla, J. V. Arthur, R. Alvarez-Icaza, A. S. Cassidy, J. Sawada, F. Akopyan, B. L. Jackson, N. Imam, C. Guo, and Y. Nakamura, "A million spiking-neuron integrated circuit with a scalable communication network and interface," *Science*, vol. 345, no. 6197, pp. 668–673, 2014.

[15] F. Akopyan, J. Sawada, A. Cassidy, R. Alvarez-Icaza, J. Arthur, P. Merolla, N. Imam, Y. Nakamura, P. Datta, and G.-J. Nam, "Truenorth: Design and tool flow of a 65 mw 1 million neuron programmable neurosynaptic chip," *IEEE Transactions on Computer-Aided Design of Integrated Circuits and Systems*, vol. 34, no. 10, pp. 1537–1557, 2015.

[16] N. Shukla, A. Parihar, E. Freeman, H. Paik, G. Stone, V. Narayanan, H. Wen, Z. Cai, V. Gopalan, and R. Engel-Herbert, "Synchronized charge oscillations in correlated electron systems," *Scientific reports*, vol. 4, p. 4964, 2014.

[17] M. Brahlek, L. Zhang, J. Lapano, H.-T. Zhang, R. Engel-Herbert, N. Shukla, S. Datta, H. Paik, and D. G. Schlom, "Opportunities in vanadium-based strongly correlated electron systems," *MRS Communications*, vol. 7, no. 1, pp. 27–52, 2017.

[18] F. J. Morin, "Oxides which show a metal-to-insulator transition at the Neel temperature," *Physical review letters*, vol. 3, no. 1, p. 34, 1959.

[19] Z. Yang, C. Ko, and S. Ramanathan, "Oxide electronics utilizing ultra-fast metal-insulator transitions," *Annual Review of Materials Research*, vol. 41, pp. 337–367, 2011.

[20] L. A. Ladd and W. Paul, "Optical and transport properties of high quality crystals of V2o4 near the metallic transition temperature," *Solid State Communications*, vol. 7, no. 4, pp. 425–428, 1969.

[21] H.-T. Kim, B.-G. Chae, D.-H. Youn, S.-L. Maeng, G. Kim, K.-Y. Kang, and Y.-S. Lim, "Mechanism and observation of Mott transition in VO2-based two-and three-terminal devices," *New Journal of Physics*, vol. 6, no. 1, p. 52, 2004.

[22] B. Schrauwen, D. Verstraeten, and J. Van Campenhout, "An overview of reservoir computing: theory, applications and implementations," in *Proceedings of the 15th european symposium on artificial neural networks. p. 471-482 2007*, 2007, pp. 471–482.

[23] H. Jaeger, "The "echo state" approach to analysing and training recurrent neural networks-with an erratum note," *Bonn, Germany: German National Research Center for Information Technology GMD Technical Report*, vol. 148, no. 34, p. 13, 2001.

[24] W. Maass, T. Natschläger, and H. Markram, "Real-time computing without stable states: A new framework for neural computation based on perturbations," *Neural computation*, vol. 14, no. 11, pp. 2531–2560, 2002.

[25] ——, "Computational models for generic cortical microcircuits," *Computational neuroscience: A comprehensive approach*, vol. 18, pp. 575–605, 2004.

[26] W. Maass, P. Joshi, and E. D. Sontag, "Computational aspects of feedback in neural circuits," *PLoS computational biology*, vol. 3, no. 1, p. e165, 2007.

[27] S. Haykin, *Kalman filtering and neural networks*. John Wiley & Sons, 2004, vol. 47.

[28] M. S. Grewal, *Kalman filtering*. Springer, 2011.

[29] H. Jaeger, *Tutorial on training recurrent neural networks, covering BPPT, RTRL, EKF and the" echo state network" approach*. GMD-Forschungszentrum Informationstechnik Bonn, 2002, vol. 5.

[30] ——, "Adaptive nonlinear system identification with echo state networks," in *Advances in neural information processing systems*, 2003, pp. 609–616.

[31] C. Sheng, J. Zhao, H. Leung, and W. Wang, "Extended kalman filter based echo state network for time series prediction using mapreduce framework," in *2013 IEEE 9th International Conference on Mobile Ad-hoc and Sensor Networks*. IEEE, 2013, pp. 175–180.

[32] N. Shukla, A. V. Thathachary, A. Agrawal, H. Paik, A. Aziz, D. G. Schlom, S. K. Gupta, R. Engel-Herbert, and S. Datta, "A steep-slope transistor based on abrupt electronic phase transition," *Nature communications*, vol. 6, p. 7812, 2015.

[33] M. L. Alomar, V. Canals, A. Morro, A. Oliver, and J. L. Rossello, "Stochastic hardware implementation of Liquid State Machines," in *2016 International Joint Conference on Neural Networks (IJCNN)*, Jul. 2016, pp. 1128–1133, iSSN: 2161-4407.

[34] B. Schrauwen, M. D'Haene, D. Verstraeten, and J. V. Campenhout, "Compact hardware liquid state machines on FPGA for real-time speech recognition," *Neural Networks*, vol. 21, no. 2, pp. 511–523, Mar. 2008. [Online]. Available: http://www.sciencedirect.com/science/article/pii/S0893608007002353

[35] M. Jerry, A. Parihar, B. Grisafe, A. Raychowdhury, and S. Datta, "Ultra-low power probabilistic IMT neurons for stochastic sampling machines," in *2017 Symposium on VLSI Technology*, Jun. 2017, pp. T186–T187, iSSN: 2158-9682.

Testing the divergence stack memory on GPGPUs:
A modular in-field test strategy

Josie E. Rodriguez Condia[†], M. Sonza Reorda[‡]

Dip. di Automatica e Informatica, Politecnico di Torino, Torino, Italy

{†josie.rodriguez, ‡matteo.sonzareorda}@polito.it

Abstract[1]—**General Purpose Graphic Processing Units (GPGPUs) are becoming a promising solution in safety-critical applications, e.g., in the automotive domain. In these applications, reliability and functional safety are relevant factors in the selection of devices to build the systems. Nowadays, many challenges are impacting the implementation of high-performance devices, such as GPGPUs. Moreover, there is the need for effective fault detection solutions to guarantee the correct in-field operation of a GPGPU, such as in the branch management unit, which is one of the most critical modules in this parallel architecture. Faults affecting this structure can heavily corrupt or even collapse the execution of an application on the GPGPU. In this work, we propose a non-invasive Software-Based Self-Test (SBST) solution to detect faults affecting the memory in the branch management unit of a GPGPU. We propose a scalar and modular mechanism to develop the test program as a combination of software functions. The FlexGripPlus model was employed to evaluate the proposed strategies experimentally. Results show that the proposed strategies are effective to test the target structure and detect up to 98% of permanent faults.**

Keywords—**General Purpose Graphics Processing Units (GPGPUs) Software-Based Self-Test (SBST), Divergence Stack Memory**

I. INTRODUCTION

Currently, General Purpose Graphic Processing Units (GPGPUs) represent effective solutions for data-intensive applications, such as multimedia, multi-signal analysis, and high-performance computing (HPC). Moreover, these devices are also increasingly considered for safety-critical applications with substantial requirements in terms of reliability and functional safety. In the automotive field, safety-critical applications (such as sensor-fusion systems and Advanced Driver Assistance Systems (ADAS)[1]) usually require real-time execution and high reliability. For this purpose, GPGPUs are implemented in cutting-edge technologies to maximize performance and reduce power consumption. Nevertheless, some studies [2, 3] proved that the latest transistor technologies are prone to be affected by faults during the operative life of the device. The most critical challenges arise when permanent faults affect a module (caused by wear-out or aging [4]), so altering the functionality and the reliability of the device.

A parallel processor, such as a GPGPU, is particularly efficient when executing embarrassingly parallel programs. Nevertheless, real applications are far from this behavior, and most of them are composed of non-easily-parallelizable algorithms. Thus, these applications usually contain Intra-Warp Divergence (IWD), which is produced when a group of threads

(warp) has different execution paths with different instructions. In [5], the authors report that sample applications in the CUDA Software Development Kit (SDK), including IWD, use approximately 33% of the total execution time in processing these conditions. Similarly, in [6], the authors profile a divergence map of typical programs and workloads in GPGPUs. Results show that most applications might produce thousands or millions of divergence conditions during the operation of the applications.

The architecture of a GPGPU includes a particular module to manage the IWD. This specific module is often called *Divergence Management Unit* (DMU), Branch Divergence Controller, Branch Controller, or Divergence Controller. DMU is devoted to controlling the operation of multiple paths in the same group of threads. Internally, the DMU evaluates control-flow instructions and uses a stack memory to store relevant information concerning the execution paths. Most DMUs can manage divergences composed of two paths. However, other locations in the stack memory can be employed to manage more than two divergence paths. Thus, the DMU is crucial for the correct operation of an application in the GPGPU, and a fault affecting this unit can propagate through the modules and collapse the entire operation of the device and the executed application.

Currently, test engineers propose some structures and mechanisms targeting the in-field test of a GPGPU and face the new reliability and technology challenges. Current solutions are sometimes based on the addition of *Design for testability* (DfT) structures, such as *Memory Built-in Self-Test* (MBIST)[7], which can be activated during idle times of the operation and stimulate/observe the internal modules of a device detecting possible faults [8]. Other solutions are based on designing special software programs to test a target functionally. These non-invasive and flexible mechanisms are also called *Software-Based Self-Test* (SBST). The last approach is based on hybrid mechanisms, combining hardware structures and software (i.e., custom instructions) to detect [9] or mitigate [10] faults of internal modules in a device. Both solutions (DfT and hybrid) are costly when targeting small modules in a GPGPU and should be developed and included in a design before the production phase.

Some previous works demonstrated that SBST solutions [11] could be successfully integrated into safety-critical applications, such as the automotive ones [12]. Most previous works applied to GPGPUs proposed SBST solutions for some data-path modules [13], including the execution units [14, 15], the register files [16], the pipeline registers [17] and some embedded memories [18]. Moreover, solutions targeting other critical modules in the control-path have been proposed (the warp scheduler [19] and their internal memories [20, 21]). However, to the best of our knowledge, practical solutions to test the DMU using the SBST mechanism are still missing.

---

[1] This work has been partially supported by the European Commission through the Horizon 2020 RESCUE-ETN project under grant 722325.

978-1-7281-5410-7/20 $31.00 © 2020 IEEE

In the present work, we propose, for the first time, a functional test strategy based on an SBST approach targeting permanent faults in the DMU located inside a GPGPU. The proposed SBST strategy has been implemented and evaluated, resorting to the FlexGripPlus model, which is a simplified open-source version of the NVIDIA GPU architecture.

This work is organized as follows. Section II introduces a basic overview of the architecture of an NVIDIA GPGPU, as with the special emphasis on the one used to validate the proposed strategy. Section III describes the SBST strategy proposed to test a permanent fault. Section IV reports the implementation details. Section V reports the experimental results, and Section VI draws some conclusions and outlines future works.

II. THE FLEXGRIPPLUS GPGPU MODEL

FlexGripPlus is a GPGPU model fully described in VHDL [22]. It is an improved version of the original FlexGrip model developed by the University of Massachusetts FPGA [23]. This model implements the Nvidia G80 micro-architecture, and it is also compatible with the commercial programming environment (CUDA under the SM_1.0 compatibility). FlexGripPlus supports 28 instructions of either 32 or 64 bits in 64 formats. The number of parallel execution units is configurable among 8, 16, and 32.

The architecture of an NVIDIA G80 GPU (and of FlexGripPlus) is based on the SIMT (Single-Instruction Multiple-Thread) paradigm. It exploits a custom Streaming Multiprocessor (SM) core with five stages of pipeline (*Fetch, Decode, Read, Execution/Control-flow* and *Write-back*), as shown in Fig. 1. This special-purpose parallel processor executes the same instruction (warp instruction) into a group of multiple threads using the available Execution Units (EUs), or Streaming Processors (SPs), in the SM. One warp is defined as a group of 32 threads. Furthermore, one controller and a warp scheduler controller (WSC) control the submission and execution of the warps into the SPs. In the SIMT paradigm, one warp instruction is fetched, decoded, and distributed to be processed on an independent SP. The *Read* and *Write-back* stages load and store data operands from/to Register Files (RFs), shared, global, or constant memories.

The *Execution/Control-flow* stage includes one DMU, which controls and traces the IWD, which is manifested in a program when the threads of the same warp execute different instructions, so generating multiple execution paths. The DMU can handle two paths in the same level of divergence and up to *n* levels of nested divergence, where *n* represents the number of threads in a warp. The DMU is located in parallel to the *Execution/Control-flow* pipeline stage and can support inter-warp branching at the hardware level. This module manages the control-flow operations to start or retake the flow from conditional branches with multiple paths. This unit also supports up to *n-1* levels of nesting branching.

Inside the DMU, one special purpose memory stores the starting (divergence point) and ending points (convergence point) of the divergence paths of a warp. This information is stored as a stack and processed by the DMU when executing control-flow instructions. More in detail, the memory is organized as a set of 32 Line-Entries (LEs) using the format presented in Fig. 2. The number of LEs directly related to the threads in a warp and the maximum nesting divergence per warp. A divergence point is a location in a parallel program where two paths (*Taken* and *Not-Taken*) are generated to be executed by a warp. Similarly, a convergence point is a location in a program where the IWD paths finish, and the program retakes one execution path.

Each LE in the stack is composed of three fields. Those fields are the thread mask (TM), the flow ID, and the program counter

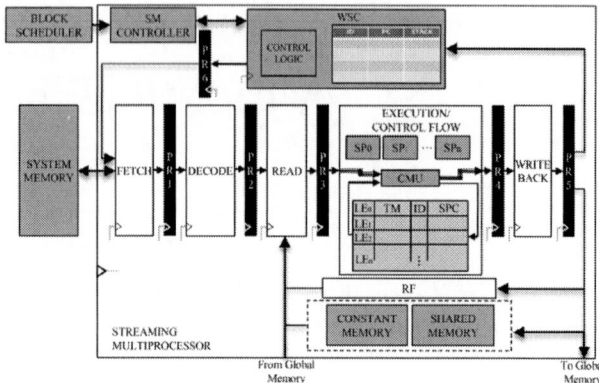

Fig. 1. A general scheme of the SM architecture in FlexGrip and the control-flow structure

TM		Flow ID	SPC	
65		34 33	32 31	0

Fig. 2. Organization format of one LE in the stack memory of the GPGPU

of the actual warp under execution (SPC). The TM stores the status of the active threads in a warp. An active logic state represents the number of active threads executing a path (Taken or Not-Taken). The flow ID represents the actual state of the execution of an IWD. This value can be "01" (for a branch condition) or "00" (for a synchronization point or embarrassingly parallel condition). The SPC can store the starting address of the paths or the synchronization point address after paths execution.

The CMU employs two LEs to manage the operations when IWD is produced. The first LE stores the synchronization point (also known as convergence point) in the SPC field and the number of active threads at the moment of starting the divergence in the TM field. The second LE stores the starting address for the not-taken path in SPC and the threads to execute this path in TM. It is worth noting that when nesting divergence is produced, the CMU uses a new set of LEs to store the status information for the further divergence.

The execution of a synchronization instruction (*SSY*) affects the address pointer in the memory, and it is moved to the next LE. When the program reaches the convergence point, the pointer returns to the previously addressed LE. During the execution, the first LE is used only for storing purposes. In contrast, the second LE is employed during the management of the divergence, and control-flow instructions can affect this LE with writing or reading operations. In this way, when the operation of the first path ends, the information in the second LE is employed to start the not-taken path until the convergence point is reached.

Two main operations can be employed to manage the LEs in the stack memory. Initially, a new LE is addressed with *SSY*, during divergence or nesting divergence generation. The return from an addressed LE to the previous one can be performed using exit control-flow instructions, such as (*NOP.S*).

III. PROPOSED METHOD TO TEST THE DIVERGENCE STACK

The proposed approach employs the functionality of the CMU and the stack memory to generate self-test routines for each LE in the memory. These self-test routines are compacted into modular functions to propose scalable and modular test solutions. The modularity provides the required controllability and observability features to inject test patterns and propagate the fault effect, respectively.

A. Controllability

The injection of test patterns is performed by forcing the execution of divergence paths targeting each one of the threads in a warp, so a sequence of divergence paths is generated to detect permanent faults in the TM bit field of each LE. As an effect of the divergence procedures, the TM field stores the threads following the not-taken path. Moreover, the SPC field stores the starting address of the not-taken path.

We propose two possible methods to control the address pointer of the LEs and inject test patterns in both fields (TM and SPC) of the LEs.

The first method (*Nesting*), see Fig. 3 (Top), can generate test patterns by using a sequence of nesting IWD routines, where the generation of each divergence produces as an effect, the movement of the address pointer to a deeper LE. The divergence is produced by comparing the *Thread.id* of each thread in a warp with a constant value. The main idea is to generate an ordered number of comparisons and follow a specific path, so causing the required test pattern in the TM field of each LE.

On each comparison, one or a group of threads is disabled, so defining a pattern to be stored into the deeper LE and generating two execution paths. This method is useful in managing the addressing of the LEs and injecting patterns into the TM field. Moreover, it can be described in the CUDA programming environment without significant modifications. The routines on each path (*Taken* and *Not-taken*) expose the presence of a permanent fault in the TM. The previous process is repeated for half the number of threads in a warp. Once, *Taken* routine finishes, the DMU submits the *Not-taken* path routine for processing purposes.

As it can be observed in the scheme in Fig. 3, the main idea is to always execute the routine in the *Taken* path, which generates new divergence paths and forces the test of other levels of LEs. The fault detection can be explained considering that once a divergence is generated, two LEs store the synchronization point and the address to start the not-taken path (as test patterns). Thus, a fault can be detected when retrieving the stored values, or when the number of threads executing a path is different from the expected one, so making the fault effects visible. The total number of possible nesting divergences is 16 when targeting any architecture with a warp composed of 32 threads. Nevertheless, this mechanism is scalable to more threads in a warp.

The *Nesting* strategy can inject test patterns on the even LEs of the stack memory. However, the odd ones are missing. The generation of test patterns for these fields requires the explicit addition of one synchronization instruction SSY before start the comparisons causing the divergence. The effect of the SSY instructions is the movement of the address pointer to the next or deeper LE in the stack memory. Then, the same previous procedure can be applied again, so testing the odd LEs. The comparison values are loaded using immediate instructions.

Nevertheless, the main issue of this strategy is the disabling state of the threads. When a thread is disabled, this cannot be turned active again until the divergence paths are executed, and a convergence point is reached. Thus, it is not possible to test or detect a permanent fault in a deeper LE location if a thread is disabled. This restriction implies that the comparisons should be performed multiple times, targeting different threads in the TM field. Thus, the strategy may suffer from considerable code length and excessive execution times.

The second method called *Sync-Trick*, see Fig. 3 (Center), exploits the functionality of *SSY* instruction to deceive the DMU when testing the stack memory. This method allocates *SSY* operations in strategically selected locations in the test program of the CMU to generate the change in the address pointer of the LEs.

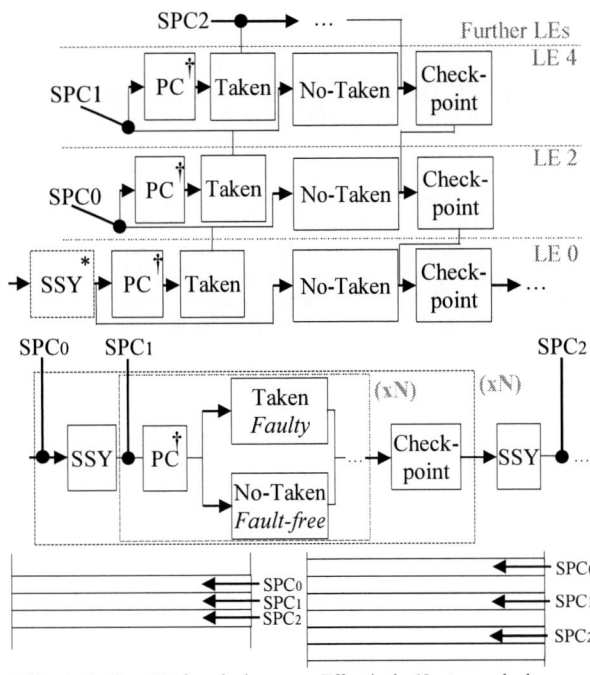

Fig. 3. General schemes of the program-flow of the proposed SBST strategies. *Nesting* method (top), *Sync-Trick* method (center), (bottom) effect in the address pointer of the stack memory of the CMU. (*)Optional function to test the odd LEs. (†) Optional functions to distribute the test functions in the system memory

More in detail, the *SSY* is explicitly located before each sequence of controlled divergence functions to test the TM of a LE. Then, this instruction forces the controller to allocate a new level of LE in the memory without the need to generate the IWD explicitly. The advantage of this mechanism is that each LE can be addressed without the need of disabling specific threads to create new addressing in the memory with the address pointer. Thus, this strategy replaces the generation of nesting divergence by the management of the address pointer in the memory. A sequence of simple IWD operations, generating the *Taken* and *Not-taken* paths, is employed to test the target LE. This process can be repeated N times to use different active threads and memory addresses (as test patterns). Then, a new SSY instruction is generated addressing a deeper LE and restarting the test procedure. It is worth noting that this mechanism is effective to move across one direction and reach deeper LEs in the memory. However, the returning phase (to a previous LE) requires the achievement of the convergence point address, which is initially stored by the *SSY* instruction. This strategy cannot be directly described in the high-level programming environment, and modifications at the assembly level are required.

B. Observability

The fault effect propagation is obtained by using the Signature per Thread (SpT) mechanism [17, 20]. This mechanism uses a set of signatures to map and to propagate the effect of a permanent fault in the stack memory into the global memory. Each SpT is updated, taking advantage of the paths in the controlled divergence routines. Thus, the same mechanism employed to perform the fault injection is used to increase its observability in the structure under test. Each SpT computes and accumulates intermediate results for each verified LE. The SpTs are finally grouped and stored in global memory for later analyses.

C. Modular test strategies

The observability and controllability methods for the TM field are complemented with some check-pointing routines, which are

978-1-7281-5410-7/20 $31.00 © 2020 IEEE

devoted to testing the SPC field. These routines are located after the convergence point. In this way, any permanent fault in the SPC is detected if the synchronization point or the starting addresses of the *Not-Taken* path present any permanent fault. The check-point routines verify, through a check-point signature, the correct flow execution of a program. In the check-point procedure, this routine compares an expected check signature value in the program with the actual value accumulated during the execution of the test program. When the comparison matches, the accumulated signature is updated, otherwise the test program finishes propagating in memory the error in the SPC field of the evaluated LE. This strategy can be applied to any of the two controllability methods (*Nesting* or *Sync-Trick*).

The check signature values are predefined before execution and are loaded through immediate instructions. A fault in the SPC field would generate an unexpected addressing in the system memory. The permanent fault is detected by changes in the execution time or the signatures stored in the global memory.

The SPC field is partially tested. This issue is mainly caused by the short length of the test program in both strategies. A control-flow routine (PC), see Fig. 3, can be included before or in one of the paths of each IWD to test the high bits in the SPC field. Moreover, the test routines are redistributed across the system memory, so generating the missing test patterns.

Some GPU instructions (in the format of 32 and 64 bits) are located before each relocated function in the memory. These instructions avoid hanging conditions by permanent faults in the SPC field. In this way, when the program counter is affected by a fault, and it jumps to any unexpected memory location, it is always possible to retake control of the program and finish the execution of the GPGPU. Nevertheless, it is expected degradation in performance by the effect of the permanent fault.

Figure 3 (Top and Center) presents the basic schemes of the operational flow for the *Nesting* and *Sync-Trick* mechanisms. In both schemes, the execution of the synchronization instruction (*SSY*) provokes a change in the addressed LE by the pointer in the stack memory. The address pointers SPC_0, SPC_1, and SPC_2 represent the effect in the stack memory when executing the functions on each method. In the *Nesting* scheme, the divergence instructions and the implicit *SSY* instructions are represented in the division of the paths.

The use of these additional functions (Check-Point and PC) is entirely optional, considering that these strategies are costly in terms of memory overhead for an in-field execution. It is worth noting that the proposed technique takes into account the operational restrictions to develop the test programs using the Stuck-at fault model. Other fault models would require the adaptation of the *Sync-trick* mechanism. However, it would be hard or impossible to follow the *Nesting* strategy.

IV. IMPLEMENTATION

Following the schemes in Fig. 3, we implemented the SBST strategy using the high-level programming environment (when possible) and combined with instructions in the assembly language (SASS) of the GPGPU. Blocks and dotted lines in Fig. 3 represents the division in the description of the SBST code as a set of functions for both methods.

The implemented code for both test methods is composed of the following functions: *i)* Initialization function, *ii)* synchronization function (*SSY*), *iii)* flow control function (*PC*), *iv)* the test pattern injection and SpT update function (*Taken* and *No-Taken*), and *v)* the check-point function (*Check-point*).

Each function is described independently and can be attached depending on the target of a test program. The initialization function defines and initializes the registers for each thread. Moreover, this function initializes the addresses to store the SpTs and check-point signatures. The functionality of other functions was introduced in the previous section.

The modular description of both SBST strategies has the advantage of allowing the fast development of multiple test programs with different test objectives. In the *Nesting* method, the modularity is used to manage the nesting divergence functions and to add or remove the optional functions targeting the test of the SPC field. In contrast, the modularity presents considerable advantages for the *Sync-Trick* method. The code description of this method is scalable and modular in such a way that it is possible to add or remove part of the description to target the individual test of LEs in the stack memory.

This modularity gives us the possibility to address any or a group of LEs in the stack memory and to generate an independent test program. The division of the test contributes to reducing the execution time of the test program during the in-field operation of a GPGPU.

The *Sync-Trick* method can employ two approaches to evaluate LEs in memory. The first approach (*Accumulative or Acc*) aims the test of a consecutive group of LEs and accumulates the signatures in memory. This approach must always start from the first LE and can finish at any of the other 31 LEs in the stack.

On the other hand, the second approach (*Individual or Ind*) targets the testing of an individual LE and then the retrieving of signature results to the host. This approach only focuses on one of the LEs in the memory and is intended to have a reduced execution time. The performance cost (execution time) of both approaches (*Acc* and *Ind*) can be calculated using the equations (1) and (2).

$$ST(Acc) = T \cdot n + Ch \cdot n + SSY \cdot (n-1) \qquad (1)$$

$$ST(Ind) = T + Ch + SSY \cdot (n-1) \qquad (2)$$

Where n represents the target LE in the stack memory. SSY, T, and Ch represent the execution time of the synchronization, test pattern injection, and check-point functions, respectively. The initialization function was not included considering that it is constant for both cases, and it is negligible in terms of duration.

From equations (1) and (2), it is clear that the cost of the Accumulative version (Acc) is higher than the *Ind* version. The cost is mainly caused due to the different approaches in each case. In the *Acc* version, the program is intended to test the number of selected LEs sequentially. In contrast, the *Ind* approach targets the test on one LE, so the test pattern and check-point functions are used once. The number of synchronization functions depends on the target level of LE in the stack memory.

On the other hand, the performance cost of the *Nesting* method is described by the expression in equation (3).

$$Ns = N \cdot Ch \cdot \sum_{i=0v1}^{m} SSY + T \qquad (3)$$

Where N represents the total number of threads in a warp, and m is the target LE to be tested. CH, SSY, and T have the same meaning from equations (1) and (2). As introduced previously, the target LE could be even or odd. Thus, the starting value of i in the summation could be 0 or 1.

Table 1 reports the results of the performance parameters for the *Nesting* method, and the *Sync-Trick* method under the *accumulative* and *individual* approaches. It is worth noting that results reported in Table 1 were obtained by simulations in the ModelSim environment using the FlexGripPlus model.

The reported results show the performance parameters for the two possible methods that can be used to test the LEs in the stack memory of the CMU. All versions present an overhead in the global memory of 64 locations (256 bytes) devoted to saving the SpTs and the Check-point signatures.

TABLE 1. PERFORMANCE PARAMETERS OF THE SBST PROGRAMS USING THE TWO APPROACHES TO DETECT PERMANENT FAULTS IN THE LES

Approach	LE	Instructions	Execution time (Clock cycles)	System memory overhead (Bytes)
Sync-Trick Ind	1	403	33,449	1,612
	2	404	34,211	1,616
	10	412	34,589	1,648
Sync-Trick Acc	1 - 2	794	66,637	3,176
	1-10	3,922	326,423	15,688
	All	12,524	1,030,473	50,096
Nesting	1	683	37,986	2,732
	1-2	1,323	83,569	5,292
	1-10	6,443	528,086	25,772
	All	19,883	2,567,209	79,532

Regarding the performance results of both versions, it can be noted that the *Sync-Trick* (*Ind*) approach maintains an average performance cost to test any LE in the stack memory. The only difference among these programs is the number of SSY instructions included to address a selected LE. Similarly, the *Sync-Trick* (*Acc*) version can test a group of LEs consecutively. However, it requires additional execution time and cannot be stopped once the test program starts.

On the other hand, Table 1 reports the required execution time to test the first and the second LEs in the stack using the *Sync-Trick Ind* (rows 2 and 3, column 4) and *Sync-Trick Acc* (row 5, column 4) approaches. The *Individual* approach requires 76 additional clock cycles to test the LEs, but it has the advantage of test each LE independently. In contrast, the Accumulative method must check both LEs consecutively. Thus, the *Ind* approach can be adapted for in-field operation by the limited number of clock cycles required during the execution.

The performance parameters show that the *Nesting* approach has a proportional relation among the number of instructions and the number of LEs to test. Similarly, the relationship between the execution time and the number of LEs to test present an increasing exponential ratio. In the end, the *Nesting* method requires more than double the execution time to test the entire stack than the *Sync-Trick* using the *Acc* approach. The execution time could be the relevant parameters to take into account when targeting the in-field operation.

V. EXPERIMENTAL RESULTS

The RT-level description of the FlexGripPlus model was employed in the experiments. The fault injector environment follows the methodology described in [17], and we injected permanent faults following the Stuck-at-Fault model. A total of 4,224 permanent faults were injected in the stack memory of the CMU of the FlexGripPlus model for each fault campaign.

The fault simulation campaign was performed using both representative benchmarks and the proposed SBST strategy. These representative benchmarks employ the CMU unit and are carefully selected to compare the detection capabilities they can achieve with the one provided by the proposed SBST approach. Descriptions and details regarding the chosen benchmarks can be found in [17, 24]. For the sake of completeness and comparison, the different versions of the SBST strategy are reported in Table 2. Moreover, both approaches were evaluated with and without the optional SPC functions in the LEs.

The last column of Table 2 reports the testable FC of the benchmarks and the proposed SBST strategy. During the analysis of the memory in the stack, a total of 192 faults were identified as untestable. These are related to the lowest bits of the SPC field of each LE, which does not affect the execution of an instruction. Thus, these faults were removed when computing the FC.

TABLE 2. FC RESULTS FOR THE REPRESENTATIVE BENCHMARKS AND THE PROPOSED SBST STRATEGY

SBST strategy or benchmark		FC (%)				
		SDC	Hang	Timeout	Total	Total testable
MxM		0.0	0.38	0.0	0.38	0.40
Sort		0.15	0.04	0.0	0.19	0.19
FFT		0.14	0.19	0.0	0.33	0.35
Edge		0.15	0.28	0.0	0.43	0.47
Sync-Trick	Ind	65.64	2.08	1.01	68.75	72.02
	Acc	64.89	2.84	1.01	68.75	72.02
	Ind + PC	83.00	8.49	2.44	93.93	98.41
	Acc + PC	82.24	9.25	2.44	93.93	98.41
Nesting		54.12	11.81	1.23	67.16	70.04
	+ PC	76.94	13.16	2.81	92.91	97.34

The *Sync-Trick* strategy presents a moderate FC for both cases (*Ind* and *Acc*). Moreover, the FC increases when adding the SP functions and the relocation of the test functions in the memory. These comprehensive approaches (*Ind+SP* and *Acc+SP*) obtain a high percentage of FC for the target structure.

An in-depth analysis of the results shows that the Individual approach allows detecting 100% of the faults in the TM of all LEs by looking at the results produced by the test procedure (Silent Data Corruption, or SDCs). In contrast, the *Acc* version makes a small percentage (0.75%) of faults in the TM field visible because they hang the GPGPU. This behavior can be explained considering that in the *Ind* approach, each LE is evaluated individually, and so all detections can be labeled as SDC. On the other hand, for the *Acc* method, a permanent fault in one LE affects the synchronization point, thus corrupting the convergence point and causing the Hang condition. More in detail, a Stuck-at-0 fault is a sensitive case during the run of the test program. A fault affecting one LE when used as synchronization causes the Hang condition.

The *Nesting* approach has a marginally lower FC than *Sync-Trick* with an increment in more than double the percentage of faults causing Hanging and Timeout. This fault effect is equivalent to the effect presented in the *Acc* version of *Sync-Trick*. In this case, the *Nesting* method generates IWD to move the address pointer among the LEs in the stack memory, testing all LEs even if a fault is detected, so the next LEs are also evaluated. The continuous evaluation generates issues when a fault affects the LE used for synchronization purposes. Thus, the test program may lose the convergence point and produces the Hang or Timeout condition. According to results, the *Nesting* strategy seems to be more susceptible to Hang and Timeout effect than the *Sync-Trick* using the *Acc* approach.

In both approaches, the addition of the relocation in memory and the SPC functions increases the testable coverage in the stack memory. However, as explained previously, these optional functions can be employed when it is possible to use the entire system memory to relocate the test functions in specific memory locations, or the application code allows this adaptation. Similarly, both SBST approaches can detect a considerable percentage of the permanent faults in the stack memory. However, a direct comparison involving the performance parameters from Table 1 shows that the *Nesting* approach consumes more than double the execution time and 37% of additional instructions. In conclusion, the *Sync-Trick* strategy seems to be a feasible candidate for in-field operations. Moreover, the *Ind* strategy can be divided into parts and adapted with the application code.

A comparison of the FC obtained by the proposed SBST strategies and the representative benchmarks shows that the FC using these specialized programs is higher and effective for this module than the FC obtained with typical applications. Thus, the FC capabilities of the representative benchmarks are lower and can be considered as almost negligible. This behavior can be explained, considering that most applications only use part of the CMU and the stack memory to manage the IWD. The main issue is that most applications employ only the first levels of stack memory to handle the divergence.

The *matrix multiplication* application generates one level of divergence. Thus, other levels inside the Stack memory are not employed, and the fault effect in not detected or propagated into the application. Similarly, the *Sort* application can generate IWD depending on the input data operands, but it remains limited to the first LE in the stack memory. However, the percentage of detection of 0.33% and 0.19% are negligible in comparison with the proposed test strategies.

The *FFT* benchmark produces two levels of IWD. This behavior slightly increases the percentage of faults detected. Nevertheless, the percentage is small. Finally, the *Edge* application causes two levels of IWD and can detect some faults as SDC and hanging conditions. However, the total coverage of all representative kernels is minimal.

The previous scenario supports the idea that executing applications and checking their results (as it is often done when using a functional test approach) is definitely not enough to verify the functionality of a crucial module in the GPGPU. Thus, special test programs are required to guarantee the correct operation of a module inside a device used in a safety-critical application.

The main advantage of the proposed method to test the stack memory is the modularity and scalability of the SBST strategy. This scalability allows the configuration and the selection of the number of LEs to be tested. Moreover, the test program can be divided into multiple parts when using the Sync-Trick approach.

It is worth noting that the implementation of the test programs required the combination of high-level descriptions ($\approx$20% of the code), when possible, and the addition of assembly functions ($\approx$80%). For all proposed approaches, the synchronization functions (*SSY*) were described in assembly language, considering that these instructions are not possible to specify at CUDA or PTx levels. Moreover, the optimizations of the compiler also remove part of the descriptions. Thus, these parts are rebuilt at the assembly level. These limitations show that the development of test programs for these complex structures in GPGPUs requires access to the assembly language of the micro-architecture to provide efficient solutions. The implementation effort could be reduced by the design of an automatic tool to include the subroutines at the assembly or binary level. Moreover, such a tool could also be employed to target other modules in the GPGPU.

Although the proposed SBST strategies targeted the test of the stack memory in the CMU of a GPGPU with the G80 micro-architecture, we still claim that the strategy can be adapted to other architectures of GPGPUs.

VI. CONCLUSIONS

We introduced and evaluated two functional test strategy (named *Sync-Trick* and *Nesting*) based on the Software-Based Self-Test (SBST) approach aimed for the in-field test permanent faults in the stack memory of the divergence management unit of a GPGPU. The experimental results show that the proposed strategies are effective in detecting up to 98% of the faults in the target structure.

Both test approaches were designed using a modular and scalable mechanism, so a set of parametric functions were created and then combined to test the target structure using different strategies (*Sync-Trick* and *Nesting*). Moreover, the modularity of the solution allows the division of the test program into parts keeping the FC, so adjusting to potential requirements of in-field operation.

As future works, we plan to evaluate the fault coverage in the divergence controller at RT level and gate-level and propose functional test approaches for other critical modules in a GPGPU.

REFERENCES

[1] W. Shi, M. B. Alawieh, X. Li, and H. Yu, "Algorithm and hardware implementation for visual perception system in autonomous vehicle: A survey," *Integration*, vol. 59, pp. 148-156, 2017.

[2] S. Hamdioui, *et al*, "Reliability challenges of real-time systems in forthcoming technology nodes," in *2013 Design, Automation & Test in Europe Conference & Exhibition (DATE)*, 2013.

[3] I. Agbo, *et al.*, "Read path degradation analysis in SRAM," in *Test Symposium (ETS), 2016 21th IEEE European*, 2016.

[4] X. Chen, Y. Wang, Y. Liang, Y. Xie, and H. Yang, "Run-time technique for simultaneous aging and power optimization in GPGPUs," in *2014 51st ACM/EDAC/IEEE Design Automation Conference (DAC)*, 2014.

[5] S. S. Baghsorkhi, M. Delahaye, S. J. Patel, W. D. Gropp, and W.-m. W. Hwu, "An adaptive performance modeling tool for GPU architectures," in *15th ACM SIGPLAN Symposium on Principles and Practice of Parallel Programming*, 2010.

[6] B. Coutinho, D. Sampaio, F. M. Pereira, and W. Meira Jr, "Profiling divergences in gpu applications," *Concurrency and Computation: Practice and Experience*, vol. 25, pp. 775-789, 2013.

[7] A. J. Becker, C. A. S. Pathirane, and R. C. Aitken, "Memory built-in self-test for a data processing apparatus," *UK Patent* US 9,449,717 B2, 2016.

[8] R. Gulati, *et al.*, "Self-test during idle cycles for shader core of GPU," *US Patent* US 10,628,274, 2020.

[9] J. E. R. Condia, P. Narducci, M. Sonza Reorda, and L. Sterpone, "A dynamic hardware redundancy mechanism for the in-field fault detection in cores of GPGPUs," in *2020 23rd International Symposium on Design and Diagnostics of Electronic Circuits & Systems (DDECS)*, 2020.

[10] J. E. R. Condia, P. Narducci, M. Sonza Reorda, and L. Sterpone, "A dynamic reconfiguration mechanism to increase the reliability of GPGPUs," in *VTS 2020: VLSI Test Symposium*, 2020.

[11] M. Psarakis, D. Gizopoulos, E. Sanchez, and M. Sonza Reorda, "Microprocessor software-based self-testing," *IEEE Design & Test of Computers*, vol. 27, pp. 4-19, 2010.

[12] P. Bernardi, M. Grosso, E. Sanchez, and O. Ballan, "Fault grading of software-based self-test procedures for dependable automotive applications," in *2011 Design, Automation & Test in Europe*, 2011.

[13] M. Abdel-Majeed and W. Dweik, "Low overhead online periodic testing for GPGPUs," *Integration*, vol. 62, pp. 362-370, 2018.

[14] S. Di Carlo, G. Gambardella, M. Indaco, I. Martella, P. Prinetto, D. Rolfo, *et al.*, "A software-based self test of CUDA Fermi GPUs," in *2013 18th IEEE European Test Symposium (ETS)*, 2013.

[15] D. Defour and E. Petit, "A software scheduling solution to avoid corrupted units on GPUs," *Journal of Parallel and Distributed Computing*, vol. 90, pp. 1-8, 2016.

[16] D. Sabena, M. Sonza Reorda, L. Sterpone, P. Rech, and L. Carro, "On the evaluation of soft-errors detection techniques for GPGPUs," in *2013 8th IEEE Design and Test Symposium*, 2013.

[17] J. E. R. Condia and R. Sonza Reorda, "Testing permanent faults in pipeline registers of GPGPUs: A multi-kernel approach," in *2019 25th IEEE International Symposium on On-Line Testing and Robust System Design (IOLTS)*, 2019.

[18] J. E. R. Condia and M. Sonza Reorda, "On the testing of special memories in GPGPUs," in *2020 26th IEEE International Symposium on On-Line Testing and Robust System Design (IOLTS)*, 2020.

[19] S. Di Carlo, J. E. R. Condia, and M. Sonza Reorda, "An On-Line Testing Technique for the Scheduler Memory of a GPGPU," *IEEE Access*, vol. 8, pp. 16893-16912, 2020.

[20] B. Du, J. E. R. Condia, M. Sonza Reorda, and L. Sterpone, "About the functional test of the GPGPU scheduler," in *2018 IEEE 24th International On-Line Testing Symposium (IOLTS)*, 2018.

[21] S. Di Carlo, J. E. R. Condia, and M. Sonza Reorda, "On the in-field test of the GPGPU scheduler memory," in *22nd IEEE International Symposium on Design and Diagnostics of Electronic Circuits and Systems (DDECS 2019)*, 2019.

[22] J. E. R. Condia, B. Du, M. Sonza Reorda, and L. Sterpone, "FlexGripPlus: An improved GPGPU model to support reliability analysis," *Microelectronics Reliability*, vol. 109, p. 113660, 2020/06/01/ 2020.

[23] K. Andryc, M. Merchant, and R. Tessier, "FlexGrip: A soft GPGPU for FPGAs," in *2013 International Conference on Field-Programmable Technology (FPT)*, 2013.

[24] B. Du, J. E. R. Condia, and M. Sonza Reorda, "An extended model to support detailed GPGPU reliability analysis," in *14th IEEE International Conference on Design & Technology of Integrated Systems in Nanoscale Era (DTIS)*, 2019.

A Model Study of Multilevel Signaling for High-Speed Chiplet-to-Chiplet Communication in 2.5D Integration

Rakshith Saligram*, Ankit Kaul, Muhannad S Bakir, Arijit Raychowdhury

School of Electrical and Computer Engineering
Georgia Institute of Technology
Atlanta USA

Abstract— The quest for high yield has motivated significant advancement in 2.5D integrated circuits, where chiplets are integrated on a silicon interposer or a package substrate with high-speed parallel communication among them. These channels for 2.5D integrated systems need to have high data bandwidth per unit length (also called shoreline-BW-density and measured in Gb/s/mm) and lower energy per bit area (measured in pJ/b). Typically, NRZ signalling is used but achieving higher data rates continues to be a major challenge. In this paper we explore PAM4 as an alternative to NRZ for signalling the channels. Simulations show that we can achieve up to 63% more energy-efficiency and 27% higher BW density for 2.5D integrated systems.

Keywords— *Heterogeneous Integration, Coplanar Microstrip, NRZ, PAM4, Channel Operating Margin*

I. INTRODUCTION

Heterogeneous 2.5D integration is one of the promising technologies that can increase system yield as well as alleviate the memory bandwidth bottlenecks for microprocessors. Several 2.5D and 3D chip stacking technologies have been recently demonstrated by both the foundries as well as academic groups, including silicon interposers, Embedded Multi-Interconnect Bridge (EMIB), Foveros, CoWoS, HIST. These technologies continue to evolve and address the challenges associated with the slowing down of Moore's Law.

The number of chiplet to chiplet interconnects per unit area that can be achieved, depends on the technology which governs the interconnect pitch. From a signalling perspective, as the pitch decreases, more of the interconnects (also known as physical IO) get packed in much smaller area leading to higher shoreline-BW-density albeit with increased crosstalk and interference. In this article, we explore the design space of fine-pitch interconnects and evaluate multiple signalling schemes for their applicability in 2.5D systems.

In this paper we first model the channel that characterizes the chiplet to chiplet connections in such integrated systems. One such prototypical system is shown in Fig. 1. We introduce the scattering parameters of the channel for different pitches and channel lengths and systematically study two signalling schemes. The highest frequency of operation for each pitch/length configuration is determined. Further, simple transceiver models are used to estimate the energy-efficiency of transmission to quantify energy/bit. The rest of the paper is

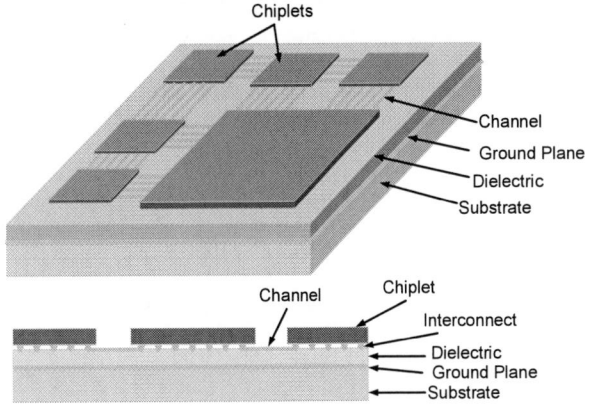

Fig. 1.   General model of Heterogeneous Integrated System

organized as follows. Section II briefs the previous works and current state of art for the integration technology. Section III describes the channel model. Section IV elaborates on the transceiver system with its different components. In section V, the channel simulation setup is explained along with the methods and metrics for simulation. Section VI describes the power estimation and explores the entire design space. Conclusion and references follow.

II. PRIOR WORK

Integration of multiple heterogenous dies on a silicon interposer achieves high interconnect density with lower power and better performance. This has motivated the recent advances in 2.5D integration and system-in-package solutions. There are few noteworthy technologies in this area. CoWoS [8] (Chip-on-Wafer-on-Substrate) has been proposed as a packaging solution comprising of ultra-thin interposers of 50μ thickness with through-silicon-vias (TSVs). Here different logic dies can be placed on the interposer connected through micro-bumps with multiple interconnect layers fabricated using TSVs. It shows better coplanarity, physical scalability with reduced TSV diameter and depth. A second generation of COWOS [9] extends the size of Si Interposer using a stitching method which can scale the width and spacing of the metal lines down to 0.4μm and 0.4μm respectively, over a large IO area.

On the other hand, EMIB [11] uses a thin die of Si with multi-layer BEOL interconnects embedded in an organic substrate. The key advantage of EMIB is there is no practical limit to the die size and eliminates the need for processing a TSV layer. The signalling rates of approximately 3Gb/s for 10mm

---

Work is funded by SRC JUMP and is a part of ASCENT Task 2776.037
*Corresponding Author: rakshith.saligram@gatech.edu

978-1-7281-5410-7/20 $31.00 © 2020 IEEE

long EMIB channel and 6-8Gb/s for 5mm long EMIB channel have been demonstrated by Intel. The IO density is ~300 channels/mm. HIST [10], unlike an interconnect bridge used in EMIB, uses a stitch-chip with high density fine pitch wires placed between the substrate and the chiplets. The contact between the stitch chip and the chiplets is established using micro bumps. Unique compressible micro-interconnects are used to reduce the effects of non-planarity in packaging. Another face-to-face die-stacking approach is the Foveros [12]. This uses TSVs to connect the bottom-most die to the package while face-to-face connections are achieved using micro-bumps. Foveros can achieve micro bump pitch of ~50μ.

Signalling in these integration technologies continues to be an area of active research. In general, the availability of high-density IO over a large area, has motivated the use of parallel IO as a high-bandwidth solution, compared to serial IO (SERDES) that has been the workhorse for die to memory interface. Parallel IO reduces the design complexity, by allowing the clock to be forwarded along with the data and individual communication channels work at a few GHz of bandwidth, which is about 10-100X lower than the most advanced SERDES links. This results in higher energy-efficiency for data-transfer while maintaining the target data-rate. To address the signalling need in chiplet to chiplet interconnects, Intel has recently proposed Advanced Interface Bus (AIB) [13] connections, which are either with lithographically printed wires on an interposer or a bridge. The AIB protocol also uses parallel IO instead of conventional SERDES. Running each IO at a much lower speed will simplify the transmitter and receiver circuity. This paper explores the design space of signalling and channel modelling in chiplet to chiplet connections. The main contributions of the paper include (1) a comprehensive EM model of the parallel IO links to quantify the intersymbol interference and the operating margins, and (2) evaluation of the conventional NRZ signaling scheme vis-à-vis PAM4 (multilevel signaling) for achieving higher bandwidth and energy-efficiency. PAM4 has not been successful in SERDES, because of the long wires in SERDES that degrade signal margins and cause high bit-error-rate (BER) at operating voltages. However, we demonstrate that in a chiplet to chiplet interconnects over short distances (which is required in the 2.5D integrated circuits), PAM4 is indeed a competitive signalling scheme.

III. CHANNEL MODELLING

With the inherent advantages of planarity, ease of fabrication and ease of integration 2.5D integrated systems have become attractive. To model the communication channels, we invoke a coplanar microstrip model for the on-chip interconnects. This is a symmetrical structure i.e., the spacing between the microstrips (pitch) is uniform and all the channels are of equal width. The generic structure of a coplanar microstrip is depicted in Fig. 2. It has a dielectric material of height h sandwiched between the common ground plane and the conductor (channel). The substrate discontinuity of the microstrip causes the dominant mode of transmission to be quasi-TEM as a result of which intrinsic impedance Z_0, phase velocity and field variation across the channel become frequency dependent. The general design constraints for microstrip lines [1] which are consistent with the interconnect designs is given by:

$$0.1 \leq \frac{w}{h} \leq 10 \tag{1}$$
$$0.1 \leq \frac{s}{h} \leq 10 \tag{2}$$
$$1 \leq \epsilon_r \leq 18 \tag{3}$$

Here, the thickness of the conductor and the ground plane t is assumed to be 2μm and the width w is 5μm. The microstrip structure though has the aforesaid advantages, comes at a cost of higher radiation due to lower isolation and thus more cross talk. The effect of E-field coupling can be observed in Fig 3 with three cases that show the variation of magnitude of electric field on the victim channel with (a) no aggressors, (b) one aggressor and (c) two aggressors. The magnitude of coupling depends on channel length and pitch. The pitch controls the crosstalk and the channel length dictates the signal attenuation. To study their effects, the pitch is varied from 5μm to 50μm and the channel length is varied from 100μm to 1000μm. Once the model is designed in HFSS, the six-port scattering parameters in the form of touchstone files are produced for further use in the channel simulation.

IV. TRANSCEIVER SYSTEM ARCHITECTURE

A. Bundle Data Clock Forwarded Channels

As seen from the previous section, each of the coplanar microstrip lines act as a channel transmitting data in form of voltages from transmitter to receiver. In order to enhance the area utilization, we propose single ended data transmission as opposed to the differential mode which uses twice the number of links. The negative effects of single ended mode like simultaneous switching and reference offset can be mitigated by adjusting the voltage amplitude of the signal. Also, to minimize the energy per bit, we do not use any equalization at the transmitter or receiver. Another large percent of link power consumption occurs in the clock recovery circuits at the receiver side. Hence, we try to eliminate this by using clock forwarding which essentially allows us to design a fully parallel IO. This is a distinguishing factor in the design of current parallel chiplet to

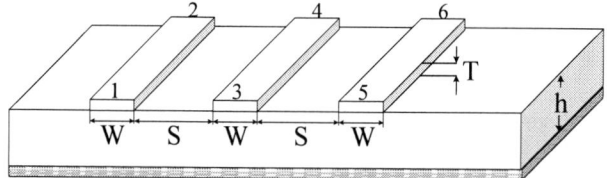

Fig. 2. Coplanar microstrip Channel Model

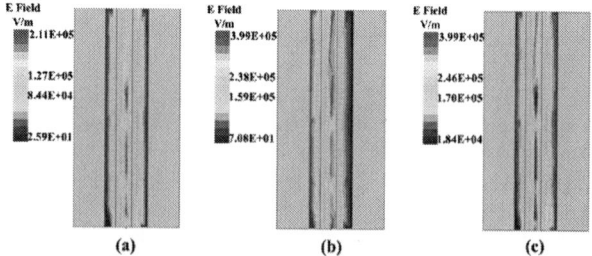

Fig. 3. E field coupling for different Scenario (a) No Aggressor Active, (b) One Aggressor Active (c) Both Aggressor Active

chiplet communication technologies and is simpler to design than traditional SERDES. This is effective because in the target designs the channel lengths are short. Thus, there can be one additional clock signal for a bundle of few data signals (8 or 16) which can be used to forward the reference clock generated on the transmitter to the receiver.

B. Signalling

In this paper, we evaluate two different signalling schemes. The circuit diagram for both are shown in Fig 4(a) and (b).

1. NRZ: The data is represented in form of single 0's and 1's. When transmitting a 0, a voltage level of 0V is sent on the channel and for transmitting a 1, a voltage of V_{dd} is sent.
2. PAM4: Here, two bits of data represent a given voltage value. The input stream of data is grouped into 2-bits and they represent the following voltage values: $00 \rightarrow 0V$, $01 \rightarrow V_{dd}/3$, $10 \rightarrow 2V_{dd}/3$, $11 \rightarrow V_{dd}$. This is the linear encoding of the data bits. There can other forms of encoding like Gray codes which can be used to further reduce bit error rates. The symbol rate in either case is half that of the NRZ or in other words, the data rate in PAM4 is twice that of the symbol rate.

C. Transmitter

1. NRZ: Since we do not use any pre-emphasis or equalization, the transmitter can be as simple as a buffer which will transmit the voltages to the channel. The only design constraint for these buffers is that they must be suitably sized to be able to drive the pad capacitance of the receiver side along with that of the channel itself.
2. PAM4: Here, two bits need to be transmitted as one value of voltage. The input data is passed through a serializer which is then input to a simple 2-bit Digital to Analog Converter (DAC). The DAC will convert it to a mapped voltage and is transmitted on to the channel by a current mode driver.

D. Receiver

1. NRZ: Much like the transmitter, the receiver can be a simple buffer which will detect the voltage on the channel and interpret it as a 0 or a 1. The buffer in this case can be viewed as a high gain voltage comparator which will compare the signal value to the trip-point voltage of the buffer in order to make the decision.
2. PAM4: The four voltage levels on the channel need to decoded back to two bits. Here, we use a simple 2-bit Analog to Digital Converter (ADC) to perform the task. Due to its speed, a flash-ADC is best suitable for the purpose. The Flash-ADC in-turn comprises of three high gain comparators which compare the signal value against the external reference voltage. The ADC output is then encoded to binary.

V. CHANNEL SIMULATION

A. Simulation Setup

The simulation setup is shown in Fig 5. The simulation is performed on the Keysight Advanced Digital Systems (ADS) platform for a typical 28nm technology node. The transmitter consists of a Pseudo Random Bit Sequence generator which is electrically encoded to voltage. Here PRBS-7 is being used. The transmitter has a transmit-resistance (R-TX) typically 50Ω in shunt with the pad capacitance. The typical pad capacitance (C_{pad}) for a 28nm node is around 5pF. For the NRZ signalling, a simple buffer is used as explained in the architecture. For PAM4, we use a simplified IBIS-AMI model along with executables generated from MATLAB SERDES toolkit which can be used in conjunction with the ADS topology. For both NRZ and PAM4, we use a V_{dd} of 1V.

The channel is modelled in the form of a 6-port S-parameter network, which uses the touchstone files generated from the HFSS models. The 6 ports represent three channels, the middle victim and the outer two aggressors which contribute to the crosstalk. On the transmit side, the "XTalk" transmitters are configured to operate at the same data rate as that of the "TX" but produces an out-of-phase signal. This emulates the worst-case cross talk scenario. On the receiving side, we first encounter the pad capacitance which is again in shunt. Most conventional channels use a termination resistance. This causes the signals to attenuate which will impact the power. Thus, to reduce the power consumption, short links typically eliminate

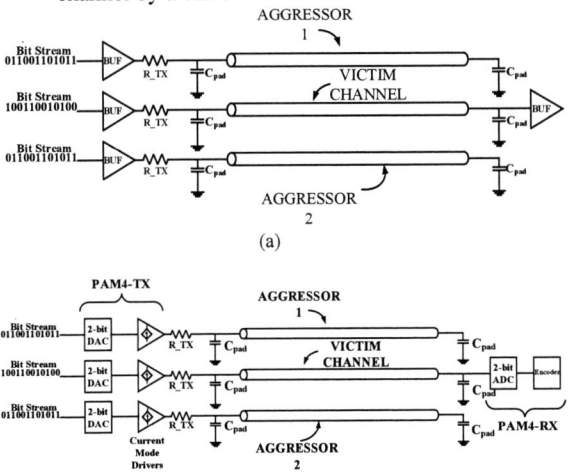

(a)

(b)

Fig. 4. Circuit of (a) NRZ Signaling (b) PAM4 Signaling

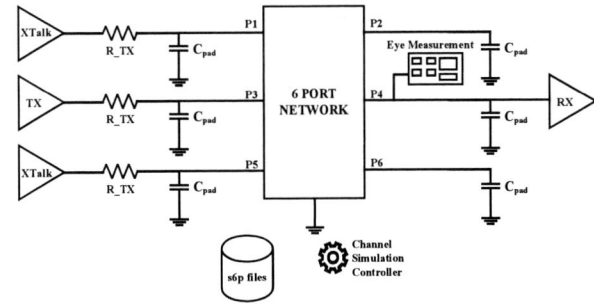

Fig. 5. Channel Simulation Setup

978-1-7281-5410-7/20 $31.00 © 2020 IEEE

the legacy termination resistance as a result of which the receiver is a capacitive load. This will cause signal reflections to the transmitter in turn affecting the quality of signal and increasing inter symbol interference (ISI). Since the channels are considerably small in length the lack of termination doesn't impact the bit-error-rate (BER) significantly. The channel simulation controller performs statistical convolution of channel impulse response with that of the data transmitted and the eye-diagram is generated at the receiver side. Finally, the receiver is a simple buffer for NRZ and IBIS-AMI model for PAM4 generated from MATLAB.

B. Simulation and Measurements

The channel simulation is performed for different pitch and channel length configurations. For a given data rate, the eye-opening increases as the pitch increases and decreases as the length increases, as can be seen for typical design parameters, in Fig. 6 for NRZ and in Fig. 7 for PAM4. In an ideal case, the eye-opening must be minimum for 1000μm length- 5μm pitch channel due to highest attenuation and crosstalk and maximum for 100μm length - 50μm pitch channel due to lowest attenuation and crosstalk. But we note that the electromagnetics of the coplanar microstrip line is much more complex than

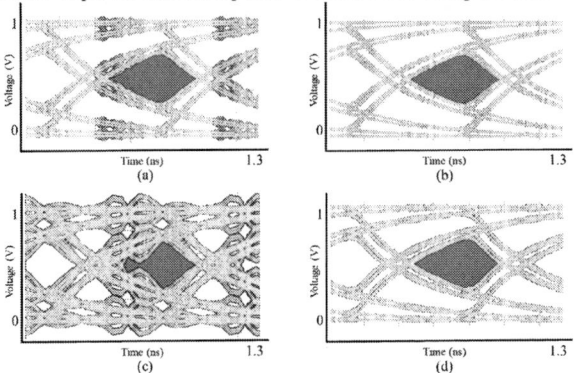

Fig. 6. NRZ Eye for (a) L=100μm, P=5μm (b) L=100μm, P=50μm
(c) L=1000μm, P=5μm, (d) L=1000μm, P=50μm

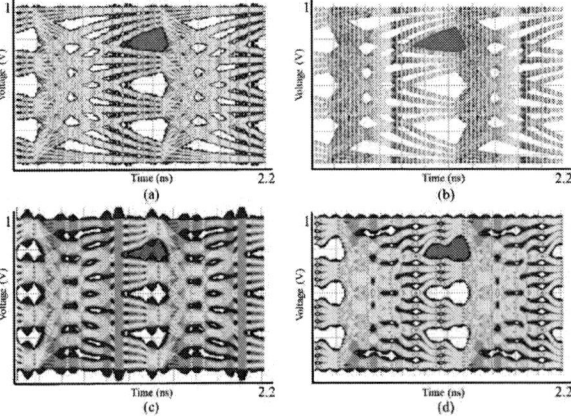

Fig. 7. PAM4 Eye for (a) L=100μm, P=5μm (b) L=100μm, P=50μm (c) L=1000μm, P=5μm, (d) L=1000μm, P=50μm

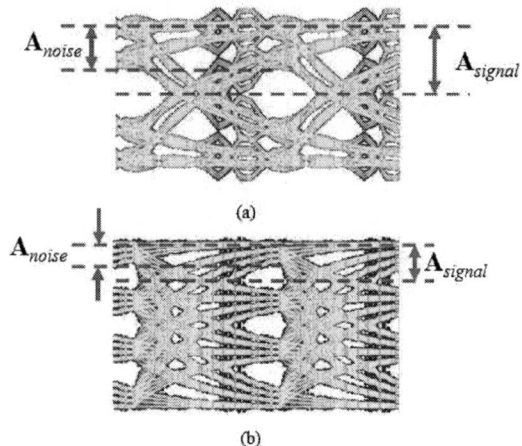

Fig. 8. COM Definition based on Eye Diagram for (a) NRZ (b) PAM4

simple linear relationships between frequency of operation and channel dimensions.

C. Channel Operating Margins and Highest Signalling Rate

The channel operating margin (COM) is a measure of channel performance which was originally developed for IEEE 802.3bj and IEEE 802.3bs Gigabit Ethernet (GbE) standards. The concept of COM has been applied for the channels under consideration. The COM is defined w.r.t the eye-diagram in Fig 8 as:

$$COM = 20log_{10}\frac{A_{signal}}{A_{noise}} \qquad (4)$$

The standard requirement for a communication channel transmitting NRZ data is that COM ≥ 3dB. For PAM4 signalling, since the amplitude of the ideal signal is 1/3rd that of NRZ, the target COM ≥ 9.5dB. In the limiting case, it will be 3dB for NRZ and 9.5dB for PAM4. For PAM4, the average of COM for all the three eyes is taken. Here in the simulation, we determine the highest data rate that can be achieved while meeting the COM requirement for every configuration of channel length and pitch. This is done by setting an optimization goal to meet the COM requirement and sweeping over a suitable frequency range. All the measurements are made for a BER of 1e-15.

Fig 9 and 10 show the intensity plot vs channel dimensions for NRZ and PAM4. The channel pitch is along the X axis and the channel length is along the Y axis. The highest data rate that can be achieved is indicated by intensity of the colour in each box.

The ideal scenario of data rate increasing with increasing channel pitch can be seen for channel length of 600μm in case of NRZ. At 40μm pitch in PAM4, the ideal trend of data rate decreasing with increasing channel length can be observed. That being said, we need to look at the general trend of the data rate as the channel dimensions are varied while considering that the maximum frequency of operation is controlled by the electromagnetics of the channel, effective dielectric constant of the substrate, characteristic impedance, resonant frequencies

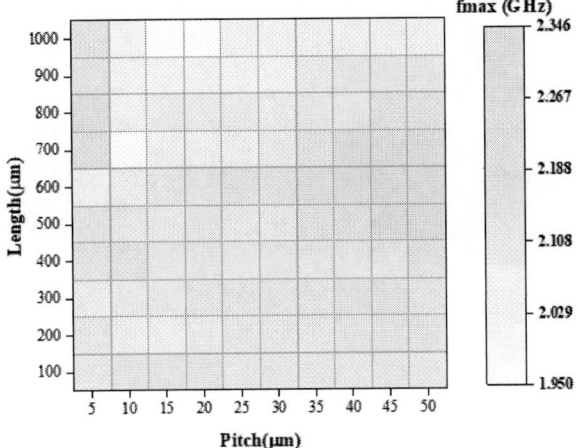

Fig. 9. Max. Freq of operation NRZ for iso-BER of 1e-15

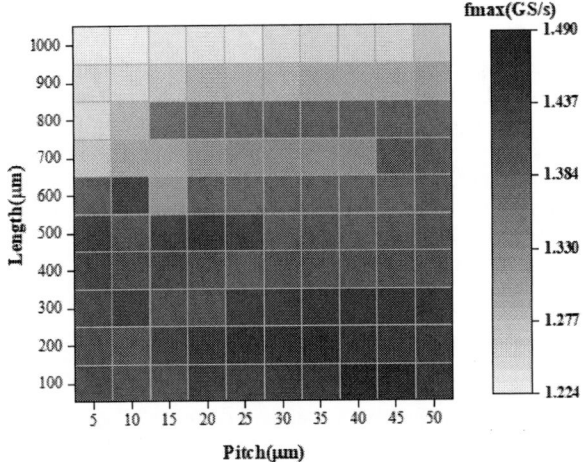

Fig. 10. Max. Freq of operation PAM4 for iso-BER of 1e-15

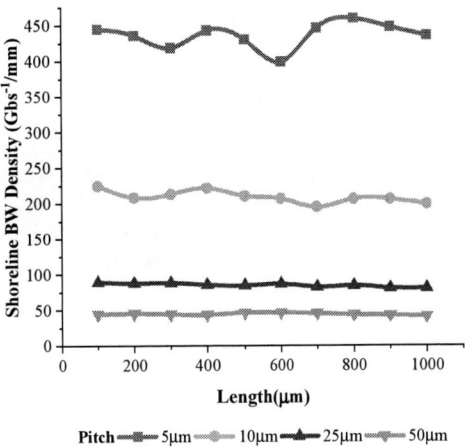

Fig. 11. NRZ Shoreline BW Density vs Channel Length

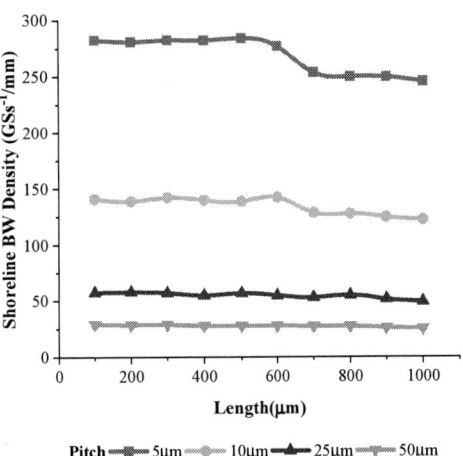

Fig. 12. PAM4 Shoreline BW Density vs Channel Length

and so on. Traditionally channels are designed by fixing most of the physical channel parameters, but here we perform a design space exploration to identify the limits of parallel IO links.

Fig 11 and 12 show the shoreline BW density vs channel length for a sample of four pitch configurations. The direct implication of the finer pitch is increased shoreline density.

VI. POWER ESTIMATIONS

A. NRZ

1. TX: Assuming a single ended voltage mode transmitter, the driver is a buffer circuit which needs to drive the wire and the pad capacitance. The magnitude of wire capacitance can be neglected compared to the pad capacitance. If C_{pad} is the pad capacitance, f_{clk} is the frequency of operation, V_{dd} is the supply voltage, the power dissipation can be modelled as

$$P_{TX} = C_{pad} f_{clk} V_{dd}^2 \qquad (5)$$

2. RX: The single ended receiver is a buffer that decodes the signal to a 0 or 1 level and has the same power expression given by (5) but with load capacitance just another buffer.

B. PAM4

1. DAC: We consider a simple capacitive binary-weighted array DAC structure. The capacitive switching will be the key component of power consumption in this structure. [2] provides a power estimation of such structures; when applied to a 2-bit DAC with equal probability of 0's and 1's gives equation (6). f_{clk} is the frequency of operation, C_0 is the capacitance of the unit capacitor, V_{ref} is the reference voltage for the conversion. A simple current mode driver comprising of two binary weighted current sources with tail currents I_T and $2I_T$ can be utilized to drive the signal. This power is given by (7):

978-1-7281-5410-7/20 $31.00 © 2020 IEEE 163

$$P_{DAC} = \frac{9}{32} f_{clk} C_0 V_{ref}^2 \qquad (6)$$

$$P_{CMD} = 3V_{DD}I_T \qquad (7)$$

2. ADC: As mentioned, due to the speed of operation and low-resolution requirements, a Flash ADC will be the most suitable candidate. The flash ADC comprises of 2^N-1 comparators and an encoder. For a N=2-bit ADC, we need 3 comparators. The power of a matching limited comparator [3] and the Wallace encoder is given by [4] as in (8) and (9) respectively. Here C_{ox} is oxide capacitance, A_{VT} is threshold voltage mismatch coefficient, V_{inpp} is peak-to-peak input voltage, C_{Cmin} is minimum required capacitance, E_{gate} is energy of a typical gate, N is number of bits.

$$P_{Comp} = \left(144 \cdot 2^{2N} C_{ox} A_{VT}^2 \frac{V_{dd}^2}{V_{inpp}^2} + C_{c\,min} V_{dd}^2 \right)(2^N - 1)f_{clk}$$
$$(8)$$

$$P_{enc} = 5 \cdot (2^N - N) \cdot E_{gate} \cdot f_{clk} \qquad (9)$$

C. Phase Locked Loop (PLL)

We consider a non-differential 5 stage VCO along with PFD from [5]. [6] provides an elaborate power estimation treating the PLL as a 2nd order continuous time system. Given the damping factor (0.707) and natural frequency (9.375MHz) with a multiplier of N= 32, the power of a PLL can be written as (10), where C_{PFD}, C_{DIV}, C_{VCO} are the total capacitances of a Phase-Frequency Detector, Frequency Divider and Voltage Controlled Oscillator respectively. The Frequency Divider circuit is a series of TSPC Flops [7] along with TG multiplexers and inverters. P_{BIAS} is the power of the bias circuitry.

$$P_{PLL} = (C_{PFD} + C_{DIV} + C_{VCO}) \cdot V_{dd}^2 \cdot f_{clk} + P_{BIAS}$$
$$(10)$$

The total power for a *2.345Gb/s* NRZ is *31.2mW* leading to an energy-efficiency of *13.323pJ/b*. For the *1.49GS/s* PAM system, the power is *14.53mW* producing an energy-efficiency of *4.876pJ/bit*.

TABLE1. TABLE SHOWING VARIOUS PARAMETERS AND THEIR VALUES

Process Parameter	Typical Value for 28nm Node
C_{ox}	45fF/μ m^2
A_{VT}	1.2mV-μ m
E_{gate}	1.2fJ
C_{cmin}	5fF
C_0	1pF
Others	**Value**
I_T	0.5mA
V_{inp-p}	1V
P_{bias}	0.5mW

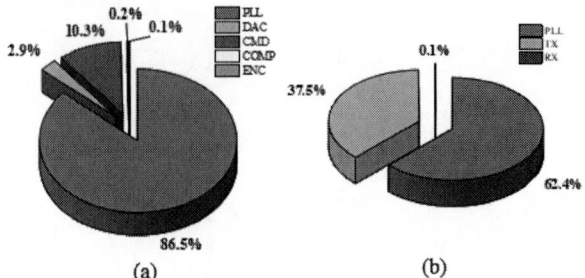

Fig 13. Power consumption of various components in (a) PAM4 (b) NRZ

Fig. 13 show the breakdown of power consumption. As expected, the PLL is the major consumer with up to 62.4% in NRZ and 86.5% in PAM. The receiver in both cases is negligible, as we do not use any equalizer or CDR. Typical process parameter values for 28nm node and other simulation variables are tabulated in Table 1.

VII. CONCLUSIONS

In this paper we explore coplanar microstrip based channels as a model for die to die interconnects for 2.5D integration. We show that higher order modulation like PAM can be applied with more than 63% energy efficiency per bit. This is enabled by the simple transceiver structures for short channel lengths. At high channel densities of up to 5μm pitch, we note that we can achieve 445Gb/s/mm of shoreline-BW-density with NRZ and 565Gb/s/mm with PAM4.

REFERENCES

[1] M. Kirschning and R. H. Jansen. "Accurate wide-range design equations for the frequency-dependent characteristic of parallel coupled microstrip lines," MTT-32, January 1984

[2] M. Saberi, et. al "Analysis of power consumption and linearity in capacitive Digital-to-Analog Converters used in successive approximation ADCs", IEEE Trans. Cir. & Sys.-I Aug 2011..

[3] Stephen O'Driscoll, er.al"Adaptive resolution ADC array for an implantable neural sensor", IEEE Trans. On Biomed. Cir. & Sys., 2011.

[4] B. Murmann, "Energy Limits in A/D Converters", SSCS Talk 2012

[5] D.Y Jeong, et.al, "Design of PLL-Based Clock Generation Circuits", IEEE Journal on Solid-State Circuits, Vol. 22, No. 2, April 1987

[6] D. Duarte, N. Vijaykrisnan, M.J. Irwin, "A Complete Phase-Locked Loop Power Consumption Model" DATE 2002.

[7] J. Rabaey, Digital Integrated Circuits: A Design Perspective, Prentice-Hall International, NJ.

[8] W.C. Chiou et. al, "An ultra-thin interposer utilizing 3D TSV technology", Symp. on VLSI Tech, 2012.

[9] W. Chris Chen et. al, "Wafer Level Integration of an Advanced Logic-Memory System Through 2nd Generation CoWoS® technology", Symp. on VLSI Tech., 2017.

[10] X. Zhang, et.al, "Heterogeneous Interconnect Stitching Technology with Compressible Micro Interconnects for Dense Multi-Die Integration", IEEE Elect. Dev. Lett. Feb 2017

[11] R. Mahajan et.al, "Embedded Multi-Die Interconnect Bridge (EMIB) – A High Density, High Bandwidth Packaging Interconnect", ECTC 2016

[12] D. B. Ingerly et.al "Foveros: 3D Integration and the use of Face-to-Face Chip Stacking for Logic Devices" IEDM 2019

[13] D. Kehlet, "Accelerating Innovation Through A Standard Chiplet Interface: The Advanced Interface Bus (AIB)", White Paper Heterogeneous Integration, Intel Corp

RAT: A Lightweight System-level Soft Error Mitigation Technique

Jonas Gava, Ricardo Reis
PPGC/PGMicro - Federal University of Rio Grande do Sul
Porto Alegre, Brazil
{jfgava, reis}@inf.ufrgs.br

Luciano Ost
Loughborough University
Loughborough, United Kingdom
l.ost@lboro.ac.uk

Abstract—**To achieve a substantial reliability and safety level, it is imperative to provide electronic computing systems with appropriate mechanisms to tackle soft errors. This paper proposes a low-cost system-level soft error mitigation technique, which allocates the critical application function to a pool of specific general-purpose processor registers. Both the critical function and the register pool are automatically selected by a developed profiling tool. The proposed technique was validated through more than 320K fault injections considering a Linux kernel, different benchmarks and two multicore ARM processors. Results show that our technique significantly reduces the code size and performance overheads while providing reliability improvement, w.r.t. the Triple Modular Redundancy (TMR) technique.**

Index Terms—**Multicore, soft error reliability, mitigation technique, fault tolerance**

I. INTRODUCTION

Multicore architectures are being adopted in many industrial segments such as automotive, medical, consumer electronics, and high-performance computing (HPC). Applications running on such architectures differ in terms of security, reliability, performance and power requirements. To achieve a substantial reliability and safety level, it is imperative to provide electronic computing systems with appropriate mechanisms to tackle systematic or transient faults, also known as soft errors or Single Event Upset (SEU). While the former originates from hardware and software design defects, soft errors are those caused by alpha particles or atmospheric neutrons [1]. The occurrence of soft errors can either corrupt the memory data, or output of a program or even crash the entire system, which depending on its criticality level can lead to life-threatening failures.

The soft error mitigation problem can be tackled both in hardware and software [2]. While hardware approaches lead to the area and power overhead, software techniques are generally implemented on a per-application basis that usually incurs in performance penalties due to the redundant computation. Such additional overhead might restrict the use of costly mitigation techniques under resource-constrained devices. Further, the adoption of soft error mitigation techniques also adds development complexity, which has a direct impact on the time-to-market. Examples of soft error mitigation techniques include, among others, Error Detection and Correction Code (EDAC) and Triple Module Redundancy (TMR).

This paper addresses the above challenges by proposing a novel lightweight system-level soft error mitigation technique, called RAT (Register Allocation Technique). The proposed technique along with the developed profiling toolset enable software engineers to isolate and allocate the most critical application function to a pool of least used general-purpose processor registers. RAT was compared against a selective TMR technique [3], considering a Linux kernel, 13 applications and a dual and a quad-core ARM processors. Results demonstrated that RAT reduces the code size and performance overheads while providing reliability improvement.

The rest of this paper is organised as follows. Section II presents basic concepts and related works in software soft error mitigation techniques. Section III describes the proposed mitigation technique. In Section IV, the experimental setup and adopted evaluation metrics are presented. In Section V the efficiency of RAT is evaluated as well as a specific case study analysing the registers criticality is presented. Finally, Section VI describes the final remarks and future works.

II. FUNDAMENTAL CONCEPTS AND RELATED WORKS

A. Fault Tolerance Taxonomy

The soft error assessment and mitigation literature is abundant, requiring a taxonomy to classify the different approaches. This work considers the definitions from [4], [5] for fault, error, and failure. A fault is an event that may cause the internal state of the system to change, e.g., a radiation particle strike. When a fault affects the system's internal state, it becomes an error. If the error causes a deviation of at least one of the system's external states, then it is considered as a failure.

The most commonplace classification for soft error assessment considers three classes: *Silent Data Corruption* (SDC) occurs when the system does not detect a fault and the outcome of the application is affected; In *Detected Unrecoverable Error* (DUE) on the other hand, the fault is detected and it is not possible to continue the execution (e.g., segmentation fault); and *Masked*, when the application outcome and the system state are the same as a faultless execution.

As mentioned before, soft error mitigation techniques can be implemented in hardware, software or a combination of both. The next Section reviews only the software-based approaches.

978-1-7281-5410-7/20 $31.00 © 2020 IEEE

B. Software-based Soft Error Mitigation Techniques

A processor-based system can be affected by two main types of soft errors: control-flow and data-flow. A control-flow error occurs when the error causes deviation from the correct program flow (e.g., incorrect branch). The data-flow error refers to the soft error caused by a bit-flip in a storage component, such as a register or memory element. They can, for instance, affect the output of a program generating an SDC, or leading to a DUE when computing a wrong memory addresses.

Aiming to mitigate both type of soft errors the following Authors have promoted few software system-level techniques, i.e., techniques that can be applied at software architectural level (e.g., application, operating system (OS)). In Benso *et al.* [6] and Nicolescu *et al.* [7] tools that apply fault tolerance techniques in C/C++ applications are proposed. Supported transformations are architecture-independent, but the language is fixed, and the compiler may remove redundant code during the compiler optimisation phases. The focus of [7] is on low-cost safety-critical applications, where the high memory and speed overheads (about 3-4 times) are not important metrics. Another similar tool is the REliable Code COmpiler (RECCO) [6], which relies on code reordering and selective variable duplication. In [8], authors use genetic algorithms to find a combination of optimisation parameters (i.e., compilation flags) that increase the reliability of the final binary and present a reasonable trade-off in terms of performance, and memory size. The proposed technique was evaluated considering an FPGA implementation that was exposed to a proton irradiation test. In [9], the authors implemented in C code two mitigation techniques: the Triple Modular Redundancy (TMR) and the Conditional Modular Redundancy (CMR). Their results have shown that both techniques do not provide a reasonable protection to a complex system executing Linux kernel. According to the authors, the OS itself is an enormous source of errors and need to be protected if employed on safety-critical systems.

The downside of aforementioned approaches is the fact that during the compiler optimisation phase, parts of the protected code (e.g., redundant functions) may be wrongly removed. One solution to overcome such restriction relies on modifying the assembly code after the compilation. A popular instruction-level mitigation technique introduced by Reis *et al.* [10] is the Swift-R, which implements TMR to recover from soft errors in the register file. Instead of duplicating instructions,

it triplicates, and change the checking points to a voter mechanism. In this direction, authors in [11] proposed the CFT-tool that modifies the assembly code by applying different data-flow and control-flow protection techniques. Although this approach does not suffer from compiler optimisation, it is architecture-dependent. The CFT-tool uses a configuration file to minimise this limitation. However, this file needs to be hand-made for each new Instruction Set Architecture (ISA). Shirvani et al. [2] propose a software implementation of EDAC, i.e., an independent task that is executed periodically. Results show their approach provides protection for code segments and can enhance the system reliability with a lower check-bit overhead w.r.t. other techniques (e.g., Hamming, Parity).

Different from the reviewed works, RAT does not involve code redundancy, and it is an architecture-independent approach. Further, RAT is a fully automated approach that is developed on the basis of LLVM backend, enabling its extension and combination with other soft error mitigation techniques, as shown in Section V.

III. REGISTER ALLOCATION TECHNIQUE (RAT)

Rather than implementing a toolset from scratch, we have adopted a flexible virtual platform (VP) that provides us with the necessary means (i.e., simulator with processor and component models, full software behaviour observability) to implement the proposed technique. RAT was implemented on the basis of OVPsim framework [12] to enable a fast design space exploration, but other VPs with similar support could also be used (e.g., gem5 [13]). The main steps of the RAT (Fig. 1) are as follows:

A. Software stack (i.e., application, operating system, drivers) source code selection.

B. Target processor architecture selection and source code compilation using Clang/LLVM 6.0.1.

C. In this step, the application is executed, and essential information are extracted (i.e., processor register file utilization and critical function). Note that the software engineer can either determine the most critical application function or use the default option of our toolset, which selects the most executed one.

D. Here, our tool extracts from the object code which type (i.e., 32 or 64-bit) and how many registers need to be reserved to the function defined as critical in the previous step. In this stage, the register pool is set following the

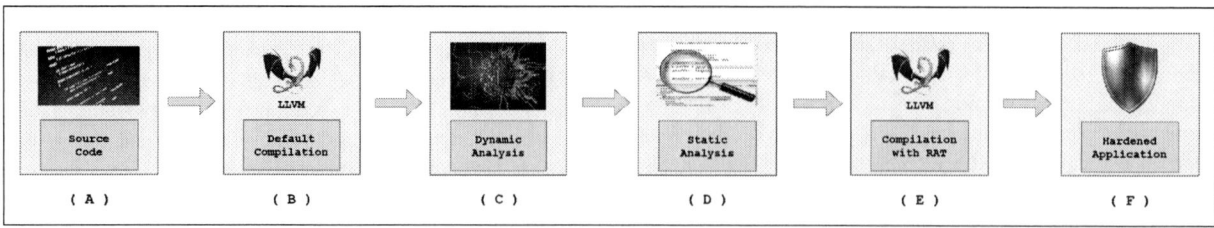

Fig. 1: Main steps of proposed register allocation technique (RAT).

strategy of allocating least used general-purpose registers for the critical function.

E. In this step, a new compilation is performed, taking into account the critical function and the register pool previously set. The underlying compilation uses a modified version of the LLVM Fast Register Allocator, which considers arguments (i.e., restrictions) that are passed to LLC (LLVM Static Compiler) through a command line (Fig. 2). Note that we do not control the use of the registers available in the pool, the compiler decides which ones to prioritise.

F. Finally, the resulting hardened binary is generated by the LLD linker.

The left-hand of Fig. 2 shows an example of a C language function that takes three integer parameters as input, performs arithmetic operations, and returns an integer value. The resulting 64-bit ARM (Aarch64) assembly code is shown in the right-hand of Fig. 2, where at the top the default register allocation is shown. In turn, at the bottom right-hand of Fig. 2 the RAT technique is applied, limiting the function register pool to "W21, W22, W23". By the calling convention, the ARMv8-A general-purpose registers with indexes from 0 to 7 are used for inputs and result. When restricting registers outside this range, the compiler only needs to insert some *MOV* instructions at the beginning and end of the function. As mentioned before, RAT is a compiler-based mitigating technique, thus it can be associate with other techniques as well. Such capacity is explored in the Section V.

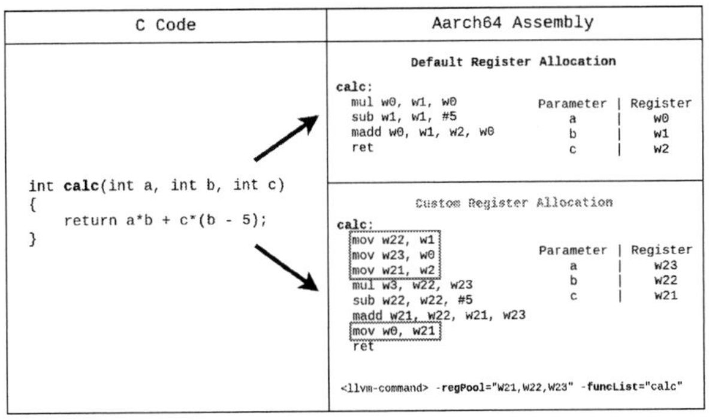

Fig. 2: Example of C code conversion to Aarch64 assembly without and with RAT flags compilation.

IV. EXPERIMENTAL SETUP AND EVALUATION METRICS

In order to demonstrate the effectiveness of RAT, we have adopted the fault injection simulator proposed in [14], which is also implemented on the basis of OVPsim. Fault analyses are obtained by injecting faults (i.e., bit-flips) in the general-purpose registers (i.e., X0-X30) of a dual-core and a quad-core ARM Cortex-A72, in a random order. Conducted experiments include 320K fault injections in a realistic software stack including unmodified Linux kernel, a standard parallelization library (OpenMP), considering 13 applications taken from the Rodinia Benchmark Suite [15] as shown in Table I. One of

the main concerns when assessing the reliability of a system is to develop a precise, well-covered and realistic approach. In this sense, this work sought to ensure that the number of fault injections has a statistical significance by applying the equations developed by [16]. This work injects 3100 faults per campaign, thus generating a 1.75% error margin with 95% confidence level.

TABLE I: Rodinia Benchmarks

#	Benchmark	Domain
A	Backprop	Pattern Recognition
B	BFS	Graph Algorithms
C	HeartWall	Medical Imaging
D	HotSpot	Physics Simulation
E	HotSpot3D	Physics Simulation
F	Kmeans	Data Mining
G	LUD	Linear Algebra
H	Myocyte	Biological Simulation
I	NN	Data Mining
J	ParticleFilter	Medical Imaging
K	PathFinder	Grid Traversal
L	SradV1	Image Processing
M	SradV2	Image Processing

Depending on the application's nature, the three categories classification described in Section II.A may be inadequate to express all the possible misbehaviours. With this in mind, the results are classified according to Cho [17], which defines five possible behaviours for a system in the presence of a fault: **Vanish**: no fault traces are left in both memory and architectural state; **Output not Affected (ONA)**: the resulting memory is not modified, however, one or more remaining bits of the architectural state is incorrect; **Output Memory Mismatch (OMM)**: the application terminates without any error indication, and the resulting memory is affected; **Unexpected Termination (UT)**: the application terminates abnormally with an error indication; **Hang**: the application does not finish requiring a preemptive removal.

Software engineers might categorize the criticality of application functions entirely differently depending on their criteria and/or system domains. For the sake of simplicity, this work assumes that most executed functions are the critical ones. Although not ideal, such an approach is entirely adequate to evaluate the benefits and drawback of the proposed mitigation technique. RAT reliability, code and performance overheads are compared against the selective TMR implementation (i.e., VAR3+) [18].

A. Reference Mitigation Technique - Selective TMR

In [3], the authors describe a set of rules for data-flow techniques that aim to detect faults affecting values stored in registers bank and memory devices. In this work, we use a triplication instead of duplication since the target is to mitigate the occurrence of soft error. The selective TMR technique implementation was based on [19] inside the Clang/LLVM 6.0.1. The VAR3+ technique was chosen due to its capability of increasing reliability while maintaining a low code and performance overhead w.r.t. previous TMR-based techniques.

In this technique, each register has a replica (rule G1), and all instructions, except for branches and stores, are replicated (D2). The replicas are checked before every load, store, or branch instruction (C3, C4, C5, C6). Some acronyms used in the following sections are RAT: reference application + register allocation technique. TMR: selective TMR technique (VAR3+). TMR+RAT: TMR + register allocation technique.

B. Evaluation Metrics

To adequately assess the soft error mitigation technique reliability, Reis *et al.* [20] introduced a metric called Mean Work To Failure (**MWTF**), which is calculated by the average amount of work that an application can perform for each error. A unit of work is a general concept whose specific definition depends on the application. The unit work is defined here as a correct program execution (i.e., Vanished fault), while the number of errors is defined as the sum of ONA, OMM, UT, and Hang results as shown in (1).

$$MWTF = \frac{Vanished}{ONA + OMM + UT + Hang} \quad (1)$$

This work also employs the *Fault Coverage* metric, which describes the percentage of faults that are either detected or masked. It is represented as the ratio of detected and masked faults (i.e., Vanished) to the total number of faults that occurred, as shown in (2).

$$F_{coverage} = \frac{UT + Vanished}{ONA + OMM + Hang} \quad (2)$$

Further, we use the Fault Coverage Increase (**FCI**) to describe the gain in the percentage of fault coverage when comparing the mitigation techniques.

V. RAT EVALUATION RESULTS

A. RAT Code and Performance Overhead

To provide relevant overhead measures, the **code size** information was extracted from the application object files, while

the **performance** figures were obtained from the gem5 full system simulator [13].

Fig. 3a shows a substantial code size overhead (e.g., up to 84.86% - benchmark C) when the TMR technique is used. In turn, the cost of the proposed technique is negligible, 0.15% in the worst case (benchmark K). Such low overhead is due to the RAT approach, which only adds *MOV* instructions at the beginning and end of the critical functions. As a consequence, the performance of applications is not jeopardised when RAT is used (i.e., less than 1% for all scenarios).

Results in Fig. 3b and 3c show that the use of the TMR can lead to up to 38.5% and 50% of performance penalty (benchmark C) when running on dual and quad-core ARM Cortex A72 processors. The reason why there is an increase in the execution time in the quad-core when compared to the dual-core is due to the increasing execution of OS thread synchronisation routines that is not linear with the number of cores. Note that the additional execution time of TMR is small for a technique that triples instructions and inserts voters into the code. This is justified by the fact that only instructions inside the application's scope are replicated, and the majority of Rodinia applications rely on external library calls. One possible solution to this problem implies replicating function calls; however, there are possible collateral damages inherent to this approach (e.g., modifying the same data structure multiple times).

B. RAT Soft Error Reliability Evaluation

1) Techniques Comparison: Fig. 4 and 5 show the reliability comparison between the three mitigation techniques. In terms of MWTF on Fig. 4, the TMR implementation provides higher reliability in 5 out of 13 cases (C, D, F, I, K), while the RAT in 4 cases (A, E, J, L), and the TMR+RAT in the other 4 cases (B, G, H, M). Results show that RAT can also provide reliability improvements of up to 40% in some cases w.r.t. TMR. Results also show that, depending on the application nature, TMR+RAT is an appropriated combination to improve system reliability. For instance, taking the benchmarks B

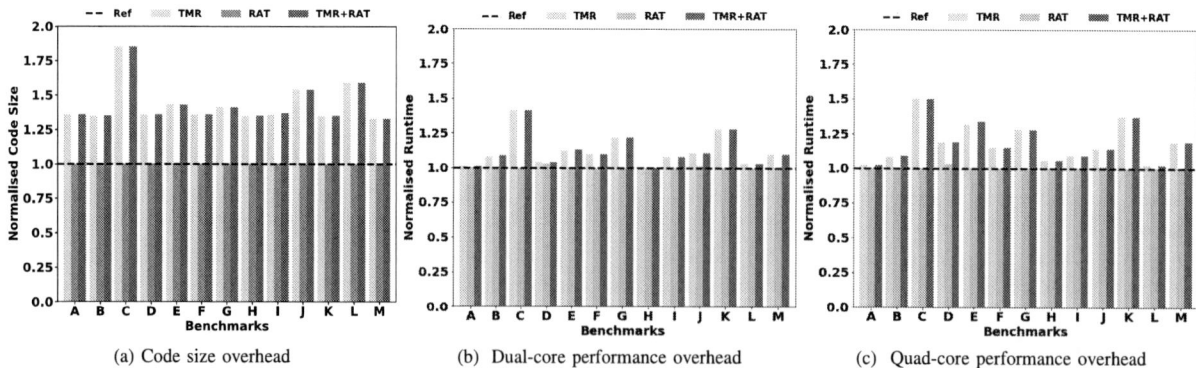

(a) Code size overhead   (b) Dual-core performance overhead   (c) Quad-core performance overhead

Fig. 3: Code size (a) and performance overhead for dual (b) and quad-core (c) ARM Cortex-A72 processor when comparing the impact of the mitigation techniques w.r.t. the original reference benchmark (**–Ref**).

and K as examples, it is possible to identify a considerable difference in the MWTF gain when comparing the two TMR implementations. While benchmark B showed a reliability improvement of 40% for `TMR+RAT`, the use of `TMR` provides an improvement of 51% for K.

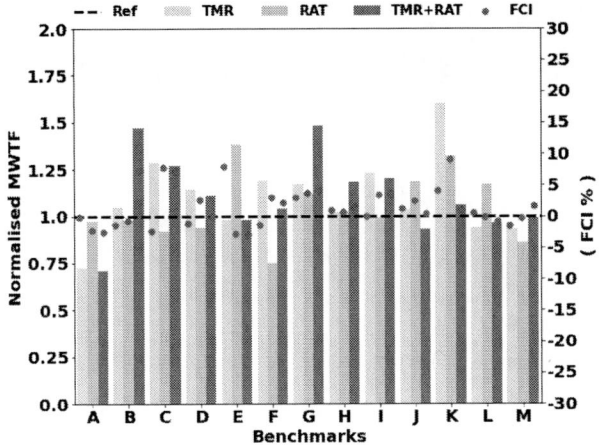

Fig. 4: Normalised reliability comparison between each technique considering the original benchmark code as reference (**–Ref**) for a dual-core ARM Cortex-A72.

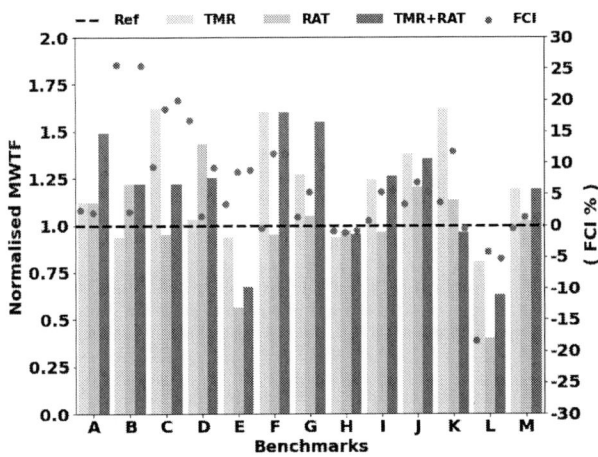

Fig. 5: Normalised reliability comparison between each technique considering the original benchmark code as reference (**–Ref**) for a quad-core ARM Cortex-A72.

Fig. 5 shows a significant increase in the FCI average compared to the results in the dual-core processor, 5.47% versus 1.48%. Note that all reliability metrics have been reduced from dual-core to quad-core, the increase was only about the reference benchmark. This behaviour occurs precisely due to the rise in the execution of thread management tasks, which have a higher susceptibility to soft errors, as mentioned earlier. The `TMR` technique obtained better reliability results in 6 of the 13 benchmarks (C, E, F, J, K, L), RAT was better in

2 cases (D, H), and `TMR+RAT` was better in 5 cases (A, B, G, I, M). Note that the applications' reliability varies from one mitigation technique to another. For that reason, we claim that engineers can use our toolset to analyse the impact of different mitigation techniques at the system-level, so they might be able to identify the most suitable one considering their application/system's constraints. Further, a more in-depth analysis is carried out, verifying the results of the fault injections in each register for a specific case study.

2) Registers Criticality Analysis: Fig. 6 shows how the 64-bit ARM (AArch64) calling convention works. The X0-X7 registers are used for input parameters and return functions; the X8 is used to hold an indirect return location address; the X9-X15 are used to hold local variables (caller saved); the X16 and the X17 are the Intra-Procedure-call scratch registers; the X18 can be used for some OS-specific purpose; the X19-X28 are callee-saved registers; the X29 is the frame pointer; while the X30 is the link register, used to return from subroutines. To better explain the RAT benefits, we chose the particlefilter benchmark (**J**) as as case study.

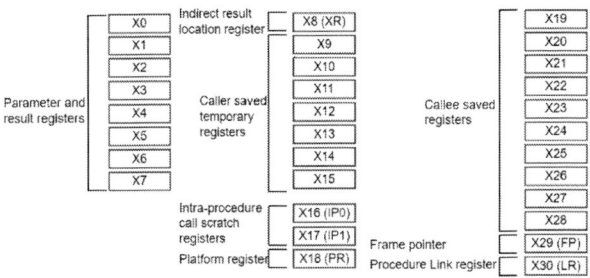

Fig. 6: Allocation of the general-purpose registers following the AArch64 calling convention. [21]

The results show that half of the registers (X0-X16) do not suffer significantly from soft errors (Fig. 7), when the particlefilter benchmark (**J**) is executed on a **dual-core** ARM Cortex-A72 processor. In contrast, the rest of the registers suffers strongly from the injected faults. Especially the callee-saved category that are used to hold long-lived values that must be preserved across calls and are used by the Linux kernel. That is, theoretically, they are registers that take a longer time to get written, but they are continuously read. However, as shown in Fig. 8, the fault masking increases when we apply the RAT technique and limit the number of registers that will be used to execute the most performed function. In general, this effect occurs because when entering the critical function, the callee-saved registers are saved in memory and return to their original values at the end of the execution. In practice, this behaviour ends up reducing the lifetime of these registers, making them more resilient to soft errors. The best examples are from the X17 and X19 registers. For the X17, we have a fault-masking rate of 70% in the reference application, and 98% when using the RAT mitigation technique. For the X19 register, we have a fault-masking rate of 36.67% in the reference application, and 58% when using the RAT technique.

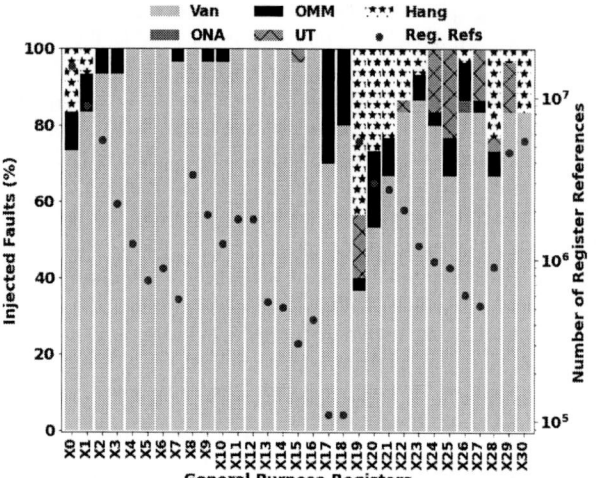

Fig. 7: Registers criticality for the **Reference** particlefilter benchmark running on a dual-core ARM Cortex-A72.

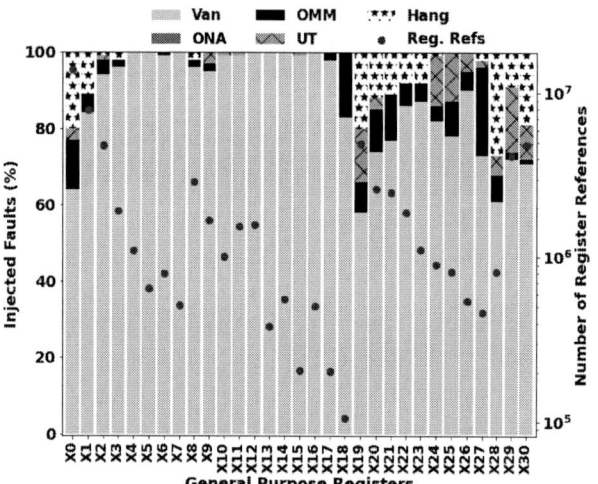

Fig. 8: Registers criticality for the **RAT** version of particlefilter benchmark running on a dual-core ARM Cortex-A72.

VI. CONCLUSION AND FUTURE WORKS

The importance of using selective and lightweight soft error mitigation techniques is increasing every year. The results show that redundancy does not always ensure reliability and other factors such as code size and performance overheads must be considered. In this regard, the proposed RAT offers a good compromise in terms of reliability improvement, code size overhead and performance penalty w.r.t. TMR. Future works include further investigation of RAT considering other processor architectures and more complex benchmarks.

REFERENCES

[1] M. Snir et al., "Addressing failures in exascale computing," *The International Journal of High Performance Computing Applications*, vol. 28, no. 2, pp. 129–173, 2014.

[2] P. P. Shirvani, N. Saxena, S. M. Ieee, E. J. Mccluskey, and L. F. Ieee, "Software-Implemented EDAC Protection Against SEUs," *IEEE Transactions on Reliability*, vol. 49, pp. 273–284, 2000.

[3] E. Chielle, F. L. Kastensmidt, and S. Cuenca-Asensi, "Overhead reduction in data-flow software-based fault tolerance techniques," in *FPGAs and Parallel Architectures for Aerospace Applications*. Springer, 2016, pp. 279–291.

[4] A. Avižienis, J.-C. Laprie, and B. Randell, "Dependability and its threats: A taxonomy," in *Building the Information Society*, Springer, 2004, pp. 91–120.

[5] S. S. Mukherjee, J. Emer, and S. K. Reinhardt, "The soft error problem: An architectural perspective," in *11th International Symposium on High-Performance Computer Architecture*, IEEE, 2005, pp. 243–247.

[6] A. Benso, S. Chiusano, P. Prinetto, and L. Tagliaferri, "A C/C++ source-to-source compiler for dependable applications," in *Proceeding International Conference on Dependable Systems and Networks. DSN 2000*, IEEE, 2000, pp. 71–78.

[7] B. Nicolescu and R. Velazco, "Detecting soft errors by a purely software approach: Method, tools and experimental results," in *Embedded Software for SoC*. Springer, 2003, pp. 39–51.

[8] A. Serrano-Cases, Y. Morilla, P. Martin-Holgado, S. Cuenca-Asensi, and A. Martinez-Alvarez, "Nonintrusive automatic compiler-guided reliability improvement of embedded applications under proton irradiation," *IEEE Transactions on Nuclear Science*, vol. 66, no. 7, pp. 1500–1509, 2019.

[9] G. S. Rodrigues, F. L. Kastensmidt, R. Reis, F. Rosa, and L. Ost, "Analyzing the impact of using pthreads versus openmp under fault injection in arm cortex-a9 dual-core," in *2016 16th European Conference on Radiation and Its Effects on Components and Systems (RADECS)*, IEEE, 2016, pp. 1–6.

[10] G. A. Reis, J. Chang, and D. I. August, "Automatic instruction-level software-only recovery," *IEEE micro*, vol. 27, no. 1, pp. 36–47, 2007.

[11] E. Chielle, R. S. Barth, A. C. Lapolli, and F. L. Kastensmidt, "Configurable tool to protect processors against see by software-based detection techniques," in *2012 13th Latin American Test Workshop (LATW)*, IEEE, 2012, pp. 1–6.

[12] Imperas, *OVPsim Simulator*, 2020. [Online]. Available: http://www.ovpworld.org.

[13] N. Binkert et al., "The gem5 simulator," *ACM SIGARCH computer architecture news*, vol. 39, no. 2, pp. 1–7, 2011.

[14] V. Bandeira, F. Rosa, R. Reis, and L. Ost, "Non-intrusive fault injection techniques for efficient soft error vulnerability analysis," in *2019 IFIP/IEEE 27th International Conference on Very Large Scale Integration (VLSI-SoC)*, IEEE, 2019, pp. 123–128.

[15] S. Che et al., "Rodinia: A benchmark suite for heterogeneous computing," in *2009 IEEE international symposium on workload characterization (IISWC)*, Ieee, 2009, pp. 44–54.

[16] R. Leveugle, A. Calvez, P. Maistri, and P. Vanhauwaert, "Statistical fault injection: Quantified error and confidence," in *Proceedings of the Conference on Design, Automation and Test in Europe*, European Design and Automation Association, 2009, pp. 502–506.

[17] H. Cho, S. Mirkhani, C.-Y. Cher, J. A. Abraham, and S. Mitra, "Quantitative evaluation of soft error injection techniques for robust system design," in *Proceedings of the 50th Annual Design Automation Conference*, 2013, pp. 1–10.

[18] J. R. Azambuja, A. Lapolli, M. Altieri, and F. L. Kastensmidt, "Evaluating the efficiency of data-flow software-based techniques to detect sees in microprocessors," in *2011 12th Latin American Test Workshop (LATW)*, IEEE, 2011, pp. 1–6.

[19] M. Bohman, B. James, M. J. Wirthlin, H. Quinn, and J. Goeders, "Microcontroller compiler-assisted software fault tolerance," *IEEE Transactions on Nuclear Science*, vol. 66, no. 1, pp. 223–232, 2018.

[20] G. A. Reis, J. Chang, N. Vachharajani, R. Rangan, D. I. August, and S. S. Mukherjee, "Software-controlled fault tolerance," *ACM Transactions on Architecture and Code Optimization (TACO)*, vol. 2, no. 4, pp. 366–396, 2005.

[21] ARM, *ARMv8-A parameters in general-purpose registers*, 2020. [Online]. Available: https://developer.arm.com/docs/den0024/latest/the-abi-for-arm-64-bit-architecture/register-use-in-the-aarch64-procedure-call-standard/parameters-in-general-purpose-registers.

A Low-Power 10 to 15 Gb/s Common-Gate CTLE Based on Optimized Active Inductors

Amin Aghighi*, Armin Tajalli*, Mohammad Taherzadeh-Sani**

*Electrical and Computer Engineering Department, University of Utah, Salt Lake City, USA

**Electrical Engineering Department, Ferdowsi University of Mashhad, Mashhad, Iran

Abstract—A new low-power common-gate continuous-time linear equalizer (CG-CTLE) is presented that exploits active matching termination to increase power efficiency. Also, a new active inductor is proposed, which introduces a higher amount of peaking as compared to the conventional active inductor topologies. The proposed two-stage CG-CTLE is designed in conventional 65-nm CMOS technology using both passive, and the proposed active inductors. These CTLEs compensate for 15 to 21 dB loss of a 12 inches electrical channel while operating within 10-15 Gb/s. Post-layout simulation results show that the CTLEs consume 6 mW and 5.4 mW using passive and active inductors, respectively.

Index Terms— SerDes, inter symbol interference (ISI), electrical channel, equalization, continuous-time linear equalizer (CTLE), common-gate CTLE, active inductor.

I. INTRODUCTION

The ever-lasting demand for higher speed and power efficiency drives the current developments for high-speed interfaces towards even higher levels of performance and speed. While modern deep sub-micrometer silicon technology facilitates the design of high-speed transceivers, off-chip interface channels have not progressed in the same manner, and suffer from limited bandwidth (BW). Consequently, the received signal is severely attenuated and affected by channel loss. Hence, designing efficient receiver front-ends that can recover the attenuated data, which is also distorted by the inter symbol interferences (ISI) is an on-going research topic [1]-[2].

High-performance wireline links employ different equalization schemes in either transmitter (TX) or receiver (RX) side to achieve the required Bit Error Rate (BER). CTLE is one of the most common equalizers in the RX side and is supposed to introduce a high-pass frequency response in order to compensate for the low-pass channel transfer function [3]-[4]. Since CTLE is the first block at the channel interface, its input impedance should match the channel characteristic impedance to prevent reflections. In this paper, two new CG-CTLEs are proposed that exploit active termination and active inductive peaking techniques to increase power and area efficiency. While a passive resistor is usually employed in conventional common-source (CS) CTLE, to terminate the 50-Ohm channel, the proposed CG-CTLEs provide a suitable input matching over the desired bandwidth without passive termination. As a result, the input energy which is dissipated through the passive termination is injected into the CG-CTLE, improving the circuit performance.

The rest of this paper is organized as follows: Section II presents the proposed active termination technique that is

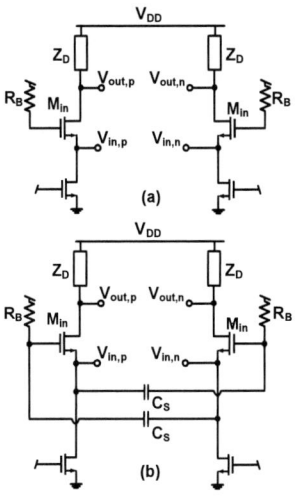

Fig. 1. (a) Conventional and (b) Proposed CG-CTLE stage.

applied to both CTLEs. Two proposed CTLEs are presented in Section III. Section IV provides post-layout simulation results, followed by the conclusion in Section V.

II. PROPOSED CG-CTLE STAGE

Conventional CG structure is widely used for providing active termination in many amplifiers such as low-noise amplifiers (LNA) in radio frequency integrated circuit design (RFIC) [5]. Fig. 1 (a) shows the conventional CG amplifier that can be used for CTLE as well. The differential input impedance of this structure is:

$$Z_{in,diff} = \frac{2}{g_m} = \frac{2}{I_D \times g_m/I_D} \tag{1}$$

where, g_m, I_D and g_m/I_D are trans-conductance, drain-source bias current and transconductance efficiency of the input transistor, M_{in}, respectively. Channel length modulation and body effect are neglected for simplicity. It is shown in [6] and [7] that there is an optimum point for the input transconductance efficiency, $(g_m/I_D)_{opt}$, which results in the lowest power consumption to achieve the target bandwidth. Assuming $Z_{in,diff} = 100\ \Omega$ necessitates a large $g_m = 20\ mS$ for input matching, which either needs large bias current or wide input devices. While the first case increases power consumption directly, the latter deviates the g_m/I_D from $(g_m/I_D)_{opt}$ that again increases the consumption due to the self-loading effect [6]. Using parallel RC-degeneration pair allows employing smaller input devices at the expense of lower DC-gain and more

978-1-7281-5410-7/20 $31.00 © 2020 IEEE

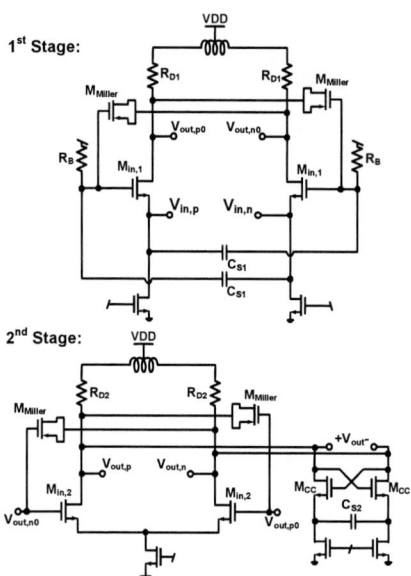

Fig. 2. Circuit schematic of the first and second stages of the proposed CTLE I.

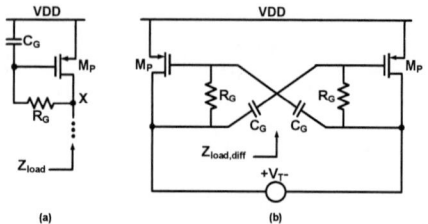

Fig. 3. (a) Conventional and (b) proposed differential active inductors.

input-referred noise. Hence, obtaining sufficiently large $g_{m,in}$ without widening input transistors alleviates the trade-off between input matching and power consumption. This is achievable by introducing a negative gain, G_{Neg}, between gate and source terminals of the input devices that increases the effective transconductance, $g_{m,eff}$ [8]. Fig. 1 (b) demonstrates how this negative gain is implemented passively using cross-coupled capacitors to prevent adding extra noise. Therefore:

$$g_{m,eff} = g_m \times \left(1 + G_{Neg}\right) \approx g_m \times (1 + \frac{C_S - C_{gs}}{C_S + C_{gs}}) \quad (2)$$

where, C_{gs} is the gate-source parasitic capacitance of M_{in}. Assuming $C_S \gg C_{gs}$ increases the effective transconductance by a factor of two with a concomitant reduction in power consumption and noise contribution as compared to the conventional CG-CTLE in Fig. 1 (a). Although this technique is previously used for LNAs to reduce the noise figure [5], it can be exploited for CTLEs in wireline applications, as well [9].

III. Proposed CG-CTLE

A. Passive inductive peaking CG-CTLE

Fig. 2 shows the first proposed CTLE, where the first stage utilizes the proposed CG structure to provide wideband input matching, and the second stage employs a CS topology with a negative impedance converter (NIC) between its output nodes. The NIC circuit provides some peaking while eliminating the low-frequency attenuation of the RC-degenerated structure. Using the positive feedback in NIC necessitates $g_{m,cc}R_D < 1$ so that the CTLE stays stable for every value of the load capacitance [10]. In addition, centered-tap symmetrical passive inductors are utilized in both stages to extend the BW and peaking of the CTLE. The gate-drain parasitic capacitances, C_{gd}, of M_{in} devices not only reduce the achievable BW but also, they degrade the input matching quality due to peaking feedthrough. Hence, the widely used negative Miller capacitance technique is used to compensate for this undesirable effect [3]. Here, M_{Miller}

devices are sized such that they partially cancel out the C_{gd} of M_{in1} and M_{in2}.

B. Proposed active peaking CG-CTLE

Although passive inductors can be employed to extend the BW and obtain gain peaking, they tend to limit the smallest realizable circuit area. Therefore, active inductors that enable the aforementioned features while obtaining small foot-print are highly desirable. Fig. 3 (a) shows the conventional active inductor which is commonly used in high-speed wireline receivers and especially in CTLEs [11], [12]. By intuition, at low frequencies where C_G can be considered as an open circuit, the transistor is operating in the diode-connected configuration for low-impedance mode ($\approx 1/g_m$). However, C_G acts as a short circuit at high frequencies to increase the impedance to $R_G||r_o$, where, r_o is the output impedance of the transistor.

$$Z_{load} = \frac{r_o}{1 + g_m r_o}\left(\frac{1 + R_G C_G S}{1 + \frac{C_G(R_G + r_o)}{1 + g_m r_o}S}\right) \quad (3)$$

This active inductor has a pole-zero pair that defines the achievable boost factor, ω_P/ω_Z [3]:

$$\omega_Z = \frac{1}{R_G C_G} \quad (4)$$

$$\omega_P = \frac{1 + g_m r_o}{C_G(R_G + r_o)} \quad (5)$$

$$G_{Conventional} = \omega_P/\omega_Z = \frac{(1 + g_m r_o)R_G}{R_G + r_o} \quad (6)$$

The conventional active inductor cannot provide the required boost factor for high-speed RXs that suffer from a severe attenuation of the interface channel. Shown in Fig. 3 (b), the proposed active inductor addresses this issue without any extra cost as compared with the conventional counterpart in Fig. 3 (a). The idea of peaking enhancement is based on increasing the high-frequency impedance of the conventional active inductor without altering its low-frequency impedance. In fact, C_G in Fig. 3 (a), bypasses M_P at high frequencies and only uses the transistor for its high output impedance, r_o. Hence, one can exploit the amplification nature of M_P to increase $Z_{load,diff}$ at high frequencies. As shown in Fig. 3 (b), instead of bypassing M_P through C_G, the cross-coupled connection provides a signal path with proper polarity at high frequencies that M_P devices can operate as an amplifier. Therefore, while the low-frequency equivalent circuit remains unchanged, positive feedback increases the load impedance at high frequencies which translates into a higher amount of peaking. The differential impedance of the proposed active inductor is given by:

978-1-7281-5410-7/20 $31.00 © 2020 IEEE

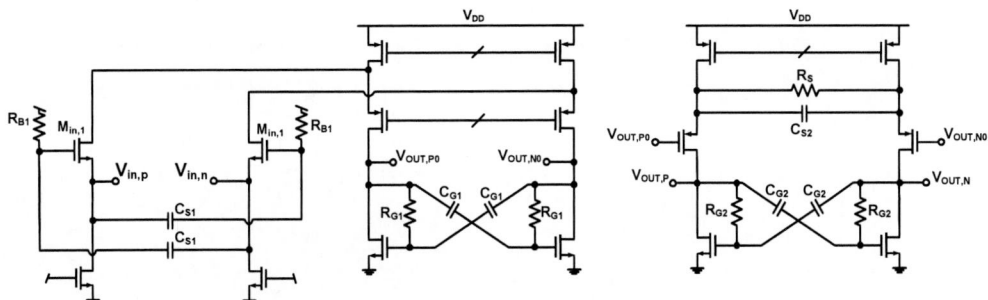

Fig. 4. Circuit schematic of the proposed CTLE II.

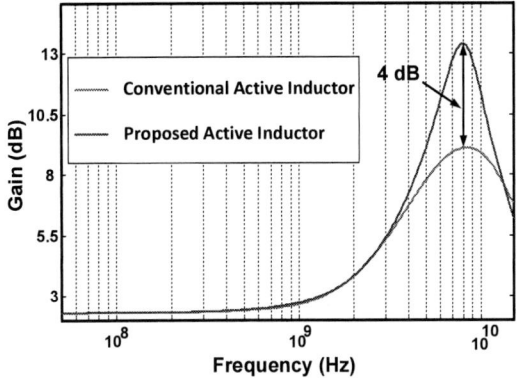

Fig. 5. Frequency response of the folded cascode stage using conventional and proposed active inductors.

$$Z_{load,diff} = \frac{2r_o}{1 + g_m r_o} \left(\frac{1 + R_G C_G S}{1 + \frac{C_G(4r_o + R_G - g_m r_o R_G)}{1 + g_m r_o} S} \right) \quad (7)$$

Similarly, the proposed active inductor has a pole-zero pair that is given by:

$$\omega_Z = \frac{1}{R_G C_G} \quad (8)$$

$$\omega_P = \frac{1 + g_m r_o}{C_G(4r_o + R_G - g_m r_o R_G)} \quad (9)$$

Comparing (8)-(9) with (4)-(5), while the zero position remains the same, the negative term in the denominator of ω_P in (9) can be exploited to increase the gain peaking. The ω_P should be a left-hand plane (LHP) pole to guarantee the stability of the system. Hence:

$$g_m r_o R_G < 4r_o + R_G \quad (10)$$

The boost factor of the proposed active inductor is given by:

$$G_{Proposed} = \omega_P / \omega_Z = \frac{(1 + g_m r_o) R_G}{4r_o + R_G - g_m r_o R_G} \quad (11)$$

Therefore:

$$\frac{G_{Proposed}}{G_{Conventional}} = \frac{r_o + R_G}{4r_o + R_G - g_m r_o R_G} \quad (12)$$

To guarantee both larger gain peaking and stability of the proposed active inductor:

$$\frac{3r_o}{R_G} < g_m r_o < 1 + \frac{4r_o}{R_G} \quad (13)$$

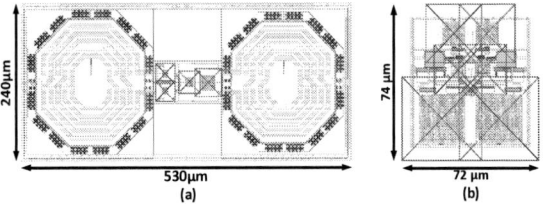

Fig. 6. Layout of the proposed (a) CTLE I, and (b) CTLE II.

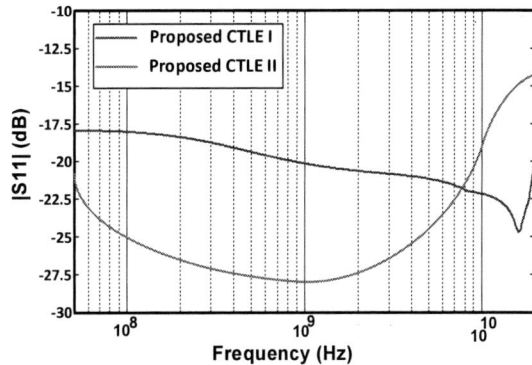

Fig. 7. Proposed CTLEs input matching.

Therefore, gain peaking can be controlled using programmable R_G to adapt the CTLE response based on the interface channel and ISI [12], [13].

Fig. 4 shows the second proposed CTLE. As it is discussed earlier, the peaking feedthrough introduced by C_{gd} of input devices can degrade the input matching. Therefore, folded-cascode topology is used for the first stage to isolate the output node and the input device that is providing the active termination. While the first stage utilizes the proposed CG structure, an RC-degenerated CS topology is employed for the second stage. Furthermore, the proposed active peaking technique is applied to both stages, while considering 25% stability margin based on (13).

The folded cascode stage of the CTLE in Fig. 4 is simulated with both conventional and proposed active inductors to verify the superior performance of the proposed technique. As shown in Fig. 5, the low-frequency gain is the same, whereas the proposed technique enhances the gain peaking by approximately 4 dB at higher frequencies. It should be noted that the proposed active inductor is implemented by only changing the C_G connections based on Fig. 3.

978-1-7281-5410-7/20 $31.00 © 2020 IEEE

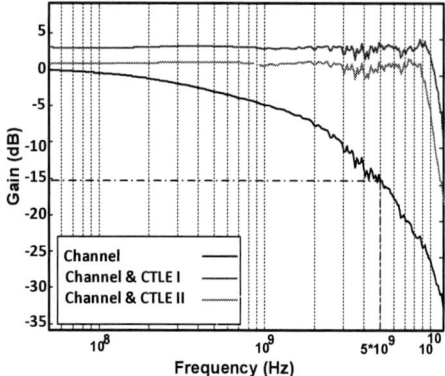

Fig. 8. Channel frequency response w/. and w/o. proposed CTLEs.

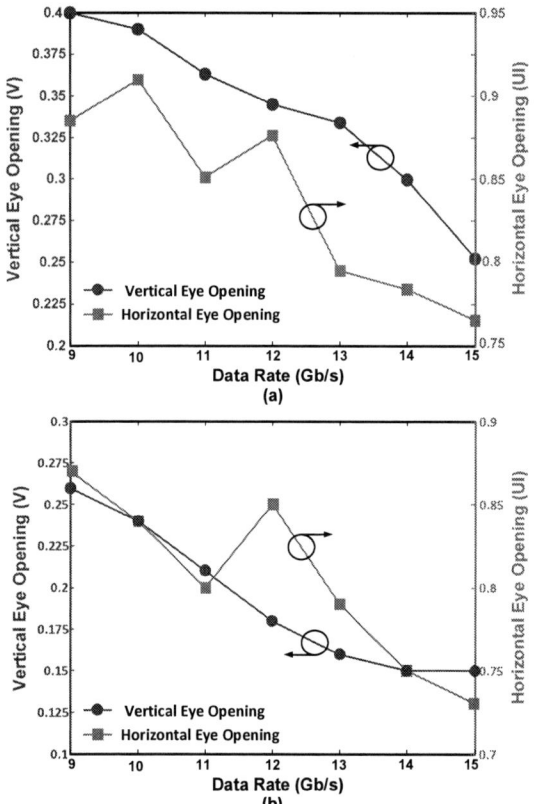

Fig. 9. Output eye height and width opening of the proposed (a) CTLE I, and (b) CTLE II for varying input data rate.

IV. POST-LAYOUT SIMULATION RESULTS AND DISCUSSION

The proposed CTLEs introduced in the previous section are designed and laid out in conventional 65-nm CMOS technology with 1.0 V supply. Fig. 6 (a) and (b) represent the layout of the proposed CTLEs in Fig. 2 and Fig. 4, respectively.

The post-layout simulation results of these CTLEs are presented here. Working at 10 Gb/s necessitates good input matching to prevent eye closure due to reflections. Fig. 7 demonstrates the input matching of the proposed CLTEs in terms of their S_{11}. As it is shown, input matching is better than -17.5 dB for both CTLEs up to 10 GHz.

A 12-inches backplane S-parameter data file given in [14] is used as the low-pass interface channel. It can be shown that for

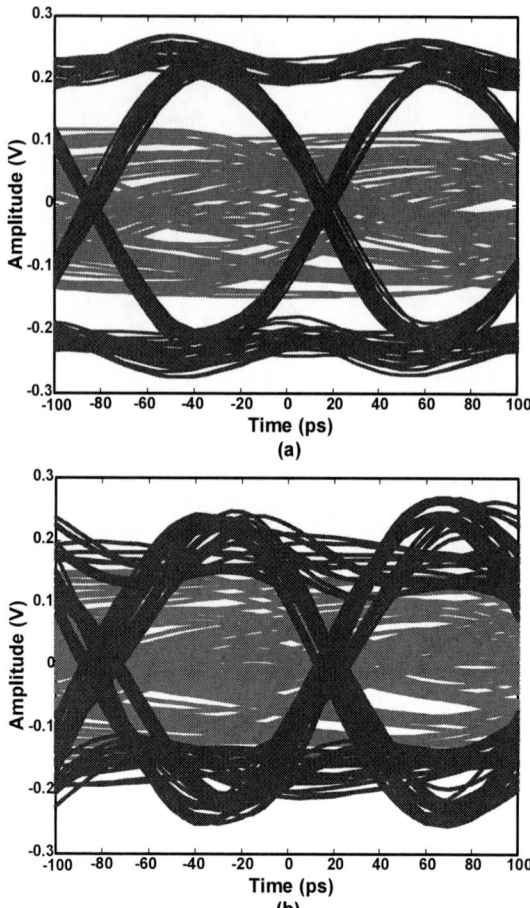

Fig. 10. Input/output (red/blue) eye diagram at 10 Gb/s for the proposed (a) CTLE I, and (b) CTLE II.

an acceptable equalization with reasonable power consumption, CTLE should compensate the low-pass channel for more BW than imposed by Nyquist frequency ($DR/2$) [15]:

$$\frac{1}{2}DR < BW < \frac{2}{3}DR \qquad (14)$$

where, BW is in GHz and DR is the input data rate in Gb/s.

Fig. 8 shows the frequency response of the interface channel and its equalized version at the output of the proposed CTLEs. The interface channel suffers from 15 dB and 20 dB loss at $5\ GHz$ and $7\ GHz$, respectively. Due to limited gain peaking of the active inductors, CTLE II is designed for smaller low-frequency gain such that it can flatten the channel response up to around 8 GHz.

By applying a random bit stream to this channel, both CTLEs are simulated over varying data rates. Fig. 9 shows the vertical (height) and horizontal (width) eye-opening for both CTLEs while operating within $9 - 15$ Gb/s. As it is expected, eye height and width for Both CTLEs have a downward trend due to the increased loss of the channel at higher data rates. At 10 Gb/s for which the proposed CTLEs are optimized, CTLE I achieves a vertical and horizontal eye-opening of 390 mV and 91% unit interval (UI), respectively. At the same speed, CTLE II with the proposed active inductors opens the closed input eye diagram by 240 mV vertically and 84% UI horizontally.

978-1-7281-5410-7/20 $31.00 © 2020 IEEE

Input and output eye diagrams at 10 Gb/s for both CTLEs are shown in Fig. 10. While the input eye is completely closed, CTLE I obtains an entirely open eye diagram at its output. The low-frequency gain of the proposed CTLEs are larger than 0 dB. Hence, output swing is larger than that of the input signal.

Since pole-zero pair in active inductors are directly related together, it restricts tuning CTLE response arbitrarily and hinders accurate compensation of the interface channel. Although CTLE II can equalize smaller losses than that of CTLE I, it still has a fairly open eye diagram while saving a huge area by replacing bulky passive inductors with proposed active inductors. While CTLE I occupies 0.127 mm^2 of die area, thanks to the proposed active inductors, CTLE II reaches an active area of 0.005 mm^2 which is approximately 25 times lower than that of CTLE I. Moreover, working with 1.0 V supply, CTLE I and CTLE II consume 6 mW and 5.4 mW, respectively. Table I compares the proposed CTLE with some of the state-of-the-art inductor-less CTLE implementations.

TABLE I
PERFORMANCE SUMMARY AND COMPARISON WITH PRIOR WORKS

	[16]	[17]		This work	
Technology	28-nm	28-nm		65-nm	
Data rate (Gb/s)	6	10	20	10	15
Channel loss (dB) @ Nyq.	28	16.2	30	15	21
Vertical eye-opening (mV)	≈540	33.5	43.5	240	150
Horizontal eye-opening (UI)	≈0.65	0.82	0.61	0.84	0.73
Power (mW)	30	4.1	7.57	5.4	5.4
FOM (fJ/bit/dB)	178	25.3	12.6	36	17
Area (mm²)	0.06	0.0014		0.005	
Measurement	Yes	No		No (Post-layout)	

V. CONCLUSION

Two common-gate CTLE circuits are presented that exploit active termination topologies to improve power efficiency while operating at 10 to 15 Gb/s. Moreover, a new active inductor is proposed and analyzed that boosts the gain peaking of the conventional topologies with almost no extra cost. The proposed CTLEs compensate for a low-pass interface channel that suffers from 15 dB and 20 dB loss at 5 GHz and 7 GHz, respectively. Using the proposed active inductor makes it possible to compensate for channel attenuation while occupying a small silicon area.

VI. REFERENCES

[1] S. Palermo, "CMOS Nanoelectronics Analog and RF VLSI Circuits: Chapter 9: High-Speed Serial I/O Design for Channel-Limited and Power-Constrained Systems," McGraw-Hill, 2011.

[2] A. Tajalli, et al., "A slew controlled LVDS output driver circuit in 0.18μm CMOS technology," *IEEE J. Solid-State Circ.*, vol. 48, no. 2, pp. 538-548, Feb. 2009.

[3] S. Gondi and B. Razavi, "Equalization and clock and data recovery techniques for 10-Gb/s CMOS serial-link receivers," *IEEE J. Solid-State Circ.*, vol. 42, pp. 1999-2011, Sep. 2007.

[4] A. Aghighi, et al., "A 10-Gb/s low-power low-voltage CTLE using gate and bulk driven transistors," *IEEE International Conference on Electronics, Circuits and Systems (ICECS)*, 2016, pp. 217-220.

[5] W. Zhuo et al., "A capacitor cross-coupled common-gate low-noise amplifier," *IEEE Trans. on Circuits and Syst. II: Exp. Briefs,* vol. 52, no. 12, pp. 875-879, 2005.

[6] A. Aghighi, et al., "CMOS Amplifier Design Based on Extended g_m/I_D Methodology," *2019 IEEE International New Circuits and Systems Conference (NEWCAS)*, Munich, Germany, 2019, pp. 1-4.

[7] J. Atkinson, et al., "Multi-Stage Current-Steering Amplifier Design Based on Extended gm/ID Methodology," *2019 IEEE 62nd International Midwest Symposium on Circuits and Systems (MWSCAS)*, Dallas, TX, USA, 2019, pp. 129-132.

[8] E. A. Sobhy, et al., "A 2.8-mW Sub-2-dB noise-figure inductorless wideband CMOS LNA employing multiple feedback," *IEEE Trans. on Microwave Theory and Tech.*, vol. 59, no. 12, pp. 3154-3161, Dec. 2011.

[9] A. Tajalli, "Symmetric linear equalizer with increased gain," US Patent No. 9'147'807, Sep. 2015.

[10] S. Badel and Y. Leblebici, "An inductorless peaking technique applied to MOS current-mode logic gates," in *Norchip Conf.*, 2004. Proceedings, 2004, pp. 36-39.

[11] H. Kimura et al., "A 28 Gb/s 560 mW multi-standard SerDes with single-stage Analog FrontEnd and 14-Tap Decision Feedback Equalizer in 28 nm CMOS," *IEEE J. of Solid-State Circ.*, vol. 49, no. 12, pp. 3091-3103, Dec. 2014.

[12] A. Tajalli et al., "A 1.02-pJ/b 20.83-Gb/s/Wire USR Transceiver Using CNRZ-5 in 16-nm FinFET," I*EEE Journal of Solid-State Circ.*, vol. 55, no. 4, pp. 1108-1123, April 2020.

[13] Tajalli, et al., "A 1.02 pJ/b 417Gb/s/mm USR link in 16nm FinFET," *IEEE VLSI Symp.*, Jun. 2019.

[14] S. Palermo, "12" Backplane S-Parameter Data" [Online]. Available: http://www.ece.tamu.edu/~spalermo/ecen689/peters_0 1_0605_B12_thr .s4p, Accessed on: Apr. 15, 2015.

[15] C. Thakkar, et. al.,, "A 10 Gb/s 45 mW Adaptive 60 GHz Baseband in 65 nm CMOS," *IEEE Journal of Solid-State Circ.*, vol. 47, no. 4, pp. 952-968, Apr. 2012.

[16] P. S. Sahni, et. al., "An Equalizer With Controllable Transfer Function for 6-Gb/s HDMI and 5.4-Gb/s DisplayPort Receivers in 28-nm UTBB-FDSOI," *IEEE Trans. on Very Large Scale Integration (VLSI) Syst.,* vol. 24, no. 8, pp. 2803-2807, Aug. 2016.

[17] J. G. Gaggatur, et. al., "A Power Efficient Active Inductor-Based Receiver Front-End for 20 Gb/s High Speed Serial Link," *AEU Int. Journal of Electronics and Comm.*, vol. 111, Nov. 2019, 152886.

PT controlled buck converter with adaptive PCCM using charge monitoring and NMOS current sensing

Yongnan Chen*, Yanhan Zeng*, Member, IEEE, Junkai Chen* and Hong-zhou Tan[†], Senior Member, IEEE

*School of Electronics and Communication Engineering, Guangzhou University, Guangzhou, China

[†]School of Electronics and Information Technology, Sun Yat-sen University, Guangzhou, China

Abstract—A pulse train controlled buck converter operating in adaptive pseudo continuous conduction mode (PCCM) is proposed and implemented in this paper. PCCM is self-adapted by using an charge monitoring and NMOS current sensor. Besides, special switch with extra timing is proposed to eliminate voltage spikes caused by PCCM. Simulation results demonstrate that the transient ripple is smaller than 24 mV for a 500 mA load current step at 1 MHz switching frequency when input and output voltages are 5 V and 2 V, respectively, providing a maximum load current of 1 A in the 0.18-μm CMOS process.

Index Terms—Buck converters, fast transient, PT control, PCCM.

I. Introduction

With the development of portable power electronic equipment, switching power converter, like DC-DC converter with the high efficiency to provide large current, has been widely applied. Conventional linear control technology such as pulse width modulation (PWM), including voltage mode and current mode [1], could hardly meet expectations both in transient response and robustness. As a non-linear technology, pulse train (PT) control has been proposed with the advantages of simple implementation, free complex compensation and fast transient response in recent years.

PT control has been widely applied in switching converters operating in discontinuous conduction mode (DCM) and continuous conduction mode (CCM) [2]. CCM is suitable for the medium and high power applications but suffers from poor transient performance. Compared with CCM, DCM needs simpler control scheme and avoids output-diode reverse-recovery problem, but results in larger current ripple and more serious electromagnetic interference(EMI) [3]. Combining the advantages of CCM and DCM, the pseudo-continuous conduction mode (PCCM) is firstly used in a boost converter proposed in [2], which is obtained in PCB instead of chip. However, spikes and glitches occur during the states switching. Besides, using a fixed and preset signal to control PCCM, the converter may mistakenly enter CCM and result in low frequency oscillation.

This work was supported by the Science and Technology Project of Guangzhou (201804010464), National Natural Science Foundation of China (61704037) and Scientific Research Project of Guangzhou University (YK2020001). Yanhan Zeng is the corresponding author (yanhanzeng@gzhu.edu.cn).

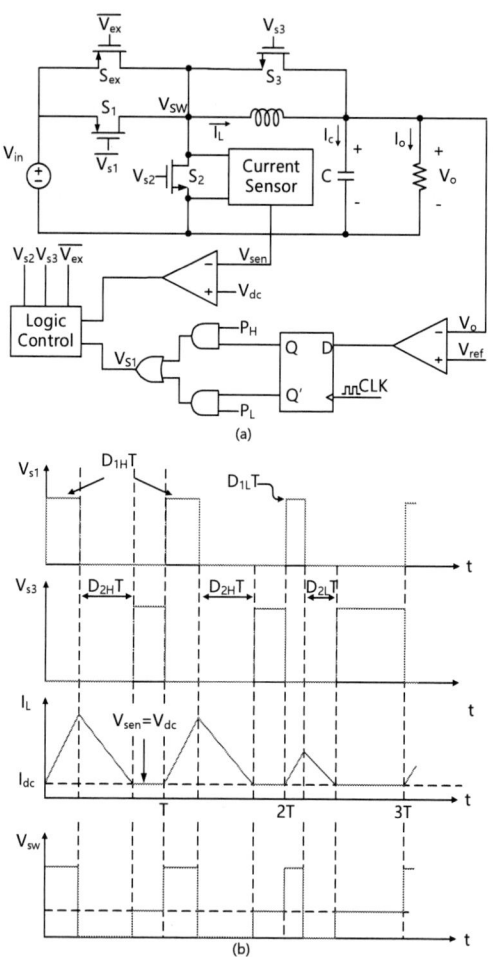

Fig. 1. PT controlled PCCM buck converter:(a) circuit schematic diagram;(b) operation waveforms.

This paper proposed a PT controlled buck converter with adaptive PCCM. By using an analog charge monitoring, both the PCCM control signal and phase duration are adaptively and continuously adjusted. The paper is organized as follows. To begin with, the operational principle and structure of the proposed converter are discussed in the section II. Then the circuit implementation is

978-1-7281-5410-7/20 $31.00 © 2020 IEEE

described in section III. In section IV, the simulation results and the comparison with other published designs are presented. Finally, the conclusion is drawn in section V.

II. PT CONTROLLED PCCM BUCK CONVERTER

The schematic diagram and operating waveforms of PT controlled PCCM buck converter are shown in Fig. 1. Assuming that the switching cycle is T, one switching cycle is composed of three parts: D_1T, D_2T and D_3T. At the beginning, S_1 is turned on and the inductor current I_L ramps up. The pulse width depends on the power control pulse P_H and P_L. If the output voltage $V_o < V_{ref}$, high-power control pulse P_H will be selected as the control signal of S_1, which will increase the energy input, leading to the increase of V_o increases. On the contrary, if $V_o > V_{ref}$, low-power control pulse P_L is selected, which will decrease the energy input, and V_o decreases. In the second part D_2T, the inductor current ramps down. When I_L decreases to a predefined I_{dc}, which is the control signal of PCCM, the third part D_3T is activated. A freewheel switch S_3 is turned on, circulating a fixed I_L in the loop formed by S_3 and the inductor. The peak inductor current in a switching cycle is i_{LP}, which is given by:

$$i_{LP} = \frac{(V_{in} - V_o) D_1 T}{L}. \tag{1}$$

Considering that the variation of inductor current should be zero in a switching cycle, D_2 is obtained by means of "voltage-second" balance principle.

$$D_2 = \frac{V_{in} - V_o}{V_o} D_1 \tag{2}$$

During freewheeling phase D_3T, input current is also zero, the same as D_2T. Only in D_1T do circuit absorbs energy from input V_{in}. From Fig. 1(b), the average input current in a switching cycle is obtained as:

$$I_{in} = I_{dc} + \frac{V_{in} (V_{in} - V_0) T D_1^2}{2V_0 L} \tag{3}$$

Thus the input power of PT controlled PCCM buck converter under high/low-power control pulses are obtained as:

$$P_H = V_{in}I_{dc} + \frac{V_{in}^2 (V_{in} - V_0) T D_{1H}^2}{2V_0 L} \tag{4}$$

$$P_L = V_{in}I_{dc} + \frac{V_{in}^2 (V_{in} - V_0) T D_{1L}^2}{2V_0 L} \tag{5}$$

From Eq. (4) and (5), it can be concluded that the input power is ranging between P_H and P_L. Therefore, the output power between P_H and P_L, can only be regulated by adjusting the combination of D_{1H} and D_{1L} in the repetition cycle. In other words, when the output power is greater than P_H or less than P_L, the output voltage can't be regulated. The PT controlled DCM buck converter can be considered as a special case of PCCM when $I_{dc} = 0$. Thus the input power of DCM can be derived as:

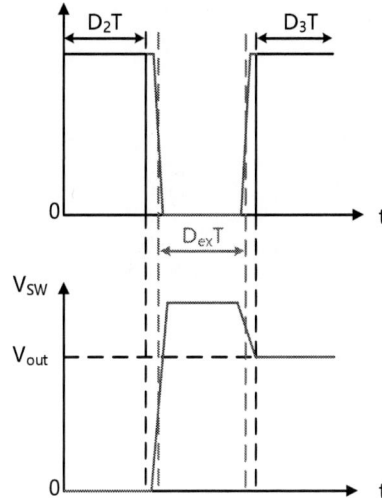

Fig. 2. Timing diagram of $D_{ex}T$.

$$P_{D\_H} = \frac{V_{in}^2 (V_{in} - V_0) T D_{1H}^2}{2V_0 L} \tag{6}$$

$$P_{D\_L} = \frac{V_{in}^2 (V_{in} - V_0) T D_{1L}^2}{2V_0 L} \tag{7}$$

Comparing (4), (5) with (6), (7), it can be concluded that buck converter operating in PCCM has a wider load range than that in DCM under the same other circuit parameters due to the existence of preset value I_{dc} of inductor current.

III. CIRCUIT IMPLEMENTATION

A. Spike decrease

As shown in Fig. 1(b), S_3 is turned on to circulate a fixed I_L during D_3T. However, because V_{sw} almost falls down to zero before D_3T, there is a large voltage between V_{sw} and V_{out}, and the conduction of S_3 causes a large current, which is flowing from V_{out} to S_3 ,creating a spike in output.

A technique to decrease the voltage ripple is proposed by making V_{sw} approach V_{out} at the beginning of D_3T. As shown in Fig. 1, an additional switch S_{ex} is used, which is controlled by an extra timing $D_{ex}T$. As shown in Fig. 2, $D_{ex}T$ begins at the end of D_2T. Thus S_{ex} is turned on, increasing V_{sw} almost to V_{in}. S_{ex} is turned off and D_3T begins when V_{sw} falls back to V_{out}. As a result, the spike almost do not occur since no large current flows from V_{out} to V_{sw}.

B. Current sensor for NMOS

PCCM buck converter needs current detection during D_2T to compare with the PCCM control signal V_{dc}. Unlike the current mode PWM control, this paper proposes a current sensor for NMOS instead of PMOS.

As shown in Fig. 3, the current sensor works at the beginning of D_2T. Both the current of S_2 and V_{sw} are

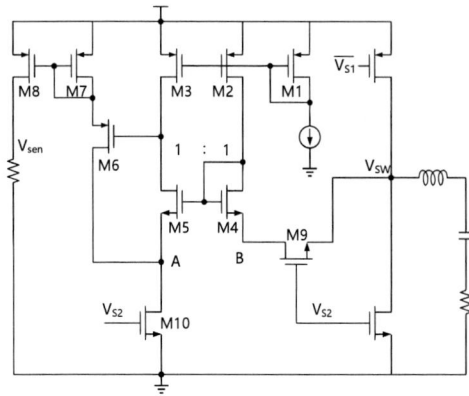

Fig. 3. The schematic of the current sensor.

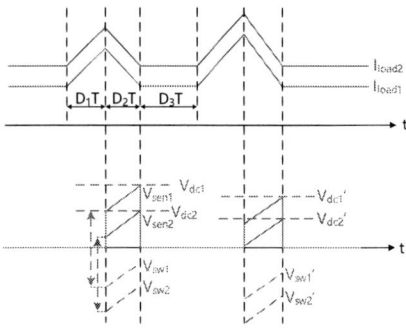

Fig. 4. Timing diagram of PCCM with load variation.

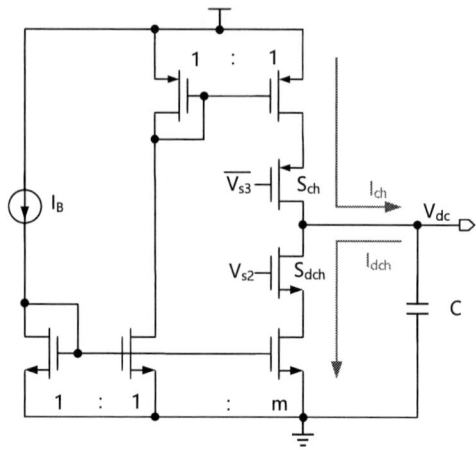

Fig. 5. The schematic of the analog charge monitoring.

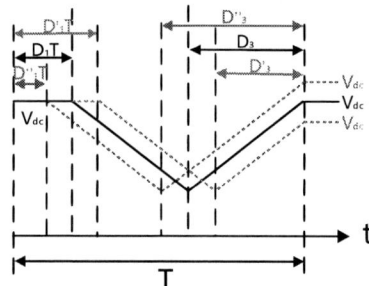

Fig. 6. Modulation processes of V_{dc} with load variation.

negative during D_2T. However, due to the DC offset of the amplifier, made up by M1-M5, the negative voltage V_B is changed to positive value in node A. As the inductor current falls down, V_{sw} increases and so as V_B and V_A. The falling of the inductor current is sensed and changed to a rising of V_{sen}, as shown in Fig. 4. Once V_{sen} increases to V_{dc}, the voltage form of PCCM control signal D_3T, is activated.

As shown in Fig. 4, the curve of V_{sen} moves down when the load current increases, thus it needs longer time to reach V_{dc}. On the other hand, the increase of load current will shorten D_2T if V_{dc} is fixed. The converter may enter CCM if the increase is significant, which results in low frequency oscillation. Therefore, V_{dc} should be adaptively changed with the load.

C. Charge monitoring

A charge monitoring [4] is proposed to adaptively adjust V_{dc}. The schematic of the circuit is depicted in Fig. 5 and its operating principle can be explained in Fig. 6.

As shown in Fig. 5, S_{ch} is turning on during D_3T and S_{dch} is turning on during D_2T. V_{dc} in Fig. 5 is charged by a constant sink current I_{ch} during D_3T and discharged by a constant source current $I_{dch} = mI_{ch}$ during D_2T. In

steady state (solid line in Fig. 6), V_{dc} keeps constant and an optimal D_3T can be achieved due to the feedback. It satisfies:

$$I_{ch} \times D_3T = I_{dch} \times D_2T \qquad (8)$$

Assuming that the capacitor charge and discharge power are Q_{in} and Q_{out}, respectively.

$$\begin{cases} Q_{in} = Q_{out}, & \text{stable state} \\ Q_{in} < Q_{out}, & \text{load decrease} \\ Q_{in} > Q_{out}, & \text{load increase} \end{cases} \qquad (9)$$

As shown in Fig. 6, if load current increases, the corresponding charge and discharge periods of inductor current ($D_1'T$ and $D_2'T$) need to be extended to deliver more power. $D_3'T$ is reduced accordingly, resulting that Q_{in} is smaller than Q_{out}. Consequently, V_{dc} is decreased to V_{dc}' (red line in Fig. 6). Similarly, when a load decrease occurs (blue line in Fig. 6), the V_{dc} value would increases. By this way, PCCM control signal V_{dc} can be adaptively adjusted based on the load current.

TABLE I
Design parameters of the proposed buck converter.

Technology	$0.18\mu m$ CMOS
Maximum load current, $I_{L,max}$	1000 mA
Switching frequency, f_s	1 MHz
Inductor, L	4.7 μH
Capacitor, C	10 μF
ESR, R_c	$10m\Omega$

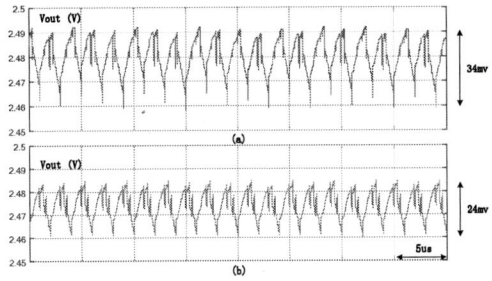

Fig. 7. Voltage ripple without spike decrease technique(a) or with decrease technique(b) when load current is 500 mA.

IV. Simulation results

The proposed buck converter with the specifications, as shown in table I, has been implemented in the 0.18 μm CMOS process. The converter provides a well regulated variable output voltage range from 1 V to 4 V and is able to supply a maximum load current of 1000 mA.

Fig. 7 shows the voltage ripple with and without proposed spikes decrease technology. The voltage ripple are 24mV and 34 mV, respectively. The ripple is decreased by 29%.

For demonstrating the fast transient response, the load current steps from 500 mA to 1000 mA, and vice versa. As shown in Fig. 8, the undershoot and overshoot are 22 mV and 8 mV, respectively. The transient recovery times of undershoot and overshoot voltages are 19 μs and 6 μs,

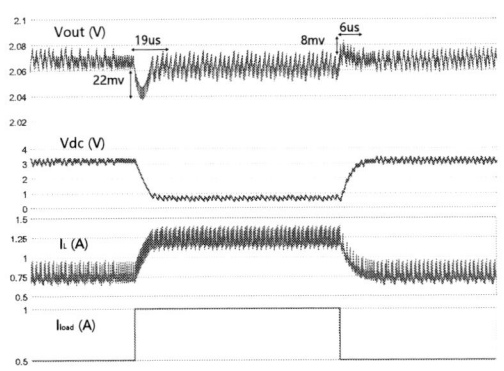

Fig. 8. Load transient response of proposed buck converter when load current steps between 500 mA to 1000 mA.

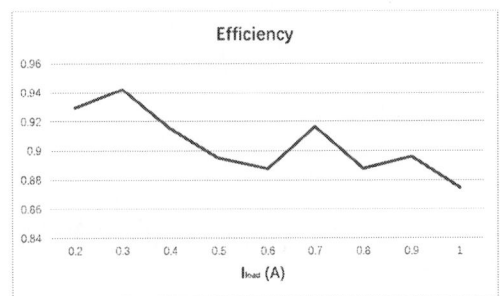

Fig. 9. Simulated efficiency at different load currents.

respectively.

Fig. 9 shows the efficiency of the proposed converter. The highest efficiency of 94% is achieved when the load is about 300 mA. To avoid the switches simultaneously turning on, an adaptive deadtime technique is used to control the duty of S_{ex}. When the load changes, the deadtime will fluctuate. Therefore, the efficiency will also fluctuate.

TABLE II
Comparison of recently published CMOS DC-DC converter.

	TCAS-I [5]	ISCAS [6]	TPE [7]	TVLSI [8]	This work
Year	2012	2014	2018	2018	2019
Process	$0.35\mu m$	$0.25\mu m$	$0.35\mu m$	$0.35\mu m$	$0.18\mu m$
Inductor	22 μH	4.7 μH	2.2 μH	4.7 μH	4.7 μH
Capacitor	22 μF	10 μF	4.7 μF	10 μF	10 μF
Frequency	1 MHz	0.5-1.5 MHz	2.5-3.1 MHz	1 MHz	1 MHz
$\triangle V_{out}$	200 mV	250 mV	47 mV	167 mV	22 mV
$\triangle I_{Load}$	450 mA	400 mA	500 mA	300 mA	500 mA
$I_{L,max}$	500 mA	500 mA	600 mA	600 mA	1000 mA
Efficiency	90%	94.4%	92%	90%	94.2%

Table II shows a comparison with previously reported works [5]–[8]. Incorporating with all the advancements described above, our design features the high efficiency and large load current. Especially, smallest transient ripple is obtained.

V. Conclusion

In this paper, a PT Controlled buck converter with adaptive PCCM is proposed and simulated in a 0.18-μm CMOS process. The operating principle, circuit implementation and simulation results have been described in detail. Particular attention has been paid to obtain spike decrease and self-adapted PCCM. Simulation results show that the overshoot/undershoot is smaller than 22 mV as the load current suddenly changes from 500 mA to 1000 mA, or vice versa. As the superiority of fast transient response and high efficiency, proposed buck converter has a wide range of application.

References

[1] Y. Zeng and H. Tan, "Fast-transient dc-dc converter using an amplitude-limited error amplifier with a rapid error-signal control," in 2019 IEEE International Symposium on Circuits and Systems (ISCAS), May 2019, pp. 1–5.

[2] G. Zhou, W. Tan, S. Zhou, Y. Wang, and X. Ye, "Analysis of pulse train controlled pccm boost converter with low frequency oscillation suppression," IEEE Access, vol. 6, pp. 68 795–68 803, 2018.

[3] F. Zhang and J. Xu, "A novel pccm boost pfc converter with fast dynamic response," IEEE Transactions on Industrial Electronics, vol. 58, no. 9, pp. 4207–4216, Sep. 2011.

[4] Y. Zhang and D. Ma, "Integrated simo dc-dc converter with on-line charge meter for adaptive pccm operation," in 2011 IEEE International Symposium of Circuits and Systems (ISCAS), May 2011, pp. 245–248.

[5] P. J. Liu, W. S. Ye, J. N. Tai, and H. S. Chen, "A high-efficiency CMOS DC-DC converter with 9-μs transient recovery time," Circuits and Systems I Regular Papers IEEE Transactions on, vol. 59, no. 3, pp. 575–583, 2012.

[6] K. I. Wu, S. H. Hung, S. Y. Shieh, B. T. Hwang, S. Y. Hung, and C. C. P. Chen, "Current-mode adaptively hysteretic control for buck converters with fast transient response and improved output regulation," in 2014 IEEE International Symposium on Circuits and Systems (ISCAS), June 2014, pp. 950–953.

[7] M. Nashed and A. A. Fayed, "Current-mode hysteretic buck converter with spur-free control for variable switching noise mitigation," IEEE Transactions on Power Electronics, vol. 33, no. 1, pp. 650–664, Jan 2018.

[8] J. J. Chen, Y. S. Hwang, J. H. Chen, Y. T. Ku, and C. C. Yu, "A new fast-response current-mode buck converter with improved i^2-controlled techniques," IEEE Transactions on Very Large Scale Integration (VLSI) Systems, vol. PP, no. 99, pp. 1–9, 2018.

Fast-transient, light-load efficient DC-DC converter using an auxiliary D-LDO

Haochang Zhi*, Yanhan Zeng*, Member, IEEE, Wei Zhou* and Hong-zhou Tan[†], Senior Member, IEEE
*School of Electronics and Communication Engineering, Guangzhou University, Guangzhou, China
[†]School of Electronics and Information Technology, Sun Yat-sen University, Guangzhou, China

Abstract—A current-mode DC-DC buck converter with a parallel auxiliary digital-LDO has been proposed and simulated in a 0.18-μm CMOS process in this paper. The proposed auxiliary digital LDO can provide both sink current and source current, which improves the transient response by rapidly injecting a positive or negative current into the output to effectively reduce the overshoot/undershoot voltage. Besides, the digital LDO is also activated and provided the out current during the light load to improve the efficiency. Simulation results demonstrate that the transient ripple is smaller than 60 mV with 800 mA load current step and 1 MHz switching frequency when the input and output voltages are 5 V and 3.3 V, respectively. Furthermore, 61% efficiency maintains under 20 mA load current, which is higher than the traditional converters.

Index Terms—Buck converters, fast transient, high efficiency.

I. Introduction

Along With the popularization of portable devices like smart watches and smart mobile phones, an urgent demand for high-performance power management converters such as DC-DC converters are given, which are able to provide large load current with high efficiency. Usage of current-mode DC-DC converter is becoming more and more popular as the simple compensation, relatively high bandwidth and intrinsic current protection. However, the transient response is limited due to the low bandwidth and the large compensation capacitor. Besides, the efficiency in light load is not as good as that in high load because of the fixed consumption.

Typical fast-transient methods are to raise the loop gain [1] or the slew-rate of the amplifier [2] during the transient period. The only fly in the ointment is that the unstable response increases the transient voltage variation and the stability should be specially considered. It is a long time to improve the transient response by auxiliary structure parallel with a DC-DC converter. [3] provide a way to obviously decrease ripple by injecting transient current from auxiliary DC-DC converter but applies another inductor. [4] in an inductor-less way but faces flexibility. Besides, applying a complex structure results in extra energy consumption and die area.

This work was supported by the National Natural Science Foundation of China (61704037), Science and Technology Project of Guangzhou (201804010464) and Scientific Research Project of Guangzhou University (YK2020001). Yanhan Zeng is the corresponding author (yanhanzeng@gzhu.edu.cn).

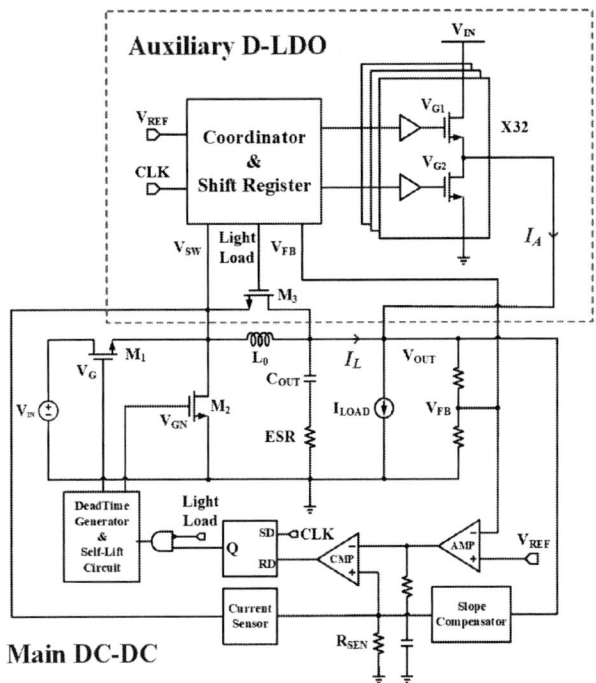

Fig. 1. The proposed system of this paper

This paper proposes a new main-auxiliary structure to improve the transient response. A digital LDO (D-LDO) without extra off-chip components, which can independently provide both source current and sink current, is used as the auxiliary structure. Proposed D-LDO is programmable and flexible, thus it can rapidly inject the non-linear current into the output during the transient period, and work as the current supply in the light load to improve efficiency.

The paper is organized as follows. To begin with, the operating principle and the system structure are discussed in section II. The circuit implementation is described in section III. In section IV, the simulation results and the comparison with other published designs are presented. Finally, the conclusion is drawn in section V.

II. The operating principle

The block diagram of the proposed buck converter with auxiliary 32-bits D-LDO is illustrated in Fig. 1. It is worth

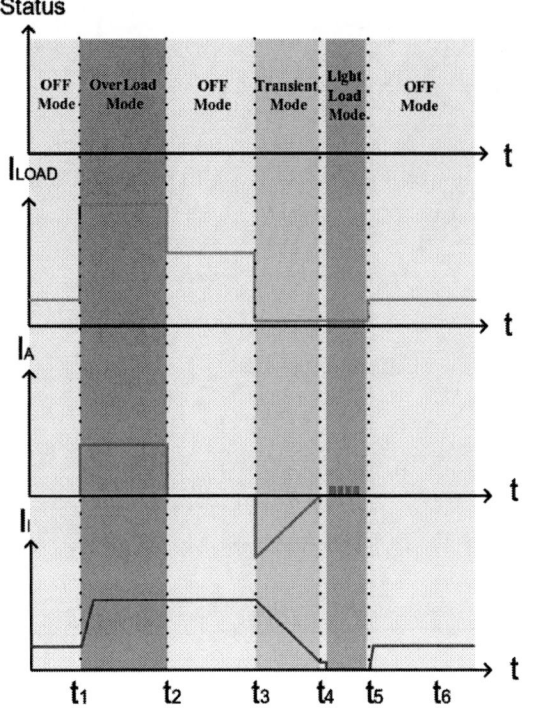

Fig. 2. Operation waveforms of different modes.

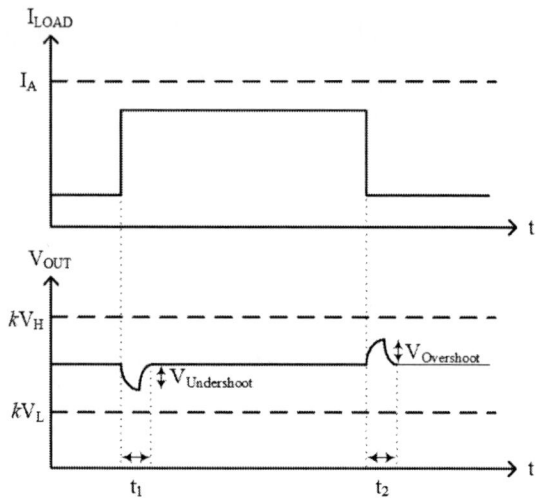

Fig. 3. Waveforms of transient

noting that both D-LDO and DC-DC converter share the same CLK. D-LDO provides different currents in both magnitude and direction based on the different status. The status is adaptively generated by the coordinator, as shown in Fig. 2.

A. Off mode

No current is provided by D-LDO in off mode so that the efficiency of the whole system is only determined by the DC-DC converter. Fig. 2 shows the operating waveforms of different modes, where I_{LOAD}, I_A and I_L are the load current, D-LDO current and inductor current, respectively. Out of consideration for the stability, it is not allowed to switch the state among light-load mode, transient mode, and over-load mode until the system returns to off mode.

B. Transient mode

During the transient period, the output ripple triggers the transient mode. During a step-up load change, D-LDO provides a sink current to charge output. All of the switches turn on when transient begins, then D-LDO cuts-off bit by bit until the system returns to off mode. During a step-down load change, D-LDO provides a source current to discharge output, as shown in Fig. 2.

$$\Delta V_{out}(i) = \begin{cases} \dfrac{T}{C}\displaystyle\sum_{j=0}^{i-1} I_{ch,j} \; - \; \Delta i_o \times i \; + \; \dfrac{1}{2}\,m_1\,i^2\,T \\[3mm] -\dfrac{T}{C}\displaystyle\sum_{j=0}^{i-1} I_{dch,j} \; + \; \Delta i_o \times i \; - \; \dfrac{1}{2}\,m_2\,i^2\,T \end{cases} \quad (1)$$

$$I_{ch,j} = \mu_n\,C_{ox}\,(N-j)\frac{W}{L}\,(2\,V_{in} - V_{out} - V_{th}\,)(V_{in} - V_{out}) \tag{2}$$

$$I_{dch,j} = \mu_n\,C_{ox}\,(N-j)\frac{W}{L}\,(2\,V_{in} - V_{th}\,)V_{out} \tag{3}$$

Considering the transient detector is able to detect the load step instantly, the expressions of the over-shoot/undershoot in $i-th$ period are obtained in Eq. (1), where $i \in [1, N], m_1 = (V_{in} - V_o)/L, m_2 = V_o/L$, Δi_o is the load step size, I_{ch} and I_{dch} are the charge and discharge current provided by D-LDO. Theoretically, if D-LDO produces a current as ΔI_{LOAD}, the undershoot/overshoot would be almost eliminated.

$$V_{Undershoot} = \frac{\Delta I_{LOAD} \times t_1}{C_{OUT}} + \Delta I_{LOAD} \times ESR \tag{4}$$

$$V_{Overshoot} = \frac{\Delta I_{LOAD} \times t_2}{C_{OUT}} + \Delta I_{LOAD} \times ESR \tag{5}$$

As shown in Fig.3, the expressions of $V_{Undershoot}$ and $V_{Overshoot}$ are obtained in Eq. (4) and Eq. (5). It shows that the overshoot voltage of V_{OUT} is directly proportional to the change of load current during the transient. Therefore, when $\Delta I_{LOAD} = I_A$, the overshoot voltage is kV_H and kV_L, where k is the ratio of two resistors connected to V_{FB}. When the transient occurs and the V_{OUT} is lower than kV_L, the coordinator will control the D-LDO to inject 32 I_1 into the output capacitor. It has been

978-1-7281-5410-7/20 $31.00 © 2020 IEEE

derived that the overshoot/undershoot would be almost eliminated by D-LDO produces a current as ΔI_{LOAD}. Therefore, the threshold voltage comparing with V_{FB} and control whether entering the transient mode is set to V_H and V_L, as shown in Eq. (6) and Eq. (7).

$$V_H = \frac{I_A \times t_2}{C_{OUT}} + I_A \times ESR \qquad (6)$$

$$V_L = \frac{I_A \times t_1}{C_{OUT}} + I_A \times ESR \qquad (7)$$

C. Light-load mode

$$\eta = \frac{V_{OUT}}{V_{IN}} \times \frac{I_{LOAD}}{I_{LOAD} + I_{STATIC}} \qquad (8)$$

Eq. (8) gives the efficiency of the D-LDO. Considering V_{IN} and V_{OUT} are 5V and 3.3V, the efficiency approximates 66% because the load current is usually much higher than the quiescent current. Unlike the high efficiency in heavy load, the efficiency of the DC-DC converter decreases in light load because the large switch losses and the operating current are fixed. In the practice design, D-LDO performs higher efficiency than the DC-DC converter when the load current is lower than 38mA, which is verified by the simulation. Thus once the load current becomes light, the system enters the light-load mode, DC-DC converter sleeps and D-LDO activates.

D. Over-load mode

The switching between the Off mode and Light-load mode is determined by the voltage of switch node V_{sw} during the M1-turn-off period, as shown in Fig. 1. When the load current is light, $V_{sw} > 0$ and system operates in Light-load mode. Once V_{sw} is lower than 0 because of the increased load, the system turns into Off mode. When the load current exceeds 1.5 A, which is the maximum load current of the DC-DC converter, and would be unstable. In this case, taking full advantage of D-LDO, the system enters over-Load mode and D-LDO provides the exceeding current.

III. CIRCUIT IMPLEMENTATION

The proposed fast-transient technique consists of a current control mode DC-DC converter and an auxiliary D-LDO.

A. DC-DC converter

As shown in Fig. 1, a typical dual-NMOS DC-DC converter is applied. M3 is used to short the inductor in light-load mode, at the same time, V_G and V_{GN} are set to ground, so that DC-DC converter is cut off. Besides, as shown in Fig. 4, a new simple dead-time generator has been used to improve efficiency. Benefiting from applying dual-NMOS with a self-lift circuit, the voltage of the switch node SW varies rapidly between over ground and under ground. As a result, an inverter with a low flip threshold is used instead of the zero-cross voltage comparator in the adaptive deadtime generation, which reduces the complexity.

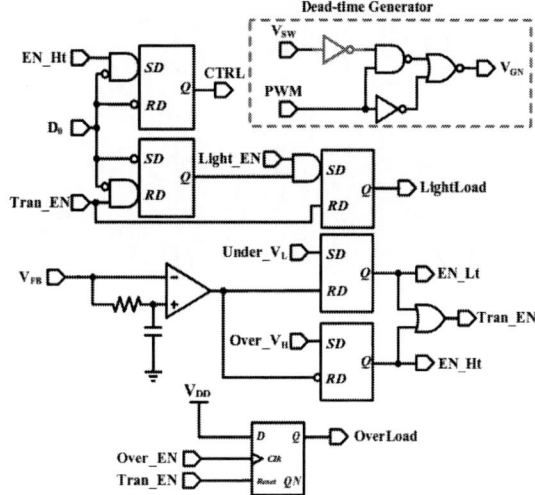

Fig. 4. The diagram of the coordinator and proposed adaptive dead-time generator.

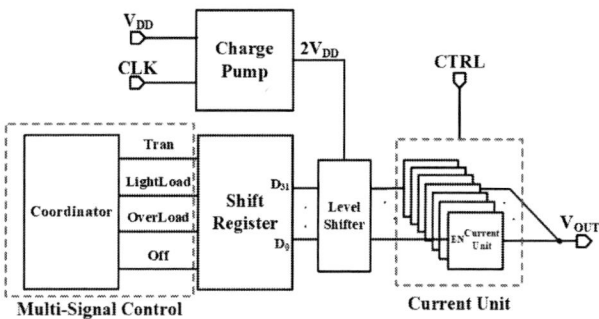

Fig. 5. The schematic of the D-LDO.

B. D-LDO

D-LDO consists of a coordinator, a shift register with multi-signal control, a level shifter, a charge pump and a class of current units, as shown in Fig. 5.

As shown in Fig. 4, the coordinator consists of D-flip-flops, detection and logic circuits. D-flip-flops record the operating mode. Detection is composed of several comparators to compare the feedback signals with the reference signals.

Four signals are used to determine the modes in the coordinator. Once the feedback voltage V_{FB} is higher/lower than V_H/V_L, the transient signal TRAN_EN is triggered. EN_Ht and EN_Lt represent the step-down and step-up, respectively. During the step-up transient period, once V_{FB} decreases to less than V_L, the coordinator enables EN_Lt. The shift register asynchronously set to the high level and begin to shifts 0 into the sequence starting from D31 bit by bit . When the lowest bit D0 is set to 0, the system returns to off mode, and then gets into light-load mode by the Light_EN signal. The signal of CTRL controls the current direction of D-LDO. D-LDO provides source current when CTRL = 1 and provides sink

978-1-7281-5410-7/20 $31.00 © 2020 IEEE

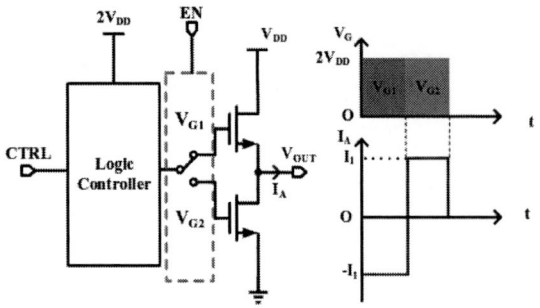

Fig. 6. Proposed Current Unit.

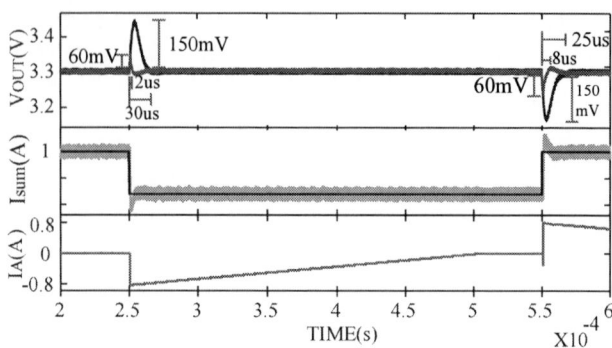

Fig. 7. Load transient response when load current changing between 200 mA and 1000 mA.

current when CTRL = 0. During the step-down transient period, the coordinator enables EN_Ht. The shift register operates in the same way while EN_Ht set CTRL to 1 and D-LDO provides source current instead of sink current.

The charge pump is used to provide a double voltage of V_{DDA}. The shift register controls the current units by bit based on the operating mode.

1) Transient mode: The shift register sets all the bits to 1 when the transient mode is triggered and begins to shift 0 bit by bit. Finally, all current units are cut-off.

2) Light-load mode: Shift register works in the normal way and is determined by the feedback comparison.

3) Over-load mode: All the bits are set to 1.

4) Off mode: All the bits are set to 0.

The current unit with dual-NMOS structure is illustrated in Fig. 6, which is powered by the double V_{DD}. In this way, both power consumption and die size are reduced. The signal EN enables the current unit and $CTRL$ controls the current direction. Each unit can provide more than 30mA sink current and not less than 40mA source current when V_{OUT}=3.3 V and V_{IN}=5 V.

TABLE I
Design parameters of the proposed system.

Technology	0.18μm BCD CMOS
Input voltage, V_{in}	5 V
Output voltage range, V_{out}	2V~4 V
Maximum load current, $I_{L,max}$	2500 mA
Switching frequency, f_s	1 MHz
Inductor, L	4.7 μH
Capacitor, C	10 μF
ESR, R_c	30$m\Omega$

IV. Simulation results

Table I shows design parameters of the proposed system. As shown in Fig. 7, when the load current steps between 1000 mA and 200 mA. The undershoot and overshoot with proposed fast transient system are both 60 mV. The transient recovery times of undershoot and overshoot voltages are 8 μs and 2 μs, respectively. The values of the converter without proposed technology increase to 150 mV and 30 μs, respectively. It is obvious that the proposed

fast transient control significantly improves the transient response, reducing undershoot/overshoot by 60% and the recovery time by 68%.

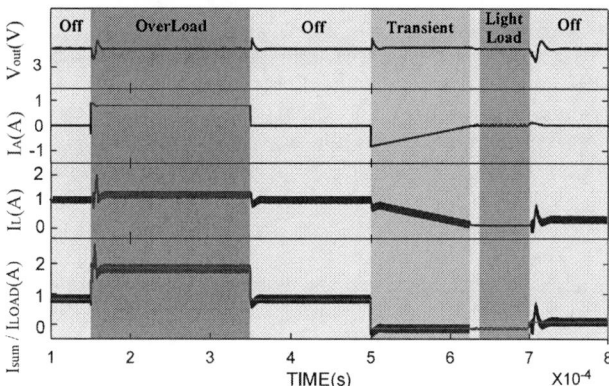

Fig. 8. Simulation of modes switch.

Fig. 8 shows how the system switches in those modes. When I_L steps from 1000 mA to 2000 mA. After I_L steps back to 1000 mA, the system returns to Off Mode, then I_L steps from 1000 mA to 50 mA, the undershoot voltage and the recovery time are 210 mV and 63 μs, respectively. After the transient, Ligth-Load detector detects the voltage of the switch node is over zero and sets DC-DC converter to sleep, multiplexer switching over, D-LDO begins to work. Finally, I_L steps from 50 mA to 200 mA, the system returns to Off Mode.

Fig. 9 shows the efficiency of the proposed converter. The highest efficiency, which is about 92% in this design, can be achieved when the load current is about 500 mA. To improve the Light-Load (20 mA-100 mA) efficiency, the proposed converter enters light-load mode. D-LDO provides the power, thus reduces the switching loss. As shown in Fig. 9, the efficiency increases by 5%~15% when the load current ranges from 10mA to 30 mA. Benefiting from the over-load mode, the maximal load current extends to 2.5A. The efficiency is deteriorated in the heavy load but still above 82%.

Table II shows a comparison with previously reported

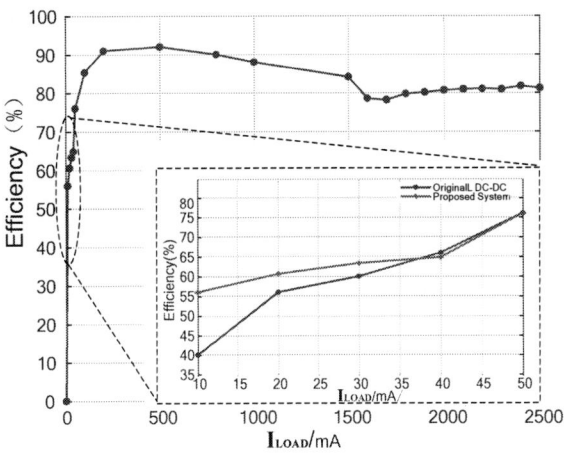

Fig. 9. The efficiency of the proposed converter

TABLE II
Comparison of recently published CMOS DC-DC converter.

	ISCAS [5]	VLSI-DAT [6]	TPE [7]	TVLSI [8]	This work
Year	2019	2018	2018	2018	2019
Process	$0.4\mu m$	$0.35\mu m$	$0.35\mu m$	$0.35\mu m$	$0.18\mu m$
Inductor	4.7 μH	4.7 μH	2.2 μH	4.7 μH	4.7 μH
Capacitor	10 μF	10 μF	4.7 μF	10 μF	10 μF
Frequency	1 MHz	0.5 MHz	2.5-3.1 MHz	1 MHz	1 MHz
$\triangle V_{out}$	35 mV	100 mV	47 mV	167 mV	60 mV*
$\triangle I_{Load}$	450 mA	450 mA	500 mA	300 mA	800 mA
Recovery Time	27.2μs	10μs	4.7μs	5μs	2/8μs*
$I_{L,max}$	1000 mA	500 mA	500 mA	600 mA	2500 mA
Efficiency	91%	94%	92%	90%	92%

*Simulation result.

works [5]–[8]. Incorporating with all the advancements described above, our design provides the maximum load current. Especially, the smallest transient ripple voltage and the shortest transient recovery time are obtained.

V. Conclusion

In this paper, a current-mode buck DC-DC converter with a parallel D-LDO is proposed to achieve a fast transient response. The operating principle, circuit implementation, and simulation results have been described in detail. Particular attention has been paid to obtain the fast-transient response by recovering the charge shortage or releasing the extra charges during the transient operation. Simulation results show that the overshoot/undershoot is smaller than 60 mV as the load current suddenly changes from 1000 mA to 200 mA, or vice versa. As the superiority of a fast transient response, a small chip area, a high load current, and a high efficiency, the buck converter proposed above have a wide range of applications, especially on the occasion of the fast load change and low costs.

References

[1] L. Chen and B. Ferrario, "Adaptive frequency compensation for DC-to-DC converter," Feb. 1 2011, uS Patent 7,880,446.

[2] J. Roh, "High-performance error amplifier for fast transient DC-DC converters," IEEE Transactions on Circuits and Systems II: Express Briefs, vol. 52, no. 9, pp. 591–595, 2005.

[3] A. Barrado, A. Lazaro, R. Vazquez, V. Salas, and E. Olias, "The fast response double buck dc-dc converter (frdb): operation and output filter influence," IEEE Transactions on Power Electronics, vol. 20, no. 6, pp. 1261–1270, Nov 2005.

[4] S. Kapat, P. S. Shenoy, and P. T. Krein, "Near-null response to large-signal transients in an augmented buck converter: A geometric approach," IEEE Transactions on Power Electronics, vol. 27, no. 7, pp. 3319–3329, July 2012.

[5] Y. Zeng and H. Tan, "Fast-transient dc-dc converter using an amplitude-limited error amplifier with a rapid error-signal control," in 2019 IEEE International Symposium on Circuits and Systems (ISCAS), May 2019, pp. 1–5.

[6] Y. Chiu, Y. Liu, and C. Hung, "A high-performance current-mode dc-dc buck converter with adaptive clock control technique," in 2018 International Symposium on VLSI Design, Automation and Test (VLSI-DAT), April 2018, pp. 1–4.

[7] M. Nashed and A. A. Fayed, "Current-mode hysteretic buck converter with spur-free control for variable switching noise mitigation," IEEE Transactions on Power Electronics, vol. 33, no. 1, pp. 650–664, Jan 2018.

[8] J. J. Chen, Y. S. Hwang, J. H. Chen, Y. T. Ku, and C. C. Yu, "A new fast-response current-mode buck converter with improved i^2-controlled techniques," IEEE Transactions on Very Large Scale Integration (VLSI) Systems, vol. PP, no. 99, pp. 1–9, 2018.

978-1-7281-5410-7/20 $31.00 © 2020 IEEE

Subthreshold-Hybrid Solutions for Thermal Sensor and Reference Circuits in Advanced CMOS

Matthias Eberlein
Institute for Integrated Circuits (IIC)
Johannes Kepler University
Linz 4040, Austria
matthias.eberlein@jku.at

Harald Pretl
Institute for Integrated Circuits (IIC)
Johannes Kepler University
Linz 4040, Austria
harald.pretl@jku.at

Abstract— **We present a family of hybrid structures, which combine subthreshold MOS operation with a single bipolar transistor. In contrast to typical BJT-based circuits, the PTAT voltage is generated from an asymmetric differential pair in weak inversion, and only the CTAT part depends on the parasitic PNP. A common feature is the signal generation and processing within a single feedback loop, which yields specifically simple solutions, with enhanced robustness towards supply and device variations. The bandgap reference, realized in 14/16nm FinFET, provides a 600mV output and > 60dB PSRR, at sub-1V power supply. We further present temperature sensors in 28nm CMOS and 14/16nm FinFET, which feature very low complexity and power. The first sensor includes a SAR ADC and achieves a precision of ± 3 °C after a 1-point trim under production conditions. The FinFET sensor variant is a scaling-friendly solution, providing enhanced resolution with a duty-cycle modulated output.**

Index Terms— *Bandgap reference, Subthreshold CMOS, temperature sensor, weak inversion*

I. INTRODUCTION

Thermal sensors and references are very basic building blocks within the backbone of nearly any SoC. Their natural affinity is due to the bipolar junction transistor (BJT), required to generate PTAT and CTAT (proportional and complementary to absolute temperature) signals with a well defined temperature coefficient. Especially bandgap references play a critical role, as they directly impact the performance of various other mixed-signal functions, like LDOs, converters, voltage scaling or sense amplifiers.

Indeed, modern technologies impose serious design challenges, since the supply voltage is below the silicon bandgap (~1.22V). Other qualities, like good supply ripple rejection or low noise, are still required, but hardly achieved with traditional techniques. The aggressively scaled MOS transistors exhibit constrained analog performance, which impedes robust and accurate circuit solutions. For bandgap references at sub-1V supply, the current-mode concept [1-3] or variants is state-of-the-art (Fig. 1a). However, it suffers from sensitivity to mismatch, large output impedance and deficient PSRR, due to the open-loop configuration.

Moreover, in latest FinFET processes the poor characteristic and spread of the parasitic PNP device add significant error, which is hard to compensate: Both macros, sensor and reference, usually have a PTAT signal created by ratioed diodes (BJT Q1/Q2). It presumes the current gain (β) to stay constant over a certain current range, which is not true for new technologies (Fig. 1b). This results in significant mismatch of the collector currents, especially for very small β (< 1) like in a typical FinFET process. As a result, the PTAT

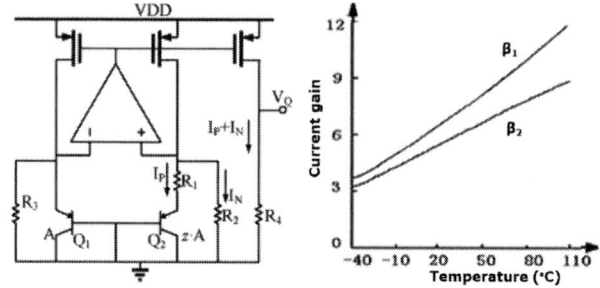

Fig. 1a (left): Current-Mode Bandgap from H. Banba [1] for sub-1V supply.
Fig. 1b (right): Typical PNP current gain at different bias in advanced CMOS.

voltage, generated from ΔVbe becomes nonlinear and affected from process spread.

A linear PTAT voltage can also be generated from MOS devices [4]. If biased in weak-inversion region, the drain current (Id) follows a strictly exponential behavior [5, 11]:

$$Id = \frac{W}{L} Io \cdot e^{Vg/\eta V_T} \cdot \left(e^{-Vs/V_T} - e^{-Vd/V_T} \right)$$

Vg, Vs and Vd are the gate/source/drain-to-bulk voltages, respectively, and V_T the thermal voltage. By driving matched devices with ratioed current densities N:1, the difference of their gate-source voltages adopts the PTAT characteristic. It is independent of MOS thresholds and also drain voltages, if the last term is made negligible (Vd > ~3V_T):

$$Vptat = Vg2 - Vg1 = \eta \cdot V_T \cdot ln(N) + Vos \qquad (1)$$

The result includes a subthreshold slope factor η and the MOS offset voltage (Vos). The latter can be handled by circuit design, in order to keep the PTAT error negligible. For the slope factor we assume for now that it is tightly controlled during manufacturing, since it directly affects the MOS leakage.

Previous work addressed the challenges of scaling and low supply by different approaches, often resulting in rather complex sub-systems. For example, the bandgap references in [2, 3] contain several OPAMPs, to establish accurate biasing and curvature correction. Likewise, a temperature sensor in [6] achieves good accuracy by using dynamic element matching (DEM) and a ΔΣ modulator, at the cost of increased silicon area. Here we focus on rather simple circuits, containing only a few components and a single feedback loop. This is major difference to other publications, where CTAT and PTAT components are processed in an open-loop configuration. In that way, the extra effort to compensate for mismatch and supply impact is avoided.

978-1-7281-5410-7/20 $31.00 © 2020 IEEE

II. A HYBRID SUB-BANDGAP REFERENCE

A. Circuit Description

Fig. 2 shows the simplified reference schematic, combining bipolar and sub-threshold MOS operation. Similar to prior art [3], the new concept generates a temperature independent voltage by adding a PTAT and CTAT current into resistor R2. But here the resistor string is placed inside the main feedback loop, which results in a low impedance output.

The PTAT voltage (Vptat) is established by an asymmetric differential pair, where the input devices (M6, M7) operate in weak inversion and with scaled current densities of N:1. The feedback loop through M10 stabilizes Vptat across Rp, which is further multiplied by the ratio R2/Rp. A single PNP transistor (Q1) supplies the CTAT-part, together with resistor divider R1/R2. Hence only a fraction of voltage drop Vbe is combined with Vptat, which enables sub-1V output and supply voltages (= "fractional bandgap"). The reference output is derived using superposition:

$$Vbg = Vptat + \frac{R2}{R1+R2} \cdot \left(Vptat \cdot \frac{R1}{Rp} + Vbe\right) \quad (2)$$

From the equation it is clear that Vbg can be adjusted by the resistor values, to achieve a flat temperature response.

This structure features certain benefits which help to keep the design simple: M10 is not constrained by matching (inside feedback loop), and neither is M11, due to the logarithmic behavior of Vbe. Also, the tail current is uncritical and may be provided through self-biasing (here with M3 replicating M10). The closed-loop concept immanently suppresses noise from various sources, since its contribution is divided by the loop gain. Likewise, the output Vbg can drive a wide range of current load, including a separate resistor to tap different reference levels. However, this comes at the cost of sensitivity towards capacitive load, which requires measures for stability (not shown). If the load varies, M10 should be decoupled from the self-bias, to avoid impact on the overall accuracy.

Further, this circuit is constrained by the asymmetry of the differential pair, which does not allow simple chopping for

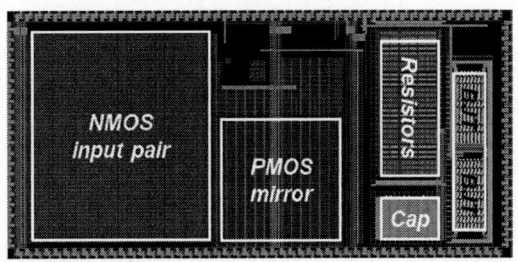

Fig. 3. Floorplan of the bandgap reference realized in 14/16nm FinFET.

offset cancellation. Yet DEM or simple device sizing are effective measures. For consistent subthreshold operation of the input NMOS, the current density is kept low by choosing W >> L. If sub-1V operation is desired, a low common-mode level for M6 and M7 is critical, to ensure a sufficiently large drain voltage. Though the cascode devices M4/M5 are not essential, they help here to stabilize Vds (at > 180mV), so that the linearity of Vptat is not compromised.

Trimming is possible by simple means, if desired. The reference output is most sensitive to the offset voltage (1), additionally to Vbe (Q1). For most applications a single-point calibration is sufficient, accepting some residual temperature gradient. Vbg can be adjusted through R2, or by tuning M11, if the BJT spread is the dominant error source.

B. Silicon Implementation

The bandgap circuit was realized in a 14/16nm FinFET node, with a reference output of Vbg = 600mV. Fig. 3 shows a layout snapshot, which occupies a silicon area of 37μm · 83μm and is constrained considerably by density rules. From CAD analyses including Monte-Carlo simulations, the expected precision across PVT was ~ ± 2.5% (3σ) without trimming. The simulated line regulation is depicted in Fig. 4, which proves accurate circuit operation down to Vdd = 0.85V. Particularly the robust DC-PSRR (typically ~ 65dB) is important, since such error can't be suppressed by filtering or calibration.

Our prototype was evaluated on a typical wafer from 12 samples and with a reduced temperature range, due to limitations in the ATE setup. Fig. 5 reports the untrimmed results, showing a total spread of -8mV/+4mV, and the overall performance is summarized in Table 1.

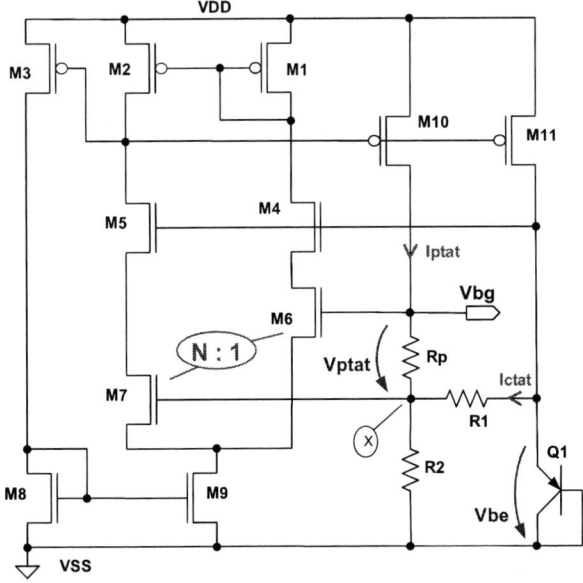

Fig. 2. Closed-loop bandgap reference with self-biasing (simplified).

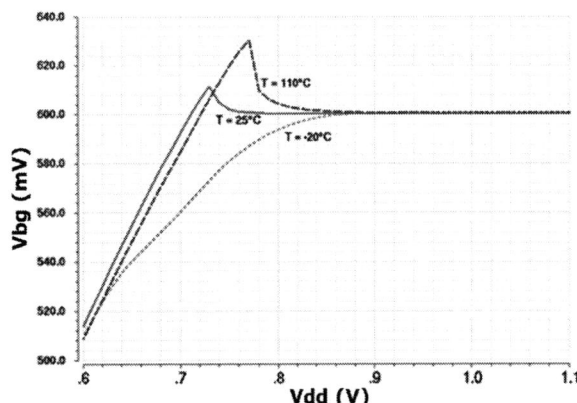

Fig. 4. Reference voltage vs. supply at different temperatures (simulated).

978-1-7281-5410-7/20 $31.00 © 2020 IEEE

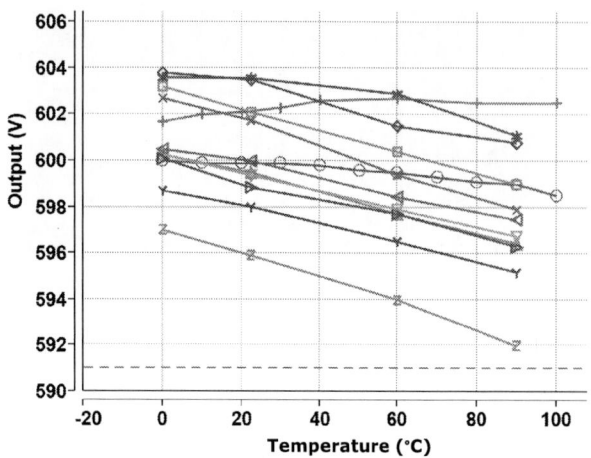

Fig. 5. Measured output voltage of the sub-bandgap reference (untrimmed).

TABLE I
HYBRID BANDGAP PERFORMANCE OVERVIEW

Process	Vref	VDD	Error[a]	DC-PSRR[b]	Power	Area
14/16nm FinFET	600 mV	0.85 ~ 1 V	~ ± 1% (3σ)	≥ 50dB	31 μW	3070 μm²

[a] from 12 samples without trimming, [b] from PVT simulation

III. A CURRENT-MODE TEMPERATURE SENSOR WITH SAR CONTROL

The hybrid architecture can be applied conveniently to temperature measurement, together with the closed-loop operation. Unlike the voltage signal processing in prior work [7], we generate here a PTAT and CTAT current, where either one can act as a reference for the other (Fig. 6). The crossing point defines the temperature T0, when Iptat = Ictat. In fact, in our structure it is profitable to generate the current difference Idiff, which is compared to zero. If either current level is modified, e.g. by a SAR algorithm, T0 will change accordingly until it matches to the silicon temperature.

A. Circuit Description

The circuit in Fig. 7 presents a compact and self-contained solution, which takes full advantage of the current-mode principle. In fact, all major functions of signal generation and processing are performed within a single feedback loop: The asymmetric pair M6/M7 defines the PTAT voltage across Rp. Here M1 does not regulate Iptat, but rather provides a replica of Ibias (M0), which feeds partly into Rp. The excess current is consumed by Q1, establishing a Vbe voltage at node A. Since V(B) follows likewise a CTAT characteristic, Iptat and Ictat (across Rc) are generated and subtracted within a single

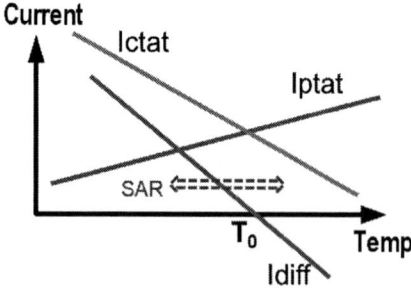

Fig. 6. Characteristic of internal signals in the current-mode sensor .

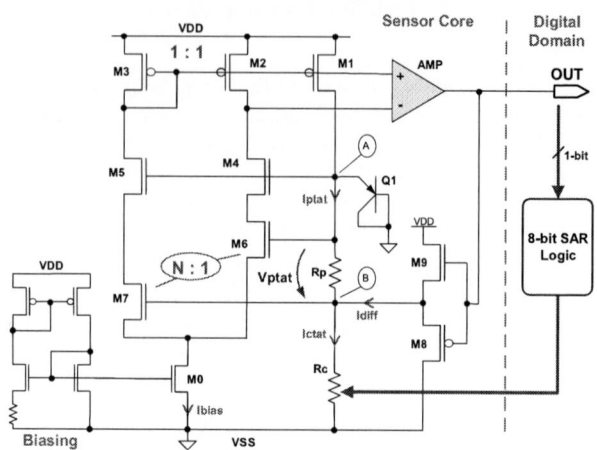

Fig. 7. Simplified circuit diagram of the thermal sensor with SAR ADC.

branch. The feedback loop via the buffer stage (AMP) and M8/M9 eventually regulates Iptat, by providing the difference current Idiff.

In this configuration the push-pull stage acts a "reverse" current comparator, which switches between ~high and ~low at node OUT whenever Idiff is crossing zero. After some buffering this signal can trigger a SAR logic, which in turn adapts Ictat until Idiff is ~0A. In our design Rc is implemented as a simple R-string DAC with 8 bits (binary weighted, not shown). After SAR conversion the temperature T0 is well defined, mainly by device ratios and physical parameters. With V(B) = Vctat = VG0 + T·tc and Vptat = η·VT·ln(N), we can extract T0 from the equilibrium condition Ictat = Iptat:

$$T[°K] = \frac{V_{G0}}{\dfrac{Rc}{Rp} \cdot \eta \cdot \dfrac{k \cdot ln(N)}{e} - tc} \qquad (3)$$

V_{G0} is the silicon bandgap, tc the Vctat temperature coefficient, and k/e the residual constants from the thermal voltage V_T. The differential pair offset is not considered here, as it can be made negligible by design (e.g. by DEM techniques). In our approach we used large devices for M6 and M7, with stable and well-matched drain-source voltages. This is accomplished through cascode devices M4/M5, which exhibit about 180mV lower threshold voltage than the input pair. Alternatively, a resistor could be added below node A, to perform the level shifting. With this design techniques the circuit is functional down to VDD ~ 1.2V, while the supply sensitivity is kept low at ≤ 0.1 LSB/100mV.

A main uncertainty remains with the process dependent parameters, which are the subthreshold slope η and tc. The latter depends on the BJT saturation current and biasing. It should be noted that this circuit is insensitive to accumulated errors like in typical open-loop designs, because the generation, subtraction and even AD-conversion happens all at one place (node B). Apart from the asymmetric differential pair, there are no critical matching requirements, including the buffer AMP and the push-pull stage. This comparator for Idiff has virtually no offset, which depends only on MOS leakage.

B. Siicon Results

While a similar sensor circuit was published earlier for a FinFET prototype [8], we present here the original version with silicon results from multiple batches. The mature

978-1-7281-5410-7/20 $31.00 © 2020 IEEE

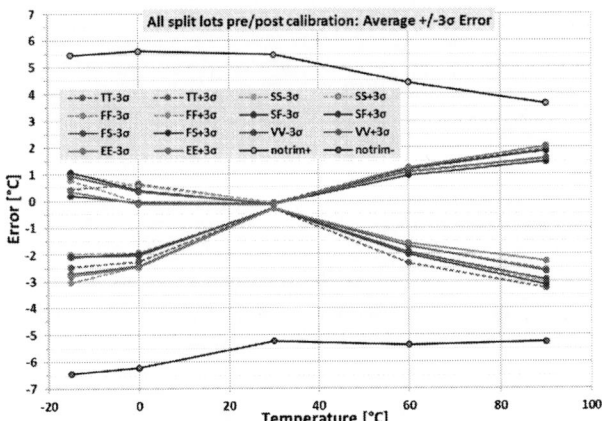

Fig. 8. Measured error (average ±3σ) of the SAR-type sensor from totally 21k samples across 8 contrary wafers: Typical spread untrimmed (solid) and with 1-point trim (dashed) for each corner lot.

structure according to Fig. 7 was implemented in a standard 28nm CMOS technology and placed at multiple locations within a complex SoC. The system firmware translates the binary values (Xb) into degree Celsius (°C) as per equation (3), which takes on the general form:

$$T[°C] = \frac{a + b \cdot X_b}{1 + c \cdot X_b} \qquad (4)$$

In contrast to other work with batch calibration, the coefficients a, b and c where extracted directly from simulation, and not from fitting data. Fig. 8 explains the evaluation results across 8 process window lots (PWL), with 2640 samples per wafer. The typical error (3σ) is around ± 6.5 °C for the untrimmed readout. This performance was in contrast with simulation results, which yield a precision of ~ ±3.5°C including mismatch. A deeper analysis of cross-wafer and process monitor data revealed, that the main root cause is from parameter η. Obviously the subthreshold slope varies much more between different lots than predicted by the standard device models.

To reach the target accuracy of ~3°C, we developed a trimming procedure, which is compatible with production conditions: The binary error value (Eb) was measured during wafer test at a controlled temperature (here: 30°C). This offset (error) is added in both nominator and denominator of formula (4), with weighting factors derived from initial silicon data. With this convenient calibration technique, which affects only the SoC firmware, the error across all PWL wafers improved to values < 3.2°C, as depicted in Fig. 8.

Notably this results already include several other uncertainties, like variations in the ATE setup/temperature (~±1°C), thermal gradients on the wafer, and also quantization error. The latter varies from 0.4 °C to 0.9 °C with increasing temperature and is significant from the low 8-bit resolution.

IV. A THERMAL SENSOR WITH DUTY-CYCLE OUTPUT

Duty-cycle modulation is an elegant solution to achieve high resolution in a thermal sensor, while keeping the interfacing particularly simple. It provides a single-wire square wave output, that can be evaluated by a CPU or a simple counter, which also reduces any self-heating effects. Such principle was presented in [9], which targets a precision in the mK-range after a 1-point trim. It includes OPAMPs and

correction techniques applied to various parts of the circuit, with dynamic error matching (DEM) across totally 8 cycles. In contrast we present here a much simpler solution, that uses chopping to achieve a moderate accuracy of ~ 3°C across process skews.

A. Circuit Description

The circuit in Fig. 9 reveals a major re-use of the previous structure, including few simplifications for the bias, startup and switching blocks. Certain adaptions facilitate especially the sub-1V capability of the sensor. In fact, the push-pull stage is replaced by an integration capacitor C1, which supplies the difference between currents Iptat and Ictat (= Idiff) dynamically. Notably the CTAT resistor takes on only two values: Rc1 or (Rc1 + Rc2). As a result, the voltage at node C increases or decreases linearly, as shown in Fig. 10. Schmitt-Trigger ST1 controls the output and alternates the polarity of Idiff through modulation of Ictat (by M8). The switching levels of ST1 (or hysteresis ΔU) effects only the output frequency, but not the duty-cycle (D), as explained from the following calculation:

$$Iptat = \frac{\eta \cdot V_T \cdot ln(N)}{Rb} \qquad (5)$$

$$Ictat_1 = \frac{V(B)}{Rc1} \qquad (Idiff > 0) \qquad (6)$$

$$Ictat_2 = \frac{V(B)}{Rc1 + Rc2} \qquad (Idiff < 0) \qquad (7)$$

Respectively, the rising and falling slopes for node C are:

$$\Delta t1 = \frac{\Delta U \cdot C1}{Ictat_1 - Iptat} \ , \quad \Delta t2 = \frac{\Delta U \cdot C1}{Iptat - Ictat_2} \qquad (8)$$

Combining (5) to (8) yields the duty-cycle of signal Out:

$$D = \frac{\Delta t1}{\Delta t1 + \Delta t2} \ = \ \frac{Iptat - Ictat_2}{Ictat_1 - Ictat_2}$$

$$= \frac{Rc1}{Rc2} \cdot \left(\frac{\eta \cdot V_T \cdot ln(N)}{V(B)} \cdot \frac{Rc1 + Rc2}{Rc1} - 1 \right) \qquad (9)$$

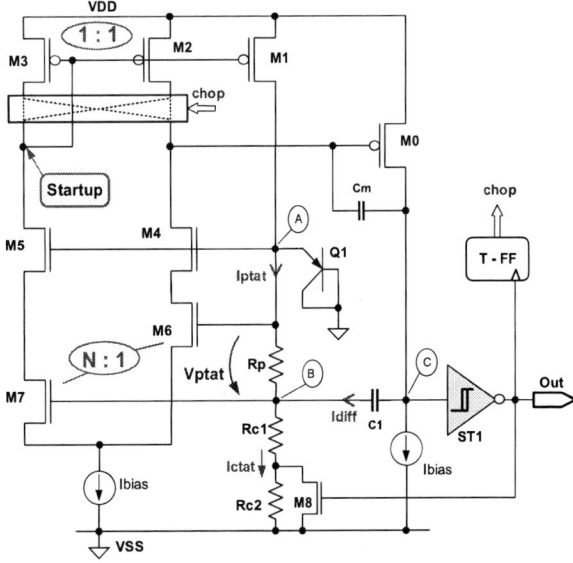

Fig. 9. Thermal sensor circuit with duty-cycle modulation (simplified).

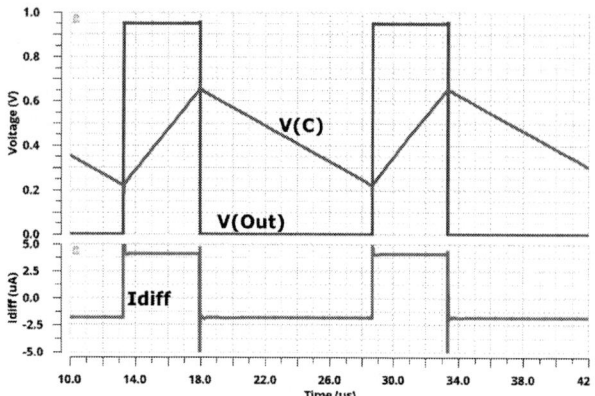

Fig. 10. Typical signals of the sensor circuit with duty-cycle output.

Similar to (3), this can be resolved to calculate the temperature by firmware, if we use $V(B) = V_{G0} + T \cdot tc$. Neglecting mismatch, the measurement depends only on device ratios, physical constants and the parameters tc and η. Fig. 11 shows the resulting transfer and frequency function (from typical simulation).

The differential amplifier in Fig. 7 is replaced here with a simple common-source stage (M0), and also PMOS cascodes were included (not shown in Fig. 9). To increase accuracy, the matching error from current mirror M2/M3 is compensated in our circuit by chopping, while the respective clock is directly derived from the output signal. For M6 and M7 we chose large device sizes, to make their mismatch contribution negligible. Hence, the precise temperature reading should be extracted from (9) using the average of only 2 consecutive duty cycles.

Comparing with [9, 10], this circuit features not only lower complexity, but also includes C1 into the feedback path. The benefit is a lower supply capability and robustness towards the impedance of related current sources (Ibias, M0). In contrast to the sensor in section III, the resolution is flexible here and depends on the duty-cycle sampling rate. Another advantage is the single-wire IO, simplifying integration into dense area.

B. Simulation Results

We chose a 14/16nm FinFET node to verify this hybrid sensor solution. Though the technology offers a PNP device, the linearity and current gain (β) is fairly constrained with values <<1 across the whole PVT range. Therefore, the combination with subthreshold NMOS in a switching feedback loop is a suitable architecture, which also helps to lower the overall design complexity. Circuit simplicity is an

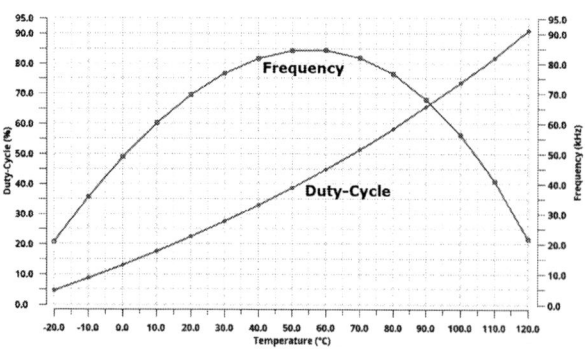

Fig. 11. Thermal sensor transfer characteristic vs. temperature (simulated).

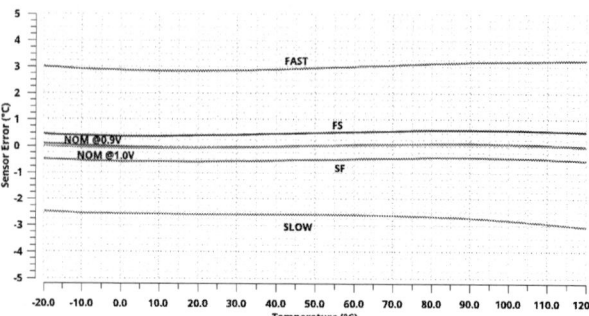

Fig. 12. Sensor error (simulated) across temperature, Vdd and process corners.

important feature specifically for advanced FinFET nodes, where analog layout becomes an extremely complicate task.

Our layout estimation (not shown) consumes an area of around 4800 μm2, and is dominated by the asymmetric NMOS pair, due to matching constraints. The floorplan utilizes the benefits of MIM-CAP device for C1, which can be placed efficiently on top of the main circuitry.

The prototype IP was simulated using an early model release across full PVT conditions, and the results are summarized in table 2. Fig. 12 explains the resulting error versus temperature (untrimmed), which stays within ± 3.2°C for the specified range of -20 °C up to 120 °C. Clearly the supply variation (at NOM corner) has negligible impact, while this IP can operate down to Vdd = 0.9 V. This data is extracted with ideal matching. Monte-Carlo simulations (chopping enabled) showed, that the additional error due to mismatch is ~1.5°C (3-sigma) and constant across the temperature range.

The device models suggest a major error contribution from the BJT, with respective variation included in the fast and slow corners. However, we expect that the spread of the subthreshold slope (η) may not be modelled precisely at this early process state, which would result in additional error, as per equation (7). In that case a soft calibration similar to the SAR-type sensor (Fig. 8) can bring the accuracy back into the target range.

TABLE II
DUTY-CYCLE SENSOR performance overview – SIMULATED

Process	T-Range	VDD	Accuracy[a]	Power	Area
14/16nm FinFET	-20 ~ 125°C	0.9 ~ 1 V	~ ± 3.5°C (3σ)	23 μW	~ 4800 μm2

[a] from PVT simulation without trimming.

V. CONCLUSION

The presented architecture realizes the profitable combination of a single PNP device with a ratioed MOS differential pair biased in weak inversion. In relation to previous work [12, 13], a unique feature is the single feedback loop, which executes the complete processing of PTAT and CTAT signals in current mode technique. This results in very compact and self-contained circuits, which offer improved supply rejection and operation at sub-bandgap levels.

We realized prototypes at various technologies down to advanced FinFET nodes, with performance validation through simulation and silicon measurements. For the temperature sensor with SAR ADC (Sec. 3), the result data is summarized in Table 3 and compared with related work. While power

TABLE III
THERMAL SENSOR PERFORMANCE COMPARISON

Silicon Data	SAR-sensor	[6] (Tang)	[7] (Bass)	[10] (Tang)
Process	28 nm	55 nm	65 nm	130 nm
VDD (V)	1.6 ~ 1.9	2.5	1.3	1.8 ~ 3.5
Current (µA)	30	14.8	86	17
3σ-Spread (°C)	± 2.6 [a]	± 1.7 [b]	± 1.7 [a]	± 0.38 [a]
Speed (ks/s)	14	0.25	0.24	0.42
Area (µm²)	3.5k	14.6k	3k	73k
# Samples	21.1k	12	25	14
Type	Hybrid	PNP	PNP	PNP
Range (°C)	-20/130 [c]	-40/125	-10/110	-10/100

[a] 1-point trim, [b] batch-trim, [c] from simulation

consumption and size numbers are competitive, the residual error after 1-point trim is relatively high with ± 2.6 °C. However, those results are extracted from a large number of samples on contrary wafers, and therefore prove the realistic capabilities under production conditions. The measurements also revealed the increased spread of the subthreshold slope, that was not reflected in device models and makes calibration indispensable. Nevertheless, the structures are highly compatible to advanced baseline technologies, and well suited for placement in dense core areas of digital SoCs.

ACKNOWLEDGMENT

The authors would like to thank the Christian Doppler Laboratory for Digitally Assisted RF Transceivers for Future Mobile Communications (CD-Lab, 4040 Linz, Austria) for supporting this work, which was carried out in close cooperation with Intel Germany and DMCE, Austria. Intel corporation also provided access to foundry PDK and silicon evaluation. We thank also Cho-Ying Lu and Zdravko Georgiev for their support during data analyses.

REFERENCES

[1] H. Banba et al., "A CMOS bandgap reference circuit with sub-1-V operation," in IEEE Journal of Solid-State Circuits, vol. 34, no. 5, pp. 670-674, May 1999.

[2] M. -D. Ker and J. -S. Chen, "New Curvature-Compensation Technique for CMOS Bandgap Reference With Sub-1-V Operation," in IEEE TCAS-II: Express Briefs, vol. 53, no. 8, pp. 667-671, Aug. 2006.

[3] Y.-W. Chen, J.-J. Horng, C.-H. Chang, A. Kundu, Y.-C. Peng and M. Chen, "A 0.7V, 2.35% 3σ-Accuracy Bandgap Reference in 12nm CMOS," in IEEE ISSCC Dig. Tech. Papers, USA, 2019, pp. 306-307.

[4] A. Annema and G. Goksun, "A 0.0025mm2 bandgap voltage reference for 1.1V supply in standard 0.16µm CMOS," 2012 IEEE International Solid-State Circuits Conference (ISSCC), San Francisco, CA, 2012, pp. 364-366.

[5] G. Tzanateas, C.A.T. Salama and Y.P. Tsividis, "A CMOS Bandgap Voltage Reference," in IEEE J. Solid-State Circuits, vol. 14, no. 3, pp. 655-657. February 1979.

[6] Z. Tang, Y. Fang, Z. Huang, X. Yu, Z. Shi and N. N. Tan, "An Untrimmed BJT-Based Temperature Sensor With Dynamic Current-Gain Compensation in 55-nm CMOS Process," in IEEE TCAS-II: Express Briefs, vol. 66, no. 10, pp. 1613-1617, Oct. 2019.

[7] O. Bass and J. Shor, "Ultra-Miniature 0.003 mm2 PNP-Based Thermal Sensor for CPU Thermal Monitoring," Dig. ESSCIRC, Dresden, 2018, pp. 334-337.

[8] C. Lu, S. Ravikumar, A. Sali, M. Eberlein and H.-J. Lee, "An 8b subthreshold hybrid thermal sensor with ±1.07°C inaccuracy and single-element remote-sensing technique in 22nm FinFET," Dig. ISSCC, pp. 318-320, Feb. 2018.

[9] A. Heidary, G. Wang, K. Makinwa and G. Meijer, "12.8 A BJT-based CMOS temperature sensor with a 3.6pJ·K2-resolution FoM," Dig. ISSCC, pp. 224-225, Feb. 2014.

[10] Z. Tang, Y. Fang, X. Yu, Z. Shi and N. Tan, "Capacitor-reused CMOS temperature sensor with duty-cycle-modulated output and 0.38°C (3σ) inaccuracy," in Electronics Letters, vol. 54, no. 9, pp. 568-570, May 2018.

[11] E. Vittoz and J. Fellrath, "CMOS analog integrated circuits based on weak inversion operation,' IEEE J. Solid-State Circuits, vol. SC-12, pp. 224-231, June 1977.

[12] K.A.A. Makinwa, "Temperature Sensor Performance Survey," [Online]. Available: http://ei.ewi.tudelft.nl/docs/TSensor_survey.xls.

[13] C. Fayomi, G. Wirth, H. Achigui and A. Matsuzawa, "Sub 1V CMOS bandgap reference design techniques: a survey". Analog Integrated Circuits and Signal Processing, vol. 62, pp 141-157, Feb. 2010.

978-1-7281-5410-7/20 $31.00 © 2020 IEEE

Temperature and Supply Voltage Monitoring with Current-mode Relaxation Oscillators

Shanshan Dai, Caleb R. Tulloss, Xiaoyu Lian, Kangping Hu, Sherief Reda, and Jacob K. Rosenstein

School of Engineering, Brown University, Providence, RI, USA

Abstract—This paper presents a new family of temperature and supply voltage sensors based on an improved current-mode relaxation oscillator. The proposed sensor circuit overcomes the finite input impedance of earlier current-mode relaxation oscillators, and it is paired with a novel opamp-free bandgap current reference. Two examples are presented: a V_{DD}-controlled oscillator, and a temperature-controlled oscillator. The temperature-controlled oscillator operates at 11.3 MHz and consumes 11.14 μW while reaching a temperature nonlinearity error less than +0.85/-0.94°C, a resolution figure of merit of 11.5 pJ·K^2, and an efficiency of 240 pJ/conv. The V_{DD}-controlled oscillator has a nonlinearity error less than +28.7/-30.0 mV from 1.2−1.8 V after two-point calibration.

Index Terms—temperature sensor, supply voltage monitor, current reference, thermal sensor, VCO, relaxation oscillator.

I. INTRODUCTION

Temperature and supply voltage monitoring are fundamental tasks in nearly all microelectronic systems, from small embedded sensors to high-performance servers. At the low-power end of the spectrum, temperature and voltage management are crucial for maximizing battery life and optimizing unpredictable loads. At the high-performance end, power dissipation in CPUs and GPUs is non-uniformly distributed, and localized hot spots can deviate significantly in temperature and supply voltage compared to other areas of the chip. Minimizing the area and energy of monitor circuits is key to supporting sensing at high spatial resolution with minimal cost and interference [1], [2].

There are several options for producing digital outputs which represent the supply voltage. One of the simplest arrangements is a voltage controlled oscillator, which is often used for supply monitoring [3]. Digital critical path monitors (CPMs) [4], [5] have very low latency and can be used to respond to power supply transients, but they are less precise for continuous monitoring, and CPMs are often combined with other complementary sensors.

Temperature sensors use a wider variety of approaches. Resistive [6]–[8] and thermal-diffusivity [9] temperature sensors are able to achieve high resolution (often < 0.1°C), but demand sophisticated frequency-locked loops or $\Sigma\Delta$-ADCs to digitize the temperature-dependent information. Their area, power consumption, and design complexity increase accordingly. Oscillator-based temperature sensors, which employ frequency [1], [10]–[12] or duty cycle modulation [13], are

This work was funded in part by grant FA8650-18-2-7851 from the Defense Advanced Research Projects Agency (DARPA). C.R.T. is also grateful for support from the Jayakumar Undergraduate Summer Research Fellowship.

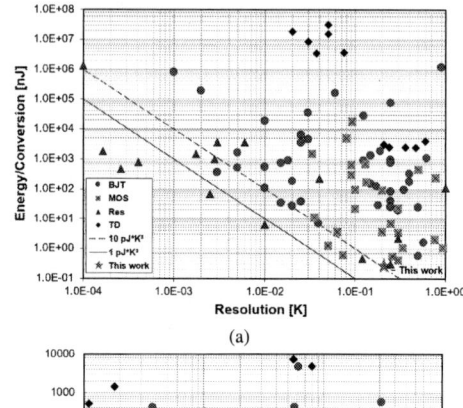

(a)

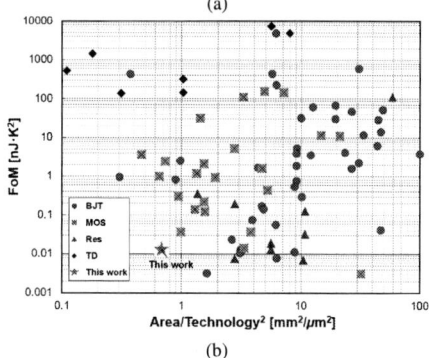

(b)

Fig. 1: CMOS smart temperature sensors [15] are compared by plotting (a) energy per conversion versus temperature resolution, and (b) an energy-resolution figure-of-merit (FoM, with unit of nJ·K^2) versus normalized circuit area.

appealing for thermal monitoring as they are straightforward to implement. Low latency temperature measurements are important to track thermal transients, which can swing 10-20 °C within 2-3 ms in smartphone SoCs [14]. Ultimately, a monitor circuit will be evaluated by a combination of factors [15] including its area, power, resolution, conversion time and accuracy, some of which are surveyed in Figure 1.

The authors previously demonstrated a low-power current-mode relaxation oscillator with an oscillation frequency that has low sensitivity to temperature and supply voltage [16]. Building on this earlier work, the key innovation presented here is to make the oscillator's reference voltage intentionally sensitive to either temperature or supply voltage, thus allowing the design of temperature sensors with low supply voltage sensitivity, or supply voltage sensors with low temperature sensitivity. Two new circuits are additionally introduced here to support the design. First, the problem of finite input impedance at the V_{ref} node was solved by adding a simple voltage buffer. Second, the bandgap reference used to generate a temperature-

978-1-7281-5410-7/20 $31.00 © 2020 IEEE

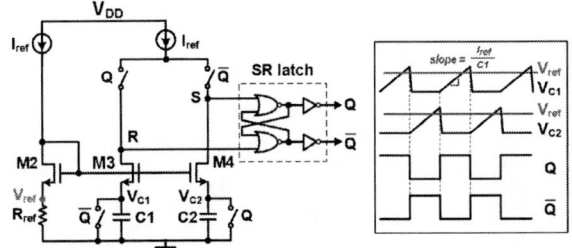

Fig. 2: A dual-phase current-mode relaxation oscillator [16], which was originally designed to be independent of temperature and supply voltage, but has been modified here to create two proposed sensor circuits.

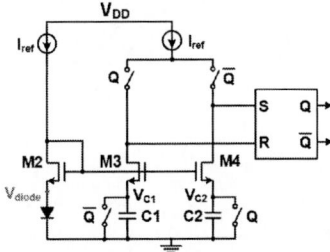

Fig. 3: Proposed temperature sensor. By replacing V_{REF} with V_{diode}, a temperature-dependent oscillation is obtained.

insensitive current was modified to function without an opamp, further simplifying the design.

The circuits proposed here have the additional feature of extremely low design complexity. There are only four key parameters (I_{ref}, C and W/L of a key transistor) in the sensor design, which can be easily scaled to achieve speed and power trade-offs for different applications, while maintaining state-of-the-art performance. With a temperature sensor core area of 0.003mm^2 and a supply voltage sensor core area of 0.005mm^2, the two sensors have conversion energies of 0.28 and 0.35 nJ/conv respectively, each achieving the lowest conversion energy in its class.

II. PROPOSED TEMPERATURE AND SUPPLY VOLTAGE MONITOR CIRCUITS

A dual-phase current-mode relaxation oscillator was previously introduced in [16], illustrated in Fig. 2. In this circuit, two capacitors C1 and C2 are alternately charged by I_{ref} and reset to ground in a feedback loop around a set-reset (SR) latch. M2, M3 and M4 form a common gate amplifier which compares the source voltage V_{C1} (or V_{C2}) against a reference voltage V_{ref}. The oscillation period can be expressed as:

$$\tau = 2\left(\frac{V_{ref}}{I_{ref}/C} + \tau_{M3,4} + \tau_{SR}\right) \quad (1)$$

where $C_1 = C_2 = C$, τ_{M3} is the comparator delay caused by finite transistor gain and the parasitic capacitance at node S (R), and τ_{SR} stands for the digital delay of SR latch. In low-power applications, current starving can be employed to reduce the dynamic power consumption of the SR latch [16].

Since the same devices and current are reused for both the comparator bias and capacitor charging, source-coupled comparators can be highly power and area efficient [17]. Here, this topology is extended as a voltage-controlled oscillator. By providing the oscillator with a V_{ref} which varies with temperature or supply voltage, a simple and efficient sensor with a digital output can be constructed. To adjust the power, area, and accuracy for different applications, one can simply modify I_{ref}, C, and the dimensions of M2-M4.

However, one obstacle to the extension of this structure is the finite input impedance of the node V_{ref}. In this paper, this drawback is solved with a small cost of one additional I_{ref}, which will be explained in detail in Section II.B.

A. Temperature sensor

Beginning from the circuit shown in Fig. 2, a low-power temperature sensor can be readily implemented by replacing R_{ref} with a forward-biased diode, as shown in Fig. 3. Biased with a constant current I_{ref}, the diode voltage V_{diode} has a complementary to absolute temperature (CTAT) coefficient, due to the exponential temperature dependence of its reverse saturation current I_s:

$$V_{diode}(T) = \frac{kT}{q}\ln\left(\frac{I_{ref}}{I_s(T)}\right) \quad (2)$$

V_{diode} is on the order of several hundred millivolts. Now the period of this relaxation oscillator becomes:

$$\tau_{temp} = 2\left(\frac{V_{diode}}{I_{ref}/C} + \tau_{M3,4} + \tau_{SR}\right) \quad (3)$$

The first term in (3) dominates the oscillation period because the comparator delay $\tau_{M3,4}$ and the digital delay τ_{SR} (limited by parasitics) are much faster than charging C to V_{diode}. Thus the oscillation period will reflect the temperature sensitivity of the diode. Excluding bias generation, the static current consumption of this temperature sensor is $2I_{ref}$.

This structure can be used to implement a µW temperature sensor, in which I_{ref} is 2.1 µA and C is 50 fF. Transistors M2-M4 have W/L=3µm/1µm, operating in the subthreshold region for high g_m/I_D (transconductance efficiency).

B. Supply-voltage sensor

A natural step towards a supply-voltage-controlled oscillator would be to create V_{ref} by simple voltage division from V_{DD}. However, one drawback of the source-coupled comparator is its low input impedance at the node V_{ref}, leading to a loading effect if V_{ref} stems from a voltage divider.

Fig. 4 addresses this limitation by adding a high-input-impedance unity-gain buffer which replicates V_{IN} at V_{ref}, in an arrangement which also reuses M2's bias current. It can be viewed as an amplifier whose inputs are at the gates of M5-M6, and whose output is the drain node of M6. This amplifier is connected as a voltage follower, so that its output is regulated to be approximately equal to V_{IN}, assuming a reasonable loop gain. The period of this voltage-controlled oscillator is:

$$\tau_{V_{IN}} = 2\left(\frac{V_{IN}}{I_{ref}/C} + \tau_{M3,4} + \tau_{SR}\right) \quad (4)$$

Now that V_{IN} is buffered, it can connect to a simple voltage division of V_{DD} as shown in the dashed box of Fig. 4. If

978-1-7281-5410-7/20 $31.00 © 2020 IEEE

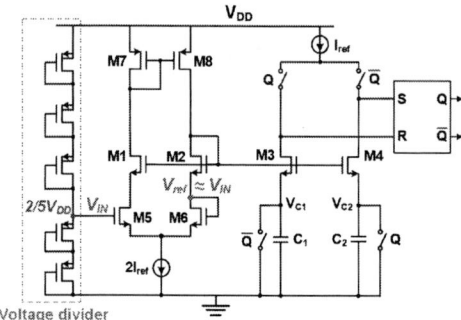

Fig. 4: Proposed voltage sensor. V_{IN} is buffered and replicated in place of V_{REF}, producing an oscillation period proportional to V_{IN}. For simplicity, the startup circuit on the gate of M1-M4 is not drawn.

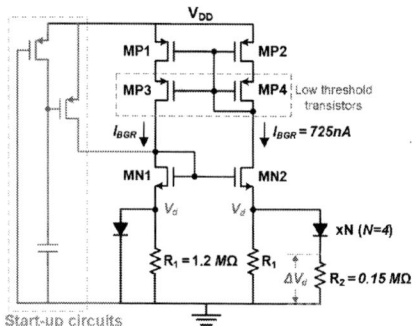

Fig. 5: A compact and easy-to-design bandgap current reference modified from [18] with removal of the operational amplifier. Copies of the reference current are created with mirrors from MP2.

$\tau_{M3,4}$ and τ_{SR} are small, the period in (4) becomes linearly proportional to the supply voltage, providing a time/frequency-encoded measurement of V_{DD}.

The primary parameters in this supply voltage sensor, such as I_{ref}, C, and the dimensions of M1-M4, are the same as those in the temperature sensor. To trade off between power, area, and accuracy, both the μW supply voltage sensor and the μW temperature sensor only require tuning I_{ref}, C, and the dimensions of M1-M4. This low design complexity is important for a modular and re-usable design.

C. Bandgap current reference

These two sensors are designed with the assumption that a constant reference current I_{ref} is available which is insensitive to both temperature and supply. In keeping with the spirit of design simplicity, our goal is to minimize the complexity of the reference generation as well as the sensor design.

One suitable bandgap circuit is shown in Fig. 5. This circuit is modified from [18], and it places a resistance R_1 in parallel with the diodes in order to reduce the minimum supply voltage compared to a classical bandgap configuration. The identical NMOS pair MN1-MN2 will regulate their source voltages to be equal. By selecting a proper ratio of R_1 to R_2 and a diode area ratio of N, one can generate a temperature-independent

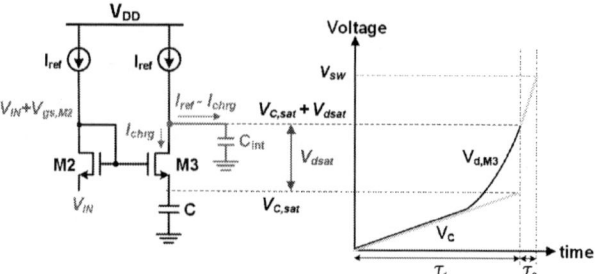

Fig. 6: The delay model and illustrated transition of the source-coupled amplifier transistors (M3 and M4).

current I_{BGR},

$$I_{BGR} = \frac{V_d}{R_1} + \frac{\Delta V_d}{R_2} = \frac{1}{R_1}\left(V_d + \frac{R_1}{R_2}\frac{kT}{q}\ln N\right). \quad (5)$$

where V_d is the voltage across the P+/N-well junction diodes, and ΔV_d is the difference between the two diodes' forward voltages, which appears across R_2. V_d has a complementary-to-absolute-temperature (CTAT) coefficient, while the second term is proportional to absolute temperature (PTAT). Similar to other bandgap circuits, the basic principle is to compensate a CTAT coefficient with a weighted PTAT coefficient, by choosing the correct value of $(R_1/R_2 \times \ln N)$ in (5). Example current and resistor values are noted in Fig. 5.

It is worth noting that the resistor temperature coefficient can affect the temperature variation of I_{BGR}. This could be addressed by implementing R_1 and R_2 with two series resistors having opposite temperature coefficients.

III. DELAY MODEL OF THE AMPLIFYING TRANSISTOR

Previous work [17] has studied the comparator delay model via the small-signal transfer function method, from which we know that the comparator delay is limited by three factors: the finite ramping speed of its input, its finite gain, and the dominant pole at the output (finite bandwidth).

In this section, we will analyze the delay of M3 from a different perspective, using the large signal model. As illustrated in Fig. 6, I_{ref} splits into two branches. One branch I_{chrg} flows into M3 to charge up the capacitor C while the remainder charges the parasitic capacitance at the drain node of M3. If M3 has a gain of A_{M3}, the voltage change at its drain node $\Delta(V_{d,M3})$ can be written as:

$$\Delta(V_{d,M3}) = A_{M3} \times \frac{I_{chrg}}{C}\,\Delta t = \frac{I_{ref} - I_{chrg}}{C_{int}}\,\Delta t. \quad (6)$$

Since the charging rate of C_{int} equals the amplified charging rate of C, we can use (6) to solve for I_{chrg}:

$$I_{chrg} = \frac{I_{ref}}{1 + A_{M3}C_{int}/C}. \quad (7)$$

Equation (7) tells us that the gain of M3 affects the current splitting. When M3 works in the linear region, $A_{M3,lin}$ is small, and the denominator is approximately 1. Thus almost all the I_{ref} flows into M3 to charge up C ($I_{ref} \approx I_{chrg}$). As

978-1-7281-5410-7/20 $31.00 © 2020 IEEE

TABLE I: Node voltages at the end of τ_1

Transistor	Source	Drain
M2	V_{IN}	$V_{IN} + V_{gs,M2}$
M3	$V_{C,sat}$	$V_{C,sat} + V_{dsat}$

M3 enters saturation region, or $A_{M3,sat}$ is large, $A_{M3,sat}C_{int}$ becomes comparable to C, and a fraction of I_{ref} begins to charge up C_{int}. At this phase, we could also say that C_{int} requires more current than C, since M3 is amplifying the changes in V_C. From the small-signal perspective, dV/dt at the output of M3 could approach $(A_{M3,sat} \times I_{ref}/C)$ only if M3 had infinite bandwidth (no C_{int}, and thus no current splitting). But in practice, C_{int} limits the bandwidth.

As shown in Fig. 6, the source voltage of M3 ramps up from zero until its drain voltage reaches V_{SW}, the switching threshold of the SR latch. We divide the total period into two phases τ_1 and τ_2 based on the operation of M3 (linear and saturation). We use the capacitor ramping voltage V_C and M3 drain voltage $V_{d,M3}$, to calculate τ_1 and τ_2 respectively, as illustrated by the two yellow segments in Fig. 6. We use $V_{d,M3}$ rather than V_C to derive τ_2, because it is simpler to obtain the intial and final voltages of $V_{d,M3}$ during τ_2.

During τ_1, M3 is in the linear region, with its drain-to-source voltage below V_{dsat}. Here M3 operates as a resistor, and initially the drain of M3 ($V_{d,M3}$) directly follows the capacitor voltage V_C. The gain $A_{M3,linear}$ is small, and the voltage is charging at a rate of I_{ref}/C based on (7).

This first phase ends when the drain-to-source voltage of M3 equals V_{dsat}, which corresponds to a capacitor voltage $V_{C,sat}$. The duration of the first phase τ_1 can be derived as:

$$\tau_1 = \frac{V_{C,sat}}{I_{ref}/C} \quad (8)$$

In order to derive τ_1, we need to find $V_{C,sat}$, the capacitor voltage at which M3 enters saturation.

At the end of τ_1, M2 and M3 form a differential amplifier, whose input voltage is the difference between the two source voltages and whose output is the difference between the two drain voltages. Their source and drain node voltages at the end of τ_1 are described in Table I. Given that the gain of M3 is $A_{M3,sat}$, $V_{C,sat}$ can be obtained from:

$$A_{M3,sat}(V_{IN} - V_{C,sat}) = V_{IN} + V_{gs,M2} - V_{C,sat} - V_{dsat} \quad (9)$$

where $(V_{IN} - V_{C,sat})$ on the left is the source voltage difference between M2 and M3, and the expression on the right is the drain voltage difference of M2 and M3, the amplifier's output. We can use equation (9) to solve for $V_{C,sat}$, and from there solve for τ_1 using equation (8).

During τ_2, M3 operates in the saturation region, amplifying the voltage difference between the two source voltages V_{IN} and V_C, and current splitting occurs. The drain voltage of M3, $V_{d,M3}$, is $(V_{dsat} + V_{C,sat})$ at the start of τ_2 and slews to V_{SW}, at a rate of $(A_{M3,sat}I_{chrg}/C)$, to trigger the SR latch, as illustrated by the second yellow segment line in Fig. 6.

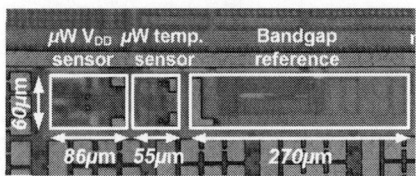

Fig. 7: Die photo of the proposed temperature and voltage sensors, fabricated in 180-nm CMOS.

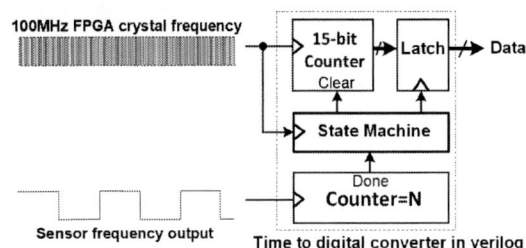

Fig. 8: A diagram of the simple frequency-to-digital converter used to digitize the sensor outputs.

From (6) and (7), the total time of the second phase τ_2 is:

$$
\begin{aligned}
\tau_2 &= \frac{V_{SW} - (V_{dsat} + V_{C,sat})}{A_{M3,sat}I_{chrg}/C} \\
&= \frac{V_{SW} - (V_{dsat} + V_{C,sat})}{A_{M3,sat}I_{ref}/(C + A_{M3,sat}C_{int})} \quad (10)
\end{aligned}
$$

Substituting the expression of $V_{C,sat}$ obtained from (9) into τ_1 and τ_2 in (8) and (10), we can reach to the overall oscillation half-period:

$$
\begin{aligned}
\tau_1 + \tau_2 &= \frac{V_{IN}}{I_{ref}/C} + \frac{V_{SW} - V_{dsat} - V_{IN}}{I_{ref}/C_{int}} \\
&+ \frac{V_{SW} - V_{gs,M2} - V_{IN}}{A_{M3,sat}I_{ref}/C} + \frac{V_{gs,M2} - V_{dsat}}{(A_{M3,sat} - 1)I_{ref}/C_{int}} \quad (11)
\end{aligned}
$$

The last term in (11) is negligible, because its ratio to the first term is on the order of $(\frac{1}{A_{M3,sat}} \cdot \frac{C_{int}}{C})$ assuming V_{IN} and $(V_{gs,M2} - V_{dsat})$ have the same order of magnitude. In both implemented sensor designs, $A_{M3,sat}$ is about 125, and C_{int} is <7 fF. Thus, the last term contributes less than 0.11% to the overall conversion time. Similarly, the contributions of the second and third term can be justified numerically by substituting $A_{M3,sat}$ and C_{int} into their ratios to the first term. Moreover, the contribution of the second term can be further reduced by decreasing $|V_{SW} - V_{dsat} - V_{IN}|$. This can be achieved by adjusting the transistor dimensions of NOR gates in the SR latch (V_{SW} adjustment).

The first term in (11) describes the ideal behavior, in which the oscillation period $\tau_1 + \tau_2$ is linear with V_{IN}, and V_{IN} represents temperature or supply voltage. The second and third terms highlight important sources of nonlinearity.

IV. Measurement Results

One μW temperature sensor based on Fig. 3 and one μW V_{DD} sensor based on Fig. 4 were fabricated in a standard 180-nm CMOS process, with the die photo shown in Fig. 7.

978-1-7281-5410-7/20 $31.00 © 2020 IEEE

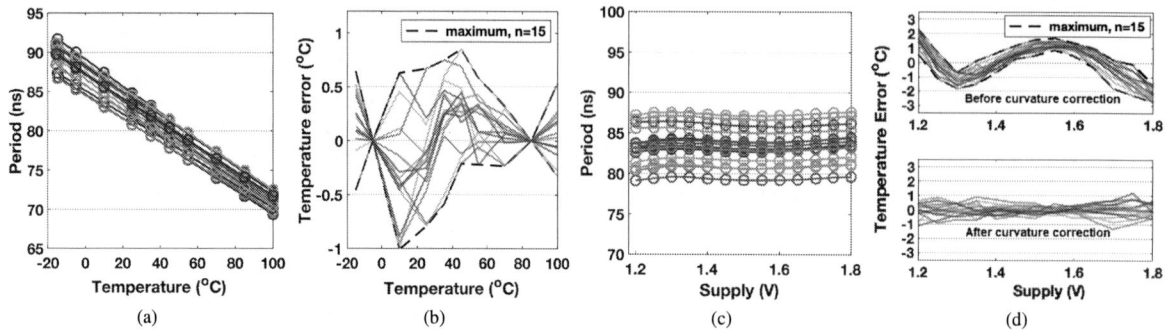

Fig. 9: Measurements of the μW temperature sensor, illustrating (a) oscillation period versus temperature, (b) temperature nonlinearity error after two-point calibration with 15 samples, (c) oscillation period versus V_{DD}, (d) supply sensitivity without calibration (upper) and supply sensitivity after curvature correction (lower).

Fig. 10: Measurements of the μW V_{DD} sensor, showing (a) oscillation period versus V_{DD}, (b) V_{DD} nonlinearity error, (c) oscillation period versus temperature, (d) temperature sensitivity before (upper) and after curvature correction (lower).

The supply voltage range for both sensors is 1.2–1.8 V. One bandgap reference circuit based on Fig. 5 is also included. The bandgap draws 2.0 μA and occupies 0.0156 mm^2, including several current mirrors to distribute I_{ref} to multiple sensors.

As depicted in Fig. 8, the experimental sensor readout is performed using a time-to-digital converter (TDC) implemented on an FPGA module (Opal Kelly XEM6310). The TDC counts reference clock cycles during N sensor cycles (N=256), which is the equivalent conversion time. To allow for comparison to other state-of-the-art designs, the TDC is also simulated in 180-nm CMOS, and its simulated power is 2.1 μW with a 0.8 V digital supply and its estimated area is 3000 μm^2, using low-power DFFs based on [19]. In more advanced process nodes the TDC power and area would decrease further.

A. Temperature sensor

The temperature sensor has an active area of 60μm $\times$ 55μm. It consumes 6.57 μA with V_{DD} = 1.3 V at room temperature. Fig. 9 (a) shows a temperature sweep of the proposed temperature sensor measured across 15 sample test chips when V_{DD} is 1.3 V, in which the periods are linear with temperature. The peak-to-peak temperature nonlinearity error is +0.85/-0.94°C after two-point linear calibration, as plotted in the upper panel of Fig. 9 (b). In the lower panel of Fig. 9 (b), measured on 15 samples, the mean voltage sensitivity is 2.28°C/V after the removal of the systematic non-linearity when V_{DD} varies from 1.2 V to 1.8 V. The RMS temperature resolution is 210 mK via 1000 consecutive temperature readings at

room temperature. Each reading was conducted by counting a 100 MHz reference clock for 256 sensor cycles as shown in Fig. 8, yielding a conversion time of 21.4 μs.

B. V_{DD} sensor

The V_{DD} sensor core occupies an area of 60μm $\times$ 86μm. Its current consumption is 8.34 μA, measured with V_{DD} = 1.3 V at room temperature. Fig. 10 (a) shows a supply voltage sweep for the V_{DD} sensor from 1.2 V to 1.8 V at room temperature. As shown in the lower panel of Fig. 10 (b), the period has a peak-to-peak nonlinearity error of +28.7/-30.0 mV after two-point calibration. Based on 1000 consecutive V_{DD} readings at 1.3 V V_{DD}, with a conversion time of 256 sensor cycles (22.4 μs), the corresponding RMS V_{DD} resolution is 0.94 mV.

Table II compares the temperature sensor performance with recent state-of-the-art demonstrations. The TDC readout utilized is similar to [1], [2], [10]–[12] in Table II, where the sensing frequency needs to be measured by a reference frequency. The implementation principle is to count the faster frequency during a fixed cycles of the slow clock. Table III compares the proposed supply voltage sensor with two state-of-the-art dedicated supply voltage monitors. Compared with the other state-of-the-art temperature and V_{DD} sensors, these circuits have very competitive conversion time, design area, and conversion energy, along with the additional advantage of a simple design methodology which can be easily scaled to achieve a desired balance between speed and power.

978-1-7281-5410-7/20 $31.00 © 2020 IEEE

TABLE II: Temperature Sensor Performance Overview

	Without readout TDC					With readout TDC					
	ISSCC14 [20]	VLSIC16 [21]	JSSC17 [22]	TCASii18 [13]	This work	JSSC16 [1]	TCASi17 [11]	ISSCC17 [10]	JSSC18 [2]	JSSC2019 [12]	This work
Sensor type	MOS	BJT	BJT	BJT	MOS	MOS	MOS	MOS	Resistor	MOS	MOS
Readout Type	SAR+$\Sigma\Delta$	Comparator output	$\Sigma\Delta$	Duty-cycle	OSC	OSC	OSC	OSC	RC constant	OSC	OSC
Tech (nm)	160	28	160	180	180	65	180	180	65	180	180
Area (mm^2)	0.085	0.0038	0.16	0.1	0.019	0.004	0.45	0.009	0.01	0.074	0.022†
Supply voltage (V)	0.85−1.2	1.1−2	1.5−2	1.6−2.0	1.2−1.8	0.85−1.05	0.55−0.85	0.8−1.8	1.1−1.4	0.7−1.5	1.2−1.8
Power (μW)	0.6	17.6	6.9	0.864	11.14	154	0.075	0.075	12.8	0.011	13.24 †
Temp. range (°C)	-40−125	-20−130	-55−125	-30−120	-15−100	0−100	0−100	-20−100	-40−110	-20−80	-15−100
Supply sens. (°C/V)	0.45	0.012	0.01	0.7	2.28	34	1.67	0.13	3	3.8	2.28
Inaccuracy (°C)	±0.4 (3σ)	±0.54 (1σ)	±0.06 (3σ)	±0.85 (3σ)	+0.85/-0.94 (p-p)	±0.9 (p-p)	+0.62/-1.33 (p-p)	+0.19/-0.22 (p-p)	+3/-1.4 (3σ)	+1.3/-1.2 (p-p)	+0.85/-0.94 (p-p)
Calibration	1-point	1-point	1-point	1-point	2-point	2-point	2-point	2-point	2-point	2-point	2-point
Number of chips	16	630	20	12	15	7	3	16	20	9	15
Resolution (mK)	63	500	15	39	210	300	100	73	150	145	210
Conv. time (sec)	6m	30μ	5m	8.1m	21.4 μ	22 μ	254m	8m	80μ	839m	21.4 μ
Energy/conv. (nJ)	3.6	0.86	34.5	7.0	0.24	3.4	19.05	0.6	1.024	8.9	0.28
FoM ‡(pJ· K^2)	14.3	132	7.8	10.6	11.5	300	190	3.2	20	190	13.7†
Require Ext. Clk$_{REF}$	Yes	Yes	Yes	Yes	Yes	No	Yes	Yes	Yes	No	Yes

$\dagger$ Including the bandgap reference, plus the estimated TDC from layout and simulation. $\ddagger$ FoM = Power × Conversion time × Resolution^2

TABLE III: V_{DD} Sensor Performance Overview

	VLSI-DAT13 [5]	ISCAS19 [23]	This work
Sensor type	Ring Osc	Gate Leakage Osc	Relax. Osc
Tech (nm)	90	65	180
Area (mm^2)	0.184	0.0047	0.024
Supply voltage (V)	0.8−1.2	0.75−1.0	1.2−1.8
Power (μW)	1940	0.1	15.54†
Temp. range (°C)	NA	NA	-15−100
Temp. sens. (mV/°C)	NA	NA	0.50
Inaccuracy (mV)	4.81 (p-p, simulation)	+12.1/-8.4 (p-p, 1 sample)	+28.7/-30.0 (p-p, 15 samples)
Resolution (mV)	3.74	2.4	0.94
Conv. time (sec)	1μ	6.15 sec	22.4μ
Energy/conv. (nJ)	1.94	615	0.35

$\dagger$ With estimated TDC by layout and simulation, and the bandgap reference.

V. CONCLUSION

This paper has introduced novel circuits for temperature and supply voltage monitoring, based on a two-phase current-mode relaxation oscillator. Future work may involve combining both temperature and supply voltage sensing in a single oscillator to save power and area. Such an arrangement could be especially desirable for applications which can tolerate moderately slower conversion in exchange for ultra low power consumption.

REFERENCES

[1] Anand, T., Makinwa, K.A. and Hanumolu, P.K., "A VCO based highly digital temperature sensor with 0.034°C/mV supply sensitivity", *IEEE Journal of Solid-State Circuits*, 51(11), pp. 2651-2663, 2016.

[2] Mordakhay, A. and Shor, J., "Miniaturized, 0.01 mm^2, Resistor-Based Thermal Sensor With an Energy Consumption of 0.9 nJ and a Conversion Time of 80 μs for Processor Applications", *IEEE Journal of Solid-State Circuits*, 53(10), pp. 2958-2969, 2018.

[3] Chen, S.W., Chang, M.H., Hsieh, W.C. and Hwang, W., "Fully on-chip temperature, process, and voltage sensors", in *IEEE International Symposium on Circuits and Systems (ISCAS)*, 2010.

[4] Vezyrtzis, C., Strach, T., Pierce, I., Chuang, J., Lobo, P., Rizzolo, R., Webel, T., Owczarczyk, P., Buyuktosunoglu, A., Bertran, R. and Hui, D., "Droop mitigation using critical-path sensors and an on-chip distributed power supply estimation engine in the z14^{TM} enterprise processor", in *IEEE International Solid-State Circuits Conference (ISSCC)*, 2018.

[5] Hsu, C.-H., Huang, S.-Y., Kwai, D.-M., Chou, Y.-F., "Worst-case IR-drop monitoring with 1GHz sampling rate", in *International Symposium on VLSI Design, Automation, and Test, VLSI-DAT*, pp. 1-4, 2013.

[6] Pan, S., Luo, Y., Shalmany, S.H. and Makinwa, K.A.,"A Resistor-Based Temperature Sensor With a 0.13 pJ · K^2 Resolution FoM", *IEEE Journal of Solid-State Circuits*, 53(1), pp. 164-173, 2018.

[7] Park, H. and Kim, J., "A 0.8-V Resistor-Based Temperature Sensor in 65-nm CMOS With Supply Sensitivity of 0.28°C/V", *IEEE Journal of Solid-State Circuits*, 53(3), pp. 906-912, 2018.

[8] Choi, W., Lee, Y., Kim, S., Lee, S., Jang, J., Chun, J., Makinwa, K.A. and Chae, Y., "A Compact Resistor-Based CMOS Temperature Sensor With an Inaccuracy of 0.12°C (3σ) and a Resolution FoM of 0.43 pJ · K^2 in 65-nm CMOS", *IEEE Journal of Solid-State Circuits*, 2018.

[9] Sonmez, U., Sebastiano, F. and Makinwa, K.A., "Compact thermal-diffusivity-based temperature sensors in 40nm CMOS for SoC thermal monitoring", *IEEE Journal of Solid-State Circuits*, 52(3), pp.834-843, 2017.

[10] Yang, K., Dong, Q., Jung, W., Zhang, Y., Choi, M., Blaauw, D. and Sylvester, D., "A 0.6 nJ −0.22/+0.19°C inaccuracy temperature sensor using exponential subthreshold oscillation dependence", in *IEEE International Solid-State Circuits Conference (ISSCC)*, pp. 160-161, 2017.

[11] Wang, X., Wang, P.H.P., Cao, Y. and Mercier, P.P., "A 0.6 V 75nW All-CMOS Temperature Sensor With 1.67m°C/mV Supply Sensitivity", *IEEE Transactions on Circuits and Systems I: Regular Papers (TCASi)*, 64(9), pp. 2274-2283, 2017.

[12] Someya, Teruki, AKM Mahfuzul Islam, Takayasu Sakurai, and Makoto Takamiya, "An 11-nW CMOS Temperature-to-Digital Converter Utilizing Sub-Threshold Current at Sub-Thermal Drain Voltage", *IEEE Journal of Solid-State Circuits*, 54(3), pp. 613-622, 2019.

[13] Wang, B., Law, M.K., Tsui, C.Y. and Bermak, A., "A 10.6 pJ·K^2 Resolution FoM Temperature Sensor Using Astable Multivibrator", *IEEE Transactions on Circuits and Systems II: Express Briefs (TCASii)*, 65(7), pp. 869-873, 2018.

[14] Said, M., Chetoui, S., Belouchrani, A. and Reda, S., "Understanding the Sources of Power Consumption in Mobile SoCs", in *IEEE Ninth International Green and Sustainable Computing Conference (IGSC)*, 2018.

[15] K. A. A. Makinwa, *Smart Temperature Sensor Survey*, [Online]. Available: http://ei.ewi.tudelft.nl/docs/TSensor_survey.xls

[16] Dai, S. and Rosenstein, J.K., "A 14.4 nW 122KHz dual-phase current-mode relaxation oscillator for near-zero-power sensors", in *IEEE Custom Integrated Circuits Conference (CICC)*, 2015.

[17] Jiang, H., Wang, P.H.P., Mercier, P.P. and Hall, D.A., "A 0.4-V 0.93-nW/kHz Relaxation Oscillator Exploiting Comparator Temperature-Dependent Delay to Achieve 94-ppm/°C Stability", *IEEE Journal of Solid-State Circuits*, 53(10), pp. 3004-3011, 2018.

[18] Banba, H., Shiga, H., Umezawa, A., Miyaba, T., Tanzawa, T., Atsumi, S. and Sakui, K., "A CMOS bandgap reference circuit with sub-1-V operation", *IEEE Journal of Solid-State Circuits*, 34(5), pp. 670-674, 1999.

[19] Piguet, C., "Logic synthesis of race-free asynchronous CMOS circuits", *IEEE Journal of Solid-State Circuits*, 26(3), pp. 371-380, 1991.

[20] Souri, Kamran, Youngcheol Chae, Frank Thus, and Kofi Makinwa, "A 0.85 V 600nW all-CMOS temperature sensor with an inaccuracy of ±0.4°C(3σ) from −40 to 125°C", in *IEEE International Solid-State Circuits Conference(ISSCC)*, pp. 222-223. IEEE, 2014.

[21] Eberlein, Matthias, and Idan Yahav, "A 28nm CMOS ultra-compact thermal sensor in current-mode technique", in *IEEE Symposium on VLSI Circuits (VLSIC)*, pp. 1-2, 2016.

[22] Yousefzadeh, B., Shalmany, S.H. and Makinwa, K.A., "A BJT-Based Temperature-to-Digital Converter With ±60 mK (3σ) Inaccuracy From −55°C to +125°C in 0.16-μm CMOS", *IEEE Journal of Solid-State Circuits*, 52(4), pp. 1044-1052, 2017.

[23] Kobayashi, A., Hayashi, K., Arata, S., Murakami, S., Xu, G., Niitsu, K., "A 65-nm CMOS 1.4-nW self-controlled dual-oscillator-based supply voltage monitor for biofuel-cell-combined biosensing systems", in *IEEE International Symposium on Circuits and Systems (ISCAS)*, pp. 1-5, 2019.

978-1-7281-5410-7/20 $31.00 © 2020 IEEE

Design, Implementation and Analysis of Efficient Hardware-based Security Primitives

N. Nalla Anandakumar
Hardware Security Research Lab
Society for Electronic Trans. and Security (SETS)
Chennai, India
Email: nallananth@gmail.com

Somitra Kumar Sanadhya
Department of CSE
IIT Ropar
Rupnagar, India
Email: somitra@iitrpr.ac.in

Mohammad S. Hashmi
School of Engineering and Digital Sciences
Nazarbayev University
Nur-Sultan, Kazakhstan
Email: mohammad.hashmi@nu.edu.kz

Abstract—Hardware-based security primitives play important roles in protecting and securing a system in Internet of Things (IoT) applications. The main primitives are physical unclonable functions (PUF) and true random number generator (TRNG) studied in this paper. Efficient FPGA implementation are proposed in the work along with relevant security analysis using prevalent metrics. Finally, an application of designed TRNG and PUF is proposed for implementing an authenticated key agreement protocol.

Index Terms—PUF, TRNG, authenticated key agreement protocol, FPGA

I. INTRODUCTION

Internet of Things (IoT) is a vast and rapidly growing technology right now in the world of innovation. Billions of new electronic devices are going to be connected to the internet in wide-ranging applications. With this massive increase in adoption and utilization of new technology, security vulnerabilities are growing exponentially as well. Traditionally, conventional cryptographic primitives are used in order to provide security to IoT devices. The security of the cryptographic protection relies on the secrecy of the key. Typically, secret keys, which are used as device identification (IDs), are stored in non-volatile memories (NVMs), and combine cryptographic primitives to implement information encryption and authentication. However, through such traditional technique, secret keys are vulnerable to various kinds of attacks and can be easily obtained or cloned. Furthermore, maintaining such secrets in NVMs is difficult and expensive. In addition, random key generation and key exchange are also very challenging in secure IoT applications. Hardware-based security primitives such as PUFs and TRNGs can overcome these limitations and provide true random numbers in order to establish security and trustworthiness in IoT systems [1].

Physically Unclonable Function (PUF) promises to be a critical hardware security primitive to provide an alternative method to create unique signatures (IDs) from complex physical characteristics of ICs rather than storing the IDs in non-volatile memories. Eventually these IDs can be used to authenticate devices and also to generate secret keys for cryptographic functions. A True Random Number generator (TRNG) is another important hardware security primitive that

generates high entropy random numbers (keys) from a physical process for use in key exchange/agreement, encryption, and digital signature, etc. The IoT infrastructure adopts a large number of these hardware-based security primitives in order to securely exchange data in an effective and resource efficient manner.

II. THESIS PROBLEM STATEMENT

After an extensive research survey on PUF primitive designs, we identify that the existing state-of-art techniques have severe limitations in most of the performance metrics namely reliability, uniqueness, and randomness. Another problem facing the wide scale deployment of hardware-based security primitives such as PUFs and TRNGs for IoT applications is that there is a high demand for low-cost resource efficient solutions. However, most of the current state of the art PUF and TRNG primitives are expensive for low area implementations. It is, therefore, imperative to investigate and propose novel solutions to address these pressing problems in the existing design approaches. This thesis attempts to address some concerns regarding design, development and implementation of highly efficient PUFs and TRNG for FPGAs with enhanced performance in IoT security systems

III. THESIS CONTRIBUTIONS

In this thesis, we have presented several novel designs and architectures for PUF and TRNG primitives and their practical usability in key agreement protocol. All of our contributions are FPGA compatible. In summary, the major contributions of this work are as follows:

First, we have developed three major types of area efficient PUF designs and improving their qualities [2]. One is a memory based PUF: RS-Latch based design. The second and third are delay based PUF: Ring oscillator and Arbiter based designs. For example, our RS Latch based PUF can be performed in 142 msec using 91 slices on a Virtex-5 FPGA. These three designs have been thoroughly tested on FPGA devices. The enhancement in performance is achieved through the incorporation of various techniques such as internal variations of FPGA Look-Up Tables (LUTs) in terms of coarse and fine Programmable Delay Lines (PDLs), Temporal Majority Voting (TMV) scheme, and hard macro techniques for routing and placements of PUF units. Performance metrics of these designs have been presented and compared to the state of the art PUFs.

Second, we present an area efficient hybrid PUF design on FPGA [3]. Our approach combines units of conventional RS Latch-based PUF and Arbiter-based PUF which is then augmented by the PDLs and TMV for performance enhancement. The proposed design of Hybrid RS Latch-Arbiter PUF is shown in Fig. 2. The measured results on the FPGA demonstrate PUF signatures exhibits good uniqueness, reliability, and uniformity with no occurrence of bit-aliasing.

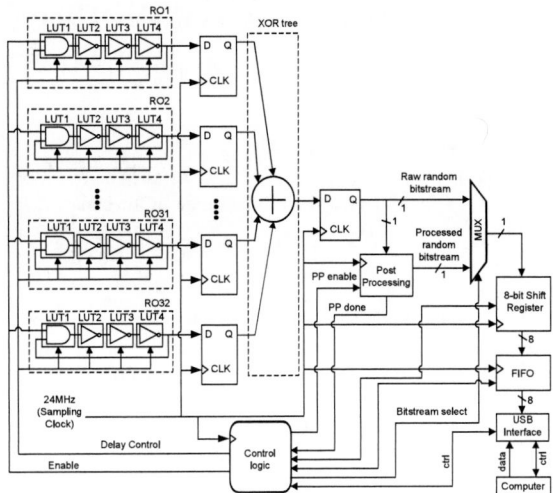

Fig. 2. Architecture of the proposed TRNG [4]

secret key and key exchange. For further details, one may refer to [1], [5]. Our implementation shows that the entire protocol can be performed in 151 msec using 15495 slices on a Virtex-5 FPGA [1]. Therefore, our proposed ECMQV design can be used as a cryptographic accelerator for IoT security applications in data center security (data center is central to the IoT as it processes data from millions of devices), intelligent automation, smart grid security.

IV. CONCLUSION

In this work, we have presented several novel designs and architectures for PUF and TRNG primitives and their practical usability in key agreement protocol. All of our contributions are FPGA compatible. Moreover, If PUFs and TRNGs are ever to be used in high security systems then they should resist side channel attacks (SCA). So far, very little has been done in this area. Studying SCA resistance of certain implementations of PUFs or proposing a SCA resistant implementation of the same could be a possible extension of our current work.

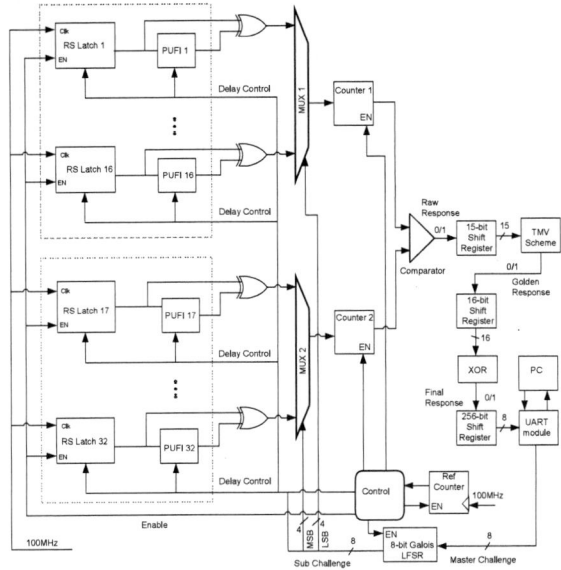

Fig. 1. The proposed design of Hybrid RS Latch-Arbiter PUF [1]

Third, we design and developed ROs based true random number generation on FPGA [4]. The proposed TRNG architecture is shown in Fig. 2. The programmable delay of FPGA LUTs has been used to achieve random jitter and to reduce correlation between several equal length oscillator rings, and thus improve the randomness qualities. Our RO based TRNG can be performed in $6.4\mu s$ using 193 slices on a Virtex-5 FPGA. Our proposed implementation achieves high entropy rate and successfully passes all NIST statistical tests.

Finally, we presented a practical design for an area efficient authenticated key agreement protocol between two IoT devices using BEC, PUF and TRNG [1], [5]. In this context, A novel hardware architecture of binary Edwards curve (BEC) point multiplication using mixed w-coordinates of the Montgomery laddering algorithm has been developed. Subsequently, an FPGA design of elliptic curve based key agreement protocol (ECMQV) using PUF and TRNG is presented. The key agreement protocol uses PUF for the unique long term secret key generation, TRNG for short term random secret key generation, BEC for generating the public key corresponding to the secret key, and ECMQV for generating the shared

REFERENCES

[1] N. N. Anandakumar, "Design, Implementation and Analysis of Efficient Hardware-based Security Primitives," Theses, IIIT-Dehi, 2020. [Online]. Available: https://repository.iiitd.edu.in/jspui/handle/123456789/798

[2] N. N. Anandakumar, M. S. Hashmi, and S. K. Sanadhya, "Compact Implementations of FPGA-based PUFs with Enhanced Performance," in *30th International Conference on VLSI Design and 16th International Conference on Embedded Systems (VLSID)*, Jan 2017, pp. 161–166.

[3] N. N. Anandakumar, M. S. Hashmi, and S. K. Sanadhya, "Efficient and Lightweight FPGA-based Hybrid PUFs with Improved Performance," *Microprocessors and Microsystems*, vol. 77, p. 103180, 2020.

[4] N. N. Anandakumar, S. K. Sanadhya, and M. S. Hashmi, "FPGA-Based True Random Number Generation Using Programmable Delays in Oscillator-Rings," *IEEE Transactions on Circuits and Systems II: Express Briefs*, pp. 1–1, 2019.

[5] N. N. Anandakumar, M. P. L. Das, S. K. Sanadhya, and M. S. Hashmi, "Reconfigurable Hardware Architecture for Authenticated Key Agreement Protocol Over Binary Edwards Curve," *ACM Trans. Reconfigurable Technol. Syst.*, vol. 11, no. 2, pp. 12:1–12:19, Nov. 2018.

Multiple-NoC Exploration and Customization for Energy Efficient Traffic Distribution

Sonal Yadav[1,2], Vijay Laxmi[1], Manoj Singh Gaur[1,3]

Dept. of CSE, [1]Malaviya National Institute of Technology, Jaipur, IN, [2]Indian Institute of Information Technology Kota, IN, [3]IIT Jammu, IN

sonal.cse@iiitkota.ac.in, {vlaxmi, gaurms}@mnit.ac.in

Motivation. As more on-chip computation resources are available, the Networks-on-Chip (NoC) power consumption is expected to increase in the future many-core era. Many of these would have to be switched off while inactive to keep power consumption and chip temperature low. However, the NoC infrastructure must be kept alive to serve shared caches and memory accesses. A recent study shows that the proportion of NoC power consumption becomes appreciable in comparison to computation counterpart. A 32 core chip at $45nm$ substantially raises NoC power ($\sim 42\%$) among the remaining on-chip active resources (cores, shared caches, memory controllers, PCIe controllers) [1]. Therefore, low power becomes the primary objective for modern NoC designs.

The static power comprises of more than 74% of the total NoC power budget at high network utilisation on $22nm$ technology [2]. This fraction of power further increases on low network utilisation. This figure is expected to increase in future technology generations. As the power of a processor is directly affected by hardware design of the circuit, the industry demands the static power efficient manycore processors. Reducing static power consumption becomes an urgent issue as it raises dark silicon more rapidly. Modern architects are looking for a static power efficient customised NoC.

Multiple Networks-on-Chip (Multi-NoCs) offer independent parallel data flows through more than one NoC interconnects. Freedom of different hardware customizations along with flexibility in fine-grain traffic distribution makes multi-NoCs more power and energy efficient as compared to bandwidth equivalent single-NoC. Many multi-NoCs customizations are made to address the modern research challenges of dark silicon [3], [4], hardware accelerators [5], reconfiguration [6], [7], fault tolerance [8], security [9], and power gating [10], [11]. Multi-NoCs research prototypes are MIT's RAW [12] and TRIPS [13] chips. In commercial chip product series, Tilera [14] and Adapteva Epiphany [15] processors have implemented multi-NoCs.

Multi-NoCs are primarily implemented for application-specific processors e.g. SoC[1], MPSoC[2], and FPGAs[3]. Contrary, general-purpose processors are less explored for multi-NoCs implementations. CMPs[4] run a wide variety of applications with unpredictable low/high heterogeneous runtime traffic variations. For these processors, it is difficult to design a static power efficient customised multi-NoC that should dynamically adapt traffic distribution according to runtime variations of fine-grain messages from computation bound to communication and memory-bound applications. We have addressed a difficult challenge to design energy efficient multi-NoC for CMPs. To attain it, NoC power consumption should be proportional to the network demand without compromising communication delays.

Thesis Contributions. In our thesis, multiple NoCs itself hardware implementation is customised for static power along with fine-grain traffic distribution exploration for improving energy efficiency of CMP's runtime traffic. Our novel contributions are as follows:

1) **Customised Multi-NoC Architecture to Attain Power Efficiency.** On-chip networks research efforts are focused to match the speed of core execution. Injecting bulk of traffic in a network could not help to achieve speed up as network start to face traffic congestion in case of higher network traffic that further slow down the network. This fact is also applied with multiple NoCs. These networks carry parallel traffic flows through external **N**etwork **I**nterface (NI) links and internal network links. We customize multi-NoCs through introducing single NI link. Such customisation reduces the static power without impacting traffic distribution flexibility of multiple NoCs. Initially, the warm-up phase takes a minor time unless the buffers of the second network get filled. Such an initial delay of the network can be compensated by efficient traffic flow. As network faces less congestion, it fastly traverses flits. The advantage comes from half of the external NI links that reduce the number of I/O ports of the router, the number of **V**irtual **N**etworks (VNs) and **V**irtual **C**hannels (VCs), routing logic and control logic overhead, size of the crossbar, etc. In the customised architecture, multiple networks originate from the router using network links, so we have placed traffic distribution hardware unit along with the routing unit of the router rather than conventional network interface placement. This customization is scalable with network size and does the following improvements–

   - The synthesis results show 42% and 36% improvement in static power, and 30% and 25% area efficiency in customised multi-NoC over conventional multi-NoC and single-NoC respectively at $32nm$ technology.
   - Experiment on PARSEC[5] benchmarks finds 45% efficiency in total router power at $(65nm, 1GHz)$. The power efficiency approaches 58% as technology shrink to $32nm$ at $1GHz$ frequency. In contrast, the power efficiency limits to 40% as frequency increases to $2.5GHz$ at $65nm$ technology.

2) **Target Multi-NoC Architecture.** The architecture and placement exploration of the traffic distribution hardware unit is unexplored prior to our work in multi-NoC. As a flit is demultiplexed to any of the multiple NoC networks through traffic distribution hardware unit we named it as **N**etwork **D**emultiplexer (Net-Demux). Conventional multi-NoCs design place Net-Demux at network interface. Our customised multi-NoCs place Net-Demux along with routing unit of the router. At the network interface, Net-Demux is placed in the datapath whereas at the router the placement is possible at the control unit. The placement at the control unit of the router is beneficiary over network interface placement as the datapath placement of NI has more area and power overhead for the implementation

978-1-7281-5410-7/20 $31.00 © 2020 IEEE

of Net-Demux. Whereas, the router microarchitecture is the pipelined architecture. The critical path delay is dominated by the crossbar of the router. The placement of the Net-Demux does not lead the critical path delay of the crossbar of the router. The Net-Demux placement at routing unit leads to the opportunity of another control unit placement on the router. We find it at switch allocator of the router that decides the output NoC for the flit. The Net-Demux placement at the switch allocator is more efficient compared to routing unit placement. This is derived through digital circuit characteristics of input signal ordering. It is beneficial to postpone the introduction of input signals with a high transition rate. As the input and output of network demultiplexer are driven by different modules of the circuit, placement changes the switching of the circuit through variation in the average number of signal transitions per cycle. The output of routing logic becomes the input of Net-Demux where signal transition rate is higher as compared to switch allocator placement of Net-Demux. Placement reorders the input signals, and hence signal transition rate for Net-Demux hardware unit. The hardware implementation synthesis results and PARSEC benchmark–

- Experiments show that switch allocator, routing unit, and network interface placements are 46%, 40% and 30% energy efficient over single-NoC on PARSEC benchmarks.
- The placement at switch allocator and routing unit are respectively 33% and 26% more energy efficient over conventional network interface placement.
- The placement at switch allocator improves the critical path delay by 33% over conventional NI placement.

3) **Case Study: Message Distribution Problem of Multi-NoC in CMPs.** Static traffic distribution is the simplest offline defined message distribution method irrespective of runtime changes in fine-grain message volume. We theoretically and experimentally prove that static message distribution is `NP-hard` problem. We have proposed a case study for different combinations of static traffic distribution to find its impact on performance and energy efficiency of multi-NoCs. We observe that it is hard to find the best solution from all the feasible choices. Efficient static message distribution can be determined for a single benchmark, but the same distribution policy is not necessarily the best solution for the rest of the applications. We have following conclusions with the case study using general-purpose applications of PARSEC benchmark–

- Communication intensive benchmarks like swaptions (in power), x264 (in energy) and rtview (in throughput) show a significant approximately $5x$ variations.
- Power varies between 3% to 78%, throughput vary 1% to 500% and energy vary 9% to 560% with the variation in static message distribution.

4) **Runtime Adaptive Fine Grained Message Distribution for improved Runtime Utilisation of Multi-NoCs.** Most of the multi-NoCs uses static communication infrastructure. A worst-case traffic scenario configured once at design time without taking into account the traffic dynamics of general-purpose workloads. Static traffic distribution unable to capture runtime fine-grain traffic variations. It results in a runtime under-utilisation of one network while another network is suffering from runtime over-utilisation by traffic. We examine inherent dynamic behavior and structural traffic variability at runtime and propose a runtime adaptive message distribution

for customised multi-NoCs. During switching between cache states transitions, the criticality of messages is devised through analysis of the volume and quality of identified fine-grain control messages. We capture the runtime dynamics of the traffic to adapt fine-grain message distribution between multiple networks of multi-NoCs.

- Runtime adaptive traffic distribution attains energy saving up to 36% and 14% compared to the single-NoC and static traffic distribution respectively for communication-intensive PARSEC benchmarks suite.
- The link utilisation improves 16% and 21% over static message distribution and single NoC.

We have implemented customised multiple NoC in GARNET[6] that is integrated with Gem5 simulator. The results are evaluated using PARSEC benchmarks in full-system simulation. The area and power of hardware implementation is estimated through synthesis using Synopsis Design Compiler.

ACHIEVEMENTS

IEEE-iSES 2018 (TCVLSI Best Paper Award), DATE 2019 PhD-Forum (Travel Grant). (For publications refer to Google Scholar (NKe86HMAAAAJ))

REFERENCES

[1] J. Zhan et al, "NoC-sprinting: Interconnect for fine-grained sprinting in the dark silicon era," in 51st ACM/EDAC/IEEE DAC, San Francisco, CA, pp. 1-6, 2014.
[2] R. Parikh et al, "Power-aware NoCs through routing and topology reconfiguration," 51st ACM/EDAC/IEEE DAC, San Francisco, CA, pp. 1-6, 2014.
[3] S. Hesham et al, "A call-up for circuit-switched NoCs in the Dark-Silicon Era," IEEE Nordic Circuits and Systems Conference (NORCAS): NORCHIP and Int. Symp. of SoC, Linkoping, pp. 1-6, 2017.
[4] M. Shafique et al, "Computing in the Dark Silicon Era: Current Trends and Research Challenges," in IEEE Design & Test, vol. 34, no. 2, pp. 8-23, April 2017.
[5] Z. Li et al, "The runahead network-on-chip," 2016 IEEE Int. Symp. on HPCA, Barcelona, 2016, pp. 333-344.
[6] H. Lu et al, "ShuttleNoC: Boosting on-chip communication efficiency by enabling localized power adaptation," in 20th ASP-DAC, IEEE, Chiba, pp. 142-147, 2015.
[7] H. Lu et al, "ShuttleNoC: Power-adaptable Communication Infrastructure for Many-core Processors," in IEEE TCAD, July 2018.
[8] Y.J. Yoon et al, "Virtual Channels and Multiple Physical Networks: Two Alternatives to Improve NoC Performance," in IEEE TCAD, vol.32, no.12, pp. 1906-1919, Dec. 2013.
[9] J. Sepúlveda, D. Flórez, and G. Gogniat, "Reconfigurable security architecture for disrupted protection zones in NoC-based MPSoCs," in Proc. of 10th Int. Symp. on ReCoSoC, Bremen, pp. 1-8, 2015.
[10] R. Das et al, "Catnap: energy proportional multiple network-on-chip," in Proc. of the 40th ISCA, ACM, New York, USA, pp. 320-331, 2013.
[11] J. Wu, D. Dong, and L. Wang, "NoC Power Optimization using Combined Routing Algorithms," in Proc of ICIC, Wuhan, China, 2017.
[12] M. Taylor et al., "The Raw Microprocessor: A Computational Fabric for Software Circuits and General-Purpose Programs," IEEE Micro, vol. 22, pp. 25-35, 2002.
[13] P. Gratz et al, "On-Chip Interconnection Networks of the TRIPS Chip," IEEE Micro, vol. 27, pp. 41-50, 2007.
[14] D. Wentzlaff et. al, "On-Chip Interconnection Architecture of the Tile Processor," IEEE Micro, vol. 27, pp. 15-31, 2007.
[15] A. Varghese et al, "Programming the Adapteva Epiphany 64-Core Network-on-Chip Coprocessor," in Proc. of the 29th IEEE-PDPSW, Phoenix, AZ, pp. 984-992, 2014.

NOTES

[1] System-on-Chip
[2] Multi-Processor System-on-Chip
[3] Field-Programmable Gate Arrays
[4] Chip Multi-Processor
[5] Princeton Application Repository for Shared mEmory Computers
[6] A detailed cycle-accurate network-on-chip model inside Gem5 full-system simulator

978-1-7281-5410-7/20 $31.00 © 2020 IEEE

Design Automation for Side Channel Resistant Lightweight Cryptography

Rajat Sadhukhan
Department of Computer Science and Engineering
Indian Institute of Technology Kharagpur
Kharagpur, India
rajatssr835@gmail.com

Debdeep Mukhopadhyay
Department of Computer Science and Engineering
Indian Institute of Technology Kharagpur
Kharagpur, India
debdeep.mukhopadhyay@gmail.com

Abstract—The scaling of devices in IoT era has opened doors to broad range of privacy and security concerns making it vulnerable to various side channel attacks (SCA) and differential fault attacks (DFA). The generic EDA flows provide powerful solutions for simulation, verification, power, performance and area (PPA) optimizations. But these flows are not enabled to detect side channel leakages or provide countermeasures or handle huge search space under both lightweightedness and security dimension paradigm. So, we propose to augment classical EDA flow with security aware notion addressing the challenges wrt primitive design, SCA and DFA thereby helping crypto-designers to reduce overall design cycle-time and early detection of security flaws in lightweight design and provide countermeasure.

Index Terms—Floorplan, ML-based S-Box Synthesis, Lightweight Cryptography, VLSI-CAD

I. ML BASED DETERMINISTIC MODEL FOR POWER EFFICIENT S-BOX SYNTHESIS

We propose a methodology focuses on efficient synthesis of non-linear layer(S-Boxes) in ciphers [1]. We first develop ML-based classifier to determinine if an S-Box is power efficient or not followed by a deterministic model using results obtained from classifier model to predict the dynamic power of an S-Box. Our ML-based tool flow along with traditional probabilistic and simulation based approaches is depicted in Fig.1. We have shown the results of the feature importance assessment in Fig.2(left) determining importance of each feature scaled to an interval of [0,1]. With an 84% accuracy we are able scale-up the efficiency in terms of speed to 14× when compared to standard synthesis tools [2]. The learning curve (Fig.2-right) conventionally depicts improvement in performance as we increase the number of training examples. Utilizing classifier model and correlation analysis we develop a dynamic power predictor function [1] to determine dynamic power consumption of S-Box as:

$$X = \Delta_N * gc(N) + \Delta_A * gc(A) + \Delta_O * gc(O) + \alpha_F * lc(factor)$$

Where Δ_N, Δ_O and Δ_A is the dynamic power of the corresponding NOT, OR and AND gate respectively. $gc(.)$ is the gate count, $lc(.)$ is literal count and α_F is the activity factor of every literal. Our obtained model has an accuracy of 80% [1](Fig.3) and can instantly predict the dynamic power consumption value of the S-Boxes bypassing actual synthesis step.

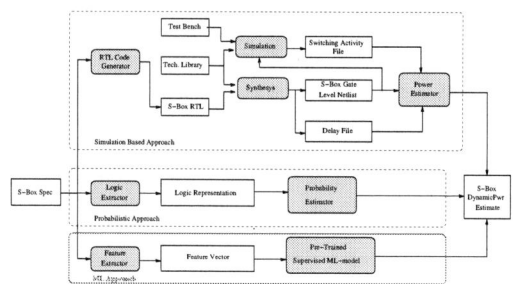

Fig. 1. ML-Based S-box Synthesis Design Flow [2]

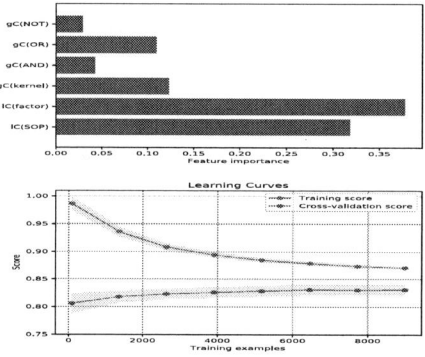

Fig. 2. Feature Importance Graph (Top) [1] and Learning Graph for 4 × 4 Optimal S-boxes (Bottom) [1]

II. PRE-SILICON POWER ANALYSIS OF CRYPTO-DESIGNS

Side channel analysis carried out at post-silicon stage incurs significant loss if the post-silicon is found to be vulnerable. Hence, we propose CAD-based methodology to perform power-analysis on pre-silicon gate-level netlist of cryptographic algorithm [3]. We first build a new leakage model of the critical component present in crypto design by approximating its dynamic power consumption from the toggle count information and then using the power model we perform power attack to recover key using power traces of crypto design. Finally, we integrated hybrid first order side channel testing methodology by Roy et al. to the traditional tool flow that

978-1-7281-5410-7/20 $31.00 © 2020 IEEE

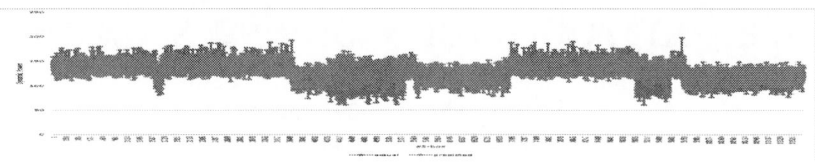

Fig. 3. Actual and Predicted Dynamic Power for 10,000 S-boxes [1]

provide us a completely automated and analytical CAD tool flow which can detect and quantify side channel vulnerability from simulated pre-silicon power analysis. The modified tool flow as shown in Fig.4 will essentially predict the success rate of first order attack of the given power trace. We have also compared our model with Hamming distance and Hamming weight leakage model and the result is illustrated in Fig.5. Our model can guess the right key with lesser number of traces and with better correlation when compared to Hamming distance and Hamming weight models.

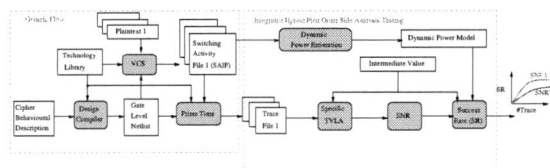

Fig. 4. Proposed SCA-aware Tool Flow for Trace Collection and Hybrid Testing Integrated with Generic Flow [3]

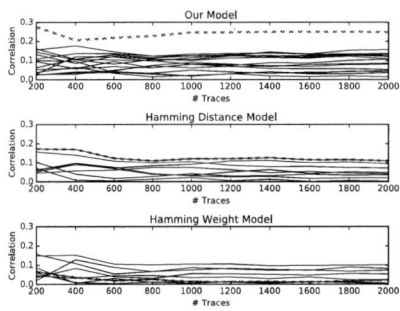

Fig. 5. Correlation Coefficient vs. Tracecount (Red Dotted Line- Right Key Guess, Black Lines- Wrong Key Guess) [3]

III. DFA-AWARE FLOORPLAN

We propose "DFA-aware" physical design floorplan methodology [4] as shown in Fig.6 , that effectively mitigates the threat posed by DFA. Fault injection can be modeled by a common fault model, known as the clustered fault model where we assume that the fault occurs in clusters. Thus if the fault injection creates a fault in a location, the neighboring region within a radius R is faulty. In this methodology mixed integer linear programming (MILP) is used to generate floorplan constraints, where exploitable fault should be R distance apart

while unexploitable fault should be within distance of R. We validate our techniques by benchmarking on AES, where blocks of same color are placed far apart depending on the fault model (Fig.7).

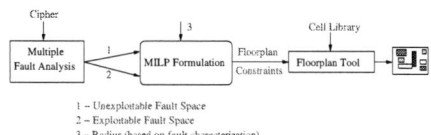

Fig. 6. DFA Aware Floorplanning Flow [4]

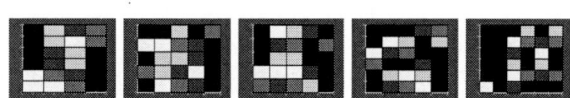

Fig. 7. Security-aware Floorplanning(From Left: Security-Unaware, 1 Diagonal Fault, 2 Diagonal Fault, 3 Diagonal Fault, 4 Diagonal Fault) [4]

IV. CONCLUSION

In our work we constructed primitive library of cryptographically strong and power efficient 4×4 S-Boxes using ML-based synthesis. We proposed methodology to perform power analysis on pre-silicon gate-level netlist of crypto-algo to early detect design vulnerability. On physical design side we suggestedd DFA-awre floorplan methodology to place the macros subjected to fault aware constraints. As future scope we will generate a system model taking the primitive library and different design strategies as input, adhering to classical and physical security along with PPA optimizations.

REFERENCES

[1] R. Sadhukhan, N. Datta, and D. Mukhopadhyay, "Power efficiency of s-boxes: From a machine-learning-based tool to a deterministic model," *IEEE Transactions on Very Large Scale Integration (VLSI) Systems*, pp. 1–13, 2019.

[2] R. Sadhukhan, N. Datta, and D. Mukhopadhyay, "A Machine Learning Based Approach to Predict Power Efficiency of S-Boxes," in *32nd International Conference on VLSI Design and 2019 18th International Conference on Embedded Systems, VLSID 2019, Delhi, India, January 5-9, 2019*, 2019, pp. 531–532.

[3] R. Sadhukhan, P. Mathew, D. B. Roy, and D. Mukhopadhyay, "Count your toggles: a new leakage model for pre-silicon power analysis of crypto designs," *J. Electronic Testing*, vol. 35, no. 5, pp. 605–619, 2019.

[4] M. Khairallah, R. Sadhukhan, R. Samanta, J. Breier, S. Bhasin, R. S. Chakraborty, A. Chattopadhyay, and D. Mukhopadhyay, "DFARPA: differential fault attack resistant physical design automation," in *2018 Design, Automation & Test in Europe Conference & Exhibition, DATE 2018, Dresden, Germany, March 19-23, 2018*, 2018, pp. 1171–1174.

Optimization Tools for ConvNets on the Edge

Valentino Peluso, Enrico Macii, Andrea Calimera

Politecnico di Torino, 10129 Torino, Italy

valentino.peluso@polito.it

Abstract—The shift of Convolutional Neural Networks (ConvNets) into low-power devices with limited compute and memory resources calls for cross-layer strategies spanning from hardware to software optimization. This work answers to this need, presenting a collection of tools for efficient deployment of ConvNets on the edge.

Index Terms—Artificial Intelligence of Things, Deep Learning, Convolutional Neural Networks, Optimization.

I. INTRODUCTION

The advancement in low power design together with the proliferation of connected devices fueled the Internet-of-Things (IoT), enabling ubiquitous sensors to sample and transmit data over the internet. Thanks to the recent breakthroughs in Artificial Intelligence (AI), Convolutional Neural Networks (ConvNets) in particular, today machines can learn trends from collected data and extract meaningful information to make decisions autonomously. Embedding AI in IoT end-nodes is the premise of a new paradigm—Artificial Intelligence of Things (AIoT)—where sensors evolve from passive data collectors to active intelligent devices. This shift will enable the design of more efficient, scalable, and secure systems.

The porting of ConvNets on IoT devices poses several issues due to the complexity of current models. The quest for high accuracy has generated networks with billions of parameters and millions of operations, thus preventing the deployment on low-power devices [1]. To tackle this problem, the design of portable ConvNets needs a proper understating of the hardware resources available from the early stage of the training process [2]. Specifically, the optimization of ConvNets should encompass a multi-objective problem formulation, involving extra-functional metrics like energy, power, and memory, besides accuracy. Rather than improving accuracy, the goal is to identify the trade-off frontiers in the design space to pick the best solution meeting the resource constraints. In practice, this can be achieved with compression methods that exploit the intrinsic resiliency of ConvNets to identify and remove those parts with less contribution to accuracy [3].

The search for optimality, however, gets challenging due to several reasons. Just like for training, the lack of a closed-form solution to describe the dynamics of the learning flow makes the optimization loop slow and uncertain. Moreover, the high number of dimensions to explore introduces an additional level of complexity overlooked by most of the existing works.

The optimization problem recalls the design of digital integrated circuits (ICs), where multiple conflicting constraints should be addressed, like area, power, and performance. Which of these dimensions gets the highest priority depends on the use-cases, the cost requirements, and in general on the design specifications. For instance, some use-cases require fast processing, whereas those applications with relaxed timing constraints enable to reduce power and area. For this reason, existing EDA tools for IC design integrates a collection of computer-aided methodologies, each of them targeting a specific goal. Designers can therefore adopt the most suited tools depending on their needs.

Applying the same approach to the optimization of ConvNets seems a natural choice and it is exactly the topic of this dissertation. Indeed, a one-size-fits-all solution does not exist due to the diversity of hardware backends and fields of application. Rather, dedicated solutions are needed for the analysis and optimization of energy, power, and memory.

Moreover, compression methods originally applied at design-time can operate at run-time to build ConvNets that modulate their resource requirements depending on external triggers raised at the application level [4] (e.g. the battery level or the severity of the task). Once again, whereas the dynamic control of resources is a standard in hardware design [5], it is a less explored field in the optimization of ConvNets. For this reason, we devoted special attention to this topic, showing that dynamic knobs extend the achievable trade-offs.

II. ENERGY: SCALABLE-EFFORT CONVNETS

State-of-the-art ConvNets are trained and implemented as static N-ways classifiers (N as the number of classes used for training) that expend equal effort no matter the surrounding context and the level of accuracy required by the application. Instead, human reasoning, which inspired ConvNets, operates differently. Our reasoning follows a hierarchical organization of knowledge that defines different abstraction levels and connect concepts through their semantic relations, spanning from high-level categories to fine-grain labels. This organization enables the reasoning process to move towards the proper abstraction level depending on the context. Intuitively, the abstraction level determines the effort needed to solve a task: to classify objects with high-level categories, like animals and vehicles is easier than assigning fine-grain labels (e.g. the kind of vehicles or animals). The abstraction level is typically driven by some external triggers, like criticality and relevance of the problem, our level of tiredness, and even the amount of time available to accomplish the task. This scalable mechanism allows our brain to adapt to the context and reach the effort-accuracy trade-off that minimizes energy waste. Having such ability implemented on IoT end-nodes represents an important step towards the implementation of more efficient systems.

To this purpose, we presented Scalable-Effort ConvNets [6], a practical strategy for the implementation of this brain-inspired paradigm with embedded ConvNets. We first demonstrated that ConvNets intrinsically learn the hierarchical organization of knowledge. They are capable to shift from fine-grain labels (used as supervision during training) to high-level categories, without the need for additional training iterations. This capability introduces an additional dimension in the optimization space: the abstraction level of the classification. Second, we presented a run-time adaptation strategy that makes use of arithmetic precision as a knob to modulate the computational effort: depending on the required level of abstraction, Scalable-Effort ConvNets reduce their precision to reach an accuracy target with minimal computational resources. The idea is to apply an algorithmic relaxation of the multiply&accumulate arithmetic in a way that allows using a unique set of weights to implement different precision options. This approach alleviates the overhead of storing multiple weight-sets for each level of abstraction.

The collected results on state-of-the-art models for image classification demonstrated that Scalable-Effort ConvNets exploit multilevel abstraction to reach better energy-accuracy trade-off than standard flat ConvNets that use single precision, achieving up to 60% energy savings or 40% higher accuracy.

III. POWER: VOLTAGE-SCALED CONVNETS

Power consumption is a major bottleneck in embedded System-on-Chips (SoCs), due to the integration of high-performance multi-core CPUs in small form-factor.

While power management of embedded systems is a well-established topic in the literature [7], its intersection with ConvNets is a less explored field. Despite the increasing number of compression pipelines to deploy embedded ConvNets, the existing solutions focus on performance optimization at nominal operating conditions, neglecting the power and thermal constraints of the hosting hardware. Meeting these constraints prevents the execution of intensive workloads (e.g. inference) at maximum performance for long runtime.

The problem calls for novel solutions playing with the availability of hardware knobs for dynamic power management [8]. Among them, Dynamic Voltage Frequency Scaling (DVFS) represents an effective knob to control the thermal stability of the system, clearly at the cost of performance degradation [9].

To enable a proper understating of the achievable performance-power trade-offs, we developed a characterization framework to assess the performance of embedded ConvNets under thermal-aware DVFS [10]. Results collected on a commercial SoC powered by an ARM Cortex-A15 enable three key achievements: (i) quantify the thermal headroom of ConvNets to identify applications that can be critical for power-constrained devices; (ii) demonstrate that the thermal profile of ConvNets depends on the network topology, revealing the need thermal-aware training of ConvNets; (iii) identify the optimal operating points of voltage-scaled ConvNets to guide designers towards the development of specialized control policies, showing that proactive policies guarantee up to 16% faster inference than reactive ones.

IV. MEMORY: PRUNE AND QUANTIZE

Memory is a primary concern for the deployment of ConvNets on tiny sensor-nodes powered by microcontroller units (MCUs). Commercial MCUs, like the ARM Cortex-M series, host less than 1MB of memory, making the deployment of even compact ConvNets extremely challenging.

To reduce the size of ConvNets, compression techniques like pruning and quantization are standard strategies today. The latest advances show that a joint combination of the two achieves state-of-the-art. However, these kinds of methods are accuracy-driven, namely, they seek for the optimal setting that guarantees the highest memory compression with the lowest accuracy loss. Moreover, they leverage algorithmic optimizations whose implementation is impractical on MCUs. Among them, quantization below the 8-bit mark (e.g. from 7- to 2-bit) remains a theoretic study as it asks for custom hardware components not available in low-power cores, e.g. variable bit-width integer arithmetic units. In this context, it is questionable whether accuracy-driven, unconstrained compression methods can meet the needs of real-life applications.

To answer these needs, we presented Prune and Quantize (PaQ) [11], a memory-driven heuristic integrating a smart selection of techniques tailored to meet the hardware specifications. Through an efficient exploration of the memory-accuracy space, PaQ enables three main achievements: (i) demonstrate that the implementation of practical ConvNets is governed by the memory constraint, and not just accuracy; (ii) enumerate the optimal configurations in the memory-accuracy space when the optimization is conducted under tight memory constraints; (iii) quantify the accuracy distance between theoretical configurations and the closest implementations that can be deployed on low-power MCUs, showing that 8-bit models reach up to 3x compression ratios with only 1% less accuracy than arbitrary bit-width models.

REFERENCES

[1] X. Xu *et al.*, "Scaling for edge inference of deep neural networks," *Nature Electronics*, vol. 1, no. 4, pp. 216–222, 2018.

[2] D. Marculescu *et al.*, "Hardware-aware machine learning: modeling and optimization," in *Proceedings of the International Conference on Computer-Aided Design*, 2018, pp. 1–8.

[3] L. Deng *et al.*, "Model compression and hardware acceleration for neural networks: A comprehensive survey," *Proceedings of the IEEE*, vol. 108, no. 4, pp. 485–532, 2020.

[4] V. Peluso *et al.*, "Weak-mac: Arithmetic relaxation for dynamic energy-accuracy scaling in convnets," in *2018 IEEE International Symposium on Circuits and Systems (ISCAS)*. IEEE, 2018, pp. 1–5.

[5] D. J. Pagliari *et al.*, "Energy-efficient digital processing via approximate computing," in *Smart Systems Integration and Simulation*. Springer, 2016, pp. 55–89.

[6] V. Peluso *et al.*, "Scalable-effort convnets for multilevel classification," in *Proceedings of the International Conference on Computer-Aided Design*, 2018, pp. 1–8.

[7] Y. Chen *et al.*, "Battery-aware design exploration of scheduling policies for multi-sensor devices," in *Proceedings of the 2018 on Great Lakes Symposium on VLSI*, 2018, pp. 201–206.

[8] V. Peluso *et al.*, "Beyond ideal dvfs through ultra-fine grain vdd-hopping," in *IFIP/IEEE International Conference on Very Large Scale Integration-System on a Chip*. Springer, 2016, pp. 152–172.

[9] G. Santoro *et al.*, "Design-space exploration of pareto-optimal architectures for deep learning with dvfs," in *2018 IEEE International Symposium on Circuits and Systems (ISCAS)*. IEEE, 2018, pp. 1–5.

[10] V. Peluso *et al.*, "Performance profiling of embedded convnets under thermal-aware dvfs," *Electronics*, vol. 8, no. 12, p. 1423, 2019.

[11] M. Grimaldi *et al.*, "Optimality assessment of memory-bounded convnets deployed on resource-constrained risc cores," *IEEE Access*, vol. 7, pp. 152 599–152 611, 2019.

Memory and Energy Efficient Method Toward Sparse Neural Network Using LFSR indexing

Foroozan Karimzadeh* and Arijit Raychowdhury*

*School of Electrical and Computer Engineering, Georgia Institute of Technology, Atlanta, Georgia 30332

Email: fkarimzadeh6@gatech.edu, arijit.raychowdhury@ece.gatech.edu

Abstract—**Deep Neural Networks (DNNs) require enormous computational power and storage memory. This impose a critical challenge to their efficient deployment on resource-constrained computing platforms such as edge devices. In this paper, we present a novel pruning algorithm and its hardware implementation to reduce the required memory-footprint and power usage of DNNs to enable them to be deployable on edge and mobile devices. we demonstrated a hardware-friendly pruning method where the locations of non-zero weights are derived from a Linear Feedback Shift Registers (LFSRs) in real-time. The results show a total power and area savings up to 49.97% and 50.20% for VGG-16 network on down-sampled ImageNet, respectively.**

I. INTRODUCTION

Ever increasing network of Internet of Thing (IoT) and edge devices requires the computation to be done on or close the source of the data. Edge computing has several benefits such as decreased latency and required energy, improved response times and better bandwidth availability. Accurate and fast computing on the edge plays an important role on different resource constrained applications such as autonomous driving and health-care monitoring. DNN has shown a great performance improvement especially in applications such as computer vision and healthcare data analysis [1]. However, their large and over-parameterized networks unable them to be used on resource-constrained edge devices.

To overcome this problem, various network compression techniques have been proposed to sparsify large DNNs, and enable them to fit in on-chip SRAM. However, sparse network add irregularity to the network which impose another level of complexity. Platforms such as GPU and DNN accelerators cannot take advantage of these sparse networks due to the lack of structure [2]. To overcome this problem, we propose a novel methodology to sparse large DNNs to reduce the memory footprint and energy usage while preserving the accuracy. We used an on-die linear feedback shift register (LFSR), to generate a pseudo random sequence (PRS). The generated PRS is utilized to prune the network. In addition, during inference we use the LFSR with the same seed to generate the indices in real-time to perform multiplication between the sparse weight matrix and input/activation vector. The advantage of our proposed hardware-friendly method is that we no longer need to store the sparse weight addresses – thereby reducing the memory foot-print significantly.

In this section, we first describe the pruning algorithm to sparse the network during the training step. Then the proposed hardware architecture for inference will be explained.

A. Training pipeline of the proposed method

The algorithm consists of four steps (figure 1): (1) generating pseudo-random sequences (PRS) using two LFSRs (2) training the network with regularization (3) Pruning the weights (4) retraining the sparse network [3].

We first generate a PRS using two LFSRs, one for row indices indicating the address of the element in the input vector and another one for column indices which encode the address of the output vector. LFSR, consists of an array of flip-flops with an initial state called input seed (s), followed by linear feedback performed by several exclusive-or (*XOR*) gates (c_i). The output of LFSRs can be mathematically described through n^{th} order characteristic polynomials as follow:

$$x^n + c_{n-1}x^{n-1} + ... + c_1x + 1 \qquad (1)$$

We then use these generated addresses for the weights that we want to keep during training and regularize the rest of the weight values to make them zero. As an example, a fully connected (*FC*) layer of DNN performs the following function:

$$a = ReLU(Z) \quad \text{where} \quad Z = W^Tx + b \qquad (2)$$

Where W is a weight matrix, x is an input vector, b is a vector of bias values. For simplification, b can be merged with W by appending an additional column to the end of matrix W. We used L2 regularization which a regularizer component is added to the cost (J) , as shown in Eq.3.

$$J(W^{[l]}, b^{[l]}) = \frac{1}{m}\sum_{c=1}^{m}L(\hat{y}^{(i)}, y^{(i)}) + \frac{\lambda}{2m}\sum_{l=1}^{L}||W_{ij}^{[l]}||_F^2 \quad (3)$$

Where i and j are correspond to the remaining indices (i.e. not chosen from LFSR 1 and LFSR 2) for rows and columns, respectively. L is the layer's number in the network. λ is the regularization parameter and can be tuned. Larger λ will more penalized the weights values and make them closer to zero.

After training step, in order to make sure that the regularized weights are exactly zero, we use a mask to make them exactly equal to zero. Finally, we retrain the network to better compensate the deleted weights and connections.

For inference hardware implementation, The generated LFSR #1 indices is used for the input to be multiplied to

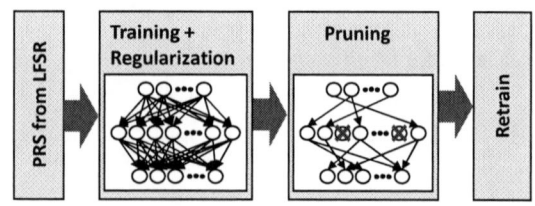

Fig. 1: Training pipeline for the proposed method.

978-1-7281-5410-7/20 $31.00 © 2020 IEEE

the corresponding weight in sparse matrix weight (figure 2). LFSR generate the PRS with values between 1 to $2^N - 1$. To avoid redundant clock cycles when the generated number is greater that the number of neurons and keep the values between number of neurons, the generated PRS is multiplied to the length of neurons and the most significant bits (MSBs) are selected. Finally, the result is stored in the output buffer with the address generated by LFSR #2.

II. RESULTS

The results are demonstrated on three networks: Lenet300-100, LeNet-5 and VGG-16 using MNIST, CIFAR-10 and down-sampled ImageNet data-sets. Training is done using Tensorflow platform on Nvidia GTX 1080 Ti GPUs.

We compared our results with a state-of-the-art pruning algorithm proposed in [2] where the pruning is done based on a threshold during training. For the baseline algorithm, the sparse weight matrix is compressed and saved in memory in three vectors: the non-zero weights' value (S), non-zero weights' location (I) and a pointer, pointing to each column of the weight matrix (P).The accuracy versus sparsity rates of our method is compared with the baseline (figure 3). The results show that our method can sparse the network more than 80% while preserving the accuracy even in complex network and datasets such as VGG-16 on down-sampled ImageNet.

To evaluate our method during inference, we have synthesized baseline and our method with 65nm CMOS technology to measure hardware metrics. The bit-width precision is considered to be 8 bits for weights and activations and 4, 8 bits for indices. The pre-layout analysis demonstrates up to 64% reduction in required memory footprint compared to the baseline pruning technique (Figure 4). Moreover, the power and area of the overall system (memory, multiplier, accumulator and input/output buffers) are measured for VGG-16 network (figure 5). We observe a maximum of 49.97% power savings and 50.20% area savings across varying sparsity versus baseline designs.

REFERENCES

[1] R. Boostani, F. Karimzadeh, and M. Nami, "A comparative review on sleep stage classification methods in patients and healthy individuals," *Computer methods and programs in biomedicine*, vol. 140, pp. 77–91, 2017.

[2] S. Han, H. Mao, and W. J. Dally, "Deep compression: Compressing deep neural networks with pruning, trained quantization and huffman coding," *arXiv preprint arXiv:1510.00149*, 2015.

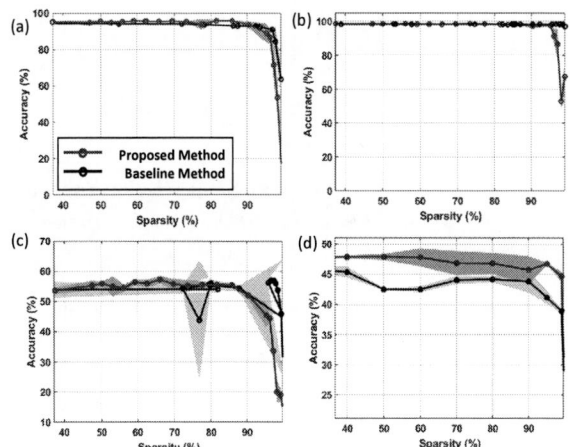

Fig. 3: Accuracy (%) versus different sparsity rates. (a) LeNet300-100 on MNIST, (b) LeNet5 on MNIST, (c) LeNet5 on Cifar10, (d) VGG-16 on down-sampled ImageNet.

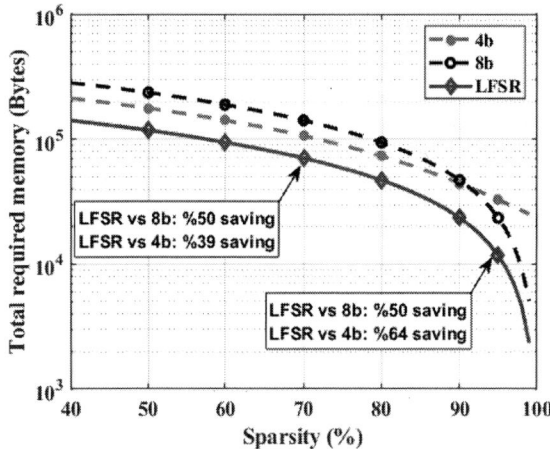

Fig. 4: Total required memory for our method and the baseline method with 4 and 8 bit precision at different levels of sparsity.

[3] F. Karimzadeh, N. Cao, B. Crafton, J. Romberg, and A. Raychowdhury, "Hardware-aware pruning of dnns using lfsr-generated pseudo-random indices," *arXiv preprint arXiv:1911.04468*, 2019.

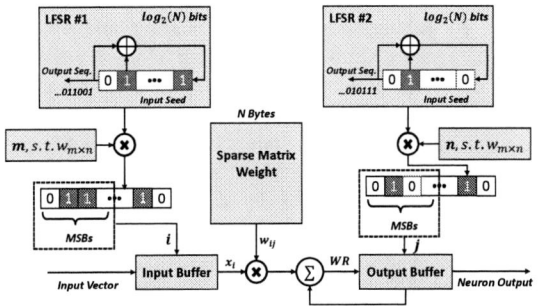

Fig. 2: The proposed method high-level architecture.

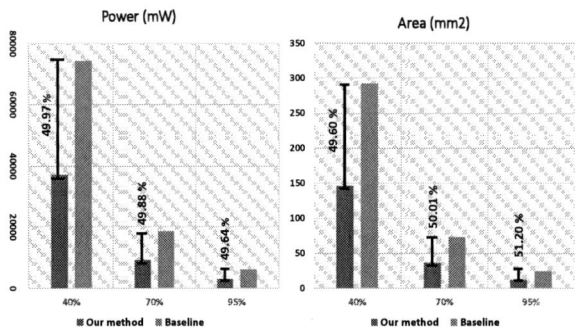

Fig. 5: Measured Power (mW) and area (mm^2) of the overall system for our proposed and baseline method on VGG-16.

Online Reward-Based Training of Spiking Central Pattern Generator for Hexapod Locomotion

Ashwin Sanjay Lele, Yan Fang, Justin Ting, Arijit Raychowdhury
School of Electrical and Computer Engineering, Georgia Institute of Technology

Abstract—Online learning in legged robot under stringent performance and energy constraints thwarts the application of conventional reinforcement learning and optimization algorithms. The integration of complex sensors and data pre-processing required in using these algorithms makes this more challenging. Spiking neural networks allow local learning and low computing power opening new possibilities neuromorphic paradigm to such tasks. Central pattern generation based learning to walk in hexapod robots perfectly matches the temporal learning in SNNs allowing end-to-end learning. We propose a stochastic reinforcement-based algorithm allowing the hexapod to learn using the reward generated by the gyro sensors and camera-based visual inputs. The system is implemented on a Raspberry pi to demonstrate convergence to bio-observed gait patterns.

I. INTRODUCTION

Central pattern generators are neural circuits in the brain generating temporally correlated spiking patterns actuating rhythmic muscle movements of limbs. Their activity is modulated by the spiking generated by various sensory inputs like the vestibular system in cockroaches. The CPG is also triggered by the visual inputs required in approaching and tracking prey. This biological closed-loop spiking system presents an ideal inspiration for a low-power autonomous robot. [1]

Spiking neural networks promise high energy efficiency and decentralized processing coupled with the ability to accommodate RL based learning perfectly suited for online reward-based tasks. Electronic implementations of spiking-CPG (SCPG) typically use a fully connected spiking neural network model shown in Fig. 1(a). Linear equation solving based reverse engineering approach has been demonstrated for hexapod CPGs capable of generating multiple gaits [1]. Another offline training using an evolutionary algorithm has been demonstrated in [2]. However, none of these approaches uses the autonomous learning capability of SNNs to learn to produce walking gaits without any prior knowledge in online reward-based end-to-end learning.

In this work, we demonstrate a stochastic reward-based weight update algorithm allowing a hexapod robot to learn to walk without any prior knowledge using the online rewards generated by the gyro-sensor and visual input. The rewards alter the weights exploring the correct pattern of leg movement Fig. 1(e). The gait converges to bio-observed tripod gait in most cases with some occurrences of sub-optimal slower gaits still enabling the forward motion. To the best of our knowledge, this is the first autonomous gait-learning demonstration using SNNs published in [3].

II. METHODOLOGY: ALGORITHM AND HARDWARE

The SCPG network consists of six fully connected neurons, firing of a neuron causes movement of the corresponding leg. The neurons follow leaky-integrate and fire (LIF) behaviour. Two neurons namely, input neuron (N_{in}) and gyro-driven neuron (N_{gyro}) trigger the SCPG. The ideal gait pattern requires alternate neurons fire in one step causing tripod gait Fig. 1(e).

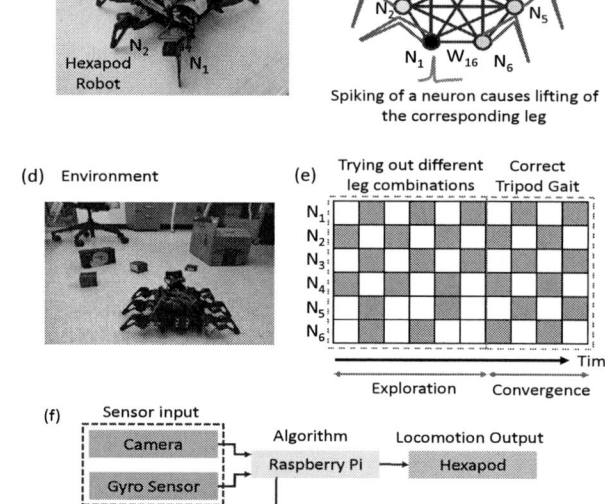

Fig. 1. (a,b) Hexapod robot with labeled neurons for mimicking the CPG. Spiking of a neuron causes motion of the corresponding leg (c) Office environment for demonstration (d) SCPG spiking as the algorithm progresses. The random spiking in the exploration phase gets latched to the correct tripod gait showing convergence (f) Hardware implementation of the system

The full system consists of an SCPG, gyro-sensor and optical camera. In every exploration step, the system is initialized by taking the gyroscope reading and capturing the image. Input neuron triggers a combination of SCPG neurons and corresponding legs to move. During the step, gyro-sensor evaluates the balance and camera captures an image after the step. Higher stability generates higher gyro-reward. The vision-based reward is calculated by applying the light-weight odometry method described in [4] to the initial and final images to estimate the magnitude and direction of the movement. The total reward is modulated by a random number and the weights are stochastically updated.

978-1-7281-5410-7/20 $31.00 © 2020 IEEE

Gy-521 MPU-6050 gyro sensor and piCamera provide sensory inputs to the Raspberry pi 3 Model B+ processing unit. Adeept RaspClaws Hexapod Spider Robot is used as the locomotion platform.

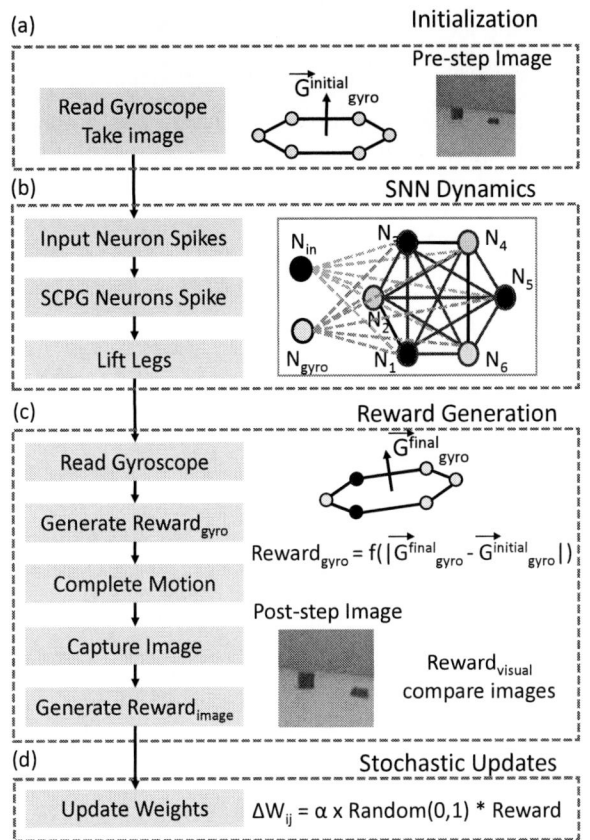

Fig. 2. Block diagram of the algorithm (a) Every step is initialized with a gyroscope reading and an image of the surroundings (b) LIF neuronal dynamics compute the spiking of the neurons (c) Sensory inputs are used to calculate the awards (e) Weight update using calculated rewards

III. RESULTS

Fig. 3 shows the time evolution of the simulation. The exploration starts and oscillates between different combinations of two and four leg motions. The corresponding cumulative reward is shown in the adjacent figure. After the network latches on to the correct tripod gait, the balance and forward motion are maintained simultaneously generating high reward. Demo-1, shows the convergence to tripod gait in the 66th cycle in hardware. However, the weight updates being stochastic, in some cases, intermediate non-bio-observed gaits are also seen to result in a forward motion with balance preservation. These sub-optimal gaits corresponding to weight parameters getting stuck into local minima are seen in (demo-2). The average number of time steps required for convergence are calculated by careful tuning of rewards and learning rate. Energy cost on Intel's Loihi [5] is estimated to be $\approx 855.1nJ$ using this. After the system has learned the correct gait, the energy consumed in every step is $\approx 9.1nJ$.

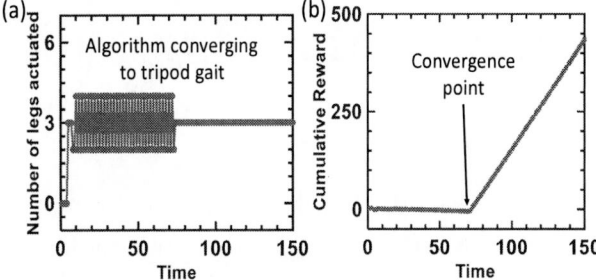

Fig. 3. The number of legs moved at a time instance oscillates around 3 and converges to 3 in a tripod gait resulting in accumulation of high constant reward (b) High positive reward is accummulated with latching to correct gait

TABLE I
COMPARISON WITH PRIOR WORK

Reference	Training Approach	Sensory Feedback	Online / Offline
[1]	Linear Programming	None	Offline
[2]	Grammar Evolution	None	Offline
[6]	Reward STDP	Olfactory + Visual	Offline
This Work	Stochastic Reward	balance + visual	Online

IV. CONCLUSION

We propose a closed-loop learning system in spiking neural network-based CPG with online reward generation to train a hexapod to learn to walk. The low energy consumption validates its potential in edge-robotics. The gait converges to biological tripod gait in most cases while converging to non-bio-observed sub-optimal gaits in some cases.

V. ACKNOWLEDGEMENT

This work was supported by CBRIC, a center in JUMP, a Semiconductor Research Corporation program by DARPA.

REFERENCES

[1] H. Rostro-Gonzalez, P. A. Cerna-Garcia, G. Trejo-Caballero, C. H. Garcia-Capulin, M. A. Ibarra-Manzano, J. G. Avina-Cervantes, and C. Torres-Huitzil, "A cpg system based on spiking neurons for hexapod robot locomotion," *Neurocomputing*, vol. 170, pp. 47–54, 2015.

[2] A. Espinal, H. Rostro-Gonzalez, M. Carpio, E. I. Guerra-Hernandez, M. Ornelas-Rodriguez, and M. Sotelo-Figueroa, "Design of spiking central pattern generators for multiple locomotion gaits in hexapod robots by christiansen grammar evolution," *Frontiers in neurorobotics*, vol. 10, p. 6, 2016.

[3] A. S. Lele, Y. Fang, J. Ting, and A. Raychowdhury, "Learning to walk: Spike based reinforcement learning for hexapod robot central pattern generation," in *2020 2nd IEEE International Conference on Artificial Intelligence Circuits and Systems (AICAS)*, pp. 208–212, IEEE, 2020.

[4] M. J. Milford and G. F. Wyeth, "Mapping a suburb with a single camera using a biologically inspired slam system," *IEEE Transactions on Robotics*, vol. 24, no. 5, pp. 1038–1053, 2008.

[5] M. Davies, N. Srinivasa, T.-H. Lin, G. Chinya, Y. Cao, S. H. Choday, G. Dimou, P. Joshi, N. Imam, S. Jain, *et al.*, "Loihi: A neuromorphic manycore processor with on-chip learning," *IEEE Micro*, vol. 38, no. 1, pp. 82–99, 2018.

[6] E. Arena, P. Arena, R. Strauss, and L. Patané, "Motor-skill learning in an insect inspired neuro-computational control system," *Frontiers in Neurorobotics*, vol. 11, p. 12, 2017.

Device Modeling and Circuit Design for Scalable Beyond-CMOS Computing

Xuan Hu[1], Naimul Hassan[1], Wesley H. Brigner[1], Maverick Chauwin[1,2], and Joseph S. Friedman[1]

[1]Electrical & Computer Engineering, The University of Texas at Dallas, Richardson, TX, United States
[2]Department of Physics, École Polytechnique, France

Abstract—Emerging technologies provide potential solutions to overcome the limitations of modern CMOS technologies. Specifically, as power density limitations impede further CMOS scaling, emerging technologies including spintronics, memristors, ambipolar transistors, and other beyond-CMOS devices are promising replacements for conventional CMOS transistors due to features such as non-volatility, low energy consumption, high operation speed, or high logical expressiveness. Specifically, we have evaluated spintronic technologies such as domain wall-magnetic tunnel junctions (DW-MTJs) and magnetic skyrmions that are particularly exciting for highly-efficient non-volatile information processing. Additionally, we have explored unconventional electronic switching devices including ambipolar transistors and memristors as replacements to CMOS and for hybrid emerging technology-CMOS computing systems.

Keywords—emerging technology, spintronics, compact modeling, ambipolarity, neuromorphic computing

I. INTRODUCTION

Beyond-CMOS computing is being explored with novel devices, architectures, and computing paradigms. For the development of these emerging technologies, it is critical to be able to directly-cascade devices without extra circuitry in scalable computing systems without CMOS. These beyond-CMOS devices and circuits exhibit the versatility and scalability of emerging technologies.

II. SPICE-COMPATIBLE MODELING OF BEYOND-CMOS COMPUTING DEVICES

As the fabrication processes for emerging technologies are relatively immature, it is important to develop device models that are able to accurately and efficiently simulate medium-to-large scale circuits. These simulation results can be used to compare the efficiency of these emerging technologies and provide feedback for system prototype fabrication.

A. Modeling & Circuit Design of Ambipolar Carbon Nanotube Field-Effect Transistors

Ambipolar semiconductor transport provides exciting opportunities for computing and has been experimentally demonstrated in various materials including carbon nanotubes (CNTs), silicon nanowires (SiNWs), and transition metal dichalcogenide (TMD) monolayers, enabling ambipolar transistors in which the majority carrier type can be switched dynamically. This flexibility enables compact circuit design by integrating more functionalities into fewer devices, but leads to fundamentally new challenges for device modeling.

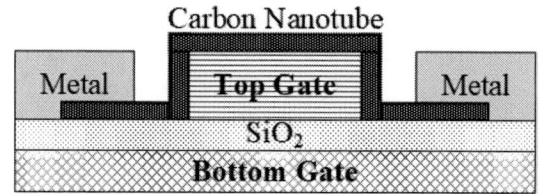

Fig. 1. Cross-section diagram of dual-gate ambipolar field-effect transistor (adapted from [1]).

Fig. 1 demonstrates the cross-section diagram of a dual-gate ambipolar carbon nanotube field-effect transistor (DG-A-CNTFET), in which both the top and bottom gates can be used to modulate the current through the CNT [1]. In addition, the bottom gate can be used to switch the transistor polarity, permitting the CNTFET to operate as an ambipolar device. We have therefore developed a closed-form SPICE-compatible model [2], [3] for dual-gate ambipolar transistors with the ability to dynamically switch the transistor polarity and interchange the terminals. This has enabled the first transient simulations of ambipolar CNTFET-based logic circuits in which the transistor polarity switches during the simulation, permitting the development of large-scale computing systems that fully exploit ambipolar transport.

B. SPICE-Only Model for Spin-Transfer Torque Domain Wall MTJ Logic

The spin-transfer torque domain wall (DW) magnetic tunnel junction (MTJ) enables spintronic logic circuits that can be directly cascaded without deleterious signal conversion circuitry and is one of the only spintronic devices for which cascading has been demonstrated experimentally [4], [5]. Fig. 2 shows the diagram of the DW-MTJ device: the current flow through the bottom soft ferromagnetic track shifts the position of the DW which leads to the resistance switching of the MTJ formed by

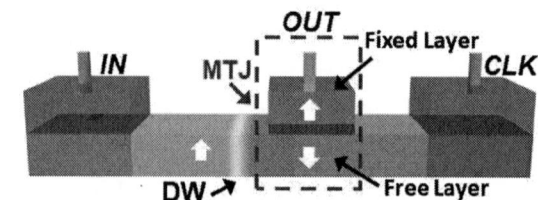

Fig. 2. Device diagram of DW-MTJ.

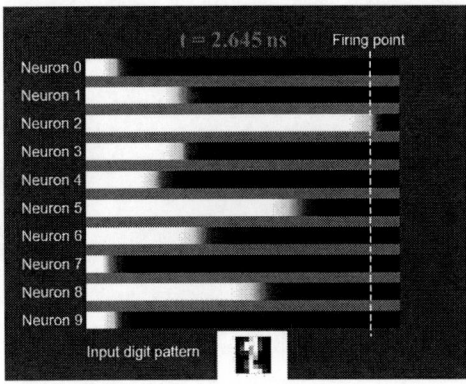

Fig. 3. Simulation snapshot of the output neuron layer for a hand-written digit recognition task composed of ten DW-MTJ neurons.

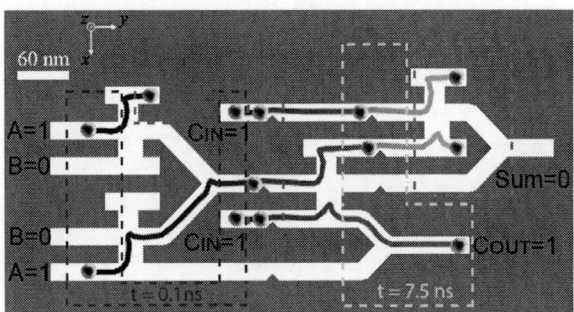

Fig. 4. Micromagnetic simulation of skyrmion-based one-bit full adder.

the combination of the soft ferromagnetic layer, the fixed layer, and the tunnel barrier.

To accelerate the design and simulation of large-scale DW-MTJ systems, we proposed a SPICE-only compact model [6] that is directly compatible with modern circuit design CAD tools. Compared to the previous simulation techniques, the proposed model enables accurate and rapid simulation with 10,000x decrease in simulation speed.

III. NEUROMORPHIC AND BOOLEAN CIRCUIT DESIGN WITH SPINTRONIC DEVICES

The increasing leakage and instability of scaled CMOS memories inhibit their application in further neuromorphic or other non-von Neumann computing systems, whereas spintronic devices have been proven to be suitable and efficient for neuromorphic computing and versatile logical systems. It is therefore important to develop spintronic computation and signal processing circuits in addition to CMOS, as conventional transistors do not effectively emulate neurobiological behavior.

A. Artificial Neuron with DW-MTJ

In light of the non-volatility, low-energy consumption, scalability, and its natural analog to biological neuron behavior, the DW-MTJ in Fig. 2 is used to implement a leaky integrate-and-fire (LIF) artificial neuron [7]. Fig. 3 shows a snapshot of the hand-written digit recognition simulation using the proposed LIF neuron, in which the system correctly identifies 94% of the digits in the task simulated. This system, in concert with artificial synapses with the same device [8], enable purely-spintronic neuromorphic computing systems without CMOS.

B. Scalable Skyrmion Logic System

Magnetic skyrmions are topologically-stable magnetic whirls that have been proposed as highly-efficient information carriers for computing systems that exploit the fact that these quasiparticles are small, non-volatile, and can be propagated with minimal energy. Reversible computing, meanwhile, is a computing paradigm that conserves information, leading to theoretical proposals for non-dissipative computing systems based on conservative logic that consume zero energy. By combining the concept of conservative logic with the features of skyrmions, we have proposed a nanoscale conservative skyrmion logic system and demonstrated the ability to perform cascaded logical operations [9]. Fig. 4 shows the schematic of a skyrmion-based one-bit full adder. This system includes a synchronization circuit that enables the realization of large-scale non-volatile spintronic computing systems with high operating speed and low energy consumption.

IV. CONCLUSIONS

By exploiting low-energy switching, reconfigurability, non-volatility, and unique operating mechanisms, these emerging technologies provide the potential to revolutionize information processing for the next generation of computing systems. However, the development and application of these novel devices in large-scale systems are impeded by the lack of proper simulation techniques and inefficient circuit topologies. Here we described the models for different emerging technology devices which enable rapid and accurate simulations in different scales of systems. We also discussed the characteristics and efficient applications of novel spintronic devices in various paradigms for next-generation computing systems.

REFERENCES

[1] Y. Lin, J. Appenzeller, and P. Avouris, "Novel carbon nanotube FET design with tunable polarity", *IEDM*, pp. 687-690, 2004.

[2] X. Hu and J. S. Friedman, "Closed-Form Model for Dual-Gate Ambipolar CNTFET Circuit Design", *IEEE ISCAS*, May 2017.

[3] X. Hu and J. S. Friedman, "Transient Model with Interchangeability for Dual-Gate Ambipolar CNTFET Logic Design", *IEEE/ACM Nanoarch*, July 2017.

[4] J. A. Currivan, Y. Jang, M. D. Mascaro, M. A. Baldo, and C. A. Ross, "Low energy magnetic domain wall logic in short, narrow, ferromagnetic wires," *IEEE Magn. Lett.*, vol. 3, 2012, Art. no. 3000104.

[5] J. A. Currivan-Incorvia et al., "Logic circuit prototypes for three-terminal magnetic tunnel junctions with mobile domain walls," *Nature Commun.*, vol. 7, Jan. 2016, Art. no. 10275.

[6] X. Hu, A. Timm, W. H. Brigner, J. A. C. Incorvia, and J. S. Friedman, "SPICE-only model for spin-transfer torque domain wall MTJ logic," *IEEE Trans. Electron Devices*, vol. 66, no. 6, pp. 2817–2821, Jun. 2019.

[7] N. Hassan et al., "Magnetic domain wall neuron with lateral inhibition," *J. Appl. Phys.*, vol. 124, no. 15, 2018, Art. no. 152127.

[8] O. Akinola, X. Hu, C. H. Bennett, M. Marinella, J. S. Friedman, and J. A. C. Incorvia, "Three-Terminal Magnetic Tunnel Junction Synapse Circuits Showing Spike-Timing-Dependent Plasticity", *Journal of Physics D: Applied Physics*, vol. 52, no. 49, 2019, Art. no. 49LT01.

[9] M. Chauwin*, X. Hu*, F. Garcia-Sanchez, N. Betrabet, A. Paler, C. Moutafis, and J. S. Friedman, "Skyrmion Logic System for Large-Scale Reversible Computation", *Physical Review Applied*, vol. 12, no. 6, p. 064053 (2019).

AUTHOR INDEX

Aghighi, Amin 100, 105, 171
Ajayi, Tutu.................................141
Aksanli, Baris 10
Anandakumar, N. N. 198
Atienza, David 94
Bakir, Muhannad S.........................159
Ben-Hur, Rotem 28, 64
Bertozzi, Davide 58
Blaauw, David 141
Boemmels, Juergen 34
Bragaglio, Moreno 111
Brigner, Wesley H.........................210
Calhoun, Benton...........................141
Calimera, Andrea..........................204
Carloni, Luca P. 7
Catthoor, Francky 34
Chauwin, Maverick 210
Chen, Chien-Hen...........................141
Chen, Junkai 176
Chen, Yongnan 176
Cherivirala, Yaswanth K...................141
Chowdhury, Siddhartha 70
Condia, Josie E. R. 153
Cops, Wim.................................52
Cordova, David 52
Crafton, Brian 123
Dai, Shanshan.............................192
Deng, Marina..............................76
Deval, Yann...............................52
Dreslinski, Ronald........................141
Dubrova, Elena............................117
Eberlein, Matthias 186
Eliahu, Adi...............................28
Fang, Yan 208
Farhang-Boroujeny, Behrouz 105
Fayazi, Morteza...........................141
Friedman, Joseph S........................210
Fujita, Masahiro..........................22
Gaillardon, Pierre-Emmanuel 34
Ganguly, Samiran 147
Gaur, Manoj S.............................200
Gava, Jonas 165
Germiniani, Samuele 111
Ghosh, Avik W. 147
Giacomin, Edouard.........................34
Goossens, Sven 16
Gupta, Saransh 10
Gupta, Shourya 141
Hao, Yilun 10
Hashmi, Mohammad S. 198
Hassan, Naimul 210
Hu, Kangping 192
Hu, Xuan..................................210
Hu, Yinghua 129

Imani, Mohsen 10
Islam, Sheikh A. 135
Jenihhin, Maksim 16
Kahng, Andrew B............................1
Kamineni, Sumanth 141
Karimzadeh, Foroozan 206
Katkoori, Srinivas 135
Kaul, Ankit 159
Koufopavlou, O. 46
Krishna, Tushar 123
Krstic, Milos 58
Kumar, Abhishek 76
Kvatinsky, Shahar 28, 64
Kwon, Kyumin 141
Lai, Xinhui 16
Lapuyade, Herve 52
Larrieu, Guilhem 76
Laxmi, Vijay 200
Le Beux, Sébastien 76
Lecestre, Aurélie.........................76
Lele, Ashwin S. 208
Levisse, Alexandre........................94
Lian, Xiaoyu 40, 192
Lim, Sung-Kyu 123
Macii, Enrico 204
Maes, Roel 16
Maneux, Cristell 76
Marc, François............................76
Marchand, Cedric 76
Miyasaka, Yukio 22
Moraitis, Michail 117
Moraitis, S. 46
Morris, Justin 10
Muduli, Sujit K. 88
Mukherjee, Chhandak 76
Mukhopadhyay, Debdeep 70, 202
Murali, Gauthaman 123
Nazarian, Shahin 129
Ngo, Kalle 117
Nodenot, Nicolas 52
Nuzzo, Pierluigi 129
O'Connor, Ian 76
Ost, Luciano..............................165
Paul, Kolin 16
Peled, Natan 64
Peluso, Valentino 204
Piccin, Yohan 52
Poittevin, Arnaud 76
Polnati, Srivarsha 135
Pravadelli, Graziano 111
Pretl, Harald 186
Ramkumar, Ranganathan 10
Rawat, Mayank 88
Raychowdhury, Arijit 123, 159, 206, 208

AUTHOR INDEX

Reda, Sherief 40, 192
Reis, Ricardo 165
Reorda, M. S..153
Rivet, Francois 52
Ronen, Ronny 28, 64
Rosenstein, Jacob K. 40, 192
Rosing, Tajana 10
Roy, Debapriya B..70
Ryckaert, Julien 34
Sadhukhan, Rajat 202
Sah, Love K. 135
Saligane, Mehdi 141
Saligram, Rakshith 159
Sanadhya, Somitra K..198
Seitanidis, D...46
Selimis, Georgios 16
Sharma, Tannu 82
Shukla, Nikhil...147
Simon, William A...94
Spetalnick, Samuel 123
Stevens, Kenneth S...82
Subramanyan, Pramod..88
Sylvester, Dennis 141
Tabib-Azar, Massood 100
Taherzadeh-Sani, Mohammad..171
Tajalli, Armin..100, 105, 171
Tan, Hong-Zhou 176, 181
Theodoridis, G...46
Ting, Justin..208
Tulloss, Caleb R. 192
Veronesi, Alessandro 58
Wentzloff, David D. 141
Yadav, Sonal...200
Yang, Kaixin...129
Yu, Jeffrey 10
Zapater, Marina...94
Zeng, Yanhan 176, 181
Zhi, Haochang 181
Zhou, Wei 181

IEEE
445 Hoes Lane
Piscataway, NJ 08854-4141

ISBN 978-1-7281-5410-7